AF856168

Chemical Degradation Methods for Wastes and Pollutants

Environmental Science and Pollution Control Series

1. Toxic Metal Chemistry in Marine Environments, *Muhammad Sadiq*
2. Handbook of Polymer Degradation, *edited by S. Halim Hamid, Mohamed B. Amin, and Ali G. Maadhah*
3. Unit Processes in Drinking Water Treatment, *Willy J. Masschelein*
4. Groundwater Contamination and Analysis at Hazardous Waste Sites, *edited by Suzanne Lesage and Richard E. Jackson*
5. Plastics Waste Management: Disposal, Recycling, and Reuse, *edited by Nabil Mustafa*
6. Hazardous Waste Site Soil Remediation: Theory and Application of Innovative Technologies, *edited by David J. Wilson and Ann N. Clarke*
7. Process Engineering for Pollution Control and Waste Minimization, *edited by Donald L. Wise and Debra J. Trantolo*
8. Remediation of Hazardous Waste Contaminated Soils, *edited by Donald L. Wise and Debra J. Trantolo*
9. Water Contamination and Health: Integration of Exposure Assessment, Toxicology, and Risk Assessment, *edited by Rhoda G. M. Wang*
10. Pollution Control in Fertilizer Production, *edited by Charles A. Hodge and Neculai N. Popovici*
11. Groundwater Contamination and Control, *edited by Uri Zoller*
12. Toxic Properties of Pesticides, *Nicholas P. Cheremisinoff and John A. King*
13. Combustion and Incineration Processes: Applications in Environmental Engineering, Second Edition, Revised and Expanded, *Walter R. Niessen*
14. Hazardous Chemicals in the Polymer Industry, *Nicholas P. Cheremisinoff*
15. Handbook of Highly Toxic Materials Handling and Management, *edited by Stanley S. Grossel and Daniel A. Crowl*
16. Separation Processes in Waste Minimization, *Robert B. Long*
17. Handbook of Pollution and Hazardous Materials Compliance: A Sourcebook for Environmental Managers, *Nicholas P. Cheremisinoff and Madelyn Graffia*
18. Biosolids Treatment and Management: Processes for Beneficial Use, *edited by Mark J. Girovich*

19. Biological Wastewater Treatment: Second Edition, Revised and Expanded, *C. P. Leslie Grady, Jr., Glen T. Daigger, and Henry C. Lim*
20. Separation Methods for Waste and Environmental Applications, *Jack S. Watson*
21. Handbook of Polymer Degradation: Second Edition, Revised and Expanded, *S. Halim Hamid*
22. Bioremediation of Contaminated Soils, *edited by Donald L. Wise, Debra J. Trantolo, Edward J. Cichon, Hilary I. Inyang, and Ulrich Stottmeister*
23. Remediation Engineering of Contaminated Soils, *edited by Donald L. Wise, Debra J. Trantolo, Edward J. Cichon, Hilary I. Inyang, and Ulrich Stottmeister*
24. Handbook of Pollution Prevention Practices, *Nicholas P. Cheremisinoff*
25. Combustion and Incineration Processes: Third Edition, Revised and Expanded, *Walter R. Niessen*
26. Chemical Degradation Methods for Wastes and Pollutants: Environmental and Industrial Applications, *edited by Matthew A. Tarr*

Additional Volumes in Preparation

Chemical Degradation Methods for Wastes and Pollutants

Environmental and Industrial Applications

edited by

Matthew A. Tarr

University of New Orleans
New Orleans, Louisiana, U.S.A.

MARCEL DEKKER, INC. NEW YORK • BASEL

Although great care has been taken to provide accurate and current information, neither the author(s) nor the publisher, nor anyone else associated with this publication, shall be liable for any loss, damage, or liability directly or indirectly caused or alleged to be caused by this book. The material contained herein is not intended to provide specific advice or recommendations for any specific situation.

Trademark notice: Product or corporate names may be trademarks or registered trademarks and are used only for identification and explanation without intent to infringe.

Library of Congress Cataloging-in-Publication Data
A catalog record for this book is available from the Library of Congress.

ISBN: 0-8247-4307-5

This book is printed on acid-free paper.

Headquarters
Marcel Dekker, Inc., 270 Madison Avenue, New York, NY 10016, U.S.A.
tel: 212-696-9000; fax: 212-685-4540

Distribution and Customer Service
Marcel Dekker, Inc., Cimarron Road, Monticello, New York 12701, U.S.A.
tel: 800-228-1160; fax: 845-796-1772

Eastern Hemisphere Distribution
Marcel Dekker AG, Hutgasse 4, Postfach 812, CH-4001 Basel, Switzerland
tel: 41-61-260-6300; fax: 41-61-260-6333

World Wide Web
http://www.dekker.com

The publisher offers discounts on this book when ordered in bulk quantities. For more information, write to Special Sales/Professional Marketing at the headquarters address above.

Copyright © 2003 by Marcel Dekker, Inc. All Rights Reserved.

Neither this book nor any part may be reproduced or transmitted in any form or by any means, electronic or mechanical, including photocopying, microfilming, and recording, or by any information storage and retrieval system, without permission in writing from the publisher.

Current printing (last digit):

10 9 8 7 6 5 4 3 2 1

PRINTED IN THE UNITED STATES OF AMERICA

Preface

Human activities have a large and important impact on the environment. Naturally occurring elements or compounds are often concentrated and redistributed in the environment through industrial processes, power production, and consumer activity. For example, lead, which is found in naturally occurring mineral deposits, has become a major pollutant through its use in batteries, paints, and gasoline additives. In addition, the production of non-natural or anthropogenic substances, such as halogenated solvents, can also result in the eventual release of often toxic and biorecalcitrant substances into the environment. Wide-scale redistribution of pollutants by humans dates at as far back as the ancient Greek and Roman civilizations (2000–2500 years ago), during which time extensive smelting activities resulted in significant atmospheric pollution by heavy metals such as lead. In fact, heavy-metal contamination of Arctic and Antarctic ice has revealed evidence of global pollution from smelting and other human activities since these ancient times.

Most certainly the people of ancient Greek and Roman times were not aware of the extent of their pollution. In fact, only in the late twentieth century did widespread awareness and understanding of the degree of anthropogenic pollution begin to develop. Unfortunately, large releases of contaminants into the environment transpired without either knowledge of or concern for the consequences. Once contaminants have been introduced into the environment, subsequent clean-up is extremely difficult, time consuming, and costly. Due to the existence of many contaminated sites, significant research and development efforts have been expended to develop effective means of remediating these sites. These methods must be both economically feasible and environmentally sound. For some sites, these challenges have been successfully met, while other sites remain contaminated because of lack of acceptable (economically and/or environmentally) technologies or because the sites pose a low risk.

While cleaning up previous contamination is a high priority, developing new technologies to prevent future contamination is equally important, if not more so. Without environmentally acceptable industrial processes, power production, and consumer activity, the Earth's environment will continue to be threatened. Development of inherently clean technologies as well as implementation of effective waste stream treatment are viable routes to preventing future environmental contamination.

Chemical Degradation Methods for Wastes and Pollutants focuses on chemical methods of destroying pollutants. Chemical methods can be advantageous over biological methods because they are often faster, can treat highly contaminated systems, and may be less sensitive to ambient conditions. In contrast, bacteria can be killed by contaminants or solvents and lose viability outside relatively narrow pH and temperature ranges. However, chemical methods are often more costly and labor-intensive than biodegradation technologies. Despite their limitations, both biological and chemical technologies are valuable tools that can be used successfully under appropriate conditions. Furthermore, combinations of biological and chemical treatment methods can often provide advantages over the individual systems.

The book covers several chemical technologies for remediation or waste stream treatment of predominantly organic contaminants. Although not every chemical technology has been included, ten common or potentially useful methods are covered. Each chapter presents the fundamentals behind each technology and covers selected applications and practical issues relevant to adaptation of the technique to real treatment systems.

Continued research into both fundamentals and applications of chemical treatment technologies will hopefully provide solutions to many current pollution treatment problems, both for waste streams and for contaminated sites. Only through cooperation among scientists, engineers, industry, government, and consumers can we maintain a healthy and productive environment for the future.

Finally, I would like to thank those who served as reviewers for each chapter.

Matthew A. Tarr

Contents

Contributors

Fernando J. Beltrán Departamento de Ingenieria Quimica y Energetica, Universidad de Extremadura, Badajoz, Spain

Enric Brillas Laboratori de Ciencia i Tecnologia Electroquimica de Materials, Departament de Quimica Fisica, Universitat de Barcelona, Barcelona, Spain

Pere-Lluís Cabot Laboratori de Ciencia i Tecnologia Electroquimica de Materials, Departament de Quimica Fisica, Universitat de Barcelona, Barcelona, Spain

Juan Casado Departamento de Investigacion, Carburos Metálicos S.A., Barcelona, Spain

William J. Cooper Department of Chemistry, University of North Carolina–Wilmington, Wilmington, North Carolina, U.S.A.

Hugo Destaillats Department of Environmental Science and Engineering, California Institute of Technology, Pasadena, California, U.S.A.

Gerry D. Getman Commodore Solution Technologies, Inc., Marengo, Ohio, U.S.A.

Michael R. Hoffmann Department of Environmental Science and Engineering, California Institute of Technology, Pasadena, California, U.S.A.

Indira Jayaweera SRI International, Menlo Park, California, U.S.A.

Timothy L. Johnson AMEC Earth & Environmental, Inc., Portland, Oregon, U.S.A.

Leah J. Matheson MSE Technology Applications, Inc., Butte, Montana, U.S.A.

Bruce J. Mincher Radiation Physics Group, Idaho National Engineering & Environmental Laboratory, Idaho Falls, Idaho, U.S.A.

James A. Nicell Department of Civil Engineering and Applied Mathematics, McGill University, Montreal, Quebec, Canada

Pierre Pichat Laboratoire Photocatalyse, Catalyse et Environment, Ecole Centrale de Lyon, Ecully, France

Charles U. Pittman, Jr. Department of Chemistry, Mississippi State University, Mississippi State, Mississippi, U.S.A.

Michelle M. Scherer Department of Civil and Environmental Engineering, University of Iowa, Iowa City, Iowa, U.S.A.

Matthew A. Tarr Department of Chemistry, University of New Orleans, New Orleans, Louisiana, U.S.A.

Paul G. Tratnyek Department of Environmental and Biomolecular Systems, Oregon Health and Science University, Beaverton, Oregon, U.S.A.

Henry C. Wallace Ultrasonic Energy Systems Co., Panama City, Florida, U.S.A.

1
Ozone–UV Radiation–Hydrogen Peroxide Oxidation Technologies

Fernando J. Beltrán
Universidad de Extremadura, Badajoz, Spain

I. INTRODUCTION

Processes involving the use of ozone, UV radiation, and hydrogen peroxide, characterized by the generation of short-lived chemical species of high oxidation power, mainly the hydroxyl radical, are classified as advanced oxidation technologies (AOTs). Possibly, the term may be attributed to Glaze et al. [1], who pointed out that hydroxyl radical oxidation is the common feature of these processes. The importance of these processes is due to the high reactivity and redox potential of this free radical that reacts nonselectively with organic matter present in water. In practical cases, these processes present a high degree of flexibility because they can be used individually or in combination depending on the problem to be solved. For instance, for phenols or substances with high UV molar absorption coefficients, ozone or UV radiation can be used alone, respectively, without the need of any additional reagent, such as hydrogen peroxide. Another advantage of these AOTs is that they may be applied under mild experimental conditions (atmospheric ambient pressure and room temperature).

The need for the application of these AOTs is based on different social, industrial, environmental, and even academic reasons. The increasing awareness of society for the quality of drinking water has led to the establishment of maximum contaminant levels of priority pollutants in drinking water [1,2]. The preparation of ultrapure water is needed for some industrial activities such as those derived from the pharmaceutical and electronic processes.

Also, the release of wastewater into natural environmental reservoirs is another concern; recycling of wastewater is already in progress in countries where the lack of water is a national problem [4]. Finally, academic interest exists because the study of these AOTs allows testing the application of some physical and chemical laws and engineering theories (mass, energy, and/or radiation conservation equations, kinetic modeling, absorption theories, etc.) to the environmental problems of water treatment.

Because of the aforementioned reasons, the number of research works and applications based on these AOTs in the treatment of water has increased considerably during the past 20 years. Numerous publications that refer to different aspects of these processes have so far been published in journals such as *Ozone Science and Engineering*, *Water Research*, *Ozone News*, *IUVA News*, and the *Journal of Advanced Oxidation Technologies*. In addition, several books on the subject are available, such as that edited by Langlais et al. [5] on applications and engineering aspects of ozone in water treatment and that of Doré [6] on the chemistry of oxidants. Reviews are also abundant, including those of Camel and Vermont [7] on ozone involving oxidation processes, Reynolds et al. [8] and Chiron et al. [9] on the oxidation of pesticides, Legrini et al. [10] on photochemical processes, Yue [11] on kinetic modeling of photooxidation reactors, and Scott and Ollis [12] on the integration of chemical and biological oxidation processes for wastewater treatment.

In this chapter, AOTs based on ozone, UV radiation, and hydrogen peroxide are presented with special emphasis on their fundamental and application aspects. Related literature of research studies and applications, especially those appearing in the last decade, are also listed, and specific examples of laboratory and scale-up studies are described in separate sections.

II. BACKGROUND AND FUNDAMENTALS OF $O_3/UV/H_2O_2$ PROCESSES

$O_3/UV/H_2O_2$ processes are characterized by the application of a chemical oxidant (ozone and/or hydrogen peroxide) and/or UV radiation. Individual description of properties and reactivities of these oxidation technologies is necessary to understand their synergism when used in combination for the treatment of specific water pollutants or wastewaters. However, because combined processes (O_3/H_2O_2, UV/H_2O_2, or O_3/UV) are usually recommended in real situations, a general description of the processes and fundamentals of the individual and integrated $O_3/UV/H_2O_2$ technologies is also presented in the following sections.

A. General Description

Ozone- or UV-radiation-based technologies ($O_3/UV/H_2O_2$) are chemical oxidation processes applied to water treatment for the degradation of individual pollutants or the reduction of the organic load (chemical oxygen demand, COD) and improved biodegradability of wastewaters. In addition, ozone and UV radiation alone can be used for disinfection purposes; in fact, this was their first application in water treatment [13,14]. In addition, these AOTs, particularly ozonation, can be used to enhance the efficiency of other processes such as Fe–Mn removal [15,16], flocculation–coagulation–sedimentation [17,18], biological oxidation [12], or biological degradation of organic carbon in granular activated carbon [19–21].

$O_3/UV/H_2O_2$ AOTs are suitable for the treatment of water containing organic pollutants in concentrations not higher than some tens of milligrams per liter. However, these technologies can also be used to treat concentrated solutions. In addition to concentration, factors such as molecular structure of pollutant, aqueous organic matrix, pH, etc. are variables that affect the efficiency and applicability of $O_3/UV/H_2O_2$ AOTs for practical application. For wastewater treatment, $O_3/UV/H_2O_2$ AOTs are used in combination with biological oxidation processes because of the enhancement achieved on the biological oxygen demand (BOD). In fact, another feature of $O_3/UV/H_2O_2$ AOTs is that they steadily transform high molecular weight substances into more oxygenated lower molecular weight substances, which involves an increase of BOD [22,23]. Examples of studies on wastewater treatment that give a general view of the application of $O_3/UV/H_2O_2$ AOTs are those of Rice and Browning [24] and, more recently, by Rice [25] on the use of ozonation, or Zhou and Smith [26], Rivera et al. [27], and Kos and Perkowski [28] for combined oxidation involving UV radiation.

$O_3/UV/H_2O_2$ AOTs, together with other processes treated in different chapters (such as Fenton oxidation), can be named ambient (temperature and pressure), advanced oxidation technologies, in contrast with other AOTs such as hydrothermal oxidation processes that require pressures and temperatures above 1 MPa and 150°C, respectively, and which are more suitable for the treatment of concentrated wastewaters. It is evident that appropriate ranges of concentrations for the different oxidation technologies cannot be exactly established but some recommended values have been reported [29]. Fig. 1 shows some possible recommended ranges of concentrations for these types of AOTs.

$O_3/UV/H_2O_2$ AOTs generally involve two oxidation/photolysis routes to remove foreign matter present in water. Thus, ozone, hydrogen peroxide, and/or UV radiation can react individually or photolyze directly the organic in water. However, when used in combination, they can degrade pollutants by

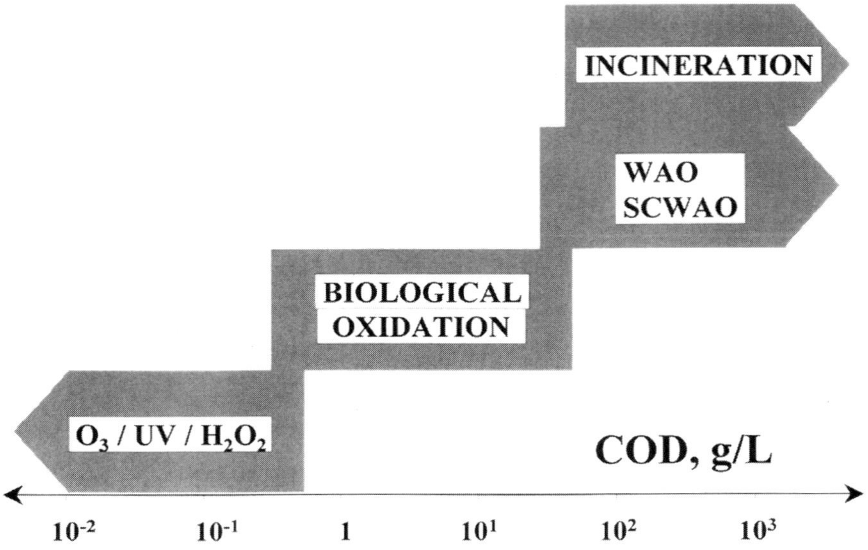

Figure 1 Oxidation process advisable according to COD of water. (WAO, wet air oxidation. SCWAO, supercritical wet air oxidation).

oxidation through hydroxyl free radicals generated in situ. Hydroxyl radicals have the largest standard redox potential except for fluorine (see Table 1).

In addition, they react very rapidly with almost all types of organic substances through reactions whose rate constants vary from 10^7 to 10^{10} M^{-1} s^{-1} [30]. Table 2 gives a list of rate constant values of these reactions.

Because of the high and similar values of the rate constants, it is said that these free radicals react nonselectively with the organic matter present in water, although, as deduced from the above range of values, there are compounds that react with them almost three orders of magnitude faster than others. Among the most common water pollutants, phenols and some pesticides are substances that react rapidly with hydroxyl radicals, whereas some organochlorine compounds are less reactive.

Another feature of these AOTs is that they are destructive types of water pollution removal processes because they eliminate compounds rather than transfer them to another medium. Thus, carbon adsorption or stripping transfers pollutants from one phase (water) to another phase such as a solid phase (carbon) or a gas phase (air). In the latter case, purification of air is required so that an additional step (i.e., carbon adsorption) is also needed, which implies higher processing costs.

Table 1 Standard Redox Potential of Some Oxidant Species

Oxidant	$E°$, V
Fluorine	3.03
Hydroxyl radical	2.80
Atomic oxygen	2.42
Ozone	2.07
Hydrogen peroxide	1.77
Permanganate ion	1.67
Hypochlorous acid	1.49
Chlorine	1.36
Chlorine dioxide	1.27
Bromine	1.09

Table 2 Rate Constants of the Reaction Between the Hydroxyl Radical and Organic Compounds in Water

Organic compound	Rate constant $\times 10^{-9}$, M^{-1} s^{-1}	Reference no.
Benzene	7.8	30
Nitrobenzene	2.9	31
2,6-Dinitrotoluene	0.75	31
Naphthalene	5	32
Phenanthrene	13.4	33
Phenol	11	30
Phenoxide ion	9.6	30
p-Nitrophenol	3.8	30
o-Chlorophenol	12	30
Maleic acid	6.0	30
Formic acid	0.13	30
Glyoxal	0.066	30
Tetrachloroethylene	2.6	30
Trichloroethylene	1.3	34
1,1,1-Trichloroethane	0.020	34
Dichloromethane	0.022	35
Chloroform	0.011	35
Lindane	5.8	35
Atrazine	2.6	35
Aldicarb	8.1	35

At first sight, however, the main drawback of $O_3/UV/H_2O_2$ AOTs is the high processing cost, mainly because both ozone and UV radiation require a continuous feed of energy for process maintenance, as well as high capital costs for ozone generators and photoreactors. However, the development of improved ozonators and UV lamp technologies has made these processes more amenable in practice as can be deduced from their actual applications (see Sec. IV).

B. Ozonation

Ozone is the basic compound for many oxidation processes included under the general term of ozonation. In these processes, ozone may be used alone or with other agents such as hydrogen peroxide, UV radiation, catalysts, ultrasound, activated carbon, etc. In this section, information concerning the individual use of ozone is given, while its combined use with hydrogen peroxide or UV radiation is reported in later sections.

1. Background and Fundamentals

Ozone is an inorganic chemical molecule constituted by three oxygen atoms. It is naturally formed in the upper atmosphere from the photolysis of diatomic oxygen and further recombination of atomic and diatomic oxygen according to the following reactions:

$$O_2 \xrightarrow{h\nu} 2O\bullet \tag{1}$$

$$O\bullet + O_2 \rightarrow O_3 \tag{2}$$

In this way, ozone forms a stratospheric layer several kilometers wide that protects life on earth by preventing UV-B and UV-C rays from reaching the surface of the planet. Ozone may arise from combustion reactions in automobile engines, resulting in pollutant gases. These gases usually contain nitric oxide that is photolyzed by sunlight in the surrounding atmosphere to yield nitrous oxide and atomic oxygen. Atomic oxygen, through reaction (2), finally yields ozone. In this sense, ozone is a contaminant of breathing air; the maximum level allowed during an 8-hr exposure is only 0.1 ppm. However, despite the importance of ozone as a tropospheric pollutant, the fate of ozone in the atmosphere is beyond the scope of this chapter.

Ozone was discovered in 1840 and the structure of the molecule as triatomic oxygen was established in 1872. The first use of ozone was reported at the end of the 19th century—as a disinfectant in many water-treatment plants, hospitals, and research centers such as the University of Paris where the first doctoral thesis on ozonation was presented [36]. Although the number of water-treatment plants using an ozonation step increased steadily during the 20th century, it was at the end of the 1970s that the use of ozone significantly increased. This increase came about when trihalomethanes and other organohalogenated compounds were identified in drinking water as disinfection by-products arising from chlorination [37]. This discovery gave rise to an enormous research effort to look for alternative oxidants to replace chlorine. Additional research aimed at discovering mechanisms of organochlorine compound formation established that these substances are formed from the electrophilic attack of chlorine on nucleophilic positions of natural humic substances present in surface water [38]. Because ozone is a powerful electrophilic agent, it was found that, generally, the application of ozone before chlorine significantly reduced trihalomethane formation. Subsequent study of ozone reactions in water led to a wide array of applications (presented in a further section) that can be summarized in the following: use as a disinfectant or biocide, use as an oxidant for micropollutant removal, and use as a complementary agent to improve other unit operations in drinking and industrial water and wastewater treatments (sedimentation, cooling water treatment, carbon adsorption, iron and manganese removal, biological oxidation, etc. [5]). The role of ozone in medical applications has also increased over the past two decades [39]. In the mid-1980s, the need to comply with environmental regulations on allowable levels of refractory substances such as pesticides [2] gave rise to another class of ozone water treatment for drinking water: ozone advanced oxidations. These processes are based on the combined use of ozone and hydrogen peroxide and/or UV radiation to generate hydroxyl radicals as indicated above [1].

Ozone is known as a very reactive agent in both water and air. The high reactivity of the ozone molecule is due to its electronic configuration. Ozone can be represented as a hybrid of four molecular resonance structures (see Fig. 2). As can be seen, these structures present negative and positively charged oxygen atoms, which in theory imparts to the ozone molecule the characteristics of an electrophilic, dipolar and, even, nucleophilic agent.

Because of this reactivity, the ozone molecule is able to react through two different mechanisms called direct and indirect ozonation. Thus, ozone can directly react with the organic matter through 1,3 dipolar cycloaddition, electrophilic and, rarely, nucleophilic reactions [40,41]. In water, only the former two reactions have been identified with many organics [42]. On the contrary, the nucleophilic reaction has been proposed in only a few cases in non-aqueous systems [43] (see examples of these mechanisms in Fig. 3).

Another group of ozone direct reactions are those with inorganic species such as Fe^{2+}, Mn^{2+}, NO_2^-, OH^-, HO_2^-, etc. [44]. These could be defined as redox reactions because in the overall process ozone acts as a true

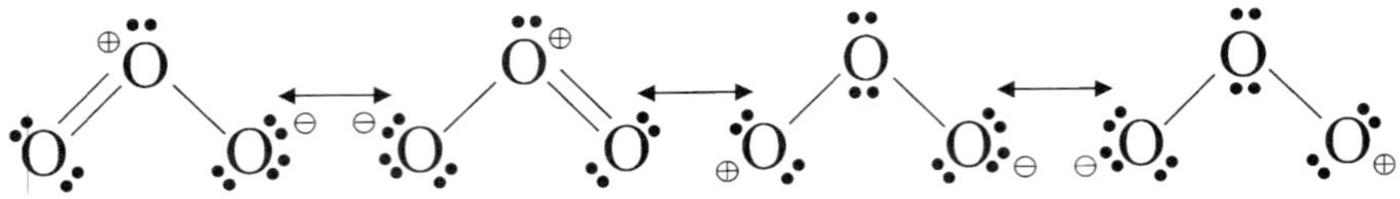

Figure 2 Resonance structures of the ozone molecule.

oxidizing agent by taking electrons whereas the other species act as true reducing agents by losing electrons. Ozone has the highest standard redox potential among conventional oxidants such as chlorine, chlorine dioxide, permanganate ion, and hydrogen peroxide (see Table 1). At acid pH, the redox reaction for ozone is as follows:

$$O_3 + 2H^- + 2e^- \longrightarrow O_2 + H_2O \qquad E^\circ = 2.07\ \mathrm{V} \qquad (3)$$

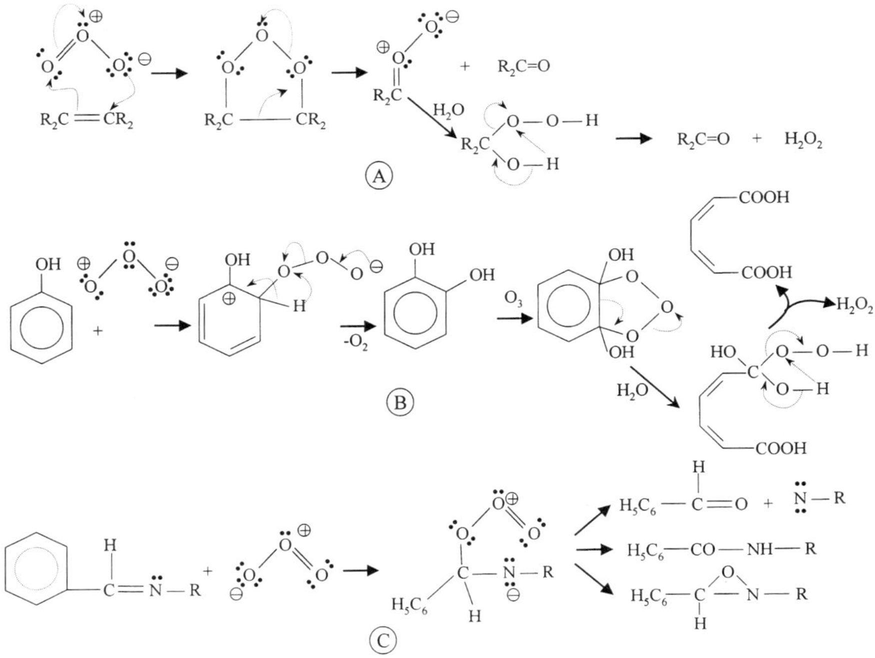

Figure 3 Direct pathways of ozone reaction with organics. (A) Criegge mechanism. (B) Electrophilic aromatic substitution and 1,3-dipolar cycloaddition. (C) Nucleophilic substitution.

However, these reactions can actually be considered as electron transfer or oxygen atom transfer reactions, as in the case of the ozone reactions with the hydroxyl and hydroperoxide ions or with the nitrite ion, respectively:

$$O_3 + OH^- \xrightarrow{k=70\ M^{-1}\ s^{-1}} HO_2^- + O_2 \quad (4)$$

$$O_3 + HO_2\bullet \xrightarrow{k=2\times10^6\ M^{-1}\ s^{-1}} HO_2\bullet + O_2\bullet^- \quad (5)$$

$$O_3 + NO_2^- \xrightarrow{k=3.7\times10^5\ M^{-1}\ s^{-1}} NO_3^- + O_2 \quad (6)$$

Reactions (4) and (5) are extremely important because they are the initiating steps of the radical mechanism leading to the formation of hydroxyl radicals when ozone decomposes.

On the other hand, the indirect type of ozonation is due to the reactions of free radical species, especially the hydroxyl radical, with the organic matter present in water. These free radicals come from reaction mechanisms of ozone decomposition in water that can be initiated by the hydroxyl ion or, to be more precise, by the hydroperoxide ion as shown in reactions (4) and (5). Ozone reacts very selectively through direct reactions with compounds with specific functional groups in their molecules. Examples are unsaturated and aromatic hydrocarbons with substituents such as hydroxyl, methyl, amine groups, etc. [45,46].

The mechanism of decomposition of ozone in water has been the subject of numerous studies, starting from the work of Weiss [47]. Among more recent studies, the mechanisms of Hoigné et al. [48] and Tomiyashu et al. [49] are the most accepted in ozone water chemistry. The main conclusion that can be drawn is that ozone stability in water is highly dependent on the presence of substances that initiate, promote, and/or inhibit its decomposition. The ozone decomposition mechanism usually assumed is given in Fig. 4 [50].

As observed from Fig. 4, ozone decomposition generates hydrogen peroxide that reacts with ozone [reaction (5)] to yield free radicals, initiating the propagation steps of the mechanism. It should be noted that hydrogen peroxide has been detected during ozonation reactions in water in the presence and absence of organics such as humic substances or aromatic compounds [51]. From this mechanism, it is also deduced that ozonation alone, or single ozonation, can be included under the group of AOTs, especially when the pH is increased. Notice that in the mechanism presented in Fig. 4 other possible reactions of ozone not shown are those corresponding to the direct pathway (see later) that leads to molecular products.

Ozone decomposition is usually a first-order process, where the apparent pseudo first-order rate constant depends on the concentration of

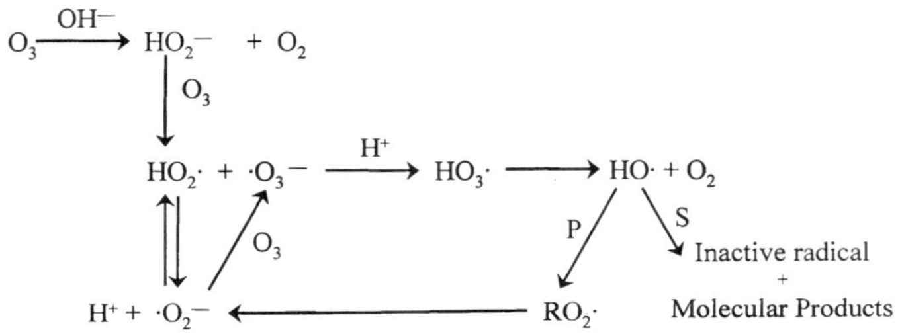

Figure 4 Scheme of ozone decomposition mechanism in water. P = promoter (e.g., ozone, methanol). S = scavenger or inhibitor (i.e., *t*-butanol, carbonate ion). I = initiators (e.g., hydroxyl ion and hydroperoxide ion).

promoters, P, inhibitors, S, and initiators, I, of ozone decomposition as was reported by Staehelin and Hoigné [48] with the equation given below:

$$-r_{O3} = \left[\sum k_{Di}C_{Mi} + \{3k_i C_{OH\_} + \sum k_{Ii}C_{Ii}\}\left(1 + \frac{\sum k_{pi}C_{pi}}{\sum k_{Si}C_{Si}}\right)\right]C_{O3} \quad (7)$$

where C_{Ii}, C_{Pi}, and C_{Si} represent the concentrations of any species i that acts as initiator, promoter, or scavenger (see also Fig. 4); C_{Mi} is the concentration of any other species i present in water other than the initiators, which react with ozone directly to yield molecular products; k_i and k_{Ii} represent the rate constants of the reactions between ozone and the hydroxyl ion and any initiator species i, respectively; k_{Pi} and k_{Si} represent the rate constants of the reactions between the hydroxyl radical and any promoter and inhibitor i of ozone decomposition, respectively; and k_{Di} represents the rate constant of the direct reaction of ozone with any other species i present in water other than the initiators. As can be deduced from Eq. (7) the half-life of ozone in water is highly dependent on the pH and matrix content of the water. For example, the half-life of ozone in distilled water can vary from about 10^{-2} sec at pH 12 to 10^5 sec at pH 2 or from 10 sec for secondary wastewater effluents to 10^4 sec for certain ground and surface waters as reported in the literature [50,52].

2. Kinetics of Ozonation

The design of ozonation contactors requires knowledge of kinetic information (see later), that is, the rate at which pollutants or matter present in water react with ozone, both directly and/or indirectly, and hence the rate of ozone absorption. Reaction rates can be calculated if rate constants of these reactions are known. Thus, the determination of rate constants represents a crucial point in contactor design. In practice, ozonation is a heterogeneous process involving ozone transfer from air or oxygen to the water phase and simultaneous chemical reactions in the aqueous medium. The kinetics of this type of processes can be established if the kinetic regime of ozone absorption is known. This process requires knowledge of the relative importance of both physical and chemical rates (diffusion of ozone and chemical reactions), which can be quantified from the dimensionless number of Hatta [53]. For any ozone–organic substance reaction in water, second-order irreversible reactions normally occur (first-order with respect to ozone and compound M) [41,44–46,54]:

$$zO_3 + M \rightarrow \text{Products} \tag{8}$$

The corresponding Hatta number, Ha, then, reduces to the following expression:

$$\text{Ha} = \sqrt{\frac{k_D C_M D_{O3}}{k_L{}^2}} \tag{9}$$

The square of this number represents the ratio between the maximum reaction rate of ozone near the water interface (film thickness) and the maximum physical absorption rate (i.e., the absorption without reaction). In Eq. (9), k_D and k_L are parameters representing the chemical reaction and physical diffusion rate constants, that is, the rate constant of the ozone–compound reaction and water phase mass transfer coefficient, respectively. Their values are indicative of the importance of both the physical and chemical steps in terms of their rates. However, two additional parameters, as shown in Eq. (9), are also needed: the concentration of the compound, C_M, and the diffusivity of ozone in water, D_{O_3}. The ozone diffusivity in water can be calculated from empirical equations such as those of Wilke and Chang [55], Matrozov et al. [56], and Johnson and Davies [57]; from these equations, at 20°C, D_{O_3} is found to be 1.62×10^{-9}, 1.25×10^{-9}, and 1.76×10^{-9} $m^2\ s^{-1}$, respectively.

The value of Ha determines the rate of the ozone reaction. Thus, for Ha < 0.3 ozone reactions are slow reactions, whereas for Ha > 3 they are fast reactions. There is also an intermediate kinetic regime defined as moderate, which is rather difficult to treat kinetically [53]. However, for most common situations, reactions of ozone in drinking water are considered as slow reactions. This does not mean that the time needed to carry out the ozonation is high (time needed to have high destruction of pollutants), but that the mass transfer rate is faster than the chemical reaction rate. For instance, in most cases, ozonation of micropollutants, which are found in very low concentrations (mg L^{-1} or μg L^{-1}), lies in this kinetic regime. In other cases, where the concentration of pollutants is higher (i.e., wastewaters containing compounds that react very fast with ozone such as phenols in high concentration), the chemical reaction rates are equal to or even much faster than the mass transfer rate and the kinetic regime is fast or instantaneous [58]. To distinguish between kinetic regimes of fast reactions, another dimensionless number, the instantaneous reaction factor, E_i, should be determined [53]:

$$E_i = 1 + \frac{zD_M C_M}{D_{O3} C^*_{O3}} \qquad (10)$$

In Eq. (10) z is the stoichiometric coefficient of the ozone-compound reaction [reaction (8)], D_M is the diffusivity of compound M in water (which can be calculated from the Wilke and Chang equation), and $C^*_{O_3}$ is the ozone solubility (or properly defined, the ozone concentration at the gas–water interface). If the parameters of Eqs. (9) and (10) are known, the kinetic regime can be established, and hence the kinetics of ozonation can be determined. Table 3 gives the kinetic equations corresponding to different kinetic regimes found in ozonation processes. As can be deduced from the equations in Table 3, the rate constant, mass transfer coefficients, and ozone solubility must be previously known to establish the actual ozonation kinetics. The literature reports extensive information on research studies dealing with kinetic parameter determination as quoted below.

Table 3 Kinetic Equations and Absorption Kinetic Regimes for Second-Order Irreversible Ozone–Organic Gas–Liquid Reactions[a]

Kinetic regime	Kinetic equation	Conditions
Very slow	$N_{O_3} = k_L a(C^*_{O_3} - C_{O_3}) = \frac{dC_{O_3}}{dt} + \sum_i r_i$	$Ha < 0.02$ $C_{O_3} \neq 0$
Diffusional	$N_{O_3} = k_L a C^*_{O_3}$	$0.02 < Ha < 0.3$ $C_{O_3} = 0$
Fast	$N_{O_3} = k_L a \frac{Ha}{\tan Ha}$	$Ha > 3$ $C_{O_3} = 0$
Fast pseudo first order	$N_{O_3} = aC^*_{O_3}\sqrt{k_D D_{O_3} C_M}$	$3 < Ha < E_i/2$ $C_{O_3} = 0$
Instantaneous	$N_{O_3} = k_L a C^*_{O_3} E_i$	$Ha > nE_i$ $C_{O_3} = 0$

[a] Equations according to film theory, see also Ref. 53. For stoichiometry, see reaction (8). N_{O_3} = ozone absorption rate, Ms^{-1}; Ha according to Eq. (9); E_i, according to Eq. (10); n = function (Ha, E_i). In the fast, pseudo first-order kinetic regime equation, a represents the specific interfacial area.

3. Ozone Solubility, Rate Constants, and Mass Transfer Coefficients

Similar to ozone decomposition, ozone solubility has been the subject of multiple studies. These studies usually propose an empirical equation for the Henry's law constant as a function of pH, ionic strength, and temperature [59,60]. For example, Sotelo et al. [60] found the following equation valid for phosphate buffer aqueous solutions at temperatures between 0 and 20°C, pH range of 2 to 8.5, and ionic strength varying from 10^{-3} to 10^{-1} M:

$$\text{He} = 1.85 \times 10^7 \exp(-2119/T)\exp(0.961) C_{\text{OH}^-}^{0.012} \quad \text{kPaM}^{-1} \tag{11}$$

where T is the absolute temperature and I is the ionic strength. Theoretically, however, He should be dependent only on temperature and the presence of ionic strength due to electrolytes in solution and independent of pH according to the following equation:

$$\log(\text{He}/\text{He}^\circ) = \sum h_i I_i \tag{12}$$

where He° is Henry's constant in ultrapure water and h is the salting-out coefficient, a function of the different ionic and dissolved gas species in water [61]. Thus, in a more recent paper, Andreozzi et al. [62] studied this problem and tried to develop an equation of this type. The authors did not arrive at this equation, but they concluded that the change in He with pH should be due to the salting-out coefficients of the different ionic species that also change with pH.

For the experimental determination of He, a mass balance of ozone in a system where ozone is absorbed in ultrapure buffered water in a semibatch reactor is usually applied:

$$k_L a\left(C_{O3}^* - C_{O3}\right) + r_{O3} = \frac{dC_{O3}}{dt} \tag{13}$$

where $k_L a$ is the volumetric mass transfer coefficient, C_{O_3} is the concentration of dissolved ozone at any time, and r_{O_3} is the ozone decomposition reaction rate. In Eq. (13), the first and second terms on the left side represent the contribution of ozone mass transfer and chemical reaction rates to the ozone accumulation rate (right side of the equation). As can be deduced, experimental results applied to Eq. (13) allows determination of the volumetric mass transfer coefficient and the ozone solubility (see also kinetic equations of Table 3). Application of Henry's law, finally, leads to the corresponding constant, He:

$$P_{O3} = \text{He} C_{O3}^* \tag{14}$$

Depending on the disappearance rate of the reacting compound or ozone, rate constants of direct ozone reactions can be obtained from both homogeneous and heterogeneous ozonation systems. Thus, for very slow reactions, homogeneous ozonation has the advantage of the absence of a mass transfer step. In these cases, the concentration of one of the reactants (ozone or compound M) can be considered constant throughout the reaction period, and the kinetics are determined by measuring the concentration of the other substance with time. When the reaction is very fast (of the order of microseconds or milliseconds) homogeneous ozonation can also be followed, but special equipment is needed to stop the reaction at very short times, for example, with stopped flow spectrophotometers [63]. For kinetic studies in these cases, heterogeneous ozonation reactions are recommended because the variation of concentration with time is much slower than in homogeneous processes. Consequently, conventional methods, such as gas or liquid chromatography or even classical spectrophotometry, can be used. For heterogeneous kinetics, the equations given in Table 3 will be needed. In Table 4, a list of rate constant values for ozone direct reactions is given together with the method of calculation. In other cases, to avoid the interferences of ozone consumption from by-products, the rate constants are deduced from competitive ozonation kinetics of two compounds: the compound whose kinetics with ozone is being determined and the reference compound. Obviously, the ozone kinetics of the reference substance must be well known. In this way, Gurol and Nekouinaini [71] and Beltrán et al. [72] have determined the rate constants of ozone fast reactions with some phenolic compounds.

Ozonation processes can also be used for determination of mass transfer coefficient. In fact, both ozone absorption in organic-free water, which is a slow gas–liquid reaction, and other ozone gas–liquid reactions have been used for this purpose. For example, Roth and Sullivan [59] and Sotelo et al. [60] determined the mass transfer coefficient from ozone absorption in organic-free water, whereas Ridgway et al. [73] and Beltrán et al. [67] carried out similar calculations from ozone absorption in water at pH 2 containing indigo and *p*-nitrophenol, respectively.

4. Kinetic Modeling

Kinetic models utilize a set of algebraic or differential equations based on the mole balances of the main species involved in the process (ozone in water and gas phases, compounds that react with ozone, presence of promoters, inhibitors of free radical reactions, etc). Solution of these equations provides theoretical concentration profiles with time of each species. Theoretical results can be compared with experimental results when these data are available. In some cases, kinetic modeling allows the determination of rate constants by trial and error procedures that find the best values to fit the

Table 4 Rate Constants of the Reaction Between Ozone and Organic Compounds in Water[a]

Organic compound	Rate constant, $M^{-1}\ s^{-1}$	pH	Method	Reference no.
Benzene	2	1.7–3	AHOK	45
Nitrobenzene	2.2	2	AHOK	64
2,6-Dinitrotoluene	5.7	2	AHOK	64
Naphthalene	3000	2	AHOK	45
Phenanthrene	2413	7	CHEK	65
Phenol	1300	2	AHOK	46
	2×10^6	7		66
Phenoxide ion	1.4×10^9	10	EX	46
p-Nitrophenol	4.5×10^6	6.5	AHEK	67
o-Chlorophenol	1600	2	CHEK	68
	2.7×10^6	7		68
Maleic acid	1000	2	AHOK	46
Formic acid	5	2–4	AHOK	46
	100	8		46
Tetrachloroethylene	<0.1	2	AHOK	45
Trichloroethylene	17	2	AHOK	45
Chloroform	<0.1	2	AHOK	45
Lindane	<0.04	2.7–6.3	AHOK	54
Atrazine	4.5	2	AHEK	69
Metoxychlor	270	2.7–6.4	AHOK	54
Aldicarb	4.4×10^4	2.1	CHOK	54
	4.3×10^5	7	CHOK	54
	4.7×10^5	7	AHEK	70

[a] AHOK = absolute rate constant by homogeneous kinetics; CHOK = competitive homogeneous kinetics; AHEK = absolute rate constant by heterogeneous kinetics; CHEK = competitive heterogeneous kinetics. EX = by extrapolation of values at lower pH.

experimental and calculated concentrations. Table 5 presents a list of studies where kinetic modeling of ozonation processes were carried out.

C. Hydrogen Peroxide Oxidation

Similar to ozone, hydrogen peroxide can react with organic matter present in water through direct and indirect pathways. In direct mechanisms, hydrogen peroxide participates in redox reactions where it can behave as an oxidant:

$$H_2O_2 + 2H^+ + 2e^- \longrightarrow 2H_2O \quad E^\circ = 1.776\ V \tag{15}$$

Table 5 Studies Dealing with AOP Kinetic Modeling Involving Ozone, Hydrogen Peroxide, and UV Radiation[a]

Compounds treated	AOP system	Reacting system	Reference no.
1,2-Dibromo-3-chloropropane	UV/H_2O_2	CMBPR	74
Chlorobutane	UV/H_2O_2	CSTPR	75
Trichloroethene, tetrachloroethene	O_3/H_2O_2	CBDR	76
VOCs	O_3/UV	PFHPR	77
Tri- and perchloroethene	O_3/H_2O_2	BHR	78
Atrazine	UV/H_2O_2	CMBPR	79
Acetone	UV/H_2O_2	CMBPR	80
Aromatic hydrocarbons	$O_3/UV/H_2O_2$	CMSBPR	81

[a] CMBPR = completely mixed batch photoreactor; CSTR = continuous stirred tank photoreactor; CBDR = continuous bubble reactor with dispersion; PFHPR = plug flow homogeneous photoreactor; BHR = batch homogeneous tube reactor; CMSBPR = completely mixed semibatch photoreactor.

or as a reductant:

$$H_2O_2 - 2e^- \longrightarrow O_2 + 2H^+ \qquad E^\circ = -0.7\ \text{V} \tag{16}$$

Indirect reactions are due to the oxidizing action of free radicals that are formed from the decomposition of aqueous hydrogen peroxide when it reacts with other inorganic compounds, such as ozone or Fe^{2+}, or when it is photolyzed.

Examples of direct reactions are mainly with inorganic compounds such as cyanides and sulfides or ozone and Fe^{2+}. Both reactions of ozone and Fe^{2+} with hydrogen peroxide represent the initiating steps of advanced oxidation processes: O_3/H_2O_2, treated later in this chapter, and the Fenton oxidation, presented in another chapter, respectively. Hydrogen peroxide, on the other hand, does not significantly react with most organic compounds, at least at appreciable rates for water treatment [6].

Hydrogen peroxide was discovered in 1818 by Tenard; the molecular structure forms an oxygen bridge, with each oxygen bonded to one hydrogen atom. In water, it is a weak acid, which dissociates to yield the hydroperoxide ion, HO_2^-:

$$H_2O_2 \leftrightarrow HO_2^- \qquad pK_a = 11.7 \tag{17}$$

It is the ionic form of hydrogen peroxide that reacts with ozone to yield free radicals as indicated before [see reaction (5)].

As far as water treatment is concerned, some of the main reactions of hydrogen peroxide are with ozone and Fe^{2+} or its photolysis with UV radiation, another advanced oxidation system commented on later. In this chapter, discussion of hydrogen peroxide reactions will be limited only to those of the O_3/H_2O_2 and UV/H_2O_2 systems.

D. UV Radiation

UV radiation is also the basis of several chemical oxidation technologies where the action of radiation and free radicals generated in the process allow for a high degree of micropollutant degradation and/or disinfection. Similar to ozonation or hydrogen peroxide oxidation, UV radiation may act on the matter present in water in two different ways: direct photolysis or indirect photolysis (e.g., free radical oxidation).

1. Background and Fundamentals

UV radiation comprises energies from about 300 kJ Einstein^{-1} (UV-A radiation, 1 einstein = 1 mol of photons), up to 1200 kJ Einstein^{-1} (vacuum UV). Table 6 shows the wavelength and energy of different UV radiation types.

For disinfection and oxidation purposes, UV-C radiation is normally used although the application of other types of UV radiation has also been reported in the literature [10]. For example, the use of UV-A or even visible radiation to treat natural organic matter present in surface water has been reported with and without the presence of catalysts [82,83]. Concerning the utilization of UV-C radiation, the most common use is 254-nm radiation due to the development of low-pressure vapor mercury lamps by Hewitt in 1901 [13]. For this reason, in this chapter the information presented mainly focuses on the use of 254-nm UV-C radiation.

Similar to ozonation processes, since the discovery of the germicidal effects of solar UV radiation by Downes and Blount in 1877 [13], UV radiation was first used for disinfection. The development of reaction mechanisms in photochemistry led to the discovery of the advantages of UV radiation as an oxidation technology. At room temperature, most molecules

Table 6 Radiation Type and Associated Energy

Radiation	Wavelength range, nm	Energy range, kJ Einstein^{-1}[a]
Infrared	>780	<155
Visible	780–400	155–300
Ultraviolet A	400–315	300–377
Ultraviolet B	315–280	377–425
Ultraviolet C	280–100	425–1198

[a] 1 Einstein $= 6.023\times 10^{23}$ photons.

reside in their lowest-energy electronic state, known as the "ground state." When UV radiation (or any other type of radiation with sufficient energy per photon) is incident upon a molecule, the radiation can be absorbed, promoting the molecule to an excited state. That is, one electron of the molecule goes to a higher-energy state or excited state. Depending on the direction of the electron spin, for most organic molecules, the excited state is either a singlet (all electron spins cancel) or a triplet state (two unpaired electrons with parallel spins). Overall the most probable transition occurs from the ground state to the singlet state. The energy difference between the ground and excited states corresponds to the absorbed energy, $h\nu$, ν being the frequency of the absorbed radiation and h the Planck constant. The molecule in the excited state has a very short lifetime (10^{-9} to 10^{-8} sec) [84], after which it returns to the ground state by one of several mechanisms (fluorescence, phosphorescence, internal conversion, collisions, etc.) or decomposes to yield a different molecule; that is, it undergoes a photochemical reaction. A simple mechanism of photochemical reaction already used in some studies [85,86] is given below:

$$M + h\nu \xrightarrow{k_a} M^* \tag{18}$$

$$M^* \xrightarrow{k_b} M \tag{19}$$

$$M^* \xrightarrow{k_c} \text{Products} \tag{20}$$

Nevertheless, mechanisms of UV radiation can be more complicated in the presence of oxygen. In this case, the electron in the excited state can be transferred to one oxygen molecule in its ground state to form the superoxide ion radical. Also, the organic molecule may first undergo homolysis in a carbon–hydrogen bond followed by reaction with oxygen to yield organic peroxyl radicals [10,87]:

$$M + h\nu \xrightarrow{k_a} M^* \tag{18}$$

$$M^* + O_2 \longrightarrow M\bullet + O_2\bullet^- \tag{21}$$

$$M - H + h\nu \longrightarrow M\bullet + H\bullet \tag{22}$$

$$M\bullet + O_2 \longrightarrow MO_2\bullet \tag{23}$$

Indirect photolysis mechanisms involve the excitation of an additional compound called a photosensitizer (PS), which in its excited state can directly oxidize the pollutant of interest. This type of mechanism was investigated by Faust and Hoigné [82] using fulvic substances as photosensitizers of phenols in natural waters. These latter mechanisms correspond to the indirect photolysis of M. In fact, Faust and Hoigné [82] reported that there are four possible routes of the excited photosensitizing action:

reactions of the photosensitizer with any compound M, with natural or added solvents, with itself, and unimolecular decay as shown below:

$$PS + h\nu \longrightarrow PS^* \tag{24}$$
$$PS^* + M \longrightarrow$$
$$PS^* + \text{Solvent} \longrightarrow \tag{25}$$
$$PS^* + PS^* \longrightarrow$$
$$PS^* \longrightarrow$$

Photosensitization for the removal of certain pollutants in photolytic processes can contribute significantly to the degradation rate. Thus, Simmons and Zepp [88] observed increases of up to 26 times of the photodegradation rates of nitroaromatic compounds due to the action of natural or commercial humic substances with solar irradiation. In another work [89], the herbicide 4-chloro-2-methylphenoxyacetic acid (MCPA) was irradiated in water with 300 nm light in the presence of different photosensitizers. This compound, which does not photolyze directly at this wavelength, could be degraded more than 95% in 5 hr when riboflavin was used as photosensitizer.

Another possible mechanism of photolysis is through the formation of secondary photooxidants that can be formed from one of the photosensitizer routes shown above. For example, a possible mechanism with humic substances as photosensitizers [90] could involve the formation of hydrogen peroxide and, subsequently, hydroxyl radicals:

$$PS + h\nu \longrightarrow PS^* \tag{26}$$
$$PS^* + O_2 \longrightarrow PS\bullet^+ + O_2^-\bullet \tag{27}$$
$$2O_2^-\bullet + 2H^+ \longrightarrow H_2O_2 \tag{28}$$
$$H_2O_2 + h\nu \longrightarrow 2HO\bullet \tag{29}$$
$$PS\bullet^+ + O_2 + RH \longrightarrow PSO_2H + R\bullet^+ \tag{30}$$
$$PSO_2H \longrightarrow PSO\bullet + HO\bullet \tag{31}$$
$$HO\bullet + M \longrightarrow \text{Products and radicals} \tag{32}$$

In addition to humic substances, nitrites and nitrates usually found in natural water also act as indirect photosensitizers to produce secondary oxidants such as hydroxyl radicals [91]. A simplified scheme of the mechanism is as follows [92]:

$$NO_2^- + h\nu \longrightarrow NO\bullet + O^-\bullet \tag{33}$$
$$NO_3^- + h\nu \longrightarrow NO_2^- + O \tag{34}$$
$$NO_3^- + h\nu \longrightarrow NO_2\bullet + O^-\bullet \tag{35}$$

$$2NO_2 \bullet + H_2O_2 \longrightarrow NO_2^- + NO_3^- + 2H^+ \quad (36)$$

$$O + H_2O \longrightarrow 2HO \bullet \quad (37)$$

$$O^- \bullet + H_2O \longrightarrow HO \bullet + OH^- \quad (38)$$

$$HO \bullet + M \longrightarrow \text{Products and radicals} \quad (39)$$

The use of nitrate to improve the photodegradation rates of pollutants has been reported. For example, Sörensen and Frimmel [92] observed that the rate of photolysis of xenobiotic compounds such as EDTA and some phenyl and naphthalene sulfonates was significantly increased in the presence of nitrates.

Finally, another possibility of photolytic reaction is due to heterogeneous processes, that is, photocatalysis. In these processes, a metal oxide surface is irradiated to yield surface hole–electron pairs [93]. For example, TiO_2 suspensions are often used for this aim to generate these species [94]:

$$TiO_2 + h\nu \longrightarrow TiO_2 + (h^+ + e^-) \quad (40)$$

The electron and hole may react at the surface with adsorbed compounds to initiate oxidation or reduction reactions:

$$e^- + O_2 \longrightarrow O_2^- \bullet \quad (41)$$

$$h^+ + H_2O \longrightarrow HO \bullet + H^+ \quad (42)$$

$$h^+ + OH^- \longrightarrow HO \bullet \quad (43)$$

$$h^+ + M \longrightarrow M^+ \bullet \quad (44)$$

Literature reports present many examples of these photocatalytic processes that will be described in detail in another chapter.

2. Kinetics of Photolysis

Once the basic mechanism of photolysis [reactions (18) to (20)] is established, the kinetics of the photochemical reaction can be studied. The kinetics of photochemical reactions is dependent on factors such as the intensity and wavelength of the incident radiation, the optical path of the radiation, and the nature of the compound irradiated and the solution in which it is present. The performance of UV radiation will also depend on the photoreactor design. For example, in a batch photochemical reactor, the rate of compound removal due to direct photolysis, assuming the mechanism of reactions (18) to (20), is as follows [95]:

$$r_{UV} = \frac{dC_M}{dt} = -\frac{1}{V}\int_V k_a\left[1 - \frac{k_b}{k_b + k_c}\right] F_M \mu q_i dV \quad (45)$$

where V is the reaction volume and μ [96] is a function of molar absorption coefficients of species present in water, ε_i, defined as follows:

$$\mu = 2.303 \sum \varepsilon_j C_i \tag{46}$$

and k_a, k_b, and k_c are the rate constants of steps (18) to (20), respectively. Notice that the first minus sign on the right side of equation (45) is due to the stoichiometric coefficient of M which is -1. As a rule, stoichiometric coefficients of reacting products are negative.

In Eq. (45) q_i, is the flux of incident radiation, which varies according to the geometrical configuration of the photoreactor and photochemical model used, and F is the fraction of absorbed radiation that M absorbs:

$$F_M = \frac{\varepsilon_M C_M}{\sum \varepsilon_j C_i} \tag{47}$$

where ε_M and ε_i are the molar absorption coefficient and molar absorptivity or optical density of M and any compound i, respectively, present in solution that also absorbs radiation. The term $[k_a(1-k_b/(k_b+k_c)]$ in Eq. (45) can be defined as the quantum yield of M, ϕ_M. The quantum yield is perhaps the most important parameter in UV radiation kinetics because it measures the fraction of the excited molecules that are transformed into products. This parameter is defined as the moles of M decomposed per Einstein absorbed (1 Einstein being 1 mol of photons, 6.023×10^{23} photons). Substances with high quantum yields that are constant over a wide range of wavelengths are usually called actinometers, and are used to measure the intensity or flux of incident radiation as shown later. In any case, compounds of high quantum yield are prone to decomposition through UV radiation. In Table 7 values of quantum yield and molar absorption coefficients for different compounds and oxidants in water are shown as examples.

Equation (45) can be solved by applying different photoreactor models. The literature reports several photochemical reactor models for both homogeneous and heterogeneous reactors [11,108,109]. In practice, annular photoreactors are often used (see Fig. 5); therefore, models for this type of reactor are considered here. For other types of reactors, attention should be given to other publications [109].

Here Eq. (45) is solved for three models: the linear source with emission in parallel planes to the lamp axis (LSPPM) model, the point with spherical emission (PSSE) model, and a semiempirical model based on Lambert's law (LLM). The first two models come from the solution of a radiation balance equation throughout the photorreactor assuming different hypotheses.

Table 7 Values of Quantum Yields and Molar Absorption Coefficients for Different Water Pollutants and Oxidants in Water

Pollutant or oxidant	ε, M^{-1} cm^{-1} (λ, nm)	ϕ, mol $Einstein^{-1}$ (λ, nm)	Reference no.
Ozone	3300 (253.7)	0.62 (253.7)	97
H_2O_2	18.7 (253.7)	0.5[d]	98,99
HO_2^-	210	0.5[d]	98,99
NO_3^-	9900 (200)		100
Phenol	516 (213–400)	0.05 (213–400)	101
2-Chlorophenol	1920 (272)	0.03 (296)	102
2-Chlorophenolate	3760 (293)	0.30 (296)	102
3-Chlorophenol	1750 (273)	0.09 (254 or 296)	102
3-Chlorophenolate	3000 (292)	0.13 (254 or 296)	102
4-Chlorophenol	1650 (278)	0.25 (254 or 296)	102
4-Chlorophenolate	2400 (296)	0.25	102
Nitrobenzene	5564 (254)	0.007 (254)	24
2,6-Dinitrotoluene	6643 (254)	0.022 (254)	24
Fluorene	16654 (254)	0.0075 (254)	73
Phenanthrene	40540 (254)	0.0069 (254)	73
Acenaphthene	1333	0.0052	73
Acenaphthalene	26941	0.004	103
1,3-Dichlorobenzene		0.06 (213–400)	101
1,3,5-Trichlorobenzene		0.043	101
Trichloroethylene	18.3 (254)	0.88 (254)	95
Atrazine	2486 (254)	0.05	104
Simazine	2512	0.06	105
CEAT[a]	3056	0.038	106
CIAT[b]	3211	0.035	106
CAAT[c]	3161	0.018	106
Alachlor	540	0.177	107
Parathion		0.0076 (240–320)[e]	89
Parathion		0.0016[f]	89

[a] CEAT = 2-chloro-4-ethylamine-1,3,5-*s*-triazine;
[b] CIAT = 2-chloro-4-isopropylamine-1,3,5-*s*-triazine;
[c] CAAT = 2-chloro-4,6-diamine-1,3,5-*s*-triazine.
[d] In the presence of scavengers.
[e] At pH 9.6.
[f] At pH 5.7.

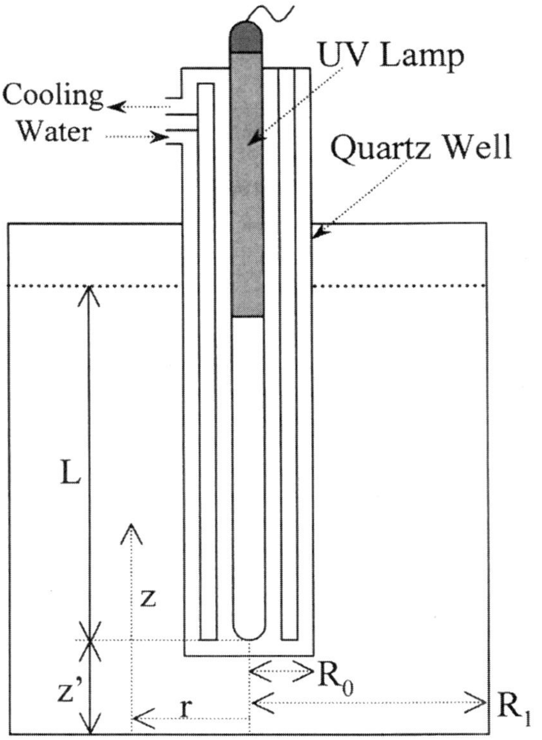

Figure 5 Scheme of annular photoreactor.

The LSPPM assumes that the lamp can be represented as a consecutive line of points, each one emitting radiation in all directions contained in a plane perpendicular to the lamp axis. An approximate equation for q_i when $r \ll L$ is as follows [110]:

$$q_i = \frac{R_o q_o}{r} \exp[-\mu(r - R_o)] \tag{48}$$

where R_o and r are defined in Fig. 5 and q_o is the flux of incident radiation at the inside wall of the photoreactor (for $r = R_o$), which is calculated from actinometry experiments. By substitution of Eq. (48) into Eq. (45) and after integration the photolysis rate of M becomes:

$$r_{UV} = \frac{dC_M}{dt} = -F_M \phi_M \frac{2\pi R_o q_o L}{V} [1 - \exp[-\mu(r - R_o)]] \tag{49}$$

The PSSE model considers each point of the lamp emitting radiation in all space directions. The expression for q_i is as follows [111,112]:

$$q_i = \frac{E_o}{4\pi} \int_{z'}^{z'+L} \frac{\exp\left(-\mu\overline{\omega}\right)}{r^2 + (z - z')} dz' \tag{50}$$

where r, z, and z' are defined in Fig. 5, E_o is the radiant energy of the lamp per unit of length, and ϖ is given by Eq. (51):

$$\overline{\omega} = \frac{(r - R_o)\sqrt{r^2 + (z - z')^2}}{r} \tag{51}$$

Substitution of Eqs. (50) and (51) in Eq. (45) leads to the photolysis rate of M:

$$r_{UV} = \frac{dC_M}{dt} = -F_M \frac{E_o \phi_M \mu}{4} \int_0^L \int_{R_o}^{R_1} \int_{z'}^{z'+L} \frac{\exp(-\mu\overline{\omega})}{r^2 + (z - z')} dr dz dz' \tag{52}$$

In this model, the solution is obtained by numerical integration of Eq. (52).

The LL model has been applied to numerous works [74,95,96,113]. Because of its simplicity, this model has a great acceptance among researchers of advanced oxidation kinetics. The model constitutes a simplification of the LSPP model. The rate of photolysis is, in this case:

$$r_{UV} = \frac{dC_M}{dt} = -F_M \phi_M I_o [1 - \exp[-\mu y_L]] \tag{53}$$

where I_o is the intensity of incident radiation and y_L is the effective path of radiation through the photoreactor.

In all these models, knowledge of parameters such as q_o (LSPP model), E_o (PSSE model), or I_o and y_L (LL model) are necessary to determine the photolysis rate of M. These parameters are determined experimentally by actinometry experiments [86]. It is noteworthy to mention that the use of these theoretical models (LSPP or PSSE models) implies that all radiation incident into the solution is absorbed without end effects, reflection, or refraction. In experimental photoreactors, it is not usual to fulfill all these assumptions because of the short wall distance of the photoreactor. For instance, to account for such deviations, Jacob and Dranoff [114] introduced a correcting equation, as a function of position. Another important disadvantage is the presence of bubbles that leads to a heterogeneous process as, for example, in the case of O_3/UV oxidation. In this case, photoreactor models should be used [109]. This is the main reason for which the LL model is usually applied in the laboratory for the kinetic treatment of photochemical reactions. In the LLM, the effective path of radiation can be considered as the correction function accounting for deviations from ideality.

E. Combined Oxidations: O_3/H_2O_2, UV/H_2O_2, and O_3/UV

The reaction of ozone and hydrogen peroxide in its ionic form and photolysis of both oxidants constitute the initiation reactions leading to a mechanism of hydroxyl radical formation in water. This mechanism is basically the same for all these advanced oxidation systems, whereas the main differences lie in the initiating steps. These oxidation technologies have been applied for the treatment of pollutants in water for more than two decades.

1. Background and Fundamentals

Photolysis of hydrogen peroxide was first studied by Baxendale and Wilson [99]. They reported that the decomposition of 1 mol of hydrogen peroxide needed one Einstein of incident 254 nm UV radiation:

$$H_2O_2 \xrightarrow{h\nu} 2HO\bullet \tag{54}$$

This quantum yield corresponds to the overall process where hydrogen peroxide is not just removed by reaction (54), its direct photolysis, but also by parallel reactions involving the hydroxyl and hydroperoxide radicals as shown below:

$$HO\bullet + H_2O_2 \longrightarrow HO_2\bullet + H_2O \tag{55}$$

$$HO_2\bullet + H_2O_2 \longrightarrow HO\bullet + H_2O + O_2 \tag{56}$$

although the latter reaction is negligible.

These authors [99] also showed that the quantum yield of both forms of hydrogen peroxide, H_2O_2 and HO_2^-, remained constant over a wide range of concentrations and UV radiation intensity. Baxendale and Wilson [99] also carried out experiments in the presence of organic substances, such as acetic acid (a well-known scavenger of the hydroxyl radicals) so that the measured rate of hydrogen peroxide disappearance corresponded to the rate of its direct photolysis [reaction (54)]. From these experiments, they found that the rate was half of that of the process in the absence of hydroxyl radical scavengers. Consequently, they concluded that the quantum yield of reaction (54) was 0.5 mol of hydrogen peroxide per Einstein. This value is called the primary quantum yield of hydrogen peroxide photolysis.

As reported by Staehelin and Holgné [115], ozone reacts only with the ionic form of hydrogen peroxide, the hydroperoxide ion, HO_2^-. These authors studied this reaction at different hydrogen peroxide concentrations and in the presence of methylmercury hydroxide, another hydroxyl radical scavenger. At pH values below the pK_a of hydrogen peroxide ($pK_a = 11.7$), these authors observed that the rate of reaction (5) increased one order of magnitude per unit increase of pH. They found a second-order reaction rate constant of $2.8 \times 10^6\ M^{-1}\ s^{-1}$. A similar behavior can be noticed with the ozone decomposition rate in organic-free water in the absence of hydrogen peroxide [reaction (4)], although the rate constant is several orders of magnitude lower ($70\ M^{-1}\ s^{-1}$) as found by the same authors and by Forni et al. [116].

Taube [97] studied the photolysis of aqueous ozone and postulated the formation of hydrogen peroxide, which he found to be formed with almost exact stoichiometry. Taube [97] reported a quantum yield for ozone of 0.62 at 254 nm. Later, Prengle [117] claimed that ozone photolysis yields atomic oxygen, which directly leads to hydroxyl radicals. To elucidate which mechanism is the correct one, Peyton and Glaze [118] later studied this reaction and concluded that hydrogen peroxide is first formed from ozone photolysis without formation of atomic oxygen. From these studies [115–118] and others reported by Staehelin and Hoigné [48] and Tomiyasu et al. [49], the mechanism of any type of advanced oxidation system involving ozone, hydrogen peroxide, and UV radiation can be established. A simplified scheme of this mechanism, applied to the oxidation of a potential pollutant M, is presented in Fig. 6.

The main reactions of the mechanism are given below:

$$O_3 + H_2O \xrightarrow{h\nu} (2HO\bullet) \longrightarrow H_2O_2 \quad (57)$$

$$H_2O_2 \xrightarrow{h\nu} 2HO\bullet \quad (58)$$

$$O_3 + OH^- \longrightarrow HO_2^- + O_2 \quad (59)$$

$$H_2O_2 \rightleftharpoons HO_2^- + H^+ \qquad pK = 11.7 \quad (60)$$

$$HO_2^- + O_3 \longrightarrow HO_2\bullet + O_3\bullet \quad (61)$$

$$HO_2^- \rightleftharpoons + H^+ + O_2^-\bullet \quad (62)$$

$$O_2^-\bullet + O_3 \longrightarrow O_2 + O_3^-\bullet \quad (63)$$

$$O_3^-\bullet + H^+ \longrightarrow HO_3\bullet \longrightarrow HO\bullet + O_2 \quad (64)$$

Reactions (58) and (61) constitute the main initiation reactions of the mechanism leading to the formation of hydroxyl radicals. The reaction between ozone and the hydroperoxide anion constitutes the faster mechanism as has been demonstrated in previous studies [119,120]. The reaction between ozone and the hydroxyl ion is negligible compared to the reaction between ozone and the hydroperoxide ion because the rate constants of these reactions differ by several orders of magnitude: $70\ M^{-1}\ s^{-1}$ for reaction (59)

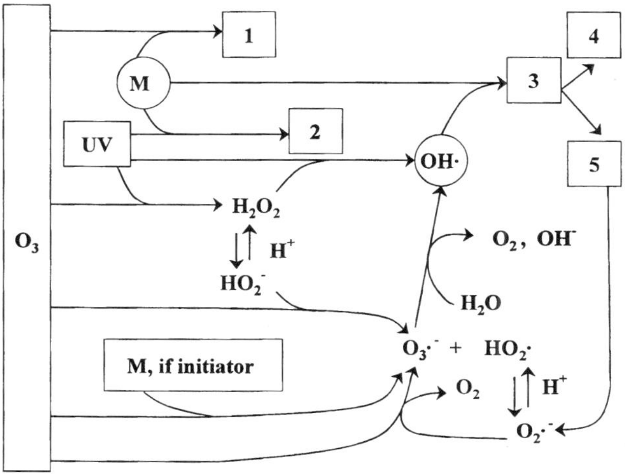

Figure 6 Scheme of $O_3/UV/H_2O_2$ oxidation processes. Key: 1: From direct ozonation. 2: From direct photolysis. 3: From free radical oxidation. 4: Intermediate pathway if M is inhibitor. 5: Intermediate pathway if M is promoter.

vs. 2.8×10^6 M^{-1} s^{-1} for reaction (51) [115]. The relative importance, however, is pH dependent. For the conditions usually applied in water treatment, reaction (61) is faster. An extensive study of these reactions can be seen in the work of Staehelin and Hoigné [115]. Reaction (61) also predominates against the direct photolysis of hydrogen peroxide [reaction (58)]. This is due to the low value of the molar absorption coefficient of hydrogen peroxide (see Table 7) and the very high rate constant value of reaction (61) (see above). Furthermore, the direct photolysis of hydrogen peroxide competes, among others, with the direct photolysis of ozone, which is also faster. This can be deduced from the values of the product between the quantum yield and molar absorption coefficient of each photolysis reaction. The product is 2046 L $Einstein^{-1}$ cm^{-1} for the case of ozone and 105 L $Einstein^{-1}$ cm^{-1} for the case of hydrogen peroxide at its most favorable conditions (see Table 7), that is, at alkaline pH when hydrogen peroxide is present in the ionic form. A thorough study of the competition between these initiation reactions can be followed from the studies of Beltrán [119,120].

For chemical structures refractory to direct reactions with ozone and UV photolysis, free radical reactions are fundamental. Among free radicals, the hydroxyl radical shows a high oxidizing power, and it is generally accepted as the main oxidant in these advanced oxidation systems.

The reactions through which hydroxyl radicals participate are shown below [121]:

* Hydrogen abstraction:

$$HO\bullet + RH \longrightarrow R\bullet + H_2O \tag{65}$$

* Electrophilic addition:

$$HO\bullet + PhX \longrightarrow HOPhX \tag{66}$$

* Electron transfer:

$$HO\bullet + RX \longrightarrow RX\bullet^{+} + HO^{-} \tag{67}$$

Organic radicals formed in these reactions may further react with oxygen (in an aerated medium as in water treatment) to yield organic peroxyl radicals that can eventually react with compounds present in the medium to release the superoxide ion radical (see route through 5 in Fig. 6; see also the work of von Sonntag and Schuchmann [122] for more details about peroxyl radical reactions). In these cases, compounds that react with the hydroxyl radical are known to be promoters of ozone decomposition because the superoxide ion radical consumes ozone at a fast rate [see reaction (63) above]. On the contrary, if the reaction between hydroxyl radical and compound M does yield inactive radicals, M is known as a scavenger or inhibitor of ozone decomposition (see route to 4 in Fig. 6). Many natural substances such as humic substances and carbonates are known to possess such a role [121]. However, the case of carbonate ion is rather special because it reacts with hydroxyl radicals to yield the carbonate ion radical:

$$CO_3^{=} + HO\bullet \xrightarrow{k=3.7\times10^{8}\ M^{-1}\ s^{-1}} CO_3^{-}\bullet + OH^{-} \tag{68}$$

$$HCO_3^{-} + HO\bullet \xrightarrow{k=2\times10^{7}\ M^{-1}\ s^{-1}} CO_3^{-}\bullet + H_2O \tag{69}$$

which is also known to react with organic compounds in water. Table 8 presents a list of rate constants of these reactions. In addition, the carbonate ion radical also reacts with hydrogen peroxide, if present in water, to regenerate the carbonate ion and the hydroperoxide ion radical that eventually leads to the hydroxyl radical in the presence of ozone. Consequently, there could be a fraction of carbonates that do not inhibit the ozone decomposition in water. Notice that in some cases when ozone is applied, hydroxyl radical oxidation is negligible or does not develop because direct ozonation is very fast. This is, for example, the case in the ozonation of phenolic compounds under neutral or alkaline conditions, where the rate constants of the direct ozone reactions vary between 10^6 and

Table 8 Rate Constants of the Reaction Between the Carbonate Ion Radical and Organic Compounds in Water

Compound	Rate constant, $M^{-1}\ s^{-1}$	pH
Benzene	$<5\times10^4$	11.7
Toluene	4.3×10^4	7.0
Phenol	1.3×10^7	7.0
Phenoxide ion	4.7×10^8	11.2
p-Chlorophenoxide ion	1.9×10^8	12.2
Aniline	5.4×10^8	7.0
Acetone	1.6×10^2	12.1–12.7
Formate ion	1.1×10^5	6.4

Source: Ref. 123.

$10^{10}\ M^{-1}\ s^{-1}$. As observed, the order of magnitude is similar to or even higher than that of the reactions with the hydroxyl radical. An extensive study of this situation can be seen in a previous paper [124].

2. Chemical Kinetics

Once the mechanism of a reaction is established, kinetic studies constitute the next step to determine the rate of pollutant degradation. Kinetic laws must be deduced experimentally following well-established rules [125]. For the degradation of a compound M through $O_3/H_2O_2/UV$ oxidation, the rate of M disappearance is given by Eq. (70):

$$r_M = r_{UV} + r_D + r_H + r_R \tag{70}$$

where the four terms of the right-hand side of Eq. (70) refer to the contribution of direct reactions (photolysis, ozonation, reactions with hydrogen peroxide) and free radical oxidation, respectively. By assuming an irreversible first-order reaction for every reactant (a global second-order reaction), the last three rates are as follows:

$$r_D = z_M k_D C_M C_{O_3} \tag{71}$$

$$r_H = z_H k_H C_M C_{H_2O_2} \tag{72}$$

$$r_R = z_R k_{HOM} C_M C_{HO} \tag{73}$$

where the values of z_M, z_H, and z_R (usually -1) are the stoichiometric coefficients of M in their corresponding reactions. Notice that the negative value of the z coefficients of M in Eqs. (71) to (73) is due to conventional stoichiometric rules because M disappears. In the case of direct photolysis, considering the LL model, the photolysis rate is expressed by Eq. (53). Once the steady state approximation is accounted for, it is easy to show that the concentration of hydroxyl radicals is given by Eq. (74):

$$C_{HO} = \frac{r_i}{\sum_j k_{sj} C_j} \tag{74}$$

where r_1 is the rate of initiation steps and the denominator is the sum of the products of concentrations of substances that inhibit the decomposition of ozone and the corresponding rate constant of their reactions with the hydroxyl radical, a term usually called the scavenging factor. When a high concentration of hydrogen peroxide is present in the reaction medium, the concentration of hydroxyl radicals, however, can be limited by the transfer of ozone into water [76]. Some investigations have also reported [76,126,127] that during ozonation, hydrogen peroxide plays a double role as initiator and inhibitor of ozone decomposition. Thus, for concentrations usually up to 10^{-3} or 10^{-2} M, the increase of hydrogen peroxide concentration leads to an increase of the ozonation rate of M, but for concentrations above these values hydrogen peroxide inhibits the ozonation rate. In these latter cases, reactions of ozone are so fast that ozone is not detected dissolved in water and the process becomes mass transfer controlled. According to absorption theories [61], a complex kinetic equation can be deduced from the solution of microscopic mass balance equations of ozone, hydrogen peroxide, and M, but a simplified equation is obtained from the macroscopic mass balance equations as previously reported for trichloroethylene and tetrachloroethylene oxidation with O_3/H_2O_2 in a semibatch system [76]. Thus, the final equation for the concentration of hydroxyl radical is as follows [76,128]:

$$C_{HO} = \frac{k_L a C^*_{O_3}}{\sum_j k_{sj} C_j} \tag{75}$$

In this latter case, one of the terms of the denominator in Eq. (5) contains the concentration of hydrogen peroxide because this also plays a role as a scavenger. Thus, from Eq. (75), it follows that an increase in hydrogen peroxide concentration leads to a decrease in hydroxyl radical concentration and hence, a decrease of r_R.

The double role as scavenger and initiator, observed for hydrogen peroxide in the O_3/H_2O_2 system, has also been reported in the UV/H_2O_2 system. It should be noted that hydrogen peroxide does not inhibit the ozone decomposition and Eq. (75) is valid only in the cases that ozone is present in the reaction mixture and the process is chemically controlled (low concentrations of hydrogen peroxide). This is because reactions of hydrogen peroxide with the hydroxyl radical release the superoxide ion radical that subsequently reacts with ozone [see reaction (63)], and eventually another hydroxyl radical is generated (see also Fig. 6). However, in the UV/H_2O_2 case, and in the absence of ozone, reaction (63) does not proceed, and hydrogen peroxide always acts as a terminator of propagation chain reactions through its reactions with the hydroxyl radical.

Table 9 presents the main reactions involved in the mechanism of $O_3/H_2O_2/UV$ radiation systems. According to this mechanism and from rate constant data, concentrations, and the intensity applied for UV radiation, the relative importance of the initiating reactions can be deduced. In previous studies, comparative studies of the reaction rates of pollutant degradation through these systems have been reported [119,120,124]. From these studies, it was shown that in UV/O_3 oxidation, initiation of free radicals through the photolysis of hydrogen peroxide [reaction (58)] is negligible if compared to reaction (61), and that reaction (59) starts to be of comparable rate to reaction (61) when the pH is higher than 10. A comparative study between the importance of direct and free radical oxidation pathways of ozonation based on the gas absorption theory is also presented [61]. Thus, depending on the kinetic regime of ozone absorption, concentration of reactants (M, hydrogen peroxide), pH, intensity of UV radiation, etc., the direct or the indirect oxidation of a given compound M will predominate.

3. Kinetic Modeling

To complete the fundamentals of these advanced oxidation systems, some discussion should be included about the kinetic modeling studies found in the literature. As stated previously, kinetic modeling allows predictions of the performance of any reacting system and represents an aid for process design (see Table 5). Kinetic modeling of $O_3/H_2O_2/UV$ systems has been studied extensively for the last decade with promising results especially for the oxidation of volatile organochlorine compounds [74–78]. For higher molecular weight compounds, with the exception of a few studies [79,81,129], the literature lacks information. This is probably because high molecular weight compounds, while reacting with ozone or hydroxyl radicals or being photolyzed, lead to intermediate compounds that also consume oxidants and/or UV radiation, thus making the kinetics very complex. In a recent work [81], a kinetic model for any $O_3/H_2O_2/UV$ radiation system to treat aromatic compounds such as phenanthrene and nitrobenzene has been proposed. In the absence of UV radiation, the authors found that the oxidation of phenanthrene, which only reacts directly with ozone, could be modeled acceptably without considering the presence of intermediates. A different situation holds for the case of the

Table 9 Reaction Mechanism for the O_3/H_2O_2/UV System: Elementary Reactions, Reaction Rate Constant, and Quantum Yield Data[a]

Reaction	Rate constant or quantum yield	Reaction	Rate constant, pK_a, or quantum yield
$O_3 + OH^- \rightarrow O_2^{\cdot -} + HO_2^{\cdot}$	$k_1 = 70$	$HCO_3^- + HO^{\cdot} \rightarrow HCO_3^{\cdot} + OH^-$	$k_{21} = 2\times10^7$
$O_3 + HO_2^- \rightarrow HO^{\cdot} + O_2^{\cdot -} + O_2$	$k_2 = 2.2\times10^6$	$CO_3^{2-} + HO^{\cdot} \rightarrow CO_3^{\cdot -} + OH^-$	$k_{22} = 3.7\times10^8$
$O_3 + H_2O + h\nu^{\bullet} \rightarrow H_2O_2^{\cdot} + O_2$	$\phi_{O_3} = 0.64$	$CO_3^{\cdot -} + O_2^{\cdot} \rightarrow CO_3^{2-} + O_2$	$k_{23} = 7.5\times10^8$
$H_2O_2 + h\nu^{\bullet} \rightarrow 2HO^{\cdot}$	$\phi H_2O_2 = 0.5$	$CO_3^{\cdot -} + O_3^{\cdot -} \rightarrow CO_3^{2-} + O_3$	$k_{24} = 6\times10^7$
$O_3 + O_2^{\cdot -} \rightarrow O_3^{\cdot -} + O_2$	$k_5 = 1.6\times10^9$	$CO_3^{\cdot -} + H_2O_2 \rightarrow HCO_3^- + HO_2$	$k_{25} = 8\times10^5$
$HO_3^{\cdot} \rightarrow HO^{\cdot} + O_2$	$k_6 = 1.4\times10^5\ s^{-1}$	$CO_3^{\cdot -} + HO_2^- \rightarrow CO_3^{2-} + HO_2^{\cdot}$	$k_{29} = 5.6\times10^7$
$O_3 + HO^{\cdot} \rightarrow HO_4^{\cdot}$	$k_7 = 3.0\times10^9$	$H_3PO_4 + HO^{\cdot} \rightarrow H_2O + H_2PO_4^{\cdot}$	$k_{27} = 2.6\times10^6$
$HO_4^{\cdot} \rightarrow HO_2^{\cdot} + O_2$	$k_8 = 2.8\times10^4\ s^{-1}$	$H_2PO_4^- + HO^{\cdot} \rightarrow OH^- + H_2PO_4^{\cdot}$	$k_{28} = 2.2\times10^7$
$H_2O_2 + HO^{\cdot} \rightarrow HO_2 + H_2O$	$k_9 = 2.7\times10^7$	$HPO_4^{2-} + HO^{\cdot} \rightarrow OH^- + HPO_4^{\cdot -}$	$k_{29} = 7.9\times10^5$
$HO_2^{\cdot -} + HO^{\cdot} \rightarrow HO_2^{\cdot} + OH^-$	$k_{10} = 7.5\times10^9$	$HO_2^{\cdot} \Leftrightarrow H^+ + O_2^{\cdot -}$	$pK_{30} = 4.8$
$HO_4^{\cdot} + HO_2^{\cdot} \rightarrow O_3 + O_2 + H_2O$	$k_{11} = 10^{10}$	$HO_3^{\cdot} \Leftrightarrow H^+ + O_3^{\cdot -}$	$pK_{31} = 8.2$
$HO^{\cdot} + HO^{\cdot} \rightarrow H_2O_2$	$k_{12} = 5\times10^9$	$H_3PO_4 \Leftrightarrow H^+ + H_2PO_4^-$	$pK_{32} = 2.2$
$HO^{\cdot} + O_2^{\cdot -} \rightarrow OH^- + O_2$	$k_{13} = 10^{10}$	$H_2PO_4^- \Leftrightarrow H^+ + HPO_4^-$	$pK_{33} = 7.2$
$HO^{\cdot} + HO_3^{\cdot} \rightarrow H_2O_2 + O_2$	$k_{14} = 5\times10^9$	$HPO_4^- \Leftrightarrow H^+ + PO_4^{3-}$	$pK_{34} = 12.3$
$HO_3^{\cdot} + O_2^{\cdot -} \rightarrow OH^- + 2O_2$	$k_{16} = 10^{10}$	$H_2CO_3 \Leftrightarrow H^+ + HCO_3^-$	$pK_{35} = 6.4$
$HO_3^{\cdot} + HO_3^{\cdot} \rightarrow H_2O_2 + 2O_2$	$k_{16} = 5\times10^9$	$HCO_3^- \Leftrightarrow H^+ + CO_3^{2-}$	$pK_{36} = 10.4$
$HO_4^{\cdot} + HO_4^{\cdot} \rightarrow H_2O_2 + 2O_3$	$k_{17} = 5\times10^9$	$H_2O_2 \Leftrightarrow HO_2^- + H^+$	$pK_{37} = 11.7$
$HO_4^{\cdot} + HO_3^{\cdot} \rightarrow H_2O_2 + O_2 + O_3$	$k_{18} = 5\times10^9$	$M + zO_3 \rightarrow$ Intermediates	$k_D = ?$
$HO_4^{\cdot} + HO^{\cdot} \rightarrow H_2O_2 + O_3$	$k_{19} = 5\times10^9$	$M + HO \rightarrow$ Intermediates	$k_{HOM} = ?$
$HO_4^{\cdot} + O_2^{\cdot -} \rightarrow OH^- + O_2 + O_3$	$k_{20} = 10^{10}$	$M + h\nu \rightarrow$ Intermediates	$\phi_M = ?$

Adapted with permission from Ref. 81. Copyright 1999 American Chemical Society.

[a] Units of rate constants and quantum yields (ϕ) in $M^{-1}\ s^{-1}$ and mol photon^{-1} unless indicated. Reactions of intermediates are not given but also take part in the mechanism. M represents any compound present in water. Notice that the quantum yield (0.5) of hydrogen peroxide photolysis corresponds to solutions irradiated in the presence of scavenger substances.

oxidation of nitrobenzene, which mainly reacts through hydroxyl radicals. In this case, the model does not lead to good results unless the mass balance equations of intermediates are included. According to the authors [81] the key factor is the presence of hydrogen peroxide, proposed to be formed from the ozonation of the starting aromatic compound. Thus, the system becomes an O_3/H_2O_2 oxidation process, regardless of the initial presence of hydrogen peroxide. This observation is fundamental to matching the experimental and predicted data. The model, however, fails when predicting the rates of UV/O_3 oxidation, probably because indirect reactions of organic peroxyl radicals, reported elsewhere as a possible route of oxidation [10], were not accounted for.

III. IN-DEPTH TREATMENT OF $O_3/UV/H_2O_2$ PROCESSES

Applied treatment of these techniques requires the necessary equipment to produce ozone and/or radiation. Hydrogen peroxide is the only reagent that can be purchased, transported to the plant, and used there. Ozone and UV radiation, however, require some capital investment, the importance of which depends on the required ozone production rate and/or the UV light flux needed for the system. These issues constitute one of the main drawbacks of these systems: They have to be generated in situ.

In addition to the generation equipment, another important element is the reactor or contactor chamber. Real reactors are similar in geometry or at least in their fundamental aspects to those of laboratory or pilot plants. In this section, a brief summary of ozone and UV contactors is given. For more detailed information related to this equipment, other literature sources should be consulted [5,11,130].

A. Ozone-Based Processes

In actual ozonation processes, contactor chambers are similar to a bubble column; the bottom chamber is equipped with many diffuser plates through which the oxidizing gas is fed. In other cases, however, the gas can be fed through injectors and intimately mixed with the water in static mixers to improve mass transfer or to avoid problems derived from the precipitation of iron or manganese oxides.

Ozone contactors supplied with diffusers are usually divided into stages (see Fig. 7) whose number depends on the major objectives of ozonation. Thus, for instantaneous or mass transfer controlled ozonations, two stages are sufficient, whereas for slow ozonation reactions a higher number is required. In order to reach plug flow behavior in the water, these

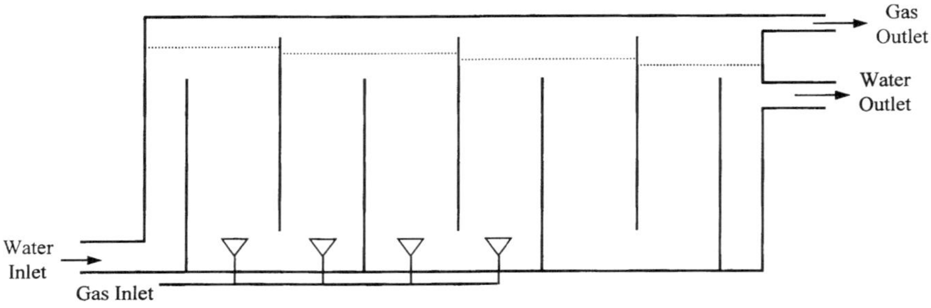

Figure 7 Conventional water–ozone contact chamber.

stages are separated by walls. This allows minimization of the residence time needed to obtain a given destruction of pollutants. In the first stages, ozone is fed through porous diffusers located at the bottom and in the last stages dissolved ozone circulates to provide a residual concentration of oxidant.

As was shown previously in some examples [15–18] in a large ozonation plant for water treatment, residual ozone in the gas exiting the ozonation stages could be sent back to the head of the water plant where it is injected in another compartment to aid flocculation, remove iron and manganese, or reduce the trihalomethane formation potential (see Fig. 8). In these cases, it is not surprising that these plants could also have a final disinfection ozonation step.

Ozonation contactors, however, are not always designed similar to bubble columns. In some cases where mass transfer represents an important problem, new contactor designs have been used, such as static mixers or the deep-U-tube contactor.

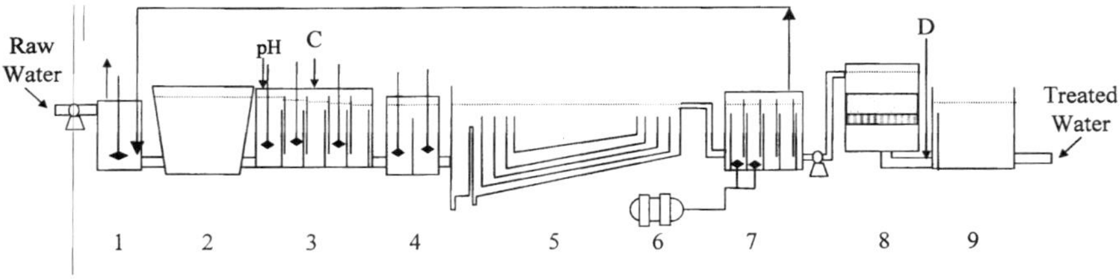

Figure 8 Water treatment plant with pre-ozonation and intermediate ozonation steps. 1: Pre-ozonation chamber. 2: Equalization basin. 3: Coagulation chamber (C = coagulant). 4: Flocculation. 5: Sedimentation. 6: Ozonators. 7: Ozonation chamber. 8. Active carbon filtration unit. 9: Final disinfection (D = disinfectant).

Static mixers are constituted by numerous perforated metal elements (i.e., sieve plates) housed in cylindrical tubes where mass transfer is significantly increased. These units are especially recommended for ozonation processes where the rate of ozone is mass transfer controlled (i.e., disinfection, decolorization, etc.) [131,132]. At the entrance of the tubes, water and the ozonated gas are simultaneously pumped and injected, respectively, yielding a gas–liquid system with a very high specific interfacial area for ozone transfer (1000 to 10,000 m^{-1}) and low energy dissipation (0.1 to 5.0 kW m^{-3}) [130]. One possible disadvantage of static mixers is the pressure drop along the unit that limits the water flow rate applied. However, in some systems, such as those with sieve plates, the large free-area percentage, about 55%, allows a moderate pressure drop of 3000 to 7000 Pa with practically 100% ozone utilization [130].

The deep-U tube is constituted by two vertical concentric tubes approximately 20 m deep through which water and gas flow (inner tube) downwards and then upward through the outer tube. The main advantage of this system is the high driving force to dissolve ozone because of the high pressure and plug flow behavior due to the high velocity [133]. Disadvantages are the necessity of perforating the soil, which may involve geotechnical problems, and royalties required because it is a proprietary system.

B. UV Radiation Based Processes

For these cases, the so-called photoreactors are usually cylindrical chambers that contain inner quartz sleeves where UV lamps are placed. These lamps are generally medium-pressure mercury vapor lamps (emitting radiation between 200 to 300 nm) of up to 60 kW each. The sleeve can have automatic devices that wipe at regular intervals any precipitate formed or particles deposited on the sleeve walls, thus avoiding the problem of reduced light transmittance [134]. When combined with ozone, UV radiation photoreactors are fed with ozone previously incorporated to the water flow [135] or with ozone fed directly into the photoreactor chamber [136]. The former methodology has the advantage that no ozone is destroyed in the gas phase by UV radiation; the latter process presents the advantage that the ozone driving force, which favors ozone dissolution in water, is enhanced by its aqueous photolysis.

IV. DEGRADATION OF POLLUTANTS

The main objective of advanced oxidation processes involving ozone, UV light, and hydrogen peroxide is the removal or degradation of pollutants from water. Environmental agencies have classified some substances as priority pollutants after a detailed study of their potential toxicity and health hazards [137]. Thus, phenols, aromatic hydrocarbons such as polycyclic aromatic hydrocarbons (PAHs), pesticides such as *s*-triazine compounds, volatile organochlorine compounds, others such as 1,4-dioxane, etc., are considered priority pollutants. As a result of industrial or agricultural activities, these substances have been found in many water environments. In this section, brief information concerning laboratory studies on $O_3/UV/H_2O_2$ oxidation is given together with a few specific examples concerning main pollutant families and general issues such as that of the bromate ion, detoxification, and biodegradability improvement of wastewater after a chemical oxidation step.

A. Laboratory-Scale Experimental Design

Laboratory work concerning the study of the advanced oxidation of pollutants focuses on aspects related to the influence of operating variables on the reaction rate (oxidant dose, UV intensity, pH, temperature, presence of free radical scavengers, etc.), kinetics and mechanism (determination of rate constants, quantum yields, etc.), identification of intermediates and final products, and kinetic modeling. Most reported studies deal with laboratory-prepared aqueous solutions, but a few deal with treatment of natural water and even wastewaters. By virtue of the number of studies published, a wide range of pollutants have been studied including different phenols, aromatic hydrocarbons, volatile organochlorine compounds, pesticides, and some other substances such as 1,4-dioxane and dissolved natural organic matter (such as humic substances). Table 10 provides a list of some of these studies published in the last decade (1990–2000). Also included in the table are the oxidation and reactor system type, the compounds treated, and the main objectives of the work.

B. Examples of Laboratory Studies

In this section, examples of laboratory studies concerning $O_3/UV/H_2O_2$ advanced oxidations of some water pollutants are discussed. They have been chosen because of the high interest that their oxidation treatment has attracted among researchers in the field. Three different types of pollutants have been chosen: phenols of different nature, *s*-triazine herbicides, and some volatile compounds, mainly chlorinated organics of low molecular weight (VOCs). Information is also given on the treatment of 1,4-dioxane, another important priority pollutant. These studies represent the scope and objectives to be reached in this type of laboratory research.

Table 10 Studies on AOP Involving Ozone, Hydrogen Peroxide, and/or UV Radiation of Organic Compounds in Water

Compound	Reacting system	Main features	Reference no.
Pyrene	O_3	BR, $C_0 = 10$–200 μg L^{-1}. Intermediate identification	138
Chlorophenols	O_3	SBBC, $C = 600$ mg L^{-1}. Effect on biodegradability and COD	139
Phenanthrene	O_3, O_3/H_2O_2, O_3/UV	SBBPR, LVP lamp 2.4×10^{14} photon s^{-1}. $C = 0.5$ mg L^{-1}	140
Phenols	O_3, O_3/H_2O_2, O_3/UV, O_3/TiO_2 UV/H_2O_2,	CBPR, LVP lamp 8 W, $C = 10$–170 mg L^{-1}. O_3/H_2O_2 better system	141
p-Chlorophenol	O_3/UV	SBPR, MVP lamp, $C_0 = 2.5$–10 mg L^{-1}, TOC changes. TOC kinetics	142
Chlorophenols	UV/H_2O_2	BPR, LVP lamp, 3,5 W. $C_0 = 3.067 \times 10^{-4}$ M. Absorbance effect on kinetics	143
Pyrene	O_3	SBBT, $C_0 = 1$–5 mM, toxicity test. Intermediates	144
Chloro- and nitrophenols	O_3/UV	SBPR, I_o, pH influence, $C = 3 \times 10^4$ M, LVP lamps 5 W maximum. Empirical kinetics	145
Trichlorobenzene	O_3, O_3/H_2O_2, O_3/UV, UV/H_2O_2	CBPR, $\tau = 5$ min. Kinetics, $C_0 = 1.2$–1.6 mg L^{-1}, LVP lamp 3.15 W	146
Toluene	O_3, O_3/H_2O_2	SFC, $C_0 = 0.75$–1.5×10^{-3} M, $C_H = 2 \times 10^{-4}$ to 0.02 M, pH = 3–11. Kinetics, reaction mechanism	147
Chlorophenols	O_3, O_3/UV, UV/H_2O_2	BPR, SBBT, LVP lamp, 0.8 W L^{-1}, $C_0 = 10^{-4}$ to 4×10^{-4} M. Reaction products, kinetics, toxicity	148
Phenol	UV/H_2O_2	BPR, LVP lamp, 6 W, $C_0 = 10^{-3}$ to 10^{-2} M. COD conversion, kinetics	149

Table 10 continued

Compound	Reacting system	Main features	Reference no.
Metol	UV/H_2O_2	BPR, $C_0 = 5\times10^{-4}$ M, pH and H_2O_2 influence, Kinetics	150
VOCs	O_3	BR, C a few milligrams per liter. Removal of chloroethylenes	151
VOCs	O_3/UV	CFPR, $C_0 = 100$–600 μg L^{-1}, LVP lamp, 60 W, $I_o = 2$–40 W m^{-2}. Kinetics	152
Herbicides	O_3	BR. Kinetics. C in μM. Rate constants O_3–herbicide reaction	153
Atrazine	O_3	CFBC, $C_0 = 10$ μg L^{-1}. Intermediates, reaction mechanism	154
VOCs	O_3/UV	CFPR, LVP lamp, 60 W, $C_0 = 100$–600 μg L^{-1}. Kinetics modeling	77
Chloroethanes	UV/H_2O_2	BPR and CFPR, LVP lamps, $I_o = 3.06$–4.44×10^{-6} einstein s^{-1}, $C_0 = 0.4$–1 μg L^{-1}, $C_H = 0.1$–10^{-5} M. Kinetic modeling	155
Atrazine	O_3, O_3/H_2O_2	CSTR simulation. Kinetic modeling. $C = 3\times10^{-8}$ M	126
Triazine herbicides	$O_3/UV/H_2O_2/Fe^{2+}$	SBBPR, LVP lamp 15 W, $C_0 = 1.9\times10^{-6}$ M	156
Dyes	O_3, O_3/UV, O_3/H_2O_2	SBBT, $C_0 = 260$–500 mg L^{-1}. Color and TOC removal, metal release	157
Pesticides	O_3, O_3/H_2O_2	BR, $C_0 = 1$–12 μg L^{-1}. Percentage removal at different conditions	158
Pesticides	O_3, O_3/H_2O_2	SBBC, $C_0 = 2\times10^{-6}$ M, O_3/H_2O_2 ratio effect	159
Atrazine	UV/H_2O_2	BPR, LVP lamp, 6.1–5.4×10^{-6} einstein s^{-1}, C_0 in μM. Kinetic modeling	79
Nonionic surfactants	O_3	SBR, $C_0 = 5.6\times10^{-4}$ M, pH 4 and 9.5. Reaction mechanism	160
RDX	O_3, O_3/H_2O_2, UV/H_2O_2, O/UV	BPR, SBPR, LVP lamp, $C_0 = 6$–170 μg L^{-1}. Kinetics, intermediate identification, reaction mechanism	162

Atrazine	O_3, O_3/H_2O_2, O_3/UV	CBPR, pH influence, mechanism, CATZ $= 6.95\times10^{-5}$ M, four LVP lamps of 15 W	163
Humic acids	O_3/UV	SBBC, kinetics, TOC rates. Intermediates	164
NOM	O_3	BR, changes on DOC	165
Humic substances	O_3	SBBC, $C_{HS} = 50$–500 mg L^{-1}. Identification of compounds	166
Fulvic acid	O_3, O_3/H_2O_2, O_3/catalyst	BR, $DOC_o = 2.84$ mg L^{-1}, $BDOC_o = 0.23$ mg L^{-1}. Changes of DOC and BDOC.	167
NOM	O_3	SBBT, $DOC_o = 3$–6 mg L^{-1}. Changes of DOC, BDOC, THMFP, AOC, intermediates	168
Low MW organics	O_3/UV	SBBPR, LVP lamp, 100 W. Compounds with 1 to 6 C atoms. Phenol oxidation. By-product identification. TOC removal	169
Polysaccharides	O_3, O_3/UV, γ radiolysis	BPR, LVP lamp, $Io = 21.7$ W m^{-2}, $C_0 = 5\times10^{-4}$ M. Comparison of treatments	170
Ethers	O_3, O_3/H_2O_2	SBBC, $C = 2\times10^{-3}$ M. By-product identification	171
Chelating agents	O_3/H_2O_2	SBT, $C_0 = 0.1$ M, $C_H = 0.25$–1.33 M. Organic removal. Ozone efficiency	172
Hydrophilic xenobiotics	UV/H_2O_2	CFPR, C_0 about 10^{-5} M. Nitrate influence, kinetics	173
PCDDs/PCDFs, 1,4-dioxane	UV/H_2O_2	BPR, MVP lamp, $I_o = 5.59\times10^{-4}$ einstein s^{-1}. Mechanism, intermediates	174
Mineral oil wastewater	O_3, O_3/H_2O_2, O_3/UV	SBBPR, LVP lamp, 17 W, 80–90% COD reduction. Kinetic modeling	175

BR = batch reactor; SBBC = semibatch bubble column; SBBPR = semibatch bubble photoreactor; CBPR = continuous bubble photoreactor system; SBPR = semibatch photoreactor; BPR = batch photoreactor; SBBT = semibatch bubble tank; CFPR = continuous flow photoreactor; CFCB = continuous flow bubble column; CST = continuous flow stirred tank; SBR = semibatch stirred reactor; SFC = stopped flow cell.

1. Phenols

Because of their extensive use in industrial activities, phenolic compounds are abundant in many wastewaters. They are present in oil refining, petrochemical, plastic, pesticide, carbon liquefaction, and food processing industrial wastewaters. In addition, phenol-like structures are part of the macromolecular natural humic substances present in water; humics are known precursors of trihalomethane compounds in drinking water chlorination [176,177]. These compounds are refractory to conventional processes such as biological oxidation because of their toxicity for microorganisms [178].

As observed in Table 10, the literature reports a large amount of information on the chemical oxidation of phenols. Subcritical oxidations such as those involving ozone, hydrogen peroxide and/or UV radiation, Fenton reagent, or wet air oxidation have been applied. In general, phenols can be photolyzed with UV radiation and react very fast with both ozone and hydroxyl radicals. The latter reactions are favored by the presence of the hydroxyl group, which activates the aromatic ring for electrophilic substitution reactions. The degree of reactivity depends on the presence of other substituent groups such as methyl, chlorine, and nitro entities [46,71].

Because of their importance in pesticide manufacturing, chlorophenols are one of the most studied types of phenols. Thus, Esplugas et al. [142] studied the direct photolysis of *p*-chlorophenol with a 250-W, medium-pressure vapor arc lamp with and without ozone. They followed the process of mineralization from total organic carbon (TOC) data. They observed that UV photolysis hardly affected the process (5% mineralization after 30 min). When ozone was fed in at a mass flow of 23 mg/min under similar conditions, they observed almost complete mineralization. The synergism between ozone and UV light was also much more effective than ozonation alone, which, under the same operating conditions, resulted in roughly 45% mineralization. For kinetic purposes, they assumed a first-order reaction for both ozonation and UV photolysis of TOC. For the latter, they used the linear spherical model to define the radiation field inside the photoreactor.

The ozonation of *o*-chlorophenol was studied by Beltrán et al. [68] in a standard agitated bubble reactor. They observed the importance of pH on this type of process and found that the degradation rate improved with increasing pH. Removal of *o*-chlorophenol was achieved in minutes by using concentrations of ozone approximately a few milligrams per liter with initial concentrations of phenols approximately 10^{-4} M. It was observed that the pH effect was not due to a higher ozone decomposition to generate hydroxyl radicals but to the direct action of the ozone molecule. Degradation rates of *o*-chlorophenol by ozonation were similar in both the presence and absence of *t*-butanol, a well-known hydroxyl radical scavenger. Moreover, the reaction rate was a function of the degree of dissociation of *p*-chlorophenol:

$$k_D = k_{D-nd}(1 - \alpha) + k_{D-d}\alpha \quad (76)$$

where $k_{D\text{-}nd}$ and $k_{D\text{-}d}$ are the rate constants of direct ozonation of non-dissociated and dissociated forms of the phenolic compound, respectively. Values of these rate constants were found to be 1300 and 1.4×10^9 M^{-1} s^{-1}, respectively, which confirmed other values reported previously [46]. The higher value of the rate constant of the ozone-ionic form reaction was because the $-O^-$ group is a more powerful activating group for electrophilic substitution reactions than is the –OH group. Fig. 9 shows the evolution of k_D with pH.

At neutral pH values, a high variation of reactivity of *o*-chlorophenol with ozone is observed with small changes in pH. These k_D values were found after gas-liquid kinetic model application to experimental data. Thus, the ozonation of *o*-chlorophenol was found to be a pseudo first-order fast gas–liquid reaction as found by Hoigné and Bader [46] in homogeneous ozonation experiments. Trends of k_D with pH for other ozone-phenol reactions are similar to that depicted in Fig. 9.

In another work, Shen et al. [143] studied the hydrogen peroxide photolyzed oxidation of some chlorophenols such as *o*-chlorophenol (CP), 2,4-dichlorophenol (DCP), and 2,4,6-trichlorophenol (TCP). They used a low-pressure mercury vapor lamp of 5.3 W maximum output. They assumed direct photolysis followed pseudo first-order kinetics, and they found that

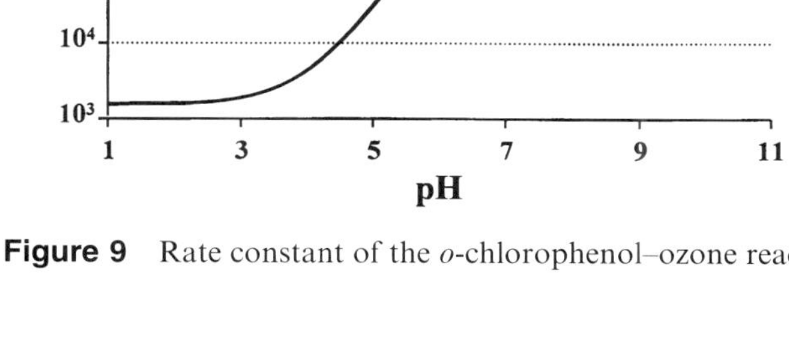

Figure 9 Rate constant of the *o*-chlorophenol–ozone reaction as a function of pH.

variation of the rate constant with pH was similar to that observed for the ozone–phenol reaction. This fact was attributed to the different molar absorption coefficients of the nonionic and ionic forms of phenols. The authors also presented the changes in the apparent pseudo first-order rate constant of hydrogen peroxide photolysis with pH. The pseudo first-order rate constants for direct photolytic processes were expressed in a similar manner to that presented in Eq. (76) for the ozone–chlorophenol rate constant. For the UV/H_2O_2 process at low pH, they found the hydroxyl radical oxidation to be more important than direct photolysis. However, they observed the opposite when the pH was increased. These effects were attributed to an increased absorptivity of phenols with pH that makes direct photolysis the predominant process. Finally, these authors observed the evolution of intermediate compounds and CO_2 formation.

In another work [148], mineralization of chlorophenols was followed by measurement of the chloride ion concentration in water during O_3, O_3/UV, and UV/H_2O_2 treatment. Compared to ozonation results, these authors did not find any improvement of chlorophenol degradation rate when UV light was used in combination with ozone. The authors also determined the toxicity of aqueous solutions of treated chlorophenols with the *Daphnia magna* 24-hr test. After 90–95% elimination of CPs they did not find any negative effect of ozonation by-products on the aquatic organism.

Finally, Sotelo et al. [179], while studying the ozonation of resorcinol and phloroglucinol, two precursors of trihalomethanes (THM) during water chlorination, found some polar intermediates that confirmed the proposed phenol mechanism reported elsewhere [42]. From the identified intermediates it was deduced that ozonation of phenols yields more oxygenated compounds that eventually could be removed in biological steps.

2. Atrazine

The widespread application of herbicides in agriculture has resulted in many polluted surface waters. As a result, numerous pesticides/herbicides have been treated in bench-scale laboratory studies with $O_3/UV/H_2O_2$ processes during the last 10 years (see Table 10). Among them, many studies focused on the treatment of atrazine and other *s*-triazine herbicides (simazine, propazine, etc.). Atrazine is a priority pollutant that similar to other individual pesticides has a very low maximum contaminant level (MCL) ($0.1\ \mu g\ L^{-1}$ for the European Environmental Commission according to Directive 80/778/EEC). In some countries atrazine cannot be used but it is still found in many surface waters. In France, for example, atrazine was banned on September 28, 2001. From applied technologies, only carbon adsorption [180] and possibly advanced oxidations can be recommended to remove some of these herbicides from water. Notice that oxidation of some herbicides could lead to unknown by-products that are difficult to remove, even with activated carbon adsorption. Here, some of these studies are discussed.

Adams and Randtke [154] and Adams et al. [181] were among the pioneers in the study of atrazine ozonation in laboratory-prepared and surface waters. These authors identified some of the ozonation by-products such as deethylatrazine, deisopropylatrazine, etc., and proposed a simplified reaction mechanism. The molecular structure of atrazine suggested that hydroxyl radical oxidation could be more effective than direct ozonation alone. Thus, the application of different advanced oxidation and photolysis technologies to eliminate atrazine was investigated in other studies [69,79, 182]. Results obtained from these studies show the paramount importance of free radical oxidation as compared to direct photolysis or direct ozonation. Degradation of the herbicide was found to be much slower in the presence of *t*-butanol due to the scavenging effect of hydroxyl radicals.

In spite of the importance of advanced oxidations, these processes yield many intermediates that eventually lead to refractory compounds such as cyanuric acid, ammeline, or the more recently reported chlorodiaminotriazine [183]. Thus, in this work [183], an extensive study of ozonation alone and ozone–H_2O_2 oxidation of atrazine under different experimental conditions has been made, with special emphasis on by-product formation with the aim of proposing a complex mechanism of reactions. Kinetic studies [69,79,153,182] have also been conducted to determine rate constants of the reactions between ozone or the hydroxyl radical and atrazine and the main intermediates, quantum yields in photochemical processes, stoichiometric coefficients, etc. Rate constants of the ozone–herbicide reactions are negligible ($< 10\ M^{-1}\ s^{-1}$), which is in agreement with the role of the hydroxyl radical in the oxidation process. Kinetic modeling of oxidation processes has also been reported. In these studies, the process efficiency of advanced oxidation is determined from the solution of molar balance equations of the main species (ozone, hydrogen peroxide, ATZ, and intermediates, carbonates, etc.) of oxidation. Finally, some other studies deal with the toxicity of strazine and by-products. Canelli [184] reported that cyanuric acid, one of the end products of atrazine oxidation, presents a rather low toxicity. Upham et al. [185] studied the toxicity of some herbicides (atrazine, alachlor, carbofuran, etc.) and their ozonated aqueous solutions. They studied the appearance of nongenotoxic mechanisms of cellular injury such as intercellular gap junctional communication (GJIC). Their results showed that for atrazine, ozonated samples were less toxic with respect to GJIC than the parent compound.

Specific studies on the application of UV and UV/H_2O_2 systems for the treatment of atrazine and intermediates such as CEAT (2-chloro-4-

ethylamine-1,3,5-*s*-triazine), CIAT (2-chloro-4-isopropylamine-1,3,5-*s*-triazine) and CAAT (2-chloro-4,6-diamine-1,3,5-*s*-triazine) have also been reported [79,106]. In these studies the influence of variables such as hydrogen peroxide concentration, pH, presence of carbonates, etc., are discussed. In addition, quantum yields at 254 nm and rate constants of the reaction between the hydroxyl radical and herbicides are given (see Tables 2 and 7). Experiments in both ultrapure and natural waters were conducted. The lamps used were low-pressure mercury vapor units of about 3×10^{-6} to 6×10^{-6} Einstein L^{-1} s^{-1}. On the whole, the oxidation rate is significantly increased by the presence of hydrogen peroxide at concentrations lower than 10^{-1} M. Above this value, the process is inhibited because hydrogen peroxide scavenges hydroxyl radicals (see Table 9), thus competing with the target compounds. Also, concentrations of carbonates equal to or higher than 10^{-2} M reduce the oxidation rates to values lower than those achieved with UV radiation alone.

3. Chlorinated Volatile Organic Compounds

Chlorinated volatile organic compounds (VOCs) are another important group of pollutants commonly studied. Literature reports include numerous works focusing on several types of advanced oxidations. These pollutants are present in water due to their use as cleaning agents, degreasers of metals, paint and ink constituents, etc. [137].

Based on their reactivity in ozone advanced oxidation processes, VOCs can be classified into two groups. Unsaturated compounds such as trichloroethylene react not only with hydroxyl radicals, but also with ozone through direct reaction and by direct photolysis [77,186]. The other group is formed by saturated halogenated hydrocarbons such as methane, ethane, or propane derivatives that do not absorb UV radiation. These compounds are not directly photolyzed and do not react with molecular ozone [126]. For VOC treatment, the ozone oxidations are not recommended, because due to their volatility an important fraction of the pollutant is transferred to the gas phase instead of being oxidized when using ozone gas processes. For example, Beltrán et al. [127] determined that volatilization during ozonation processes (O_3, O_3/H_2O_2, and O_3/UV) of trichloroethylene (TCE) and trichloroethane (TCA) in semibatch bubble columns was highly dependent on the gas flow rate. Volatilization was the only route of TCA removal from water during ozonation at gas flow rates between 2 and 20 L hr^{-1} and 500 Pa ozone partial pressure. Volatilization was reduced when ozone was applied in combination with hydrogen peroxide or UV radiation but still remained high ($>40\%$). In the case of TCE, due to the carbon–carbon double bond, the volatilization contribution was lower (i.e., 50% during ozonation alone with 20 L hr^{-1} gas flow rate). In order to overcome this problem, some studies carried out the ozonation homogeneously. Thus, Hayashi et al. [77] studied the O_3, O_3/UV (low-pressure mercury vapor lamp) oxidation of several VOCs (1,1,1 trichloroethane, trichloroethylene, and tetrachloroethylene) in a 30-mm-wide, 3-mm-thick, and 200-mm-high chamber where aqueous solutions of ozone and VOCs were fed and irradiated simultaneously. They used a kinetic model to observe the importance of some parameters such as intensity of radiation or reactor radius on degradation rates. Sunder and Hempel [78] also used homogeneous ozonation in a tubular reactor (14.8 m long, 18 mm i.d.) to study the kinetic modeling of the O_3/H_2O_2 of TCE and PCE (tetrachloroethylene). Under the conditions applied, the disappearance of VOCs was rapid (within seconds). In this work, mineralization was also determined by measuring the amount of chloride detected in water. They found a linear relation between the amount of pollutant lost and the amount of chloride ion released in water.

UV/H_2O_2 oxidation of VOCs has also been studied in detail and several studies reported kinetic models to predict the efficiency of the process. For example, Liao and Gurol [75], Glaze et al. [113], De Laat et al. [155] and Crittenden et al. [74] studied the UV/H_2O_2 oxidation of VOCs such as n-chlorobutane, 1,2-dibromo-3-chloropropane and tri- and tetrachloroethanes in batch photoreactors with low-pressure mercury vapor lamps. Effects of pH, concentration of hydrogen peroxide, UV intensity and the presence of carbonates or fulvic substances were variables studied.

In other studies [34,186], the treatment of trichloroethylene and tetrachloroethylene with UV/H_2O_2 system is reported. For example, Beltrán et al. [34] determined the rate constants of the reactions between hydroxyl radical and TCE and TCA with this oxidizing system at high concentration of hydrogen peroxide. Hirvonen et al. [186] observed the formation of chloroacetic acids during the hydrogen peroxide photo-assisted oxidation of tri- and tetrachloroethylene. These authors [186] report that formation of chloroacetic acids diminishes when hydrogen peroxide is simultaneously applied to UV radiation compared to UV radiation alone.

Other studies also give an extensive analysis of by-product identification. The work of Dowideit and Von Sonntag [187] is a good example. These authors studied the ozonation of ethylene and some methyl- and chlorine-substituted derivatives. They showed that these compounds are practically removed via Criegge mechanism (see Fig. 3) to yield a carbonyl compound and a hydroxyhydroperoxide, which, in the case of chlorine-substituted derivatives, rapidly releases HCl. It should be remembered that in the case of saturated VOCs, hydroxyl radical is the main route of oxidation regardless of the presence of ozone.

4. 1,4-Dioxane

This is another synthetic organic chemical classified as a priority pollutant of water and potential carcinogen for humans. The presence of a cyclic structure and oxygen in the carbon chain makes this compound very refractory to both aerobic and anaerobic biological oxidation. As a consequence, much research has been done on the treatment of this compound with advanced oxidation processes [174,188].

Adams et al. [188] studied the oxidation of 1,4-dioxane with ozone and hydrogen peroxide at different conditions of oxidant concentration and the presence of hydroxyl radical scavengers (carbonates, other chemical such as methanol, 1,3-dioxolane, etc.). These authors followed not only the disappearance of 1,4-dioxane but also other general parameters such as COD, TOC, and BOD. They observed that single applications of oxidants (ozone or hydrogen peroxide) were unable to remove 1,4-dioxane, but the combination of both oxidants significantly enhanced the oxidation rate to reduce COD, TOC, and 1,4-dioxane. BOD was also enhanced; that is, the aqueous solution treated with the O_3/H_2O_2 system became more amenable to biodegradation. Also, a critical mole ratio between H_2O_2 and O_3 (between 0.5 to 1) was observed above which the oxidation rate diminished. It is evident that this was because of the double character of initiator and scavenger of hydrogen peroxide as shown in other works. This was corroborated by the fact that hydroxyl radical scavengers (carbonates, etc.) decreased the oxidation rate of 1,4-dioxane, which confirms the process as an AOT.

In another work, Stefan and Bolton [174] carried out the oxidation of 1,4-dioxane with the UV/H_2O_2 system. They used a 1-kW, multivariable-wavelength, UV–vis Hg lamp (200–400 nm) with the aim of establishing the mechanism of reactions. Under the experimental conditions applied, very high oxidation rates of 1,4-dioxane were observed (i.e., 1,4-dioxane at about 9×10^{-4} M disappeared in approximately 10 min with an initial concentration of 1.5×10^{-2} M of hydrogen peroxide). Some intermediates formed were 1,2-ethanediol diformate, methoxy acetic acid, acetic acid, glycolic acid, formic acid, oxalic acid, and glyoxal. Also, reduction of TOC was observed to occur if hydrogen peroxide was present. A detailed mechanism of reactions was proposed to justify the experimental findings.

5. Natural Organic Matter and the Bromate Issue

Because trihalomethane compounds were identified in chlorinated water from drinking water-treatment plants [37], multiple research studies have addressed their efforts to identify the nature of dissolved natural organic matter and the action of oxidants to yield by-products, usually called disinfection by-products. As for natural organic matter (NOM), after the discovery of THMs, it was soon reported that humic substances present in surface water are the precursors of most of the disinfection by-products. It is known that humic substances include macromolecules possessing numerous nucleophilic positions (i.e., aromatic rings with hydroxyl substituent groups among others) prone to electrophilic attacks [176]. These positions are the first to react with chlorine or ozone and they are responsible for many disinfection by-products. Identification of some potentially dangerous ozonation by-products such as bromate (see later) and the uncertainty about others such as aldehydes has led to new research aimed to identify these compounds in ozonated water. In these studies (see Table 10), experiments were carried out to observe the influence of oxidant dose and origin of dissolved natural organic matter on the nature and evolution of by-products.

Regarding ozonation processes, the treatment with ozone leads to halogen-free oxygenated compounds (except when bromide is present), mostly aldehydes, carboxylic acids, ketoacids, ketones, etc. [189]. The evolution of analytical techniques and their combined use have allowed some researchers to identify new ozone by-products. This is the case of the work of Richardson et al. [189,190] who combined mass spectrometry and infrared spectroscopy together with derivatization methods. These authors found numerous aldehydes, ketones, dicarbonyl compounds, carboxylic acids, aldo and keto acids, and nitriles from the ozonation of Mississippi River water with 2.7–3 mg L^{-1} of TOC and pH about 7.5. They also identified by-products from ozonated–chlorinated (with chlorine and chloramine) water. In these cases, they found haloalkanes, haloalkenes, haloaldehydes, haloketones, haloacids, brominated compounds due to the presence of bromide ion, etc. They observed a lower formation of halocompounds formed after ozone–chlorine or chloramine oxidations than after single chlorination or chloramination, showing the beneficial effect of preozonation.

García-Araya et al. [191] conducted ozonation studies with and without the presence of hydrogen peroxide on humic substances extracted from different surface waters. They observed the effect of ozone dose on the absorbance at 254 nm (A_{254}), the NOM concentration, and the amount of specific ketoacids formed such as ketomalonic, pyruvic, and glyoxylic acids. They reported that hydrogen peroxide had a negligible effect on NOM reduction, and that A_{254} is more efficiently decreased with ozonation alone. Because this parameter is due to aromatic and unsaturated structures that mainly react with ozone through direct reaction, it can be postulated that hydroxyl radical advanced oxidation should not be recommended. It was also reported that hydroxyl radical oxidation is the main route for elimination of glyoxylic acid.

In another interesting work, Volk et al. [167] investigated the comparative effect of ozonation alone and combined with hydrogen peroxide or the TiO_2 catalyst. They observed the variation of UV absorbance, DOC, and biological and refractory dissolved organic carbon (BDOC and RCOD, respectively). Although no significant differences were noticed in the UV absorbance, DOC reduction was improved in combined ozonations with increases of BDOC in all cases. As reported by Richardson et al. [190], ozonation disinfection by-products can be partially removed through biofiltration. Measurements indicated that about 50% of assimilable organic carbon (AOC) was removed in biofiltration. Individual carboxylic acid removal varied from 45% to 90% whereas that of aldehydes and ketones was between 35% and 95%, respectively. Thus, biofiltration could be a recommended process to destroy some of the oxygenated disinfection by-products of water ozonation.

Also, research on the use of UV/H_2O_2 system to treat NOM has been carried out [192–194]. Backlund [192] carried out the oxidation of an aquatic humic material with a 254-nm UV lamp in the presence and absence of hydrogen peroxide. He reported significant increases in the humic removal rate with the combined system. He noticed the destruction of macromolecules to yield smaller fragments and identified some compounds such as oxalic acid, acetic acid, malonic acid, etc., which accounted for 20% and 80% of the NOM in UV and UV/H_2O_2 systems. Also, he observed that these processes do not lead to any mutagenic activity in the treated water.

Wang et al. [193,194] studied the influence of hydroxyl radical scavengers (carbonates) and type of UV–vis light on the removal of humic acids from laboratory-prepared water. They used a 450-W high-pressure Hg vapor lamp inserted in quartz or Pyrex tubes. They observed a slight oxidizing effect when the Pyrex tube was used. The major effects were observed with the quartz tube as expected because this allowed the UV light to penetrate into the aqueous solution. Again, a critical concentration of hydrogen peroxide was observed (about 3×10^{-3} M). Carbonates presented a negative effect.

In addition to organic disinfection by-products, there are also some inorganics such as bromate ion, which deserves special attention due to its potential hazard. It should be noted that the World Health Organization [3] has recommended a maximum guideline of 25 $\mu g\ L^{-1}$ for bromate ion in drinking water. For actual ozonation conditions, only waters containing high bromide levels ($>200\ \mu g\ L^{-1}$) can lead to bromate concentration higher than the above indicated upper limit. From a survey of 38 European surface waters [195], except for three cases with none or up to 700 $\mu g\ L^{-1}$, concentrations of bromide ion were found to range from 30 to 320 $\mu g\ L^{-1}$. However, the improvement of analytical techniques can lead to the reduction of guidelines, making research on bromate formation still necessary. Thus, considerable work has been conducted to establish bromate-formation mechanisms from ozonation or O_3/H_2O_2 oxidation of bromide-containing water [195,196]. Bromate ion is formed in water from the oxidation of bromide by molecular ozone. Also, it can be added to water as an impurity when sodium hypochlorite is used in a chlorination step. Formation of bromate ion from ozonation is strongly influenced by pH, presence of ammonia, carbonates, hydrogen peroxide, and organic matter in water. Hydroxyl radicals react with bromide ion to give the $BrOH\cdot^-$ radical that eventually leads to the $BrO\cdot$ radical. Also, in natural waters, the carbonate ion radical, $CO_3\cdot^-$, formed in the reaction of carbonate ions with hydroxyl radicals, reacts with hypobromite ion to yield the $BrO\cdot$ radical. This radical, after disproportionation, yields BrO_2^- that leads to bromate after oxidation with the hydroxyl or the carbonate ion radicals via $BrO_2\cdot$ disproportionation. Hence, the carbonate ion radical acts as a secondary oxidant. In AOTs involving hydrogen peroxide and ozone, bromate ion is also formed, although its concentration depends on the ozone/peroxide ratio. As the hydrogen peroxide concentration increases, formation of $HO\cdot$ radicals increases and hence the bromate ion concentration increases. However, if hydrogen peroxide is further increased, ozone is preferentially consumed by this substance so that there is less ozone available to react with bromide or hypobromide ions to yield bromate. Also, at high levels of hydrogen peroxide, hypobromide ion can be reduced to bromide ion according to the following reaction:

$$H_2O_2 + HBrO \xrightarrow{k=2\times10^4\ M^{-1}\ s^{-1}} H^+ + Br^- + H_2O + O_2 \qquad (77)$$

If water contains ammonia, hypobromous acid reacts with it to finally yield a monobromamine, and bromate formation is reduced. In systems where hydroxyl radicals are not scavenged by organic compounds, bromate can be formed in high concentration. In Table 11, a scheme of the O_3/Br^- oxidation mechanism is presented.

Based on some experimental results, application of the UV/H_2O_2 system to eliminate the bromate problem appears to be worthwhile. For example, Kruithof et al. [197], while studying the optimum conditions for disinfection and pesticide removal in a drinking water treatment plant, found that in the range 0.1–2.5 kWh m^{-3} electric energy (with medium-pressure UV lamps) and 0–25 gm^{-3} H_2O_2, bromate formation was absent. The main drawback was increased assimilable organic carbon, which would necessitate the use of activated carbon filters.

6. Wastewater Treatment

Although much of the research on AOTs has focused on drinking water treatment, the literature also reports many studies on the use of these

Table 11 Reaction Mechanism of the Bromide–Ozone–Peroxide–Ammonia–Carbonate Process for the Formation of Bromate in Water[a]

Reaction	Reaction or equilibrium
$O_3 + Br^- \longrightarrow O_2 + BrO^- \quad k = 160$	$BrOH^- \bullet \longrightarrow HO\bullet + Br^- \; k = 3.3\times10^7\ s^{-1}$
$O_3 + BrO^- \longrightarrow 2O_2 + Br^- \quad k = 330$	$BrOH^- \bullet \longrightarrow HO^- + Br\bullet \; k = 4.2\times10^6\ s^{-1}$
$O_3 + BrO^- \longrightarrow O_2 + BrO_2^- \quad k = 100$	$Br\bullet + Br^- \longrightarrow Br_2^-\bullet \; k = 10^{10}$
$O_3 + HBrO \longrightarrow O_2 + BrO_2^- + H^+ \; k \leq 0.013$	$Br_2^-\bullet + BrO^- \longrightarrow BrO\bullet + 2Br^- \; k = 8\times10^7$
$O_3 + BrO_2^- \longrightarrow BrO_3^- \quad k > 10^5$	$2BrO\bullet + H_2O \longrightarrow BRO^- + BrO_2^= + 2H^+ \; k = 4.9\times10^9$
$O_3 + HO_2^- \longrightarrow O_3^-\bullet + HO_2\bullet \; k = 2.2\times10^6$	$HO\bullet + BrO_2^- \longrightarrow BrO_2\bullet + OH^- \; k = 2\times10^9$
$HBrO + H_2O_2 \longrightarrow Br^- + O_2 + H_2O + H^+ \; k = 2\times10^4$	$2BrO_2\bullet \longrightarrow Br_2O_4 \; k = 1.4\times10^9$
$NH_3 + HBrO \longrightarrow NH_2Br + H_2O \; k = 8\times10^7$	$Br_2O_4 \longrightarrow 2BrO_2\bullet \; k = 7\times10^7$
$O_3 + NH_2Br \xrightarrow{k=40\ M^{-1}\ s^{-1}} I \xrightarrow{+2O_3\ k>>40\ M^{-1}\ s^{-1}} NO_3^- + Br^- + 3O_2 + H^+$	
$O_3\bullet$ and $HO_2\bullet \xrightarrow{\text{see reaction on Table 9}} HO\bullet$	$Br_2O_4 + OH^- \longrightarrow BrO_3^- + BrO_2^= + H^+ \; k = 7\times10^8$
$HO\bullet + Br^- \longrightarrow BrOH^-\bullet \quad k = 10^{10}$	$HBrO \rightleftharpoons BrO^- + H^+ \; pk = 9$
$HO\bullet + BrO^- \longrightarrow BrO\bullet + OH^- \quad k = 4.5\times10^9$	$H_2O_2 \rightleftharpoons HO_2^- + H^+ \; pk = 11.7$
$HO\bullet + HBrO \longrightarrow BrO\bullet + H_2O \quad k = 2\times10^9$	$HCO_3^- \rightleftharpoons CO_3^= + H^+ \; pk = 10.3$
$HO\bullet + H_2O_2$ and $HO_2^- \xrightarrow{\text{see reaction on Table 9}} HO_2\bullet$	$HCO_3\bullet \rightleftharpoons CO_3^-\bullet + H^+ \; pk = 9.6$
$CO_3^-\bullet + BrO^- \longrightarrow BrO\bullet + CO_3^- \; k = 4.3\times10^7$	$NH_4^+ \rightleftharpoons NH_3 + H^+ \; pk = 9.3$
$CO_3^-\bullet + H_2O_2$ and $HO_2^- \xrightarrow{\text{see reaction on Table 9}} HO_2\bullet + CO_3^-$ and HCO_3^-	

[a] Units of rate constants in $M^{-1}\ s^{-1}$ unless indicated. *I* represents unknown intermediates [196]. Values of rate constants and p*K*s from Ref. 196.
[b] See reactions in Table 9.

technologies for the treatment of wastewater. In these studies, attention is mainly focused on the reduction of COD, TOC, color, toxicity, and the effects on biodegradability [28,198–203]. In many cases, an integrated operation (chemical–biological oxidation) was studied [12,204–206]. Wastewater treated varied from food-processing industries, petrochemical plants, pesticides, surfactants, dyes, domestic wastewater, etc. In most cases, little reduction of COD and TOC was observed, although significant increases of BOD/COD ratio were also noticed.

Kuo [198] studied the effect of O_3/UV on the toxicity and biodegradability of an industrial pesticide wastewater. He observed increases of BOD/COD ratio up to a value of 0.4 with more than 50% reduction in toxicity (measured as EC_{50}) when using 400 g m^{-3} h^{-1} of ozone and 3 mW cm^{-2} of UV light intensity. The author also recommended the use of the integrated O_3/UV–aerobic biodegradation system for scale-up purposes.

Rodgers et al. [199] reported the use of the UV/H_2O_2 system, among others, as an alternative to reduce the toxicity of a radioactive liquid waste. Toxicity was reduced more than 50% when pH was adjusted before the oxidizing step.

Juang et al. [200] observed the effect of a UV/H_2O_2 system on the treatment of a petrochemical wastewater. They observed significant potential of this oxidizing system for detoxification and increase of biodegradability.

Beltrán et al. [201–203] studied the application of $UV/O_3/H_2O_2$ processes to different domestic and food processing wastewaters. Treatment of distillery, tomato, debittering table olive, and olive oil wastewaters were carried out under different conditions and oxidizing systems. Also, the integrated process with aerobic biological oxidation was undertaken with promising results. The results obtained with the treatment of debittering table olive wastewaters were particularly important because oxidation reduces the initial pH from 12 to 7 with significant increases in biodegradability due to the elimination of phenolic compounds present. Thus, wastewater became suitable for biological oxidation as also shown in other works [205].

V. SCALE-UP STUDIES AND ENVIRONMENTAL APPLICATIONS

Although some of the fundamentals of the mechanisms of $O_3/UV/H_2O_2$ systems are still under laboratory research, especially those concerning natural water, the beneficial effects of these oxidizing systems observed in laboratory studies were obtained in practice. At present, AOTs have been applied in many water-treatment plants. Thus, in addition to ozonation and UV disinfection, which are considered to be well established technologies, other $O_3/UV/H_2O_2$ systems have already been commercialized under different trade names like peroxone, Ultrox, perox-pure™, Rayox®, etc. Many cases concerning real applications can be seen in several publications dealing with the "state of the art" of processes in different countries like France [36], Germany [207], the United Kingdom [208], and the United States [209]. In addition to applications of ozonation, reports for UV radiation technologies are also available [210,211]. Table 12 presents some studies on the scale-up and applications of the discussed advanced oxidation systems.

As can be observed from Table 12, these studies on the application of advanced oxidations cover a wide range of uses, including degradation of pollutants [197,210,211,213,217–220], disinfection [197,215], maintenance of swimming-pools [216], and treatment of cooling waters [214], leachates [221], and domestic wastewater [222].

Table 12 Some Examples of Pilot Plant and Full Applications of $O_3/UV/H_2O_2$ AOP

Scale	AOP	Plant objective	Reference no.
Full plant	O_3	Improve flotation	212
Full plant	UV/H_2O_2	Aromatics	210
Full plant	UV/H_2O_2	PCE, TCE	210
Full plant	UV/H_2O_2	Pentachlorophenol	211
Full plant	UV/H_2O_2	Hydrazine	211
Full plant	UV/H_2O_2	1,4-Dioxane, benzene	211
Full plant	UV/H_2O_2	TCE, DCE, TCA, DCA	211
Pilot plant	UV/H_2O_2	Herbicides, disinfection	197
Pilot plant	UV/H_2O_2	Remove TCE, PCE	186
Full plant	O_3/UV	Remove propellants	213
Full plant	O_3	Cooling water	214
Pilot plant	O_3	Bactericide, fouling	215
Full plant	O_3	Swimming pool water	216
Pilot plant	O_3/H_2O_2	Aromatics	217
Pilot plant	O_3/UV	Surfactants	218
Pilot plant	UV/H_2O_2	PCE	219
Pilot plant	O_3, O_3/H_2O_2	Flocculation	18
Full plant	O_3	BAC	220
Pilot plant	O_3/H_2O_2, UV/H_2O_2, O_3/UV	Leachates	221
Pilot plant	UV/H_2O_2	Clarifier efficiency	222
Full plant	O_3/H_2O_2	Herbicide removal, bromate ion	223
Pilot plant	O_3, O_3/H_2O_2	Boiling water	224

Another set of applications includes those aimed at improving the performance of physicochemical unit operations such as coagulation–sedimentation or biological carbon filtration [18,212]. A few examples of scale-up and actual applications of $O_3/H_2O_2/UV$ systems are briefly discussed in the following two sections.

A. Experimental Design (Scale-Up Studies)

As shown in Table 12, several studies conducted in pilot plants and full-scale plants have been published. Reports are given on the performance of advanced oxidations in water-treatment plants with respect to oxidation efficiency and degree of pollutant oxidation. The control of trihalomethane and bromate ion constitutes one of the principal objectives of some studies.

Other work has been mainly concerned with the scale-up to pilot plant or full-scale installations. For example, Beltrán et al. [225] studied the scale-up of the ozonation of industrial wastewaters from alcohol distilleries and tomato-processing plants. They used kinetic data obtained in small laboratory bubble columns to predict the COD reduction that could be reached during ozonation in a geometrically similar pilot bubble column. In the kinetic model, assumptions were made about the flow characteristics of the gas phase through the column. From the solution of mass balance equations of the main species in the process (ozone in gas and water and pollution characterized by COD) calculated results of COD and ozone concentrations were determined and compared to the corresponding experimental values.

Pedit et al. [226] used a kinetic model for the scale-up of ozone/hydrogen peroxide oxidation of some volatile organochlorine compounds such as trichloroethylene and tetrachloroethylene. The kinetic model was applied to simulate the ozone/hydrogen peroxide treatment of these pollutants in a full-scale demonstration plant of the Los Angeles Department of Water and Power. The authors confirmed from the model that the reaction rate of the pollutant with ozone was several orders of magnitude lower than that with the hydroxyl radical. When considering that the natural organic matter acts as a promoter of hydroxyl radicals, the ozone utilization prediction was 81.2%, very close to the actual 88.4% experimentally observed.

In another work, Laplanche et al. [129] used rate data on the ozone/hydrogen peroxide oxidation of atrazine obtained in other studies, some of them from water-treatment plants, to check the feasibility of their kinetic model as far as the removal of atrazine was concerned. Calculated results included the variation of some intermediates. The authors demonstrated the suitability of the model and calculated results were very close to the experimental ones.

Another example of scale-up experimental design is given by Lewis et al. [227], who discussed the use of a $UV/O_3/H_2O_2$ oxidation technology in a field demonstration unit operated by Ultrox International. This unit has an effective reactor capacity of 0.568 m^3 (actual dimensions were 0.91 m long, 0.46 m wide, and 1.68 m high) divided into six chambers containing 24 UV lamps (65 W each). Each chamber also had one stainless steel sparger along the width of the reactor. The UV lamps were vertically installed and uniformly distributed throughout the reactor (four lamps per chamber). Ozone gas was diffused through the spargers and hydrogen peroxide was fed in the influent line with the contaminated water. VOCs were the main compounds followed. TCE and total VOC removal efficiencies were as high as 99% and 90%, respectively, although for some specific cases the removal efficiency was only 65%. Also, in many cases stripping contributed significantly to the removal of VOCs. Thus, 75% of TCA removal was due to stripping. Very low TOC removal was observed suggesting that only partial oxidation of compounds occurred. However, the authors did not observe the formation of intermediates using GC/MS analysis.

Duguet et al. [217] studied the removal of chloronitrobenzene compounds from a natural water in a 450 L hr^{-1} pilot plant with two ozone contactors (4 m height, 15 cm diameter). Removal efficiencies of 99% were observed by applying 8 mg L^{-1} of ozone with 3 mg L^{-1} of hydrogen peroxide after a 20-min contact time. To avoid eventual regrowth in the network, the oxidizing system was coupled in an integrated system with sand and GAC filtration units.

B. Examples of Real Applications

Here, some examples of real applications of $O_3/UV/H_2O_2$ systems are given. Applications may involve other unit operations such as coagulation, biological carbon filtration, biological oxidation in industrial wastewater treatment, or control of trihalomethane compounds. Examples taken here focus on pollutant degradation or by-product disinfection control.

An example of the combined use of ozone, hydrogen peroxide, and carbon filtration is seen in the operation of water plants located from the South and East of England to the North of London. Croll [220] presents a detailed explanation on the performance of such plants that supply the drinking water for a population of more than 3.6 million inhabitants. Water in reservoirs can have concentrations of 0.53 and 0.31 μg L^{-1} of atrazine and simazine, respectively.

1. Aromatic Hydrocarbons

As Calgon Corporation Technologies has reported [211], volatile aromatic hydrocarbons such as benzene, toluene, ethylbenzene, and xylenes have been satisfactory treated with UV processes. This company utilizes a UV commercialized process called Rayox®. The system uses high-temperature mercury lamps with electrical power between 10 and 30 kW. The system is advisable for cases where the use of ozone is limited by the stripping of contaminants like VOCs. The Rayox® system has been used for the treatment of volatile aromatic hydrocarbons that leak from underground gasoline storage tanks, as was the case in Carson City, Nevada. Here, in 1991 Calgon Carbon Corporation [211] mounted a Rayox® unit with 30 kW electrical power for the UV lamps at a total operating cost of $0.6/m^3 of treated water. The system was in operation for nearly 2 years and treated about 5700 m^3 of water containing the hydrocarbons cited above at a concentration of 5 mg

L^{-1} with an effluent concentration of less than 5 μg L^{-1}. According to the company, this system has been used later in similar situations [211].

2. Atrazine and Other *s*-Triazine Compounds

The treatment of *s*-triazine herbicides constitutes a clear application of the combined use of ozone and hydrogen peroxide. At the end of the eighties, this system was investigated to remove triazines in several pilot plants in France and the United Kingdom [228]. The successful results obtained led to the implementation of this system in some water-treatment plants, such as those owned by the Compagnie Generale des Eaux in Paris [228]. Since then, the O_3/H_2O_2system has been used in many waterworks to improve the removal of *s*-triazines. It should be noted, however, that ozone processes may not be appropriate for the removal of herbicides from water because of the potential formation of very low concentrations of harmful intermediates.

Another example can be found in the waterworks of the Anglian Water Company in England [220]. The region covered not only has an increasing population but also an intense agricultural activity requiring the use of large amounts of pesticides. This obviously has led to the appearance of high concentrations of pesticides such as atrazine, simazine, mecoprop, isoproturon, diuron, etc. In their treatment plants, the use of ozone combined with hydrogen peroxide has allowed atrazine and simazine reductions up to 79% and 85%, respectively. Another advantage observed when using this combination was the improvement of carbon filter bed life. For *s*-triazine removal, ozone was applied before coagulation (1 mg L^{-1} ozone dose) and between sand and carbon filtration with 4 mg L^{-1} ozone dose and a 0.25:0.75 O_3/H_2O_2 ratio. In full-scale plants, however, hydrogen peroxide is not usually applied because atrazine and simazine have been banned in the United Kingdom for total weed control, and ozonation alone has been proven to be effective for the removal of other herbicides such as diuron below 0.1 μg L^{-1}. It should be noticed, however, that with other surface waters results could be different because the efficiency of oxidation processes is highly dependent on alkalinity and the nature of the organic matter present in the water.

In another example [229], atrazine and one of its main by-products (deethylatrazine) have been removed in a large-scale O_3/UV plant. The plant was fed with 70 m^3 h^{-1} of groundwater (pH 7.2, 464 mg L^{-1} bicarbonate and 0.6 mg L^{-1} DOC). UV-C radiation was emitted from a medium-pressure mercury arc lamp with a maximum total electrical power of 10 kW and 1.2 kW maximum radiant power in the UV-C region. The maximum electrical power of the ozone generator was also 10 kW. Concentrations of atrazine and deethylatrazine in the plant inlet were 0.28 μg

L^{-1} and 0.6 μg L^{-1}, respectively. Due to photolysis, nitrate present in the water at a concentration of 55 mg L^{-1} was reduced to nitrite generating an additional problem (see section II.D.1). This is because nitrite is thought to be involved in the formation of nitrosamines. Experimental conditions in this plant have been investigated to provide the optimum operating conditions for the removal of herbicides and to minimize the impact of nitrate photolysis. The results obtained in this plant led to concentrations of atrazine and nitrite in the effluent lower than 0.1 and 100 μg L^{-1}, respectively. It must be said, for safe operation of the plant, that an equivalent ozone dose higher than 3 kW energy and a radiant power for the UV lamp between 20 and 35 kW were needed.

3. Volatile Organic Compounds

Likely, chlorinated VOCs are the principal priority pollutants treated in real waterworks or full-scale systems with O_3/UV/H_2O_2 processes. For example, Foelich [230] presents the case of the perox-pure™ system made up of high-intensity UV lamps (similar to those of the Rayox® system indicated previously) and hydrogen peroxide. The majority of full-scale perox-pure™ systems in operation or construction/installation treat wastewater containing organic concentrations between 10 mg L^{-1} and 1 mg L^{-1}. Although the list of chemicals treated covers different types of compounds like herbicides and aromatic hydrocarbons, most of them are VOCs. Flow rates in these systems range from 19 to 3800 m^3 day^{-1}, whereas according to Foelich [230], operating costs vary from \$0.053 m^{-3} for water containing low concentrations of TCE and DCE to about \$32 m^{-3} for the highest concentration wastewater. Initially, the perox-pure™ system was a polish operation for air stripping units but some VOCs, such as TCA, still remained in the water at high concentrations (65 μg L^{-1}). Subsequently, the sequence was reversed, and the perox-pure™ system was used before stripping. With a 40 mg L^{-1} hydrogen peroxide dose and 2 min contact time, 85% and 90% VOC and air emissions removal, respectively, were achieved with an operating cost of \$0.53 m^{-3} of water. The technology of this oxidizing system has focused on the improvement of UV lamps mainly to remove compounds difficult to oxidize such as TCA, DCA, etc.

Schwämmlein and Leitzke [231] presented another example of VOC removal with AOT in plant systems. This case was a mobile Wedeco unit able to operate with O_3/UV, O_3/H_2O_2, or UV/H_2O_2 systems. The main characteristics of the plant were 10 m^3 hr^{-1} (maximum) raw water flow, 70 g h^{-1} (maximum) ozone generation, 0.7 N m^3 h^{-1} O_2/O_3 gas flow rate (maximum), 6.6 kW maximum electrical power for UV lamps, and 4.5–5 bar absolute pressure in the absorption tank. Under favorable conditions more than 95% reduction in total VOCs (1 mg L^{-1} in the raw water) was noticed.

4. Bromate Ion

The formation of bromate ion after ozonation in real waterworks has been reviewed by Legube [195]. The use of hydrogen peroxide combined with ozone seems to be a suitable choice because of the competitive effect of hydrogen peroxide to reduce hypobromite to bromide ion, avoiding the formation of bromate. For example, Griffini and Iozzelli [223] presented data obtained in a pilot plant in Mantignano, Italy where hydrogen peroxide was added after ozonation to reduce aliphatic acid (such as oxalic, acetic, and formic acids) and bromate formation and to improve carbon filtration. This pilot plant was located in a full-scale plant fed with water from the Arno River. Because of the low bromide ion concentration in this river water (<0.07 mg L^{-1}), for pilot plant experiments, water was spiked with bromide to reach concentrations of up to 1 mg L^{-1}. For ozone and hydrogen peroxide doses of 4.5 and 6 mg L^{-1}, respectively, concentrations of brominated compounds were highly dependent on the initial bromide ion concentration. In all cases, the addition of hydrogen peroxide after ozonation allowed greater reduction of bromate and bromoform concentrations. As inferred from the results reported, bromate does not represent a serious problem in ozonated water if hydrogen peroxide is applied. Also, it should be highlighted that the use of UV/H_2O_2 system in water containing bromide ion does not lead to bromate, as observed in a large-scale test application on pesticide control in a surface water [197].

VI. ECONOMIC ASPECTS

Processes involving $O_3/UV/H_2O_2$ require different costs associated with the investment in equipment needed (ozone generator, UV lamps, photorreactor and ozonation chambers, process control buildings, etc.) and operating and maintenance costs (electrical power requirements, manpower, etc.). Because costs are largely affected by the legislation of the country where the water plant operates, it is difficult to present detailed figures about them. Aspects that also effect the economy of these treatment methods are dependent on the nature of the water to be treated and the experimental conditions applied (water flow rate, oxidant and UV doses, etc.). Thus, the presence of particulates, turbidity, and natural hydroxyl radical scavengers (such as carbonates) are factors that negatively affect the performance of AOTs. These problems deserve specific studies and lead to cost estimations that can vary considerably in each case. For these reasons, no economic figures are given here, but some references reporting data on economic aspects are provided.

For ozonation systems, Langlais et al. [5] present an extensive study on the economy of different ozonation processes. For UV processes, Bolton et al. [232] have established some figures-of-merit for the determination of electrical costs. They defined the concepts of electrical energy per order (EE/O) and electrical energy per mass (EE/M) to determine the energy requirements for reducing contaminants present at high and low concentrations, respectively. Application of these figures has aided in the determination of capital costs for some units [211] and have also been used in laboratory studies [233,234].

VII. CONCLUSIONS

The use of advanced oxidation processes based on ozone/hydrogen peroxide/UV radiation are currently very extensive in water-treatment plants or with mobile units that operate temporarily until the pollution problem is solved. The main objectives of these oxidizing systems are the removal of pollutants from drinking water and the improvement of efficiency of other unit operations, such as coagulation-flocculation or carbon filtration. Studies of these systems from laboratory studies have resulted in the improvement of real unit operations because of understanding of fundamental principles is now well established. These principles are chemical and/or photochemical reaction mechanisms, mass transfer (for ozone systems), natural organic matter interaction, determination of kinetic data, kinetic modeling, etc. From these data, together with better analytical systems, more complicated and accurate models of the mechanism, and kinetics of these processes can be developed, allowing better predictions of the behavior of advanced oxidation processes.

In addition to the systems studied in this chapter, it should be highlighted that research is still in progress. Some of the oxidants discussed here, such as ozone or UV radiation, are currently being studied at bench-scale level, including new studies combined with other agents like catalysts [235], ultrasound [236–238], or electrical discharges [239]. It is likely that these AOTs will become viable options in the near future.

VIII. NOMENCLATURE

a Specific interfacial area in a gas liquid reaction, m^{-1}
C_{HO} Concentration of hydroxyl radicals, M
C_{Ii} Concentration of any initiator compound of the ozone decomposition in radicals as defined in Eq. (7), M

C_j Concentration of any j compound, M
C_M Concentration of compound M, M
C_{O_3} Concentration of dissolved ozone, M
$C^*_{O_3}$ Concentration of dissolved ozone at the gas-water interface or ozone solubility, M
C_{Pi} Concentration of any i promoter compound of the ozone decomposition to propagate the free radical chain as defined in Eq. (7), M
C_{Si} Concentration of any i inhibitor compound of the ozone decomposition to quench the free radical chain as defined in Eq. (7), M
D_M Diffusivity of compound M in water, $m^2\ s^{-1}$
D_{O_3} Diffusivity of ozone in water, $m^2\ s^{-1}$
E^o Standard redox potential, V
E_i Instantaneous reaction factor for gas-liquid reactions, defined in Eq. (10), dimensionless
E_o Radiant energy of UV–vis lamp per unit length according to the PSSE model, $J\ m^{-1}$
F_M Fraction of absorbed energy in water that it is absorbed for any compound M, defined in Eq. (47), dimensionless
Ha Hatta number for gas-liquid reactions, defined in Eq. (9), dimensionless
He Henry's law constant for ozone in a water containing substances, Pa M^{-1}
He^o Henry's law constant for ozone in ultrapure water, Pa M^{-1}
h Planck's constant
h_i Salting-out coefficient for any i electrolyte or nonelectrolyte species present in water, see Eq. (12), M^{-1}
I Total ionic strength of an aqueous solution, M
I_i Ionic strength due to any i species, M
I_o Intensity of incident radiation in water according to the LL model, $J\ L^{-1}\ s^{-1}$
k_a Rate constant of the photolysis step (18) for any compound M, s^{-1}
k_b Rate constant of step (19) in photolysis mechanism for compound M, s^{-1}
k_c Rate constant of true decomposition step (20) for any compound M during photolysis, s^{-1}
k_D Rate constant of the direct reaction of ozone with any compound in water, $M^{-1}\ s^{-1}$
k_{Di} Rate constant of the direct reaction of ozone with any i compound in water, $M^{-1}\ s^{-1}$
$k_{D\text{-}d}$ Rate constant of the direct reaction of ozone with the dissociated form of a phenol compound in water, $M^{-1}\ s^{-1}$
$k_{D\text{-}nd}$ Rate constant of the direct reaction of ozone with the non-dissociated form of a phenol compound in water, $M^{-1}\ s^{-1}$
k_{Ii} Rate constant of the reaction between ozone and any i initiator compound to yield free radicals that initiate the free radical mechanism of the ozone decomposition, $M^{-1}\ s^{-1}$
k_i Rate constant of the reaction between ozone and the hydroxyl ion, reaction (4), $M^{-1}\ s^{-1}$
k_L Individual liquid phase mass transfer coefficient, $m\ s^{-1}$

k_{Pi} Rate constant of the reaction between the hydroxyl radical and any i promotor of ozone decomposition to propagate the free radical chain, $M^{-1}\ s^{-1}$

k_{Si} Rate constant of the reaction between the hydroxyl radical and any i inhibitor of the ozone decomposition to quench the free radical chain, $M^{-1}\ s^{-1}$

k_La Volumetric liquid phase mass transfer coefficient, s^{-1}

P_{O_3} Ozone partial pressure, Pa

PS Notation for any photosensitizing compound

q_i Flux of incident radiation in a photoreactor, $J\ m^{-2}\ s^{-1}$

r_D Rate of removal of any compound in water due to direct ozone reaction, defined in Eq. (71), $M\ s^{-1}$

r_H Rate of removal of any compound due to hydrogen peroxide direct oxidation, defined in Eq. (72), $M\ s^{-1}$

r_i Rate of initiating reaction of the free radical mechanism of an AOT process, $M\ s^{-1}$

r_M Total removal rate of any compound M due to AOT, defined in Eq. (70), $M\ s^{-1}$

r_R Removal rate of any compound due to free radical oxidation, defined in Eq. (73), $M\ s^{-1}$

r_{UV} Removal rate of any compound due to direct photolysis, defined in Eqs. (45) or (49) (LSPP model), (52) (PSSE model), or (53) (LL model), $M\ s^{-1}$

V Reaction volume, L

y_L Effective path of radiation through the photoreactor for the LL model, cm

z Stoichiometric coefficient of ozone in the direct reaction of ozone with any compound M in water, dimensionless

z_M Stoichiometric coefficient of M in the direct reaction of ozone with any compound M in water, dimensionless

z_H Stoichiometric coefficient of M in the direct reaction of hydrogen peroxide with any compound M in water, dimensionless

z_R Stoichiometric coefficient of M in the reaction of HO• with any compound M in water, dimensionless

Greek letters

α Degree of dissociation of any phenol compound, dimensionless

ε_i, ε_M Molar absorptivity coefficient for any i or M compound, $M^{-1}\ cm^{-1}$

ϕ_M Quantum yield of compound M, mol $Einstein^{-1}$

μ Parameter defined in Eq. (46), cm^{-1}

v Frequency of radiation, s^{-1}

ϖ Parameter defined in Eq. (51), dimensionless

ACKNOWLEDGMENTS

The author thanks Calgon Carbon Corporation for providing some of the information on UV technologies. The author also thanks Drs. J.F. García-Araya, F.J. Rivas, P. Álvarez, and B. Acedo of the Departamento de Ingeniería Química y Energética of the University of Extremadura at Badajoz (Spain) for their collaboration in carrying out some AOT studies quoted in this paper.

REFERENCES

1. Glaze WH, Kang JW, Chapin DH. The chemistry of water of water treatment processes involving ozone, hydrogen peroxide and ultraviolet radiation. Ozone Sci Eng 1987; 9:335–342.
2. Sayre JM. International standards for drinking water. J Am Water Works Assoc 1988; 80:53–60.
3. Organisation Mondiale de la Santé. Directives de qualité pour l'eau de boisson. Vol. 1. Recommandations. Geneve: Organization Mondiale de la Santé, 1994.
4. Rowe DR, Abdel-Magid IM. Handbook of wastewater reclamation and reuse. Boca Ratón, FL: CRC Press, Inc., 1995.
5. Langlais B, Reckhow DA, Brink DR eds. Ozone in Water Treatment: Application and Engineering. Chelsea, MI: Lewis Publ., 1991.
6. Doré M. Chimie des oxydants et traitement des eaux. Paris: Technique et Documentation-Lavoisier, 1989.
7. Camel V, Bermond A. The use of ozone and associated oxidation processes in drinking water treatment. Water Res 1998; 32:3208–3222.
8. Reynolds G, Graham N, Perry R, Rice RG. Aqueous ozonation of pesticides: a review. Ozone Sci Eng 1989; 11:339–382.
9. Chiron S, Fernández-Alba A, Rodríguez A, García-Calvo E. Pesticide chemical oxidation: state of the art. Water Res 2000; 34:366–377.
10. Legrini O, Oliveros E, Braun AM. Photochemical processes for water treatment. Chem Rev 1993; 93:671–698.
11. Yue PL. Modelling of kinetics and reactor for the water purification by photo-oxidation. Chem Eng Sci 1993; 48:1–11.
12. Scott JP, Ollis DF. Integration of chemical and biological oxidation processes for water treatment: review and recommendations. Environ Prog 1995; 14:88–103.
13. Meulemans CC. The Basic Principles of UV Disinfection in Water. Ozone Sci Eng 1987; 9:299–313.
14. Malcolm Pirnie, Inc and HDR Engineering, Inc. Guidance Manual for compliance with the filtration and disinfection requirements for public water systems using surface water sources for USEPA (contract No 68-01-6989). Denver: American Water Works Association, 1991.

15. Paillard H, Legube B, Bourbigot MM, Lefebvre E. Iron and manganese removal with ozonation in the presence of humic substances. Proceedings of the 8th Ozone World Congress, Zurich VI, 1987:C71–C96.
16. Reckhow DA, Knocke WR, Kearney M, Parks CA. Oxidation of iron and manganese by ozone. Ozone Sci Eng 1991; 13:675–695.
17. Reckhow DA, Singer PC, Trussell RC. Ozone as a coagulant aid. Proceedings of AWWA Seminar: Ozonation: Recent Advances and Research Needs, No. 20005, Denver, 1986. American Water Works Association 1986:17–46.
18. Paode RD, Chandrankanth MS, Amy GL, Gramith JT, Ferguson DW. Ozone versus ozone/peroxide induced particle destabilization and aggregation: a pilot plant. Ozone Sci Eng 1995; 17:25–51.
19. van Leewen J, Prinsloo J, van Steenderen RA, Melekus W. The effect of various oxidants on the performance of activated carbon used in water reclamation. Ozone Sci Eng 1981; 3:225–237.
20. Glaze WH, Wallace JL, Wilcox D, Johansson KR, Scalf B, Noack R, Busch AW. Pilot scale evaluation of ozone-granular activated carbon combinations for trihalomethane precursor removal. McGuire MJ, Suffet IH eds. Advances in Chemistry Series, No. 202, Treatment of water by granular activated carbon. American Chemical Society, New York 1983; 3:221–229.
21. De Laat J, Doré M, Mallevialle J. Effects of preozonation on the adsorbability and biodegradability of aquatic humic substances and on the performance of granular activated carbons filters. Water Res 1991; 25:151–164.
22. Gilbert E. Biodegradability of ozonation products as a function of COD and DOC elimination by example of substituted aromatic substances. Water Res 1987; 21:1273–1278.
23. Adams CD, Cozzens RA, Kim BJ. Effects of ozonation on the biodegradability of substituted phenols. Water Res 1997; 31:2655–2663.
24. Rice RG, Browning ME. Ozone treatment of industrial wastewater. Pollution Technology Review No. 84. Park Ridge, NJ: Noyes Data Corporation, 1981.
25. Rice RG. Applications of ozone for industrial wastewater treatment: a review. Ozone Sci Eng 1997; 18:477–515.
26. Zhou H, Smith DW. Advanced technologies in water and wastewater treatment. Can J Civ Eng 2001; 28:49–66.
27. Rivera LAC, Sillet A, Roussy J, Dumas JRD, Thomas O. Treatment of high organic loaded industrial effluents. Water Sci Technol 2000; 42:115–118.
28. Kos L, Perkowski J. Simultaneous application of H_2O_2 and UV or O_3, H_2O_2 and UV in textile wastewater treatment. Fibres Text East Eur 1999; 7:57–61.
29. Glaze WH. An overview of advanced oxidation processes: current status and kinetic models. In: Eckenfelder WW, Bowers AR, Roth JA, eds. Chemical Oxidation: Technologies for the Nineties Vol. 3 Basel: Technomic Publishing, 1994.
30. Buxton GV, Greenstock CL, Helman WP, Ross AB. Critical review of data constants for reactions of hydrated electrons, hydrogen atoms and hydroxyl radicals ($\cdot OH/\cdot O^-$) in aqueous solution. J Phys Chem Ref Data 1988; 17:513–886.

31. Beltrán FJ, Encinar JM, Alonso MA. Nitroaromatic hydrocarbon ozonation in water. 2. Combined ozonation with hydrogen peroxide or UV radiation. Ind Eng Chem Res 1998; 37:32–40.
32. Evers EL, Jayson GG. Determination by pulse radiolysis of the distribution of solubilizates between micellar and non-micellar phases. J Chem Soc, Faraday Trans I 1980; 76:528–536.
33. Beltran FJ, Ovejero G, Rivas J. Oxidation of polynuclear aromatic hydrocarbons in water. 3. UV radiation combined with hydrogen peroxide. Ind Eng Chem Res 1996; 35:883–890.
34. Beltran FJ, González M, Acedo B, Jaramillo J. Contribution of free radical oxidation to eliminate volatile organochlorine compounds in water by ultraviolet radiation and hydrogen peroxide. Chemosphere 1996; 32:1949–1961.
35. Haag WR, Yao CCD. Rate constants for reaction of hydroxyl radicals with several drinking water contaminants. Environ Sci Technol 1992; 26:1005–1013.
36. Le Paulouë J, Langlais B. State of the art of ozonation in France. Ozone Sci Eng 1999; 21:153–162.
37. Rook JJ. Formation of haloforms during chlorination of natural waters. Water Treat Exam 1974; 23:234–243.
38. Rook JJ. Possible pathways for the formation of chlorinated degradation products during chlorination of humic acids and resorcinol. In: Jolly RL, Cumming RB, Jacobs VA, eds. Water Chlorination Environmental Impacts and Health Effects. Vol. 3 Ann Arbor, MI: Ann Arbor Science, 1980.
39. International Ozone Association. Ozone in medicine. Proceedings of 12th Ozone World Congress, Vol. 3, Lille, France, 1995.
40. Bailey PS. The reactions of ozone with organic compounds. Chem Rev 1958; 58:925–1010.
41. Kuczkowski RL. Ozone and carbonyl oxides. 1,3-Dipolar Cycloaddition Chemistry. New York: John Wiley and Sons, 1984:197–277.
42. Decoret C, Royer J, Legube B, Doré M. Experimental and theoretical studies of the mechanism of the initial attack of ozone on some aromatics in aqueous medium. Environ Technol Lett 1984; 5:207–218.
43. Riebel AH, Erikson RE, Abshire CJ, Bailey PS. Ozonation of carbon-nitrogen double bonds. I. Nucleophilic attack of ozone. J Org Chem 1960; 25:1801–1807.
44. Hoigné J, Bader H, Haag WR, Staehelin J. Rate constants of the reactions of ozone with organic and inorganic compounds. III. Inorganic compounds and radicals. Water Res 1985; 19:993–1004.
45. Hoigné J, Bader H. Rate constants of the reactions of ozone with organic and inorganic compounds. I. Non dissociating organic compounds. Water Res 1983; 17:173–183.
46. Hoigné J, Bader H. Rate constants of the reactions of ozone with organic and inorganic compounds. II. Dissociating organic compounds. Water Res 1983; 17:185–194.
47. Weiss J. Investigations on the radical HO_2 in solution. J Trans Faraday Soc 1935; 31:668–681.

48. Staehelin S, Hoigné J. Decomposition of ozone in water the presence of organic solutes acting as promoters and inhibitors of radical chain reactions. Environ Sci Technol 1985; 19:1206–1212.
49. Tomiyasu H, Fukutomi H, Gordon G. Kinetics and mechanisms of ozone decomposition in basic aqueous solution. Inorg Chem 1985; 24:2962–2966.
50. Hoigné J. Chemistry of aqueous ozone and transformation of pollutants by ozonation and advanced oxidation processes. In: Hrubec J ed. The Handbook of Environmental Chemistry. Vol. 5. Part C. Quality and Treatment of Drinking Water II. Berlin: Springer-Verlag, 1998.
51. Nakamuro K, Ueno H, Nakao M, Sayato Y. Formation of hydrogen peroxide by aqueous ozonation of humic acid and aromatic hydrocarbons. Chemosphere 1990; 20:525–531.
52. Sotelo JL, Beltrán FJ, Benítez J, Beltrán-Heredia J. Ozone decomposition in water: kinetic study. Ind Eng Chem Res 1987; 26:39–43.
53. Charpentier JC. Mass transfer rates in gas liquid absorbers and reactors. Advances in Chemical Engineering. New York: Academic Press, 1981:3–133.
54. Yao CC, Haag WR. Rate constants of direct reactions of ozone with several drinking water contaminants. Water Res 1991; 25:761–773.
55. Reid RC, Prausnitz JM, Sherwood TK. The Properties of Gases and Liquids (3d ed.) 1977. New York: McGraw-Hill.
56. Matrozov V, Kachtunov S, Stepanov A, Treguhov B. Experimental determination of the molecular diffusion coefficient of ozone in water. Zh Prikl Khim 1976; 49:1070–1073.
57. Johnson PN, Davis RA. Diffusivity of ozone in water. J Chem Eng Data 1996; 41:485–1487.
58. Beltrán FJ, González M. Ozonation of aqueous solutions of resorcinol and phloroglucinol. 3. Instantaneous kinetic regime. Ind Eng Chem Res 1991; 30: 2518–2522.
59. Roth JA, Sullivan DE. Solubility of ozone in water. Ind Eng Chem Fundam 1981; 20:137–140.
60. Sotelo JL, Beltrán FJ, Benítez FJ, Beltrán-Heredia J. Henry's law constant for the ozone–water system. Water Res 1989; 23:1239–1246.
61. Danckwerts PSV. Gas–Liquid Reactions. New York: McGraw-Hill, 1970.
62. Andreozzi R, Caprio V, Ermellino I, Insola A, Tufano V. Ozone solubility in phosphate buffer aqueous solutions: effect of temperature, *tert*-butyl alcohol and pH. Ind Eng Chem Res 1996; 35:1467–1471.
63. Li KY, Kuo CH, Weeks JL. A kinetic study of ozone-phenol reaction in aqueous solutions. AIChE J 1984; 25:583–591.
64. Beltrán FJ, Encinar JM, Alonso MA. Nitroaromatic hydrocarbon ozonation in water. 1. Single ozonation. Ind Eng Chem Res 1998; 37:25–31.
65. Beltrán FJ, Ovejero G, Encinar JM, Rivas J. Oxidation of Polynuclear aromatic hydrocarbons in water. 1. Ozonation. Ind Eng Chem Res 1995; 34: 1596–1606.
66. Mehta YM, George CE, Kuo CH. Mass transfer and selectivity of ozone reactions. Can J Chem Eng 1989; 67:118–126.

67. Beltrán FJ, Gómez-Serrano V, Durán A. Degradation kinetics of *p*-nitrophenol ozonation in water. Water Res 1992; 26:9–17.
68. Beltrán FJ, Kolaczkowski ST, Crittenden BD, Rivas J. Degradation of *o*-clorophenol with ozone in water. Trans Inst Chem Eng Part B Process Saf Environ Protect 1993; 71:57–65.
69. Beltrán FJ, García-Araya JF, Acedo B. Advanced oxidation of atrazine in water. I. Ozonation. Water Res 1994; 28:2153–2164.
70. Beltrán FJ, Álvarez PM, Legube B, Allemane H. Chemical degradation of aldicarb in water using ozone. J Chem Technol Biotechnol 1995; 62:272–278.
71. Gurol MD, Nekouinaini S. Kinetic behavior of ozone in aqueous solutions of substituted phenols. Ind Eng Chem Fundam 1984; 23:53–60.
72. Beltrán FJ, Encinar JM, García-Araya JF. Oxidation by ozone and chlorine dioxide of two distillery wastewater contaminants: gallic acid and epicatechin. Water Res 1993; 27:1023–1032.
73. Ridgway D, Sharma RN, Eanlay TR. Determination of mass transfer coefficients in agitated gas liquid reactors by instantaneous reactions. Chem Eng Sci 1989; 44:2935–2942.
74. Crittenden JC, Hu S, Hand DW, Green SA. A kinetic model for H_2O_2/UV process in a completely mixed batch reactor. Water Res 1999; 33:2315–2328.
75. Liao CH, Gurol MD. Chemical oxidation by photolytic decomposition of hydrogen peroxide. Environ Sci Technol 1995; 29:3007–3014.
76. Glaze WH, Kang JH. Advanced oxidation processes. Description of a kinetic model for the oxidation of hazardous materials in aqueous media with ozone and hydrogen peroxide in a semibatch reactor. Ind Eng Chem Res 1989; 28:1573–1580.
77. Hayashi J, Ikeda J, Kusakabe K, Morooka S. Decomposition rate of volatile organochlorines by ozone and utilization efficiency of ozone with ultraviolet radiation in a bubble column contactor. Water Res 1993; 27:1091–1097.
78. Sunder M, Hempel DC. Oxidation of tri- and perchloroethene in aqueous solution with ozone and hydrogen peroxide in a tube reactor. Water Res 1997; 31:33–40.
79. De Laat J, Berger P, Poinot T, Karpel vel Leitner N, Doré M. Modeling the oxidation of atrazine by H_2O_2/UV. Estimation of kinetic parameters. Ozone Sci Eng 1997; 19:395–408.
80. Stefan MI, Hoy AR, Bolton JR. Kinetics and mechanism of the degradation and mineralization of acetone in dilute aqueous solution sensitized by UV photolysis of hydrogen peroxide. Environ Sci Technol 1996; 30: 2382–2390.
81. Beltrán FJ, Rivas FJ, Alvarez P, Alonso MA, Acedo B. A kinetic model for advanced oxidation processes of aromatic hydrocarbons in water: application to phenanthrene and nitrobenzene. Ind Eng Chem Res 1999; 38: 4189–4199.
82. Faust BC, Hoigné J. Sensitized photooxidation of phenols by fulvic acid and in natural waters. Environ Sci Technol 1987; 21:957–964.
83. Glaze WH, Kenneke JF, Ferry JL. Chlorinated by-products from TiO_2-

mediated photodegradation of trichloroethylene and tetrachloroethylene in water. Environ Sci Technol 1993; 27:177–184.
84. Owen ED. Principles of photochemical reactions in aqueous solution. In: Faust SD, Hunter JV, eds. Organic Compounds in Aquatic Environments. New York: Marcel Dekker Inc., 1971:387–423.
85. Gal E, Aires P, Chamarro E, Esplugas S. Photochemical degradation of parathion in aqueous solution. Water Res 1992; 26:911–915.
86. Beltrán FJ, Ovejero G, García-Araya JF, Rivas J. Oxidation of polynuclear aromatic hydrocarbons in water. 2. UV radiation and ozonation in the presence of UV radiation. Ind Eng Chem Res 1995; 34:1607–1615.
87. Wayne CE, Wayne RP. Photochemistry. New York: Oxford Science Publ, 1996.
88. Simmons MS, Zep RG. Influence of humic substances on photolysis of nitroaromatic compounds in aqueous systems. Water Res 1986; 20:899–904.
89. Soley J, Vicente M, Clapés P, Esplugas S. Kinetic study of 4-chloro-2-methylphenoxyacetic acid photodegradation. Ind Eng Chem Prod Res Dev 1986; 25:645–649.
90. Cooper WJ, Herr FL. Introduction and Overview. In: Zika RG, Cooper WJ, eds., in Photochemistry of Environmental Aquatic Systems, Chap. 1. Washington, DC: American Chemical Society, 1987:1–8.
91. Vaughan PP, Blough NV. Photochemical formation of hydroxyl radical by constituents of natural waters. Environ Sci Technol 1998; 32:2947–2953.
92. Sörensen M, Frimmel FH. Photochemical degradation of hydrophilic xenobiotics in the UV/H_2O_2 process: influence of nitrate on the degradation rate of EDTA, 2-amino-1-naphthalenesulfonate, diphenyl-4-sulfonate and 4,4′-diaminostilbene-2,2′-disulfonate. Water Res 1997; 31:2885–2891.
93. Litter MI. Heterogeneous photocatalysis. Transition metal ions in photocatalytic systems. Appl Catal B: Environ 1999; 23:89–114.
94. Mathews RW. Photo-oxidation of organic material in aqueous suspensions of titanium dioxide. Water Res 1986; 20:569–578.
95. Beltrán FJ, González M, Rivas J, Jaramillo J. Application of photochemical reactor models to UV radiation of trichloroethylene in water. Chemosphere 1995; 31:2873–2885.
96. Leifer A. The Kinetics of Environmental Aquatic Photochemistry: Theory and Practice. New York: American Chemical Society, 1988.
97. Taube H. Photochemical reactions of ozone in aqueous solution. Trans Faraday Soc 1957; 53:656–665.
98. Weeks JL, Meaburn MAC, Gordon S. Absorption coefficients of liquid water and aqueous solutions in the far ultraviolet. Radiat Res 1963; 19:559–567.
99. Baxendale JH, Wilson JA. Photolysis of hydrogen peroxide at high light intensities. Trans Faraday Soc 1957; 53:344–356.
100. Wagner I, Strehlow H, Busse G. Flash photolysis of nitrate ions in aqueous solutions. Z Phys Chem (NF) 1980; 123:1–33.

101. Dulin D, Drossman H, Mill T. Products and quantum yields for photolysis of chloroaromatics in water. Environ Sci Technol 1986; 20:72–77.
102. Boule P, Guyon C, Lematre J. Photochemistry and environment. IV. Photochemical behaviour of monochlorophenols in dilute aqueous solution. Chemosphere 1982; 11:1179–1188.
103. Rivas FJ, Beltrán FJ, Acedo B. Chemical and photochemical degradation of acenaphthylene. Intermediate identification. J Hazard Mater B 2000; 75:89–98.
104. Beltrán FJ, Ovejero G, Acedo B. Oxidation of atrazine in water by hydrogen peroxide combined with UV radiation. Water Res 1993; 27:1013–1021.
105. Beltrán FJ, García-Araya JF, Rivas J, Álvarez P, Rodríguez E. Kinetics of simazine advanced oxidation in water. J Environ Sci Health B 2000; 35:439–454.
106. Beltrán FJ, González M, Rivas FJ, Álvarez P. Aqueous UV radiation and UV/H_2O_2 oxidation of atrazine first degradation products: deethylatrazine and deisopropylatrazine. Environ Toxicol Chem 1996; 15:868–872.
107. Beltrán FJ, González M, Rivas J, Acedo B. Determination of kinetic parameters of ozone during oxidations of alachlor in water. Water Environ Res 2000; 72:689–697.
108. Alfano OM, Romero RL, Casano A. Radiation field modelling in photoreactors. I. Homogeneous Media. Chem Eng Sci 1986; 41:421–444.
109. Alfano OM, Romero RL, Casano A. Radiation field modelling in photoreactors. II. Heterogeneous Media. Chem Eng Sci 1986; 41:1137–1153.
110. Jacob SM, Dranoff JS. Radial scale-up of perfectly mixed photochemical reactors. Chem Eng Prog Symp Ser 1966; 62:47–55.
111. Jacob SM, Dranoff JS. Design and analysis of perfectly mixed photochemical reactors. Chem Eng Prog Symp Ser 1968; 64:54–63.
112. Costa J, Esplugas S. Fotorreactor anular continuo de mezcla perfecta. Ing Quim 1977; 100:89–95.
113. Glaze WH, Lay Y, Kang JW. Advanced oxidation processes. A kinetic model for the oxidation of 1,2-dibromo-3-chloropropane in water by the combination of hydrogen peroxide and UV radiation. Ind Eng Chem Res 1995; 34:2314–2323.
114. Jacob SM, Dranoff JS. Light intensity profiles in a perfectly mixed photoreactor. AIChE J 1970; 16:359–363.
115. Staehelin S, Hoigné J. Decomposition of ozone in water: rate of initiation by hydroxide ions and hydrogen peroxide. Environ Sci Technol 1982; 16:666–681.
116. Forni L, Bagnemann D, Hart EJ. Mechanism of hydroxide ion decomposition of ozone in aqueous solution. J Phys Chem 1982; 86:255–259.
117. Prengle HW. Experimental rate constants and reactor considerations for the destruction of micropollutants and trihalomethane precursors by ozone with ultraviolet radiation. Environ Sci Technol 1983; 17:743–747.
118. Peyton GR, Glaze WH. Destruction of pollutants in water by ozone in combination with ultraviolet radiation. 3. Photolysis of aqueous ozone. Environ Sci Technol 1988; 22:761–767.

119. Beltrán FJ. Theoretical aspects of the kinetics of competitive first order reactions of ozone in the O_3/H_2O_2 and O_3/UV oxidation processes. Ozone Sci Eng 1997; 19:13–38.
120. Beltrán FJ. Estimation of the relative importance of free radical oxidation and direct ozonation/UV radiation rates of micropollutants in water. Ozone Sci Eng 1999; 21:207–228.
121. Hoigné J. Mechanisms, rates and selectivities of oxidations of organic compounds initiated by ozonation of water. In: Rice RG, Netzer A, Kent A, eds. Handbook of Ozone Technology and Application. Ann Arbor, MI: Ann Arbor Science, 1982:341–379.
122. von Sonntag C, Schuchmann HP. The elucidation of peroxyl radical reactions in aqueous solution with the help of radiation-chemical methods. Angew Chem Int Ed Engl 1991; 30:1229–1253.
123. Neta P, Huie RE, Ross AB. Rate constants for reactions of inorganic radicals in aqueous solution. J Phys Chem Ref Data 1988; 17:1027–1284.
124. Beltrán FJ. Theoretical aspects of the kinetics of competitive ozone reactions in water. Ozone Sci Eng 1995; 17:163–181.
125. Fogler HS. Elements of Chemical Reaction Engineering. Englewood Cliffs, NJ: Prentice-Hall, 1986.
126. Beltrán FJ, García-Araya JF, Álvarez P, Rivas FJ. Aqueous degradation of atrazine and some of its main by-products with ozone/hydrogen peroxide. J Chem Technol Biotechnol 1998; 71:345–355.
127. Beltrán FJ, González M, Rivas FJ, Acedo B. Aqueous degradation of VOCs in the ozone combined hydrogen peroxide or UV radiation processes. 1. Experimental results. J Environ Sci Health A 1999; 34:649–671.
128. Beltrán FJ, Ovejero G, Rivas J. Oxidation of polynuclear aromatic hydrocarbons in water. 4. Ozone combined with hydrogen peroxide. Ind Eng Chem Res 1996; 35:891–898.
129. Laplanche A, Orta de Velasquez MT, Boisdon V, Martin N, Martin G. Mod- elisation of micropollutant removal in drinking water treatment by ozonation or advanced oxidation processes (O_3/H_2O_2). Ozone Sci Eng 1995; 17:97–117.
130. Langlais C, Wolbert D, Laplanche A, Maillard K. A new step towards a better understanding static mixers as gas liquid reactors: micropollutant oxidation by O_3/H_2O_2 system. Proceedings 13th Ozone World Congress. Berlin: International Ozone Association, 1997:III.1.1–III.1.15.
131. Zhou OS. Ozone disinfection of sewage in a static mixer contacting reactor system on a plan scale. Ozone Sci Eng 1991; 13:313–330.
132. Martin N, Galey C. Use of static mixer for oxidation and disinfection by ozone. Ozone Sci Eng 1994; 16:455–473.
133. Brodart E, Mallevialle J, Costa C, Roustan M. Tube diameter and height influence on the ozonation operating conditions with a U-tube. Ozone Sci Eng 1986; 8:235–246.
134. Cater SR, Bircher KG, Stevens RDS, Safarzadeh-Amiri A. Rayox—A second

generation enhanced oxidation process for process and groundwater remediation. Water Pollut Res J Can 1992; 1:151–168.
135. McPhee W, Wagg L, Martin P. Advanced oxidation processes for the destruction of ordnance and propellant compounds using Rayox®. In: Eckenfelder WW, Bowers AR, Roth JA, eds. Chemical Oxidation. Technologies for the Nineties. Vol. 3. Basel: Technomic Publ, 1994:249–266.
136. Zeff JD, Barich JT. UV/Oxidation of organic contaminants in ground, waste and leachate waters. Water Pollut Res J Can 1992; 1:139–150.
137. Howard PH. Handbook of Environmental Fate and Exposure Data for Organic Chemicals. Vols. 1 to 3. Chelsea, MI: Lewis Publ, 1989.
138. Corless CE, Reynolds GL, Graham NJD, Perry R. Ozonation of pyrene in aqueous solution. Water Res 1990; 24:1119–1123.
139. Hu ST, Yu YH. Preozonation of chlorophenolic wastewater for subsequent biological oxidation. Ozone Sci Eng 1994; 16:13–28.
140. Trapido M, Veressinina J, Munter R. Ozonation and advanced oxidation process (AOP) treatment of phenanthrene in aqueous solutions. Ozone Sci Eng 1994; 16:475–486.
141. Preis S, Kamenev S, Kallas J, Munter R. Advanced oxidation processes against phenolic compounds in wastewater treatment plant. Ozone Sci Eng 1995; 17:399–418.
142. Esplugas S, Yue PL, Pervez MI. Degradation of 4-chlorophenol by photolytic oxidation. Water Res 1994; 28:1323–1328.
143. Shen Y, Ku Y, Lee K. The effect of light absorbance on the decomposition of chlorophenols by ultraviolet radiation and u.v./H_2O_2 processes. Water Res 1995; 29:907–914.
144. Upham BL, Yao JJ, Trosko JE, Masten SJ. Determination of the efficacy of ozone treatment systems using a gap junction intercellular communication bioassay. Environ Sci Technol 1995; 29:2923–2928.
145. Ku Y, Su WJ, Shen YS. Decomposition of phenols in aqueous solution by the UV/O_3 process. Ozone Sci Eng 1996; 18:443–460.
146. Masten SJ, Galbraith MJ, Davies SHR. Oxidation of trichlorobenzene using advanced oxidation processes. Ozone Sci Eng 1997; 18:535–548.
147. Kuo CH, Chen SM. Ozonation and peroxone oxidation of toluene in aqueous solution. Ind Eng Chem Res 1996; 35:3973–3983.
148. Trapido M, Hirvonen A, Veressinina Y, Hentunen J, Munter R. Ozonation, ozone/UV and UV/H_2O_2 degradation of chlorophenols. Ozone Sci Eng 1997; 19:75–96.
149. De AK, Bhattacharjee S, Dutta BK. Kinetics of phenol photooxidation by hydrogen peroxide and ultraviolet radiation. Ind Eng Chem Res 1997; 36:3607–3612.
150. Andreozzi R, Caprio V, Insola A, Marotta R. The oxidation of metol (*N*-methyl-*p*-aminophenol) in aqueous solution by UV/H_2O_2 photolysis. Water Res 2000; 34:463–472.
151. Masten SJ. Ozonation of VOCs in the presence of humic and soils. Ozone Sci Eng 1991; 13:287–312.

152. Kusakabe K, Aso A, Wada T, Hayashi JI, Morooka S, Isomura K. Destruction rate of volatile organochlorine compounds in water by ozonation with UV radiation. Water Res 1991; 25:1199–1203.
153. Xiong F, Graham NJD. Rate constants for herbicide degradation by ozone. Ozone Sci Eng 1992; 14:283–301.
154. Adams CD, Randtke SJ. Ozonation by-products of atrazine in synthetic and natural waters. Environ Sci Technol. 1992; 26:2218–2227.
155. De Laat J, Tace E, Doré M. Degradation of chloroethanes in dilute aqueous solution by H_2O_2/UV. Water Res 1994; 28:2507–2519.
156. Prados M, Paillard H, Roche P. Hydroxyl radical oxidation processes for the removal of triazine from natural water. Ozone Sci Eng 1995; 17:183–194.
157. Adams CD, Fusco W, Kanzelmeyer T. Ozone, hydrogen peroxide/ozone and UV/ozone treatment of chromium and copper complex dyes: decolorization and metal release. Ozone Sci Eng 1995; 17:149–162.
158. Roche P, Prados M. Degradation of pesticides by ozonation and advanced oxidation. Ozone Sci Eng 1995; 17:657–672.
159. Lambert SD, Graham NJD, Croll BT. Degradation of selected pesticides in a lowland surface water by ozone and ozone hydrogen peroxide. Ozone Sci Eng 1996; 18:251–270.
160. Brambilla A, Bolzacchini E, Orlandi M, Polesello S, Rindone B. Reactivity of two models of non-ionic surfactants with ozone. Water Res 1997; 31:1839–1846.
161. Bose P, Glaze WH, Maddox S. Degradation of RDX by various advanced oxidation processes. 1 Reaction rates. Water Res 1998; 32:997–1004.
162. Bose P, Glaze WH, Maddox S. Degradation of RDX by various advanced oxidation processes. II. Organic by-products. Water Res 1998; 32:1005–1018.
163. Prado J, Esplugas S. Comparison of different advanced oxidation processes involving ozone to eliminate atrazine. Ozone Sci Eng 1999; 21:39–52.
164. Kusakabe K, Aso S, Hayashi JI, Isomura K, Morooka S. Decomposition of humic acid and reduction of trihalomethane formation potential in water by ozone with UV irradiation. Water Res 1990; 24:781–785.
165. Bose P, Bezbarua BK, Reckhow DA. Effect of ozonation on some physical and chemical properties of aquatic natural organic matter. Ozone Sci Eng 1994; 16:89–112.
166. Takahashi M, Nakai T, Satoh Y, Katoh Y. Ozonolysis of humic acid and its effect on decoloration and biodegradability. Ozone Sci Eng 1995; 17:511–526.
167. Volk C, Roche P, Joret JC, Paillard H. Comparison of the effect of ozone, ozone-hydrogen peroxide system and catalytic ozone on the biodegradable organic matter of a fulvic acid solution. Water Res 1997; 31:650–656.
168. Siddiqui MS, Amy GL, Murphy BD. Ozone enhanced removal of natural organic matter from drinking water sources. Water Res 1997; 31:3098–3106.
169. Takahashi N. Ozonation of several organic compounds having low molecular weight under ultraviolet radiation. Ozone Sci Eng 1990; 12:1–18.

170. Akhland MS, Schuchmann HP, von Sonntag C. Degradation of the polysaccharide alginic acid: a comparison of the effects of UV light and ozone. Environ Sci Technol 1990; 24:379–383.
171. Karpel Vel Leitner N, Papailhou AL, Croue JP, Peyrot J, Dore M. Oxidation of methyl *tert*-butyl ether (MTBE) and ethyl *tert*-butyl ether (ETBE) by ozone and combined ozone/hydrogen peroxide. Ozone Sci Eng 1994; 16:41–54.
172. Appelman EH, Jache AW, Muntean JV. Use of ozone + hydrogen peroxide to degrade macroscopic quantities of chelating agents in an aqueous solution. Ind Eng Chem Res 1996; 35:1480–1482.
173. Sörensen M, Frimmel FH. Photochemical degradation of hydrophilic xenobiotics in the UV/H_2O_2 process: influence of nitrate on the degradation rate of EDTA, 2-amino-1-naphthalenesulfonate, diphenyl-4-sulfonate and 4,4′-diaminostilbene-2,2′-disulfonate. Water Res 1997; 31:2885–2891.
174. Stefan MI, Bolton JR. Mechanism of the degradation of 1,4-dioxane in dilute aqueous solution using the UV/hydrogen peroxide process. Environ Sci Technol 1998; 32:1588–1595.
175. Andreozzi R, Caprio V, Insola A, Marotta R, Sanchirico R. Advanced oxidation processes for the treatment of mineral oil-contaminated waste-waters. Water Res 2000; 34:620–628.
176. Schnitzer M, Khan SU. Humic Substances in the Environment. New York: Marcel Dekker, 1972.
177. Croué JP, Beltrán FJ, Legube B, Doré M. Effect of preozonation on the organic halide formation potential of an aquatic fulvic acid. Ind Eng Chem Res 1989; 28:1082–1089.
178. Borja R, Martín A, Luque M, Durán MM. Kinetic study of anaerobic digestion of wine distillery wastewater. Proc Biochem 1993; 28:83–90.
179. Sotelo JL, Beltrán FJ, González M. Ozonation of aqueous solutions of resorcinol and phloroglucinol. 1 Stoichiometric and absorption kinetic regime. Ind Eng Chem Res 1990; 29:2358–2367.
180. Battaglia G. Atrazine elimination according to the drinking water regulations. Water Supply 1990; 7:161–168.
181. Adams CD, Randtke SJ, Thurman EM, Hulsey RA. Occurrence and treatment of atrazine and its degradation products in drinking water. Proceedings of American Water Works Association, 109th Annual Conference, Cincinnati, OH 1990:1–24.
182. Beltrán FJ, García-Araya JF, Acedo B. Advanced oxidation of atrazine in water. II. Ozonation combined with ultraviolet radiation. Water Res 1994; 28:2165–2174.
183. Nélieu S, Kerhoas L, Einhorn J. Degradation of atrazine into ammeline by combined ozone/hydrogen peroxide treatment in water. Environ Sci Technol 2000; 34:430–437.
184. Canelli E. Chemical, bacteriological and toxicological properties of cyanuric acid and chlorinated isocyanurates as applied to swimming pool disinfection. Am J Public Health 1974; 62:155–162.
185. Upham BL, Boddy B, Xing X, Trosko JE, Masten SJ. Non-genotoxic effects

of selected pesticides and their disinfection by-products on gap junctional intercellular communication. Ozone Sci Eng 1997; 19:351–369.
186. Hirvonen A, Tuchkanen T, Kalliokoski P. Formation of chlorinated acetic acids during UV H_2O_2 oxidation of ground water contaminated with chlorinated ethylenes. Chemosphere 1996; 32:1091–1102.
187. Dowideit P, VonSonntag C. Reaction of ozone with ethene and its methyl- and chlorine-substituted derivatives in aqueous solution. Environ Sci Technol 1998; 32:1112–1119.
188. Adams CD, Scanlan PA, Secrit ND. Oxidation and biodegradability enhancement of 1,4-dioxane using hydrogen peroxide and ozone. Environ Sci Technol 1994; 28:1812–1818.
189. Richardson SD, Thruston AD Jr, Caughran TV, Chen PH, Collette TW, Floyd TL, Schenck KM, Lynks BW, Sun G, Majetich G. Identification of new ozone disinfection by-products in drinking water. Environ Sci Technol 1999; 33:3368–3377.
190. Richardson SD, Thruston AD Jr, Caughran TV, Chen PH, Collette TW, Floyd TL, Schenck KM, Lynks BW, Sun G, Majetich G. Identification of new drinking water disinfection by-products formed in the presence of bromide. Environ Sci Technol 1999; 33:3378–3383.
191. García-Araya JF, Croué JP, Beltrán FJ, Legube B. Origin and conditions of ketoacid formation during ozonation of natural organic matter in water. Ozone Sci Eng 1995; 17:647–656.
192. Backlund P. Degradation of aquatic material by ultraviolet light. Chemosphere. 1992; 25:1869–1878.
193. Wang GS, Hsieh ST, Hong CS. Destruction of humic acid in water by UV light catalyzed oxidation with hydrogen peroxide. Water Res 2000; 34:3882–3887.
194. Wang GS, Liao CH, Wu FJ. Photodegradation of humic acids in the presence of hydrogen peroxide. Chemosphere 2001; 42:379–387.
195. Legube B. A survey of bromate ion in European drinking water. Ozone Sci Eng 1996; 18:325–348.
196. Von Gunten U, Hoigné J. Bromate formation during ozonation of bromide-containing waters: interaction of ozone and hydroxyl radical reactions. Environ Sci Technol 1994; 28:1234–1242.
197. Kruithof JC, Kamp PC, Belosevic M. UV–H_2O_2 treatment for pesticide degradation and disinfection: from challenge to full scale application. Proceedings 15th Ozone World Congress. Vol. II. London: International Ozone Association, 2001:159–173.
198. Kuo WS. Effects of photolytic ozonation on biodegradability and toxicity of industrial wastewater. J Environ Sci Health A 1999; 34:919–933.
199. Rodgers DW, Evans DW, Sheehan LV. Toxicity reduction on Ontario hydro radioactive liquid waste. Water Air Soil Pollut 1996; 90:219–229.
200. Juang LC, Tseng DH, Yang SC. Treatment of petrochemical wastewater by UV/H_2O_2 photodecomposed system. Water Sci Technol 1997; 36:357–365.
201. Beltrán FJ, González M, González JF. Industrial wastewater advanced

oxidation. 1. UV radiation in the presence and absence of hydrogen peroxide. Water Res 1997; 31:2405–2414.
202. Beltrán FJ, Encinar JM, González JF. Industrial wastewater advanced oxidation. 2. Ozone combined with hydrogen peroxide or UV radiation. Water Res 1997; 31:2415–2428.
203. Beltrán FJ, García-Araya JF, Frades J, Álvarez P, Gimeno O. Effects of single and combined ozonation with hydrogen peroxide or UV radiation on the chemical degradation and biodegradability of debittering table olive industrial wastewaters. Water Res 1999; 33:723–732.
204. Beltrán FJ, García-Araya JF, Álvarez P. Continuous flow integrated chemical (ozone)-activated sludge system treating combined agroindustrial-domestic wastewater. Environ Prog 2000; 19:28–35.
205. Rivas J, Beltrán FJ, Acedo B, Gimeno O. Two step wastewater treatment: sequential ozonation–aerobic biodegradation. Ozone Sci Eng 2000; 22:617–636.
206. Ledakowicz S, Gonera M. Optimization of oxidants dose for combined chemical and biological treatment of textile wastewater. Water Res 1999; 33:2511–2516.
207. Böhme A. Ozone technology of German industrial enterprises. Ozone Sci Eng 1999; 21:163–176.
208. Lowndes R. State of the art for ozone U.K. experience. Ozone Sci Eng 1999; 21:201–205.
209. Rice RG. Ozone in the United States of America—state of the art. Ozone Sci Eng 1999; 21:99–118.
210. Hager DG. UV-catalyzed hydrogen peroxide chemical oxidation of organic contaminants in water. Innovative hazard. Waste Treat Technol. Ser. 2. Tucson, AZ: Peroxidation Systems, 1990:143–153.
211. AOT Handbook. Markham, Ontario: Calgon Carbon Corporation Technologies, 1996.
212. Martin N, Le Corre K, Quennell S. The ozoflotation-filtration process: treatment of the Thames river water at Walton works. Ozone News 1991; 19:6–7.
213. Nebel C, Caldwell W. NASA detoxifies launch-pad water with UV and ozone. Ozone News 1992; 20:21–22.
214. Liechti PA, Kaulbach R. One year full-scale study of ozone cooling water treatment at a German electric power station. Ozone Sci Eng 1992; 14:531–544.
215. Puckorious PR. Ozone use in cooling tower systems—current guidelines—where it works. Ozone Sci Eng 1993; 15:81–93.
216. Barlow P. Ozone treatment for the 2000 Olympics aquatic facility. Ozone News 1994; 22:44–45.
217. Duguet JP, Anselme C, Mazounie P, Mallevialle J. Application of combined ozone-hydrogen peroxide for the removal of aromatic compounds from a ground water. Ozone Sci Eng 1990; 12:281–294.
218. Cline JD, Sullivan PF, Lovejoy MA, Adams CD, Collier J, Fowler R. Ozone/UV treatment to enhance biodegradation of surfactants in industrial waste-

waters: a pilot plant scale study. Proceedings of 89th Annual Meeting & Exhibition of Air & Waste Management Association. 96-RA 115.07, Nashville, TN, 1996.
219. Hirvonen A, Tuchkanen T, Ettala M, Korhonen S, Kalliokoski P. Evaluation of a field scale UV/H_2O_2 oxidation system for the purification of groundwater contaminated with PCE. Environ Technol 1998; 19:821–828.
220. Croll BT. The installation of GAC and ozone surface water treatment plants in Anglian water, UK. Ozone Sci Eng 1996; 18:19–40.
221. Wenzel A, Gahr A, Niessner R. TOC removal and degradation of pollutants in leachate using a thin-film photoreactor. Water Res 1999; 33:937–946.
222. Annachhatre AP, Bhargava A. Evaluation of Swirl-Flo™ Clarifier and UV/H_2O_2 process for domestic wastewater treatment. Environ Technol 1999; 20:285–292.
223. Griffini O, Iozelli P. The influence of H_2O_2 in ozonation treatment: experience of the water supply service of Florence, Italy. Ozone Sci Eng 1996; 18:117–126.
224. Rico B, Pérez D, Lema JM, Lucas T. Organic matter removal in boiler feed water treatment of a power plant with ozone. Ozone Sci Eng 1997; 19: 471–480.
225. Beltrán FJ, Encinar JM, García-Araya JF. Modelling industrial wastewater ozonation in bubble contactors. 2. Scale-up from bench to pilot plant. Ozone Sci Eng 1995; 17:379–398.
226. Pedit JA, Iwamasa KJ, Miller CT, Glaze WH. Development and application of a gas–liquid contactor model simulating advanced oxidation processes. Environ Sci Technol 1997; 31:2791–2796.
227. Lewis N, Topudurti K, Welshans G, Foster R. A field demonstration of the UV/oxidation technology to treat ground water contaminated with VOCs. J Air Waste Manage Assoc 1990; 40:540–547.
228. Trancart JL. Treatment of triazines in waterworks. Efficiency of O_3/H_2O_2 coupling. Ozone News 1990; 18:7–8.
229. Gahr A, Niessner R. Drinking water treatment with ozone in combination with UV-radiation; removal of pesticides and by-product formation. Proceedings of 12th Ozone World Congress, Lille, France, 1995:349–371.
230. Foelich EM. Advanced chemical oxidation of organics using the perox-pure™ oxidation system. Water Res J Can 1992; 27:169–183.
231. Schwämmlein K, Leitzke O. Field tests for the elimination of chlorinated hydrocarbon compounds in ground water by combined oxidation processes. Proceedings of 12th Ozone World Congress, Lille, France, 1995:325–336.
232. Bolton JR, Bircher KG, Tumas W, Toldman CA. Figures-of-merit for the technical development and application of advanced oxidation processes. J Adv Oxid Technol 1996; 1:13–17.
233. Arslan I, Balcioglu IA, Tuchkanen T. Treatability of simulated reactive dyebath wastewater by photochemical and non-photochemical advanced oxidation processes. J Environ Sci Health A 2000; 35:775–793.
234. Cater SR, Stefan MI, Bolton JR, Safarzadeh-Amiri A. UV/H_2O_2 treatment of

methyl *tert*-butyl ether in contaminated waters. Environ Sci Technol 2000; 34:659–662.
235. Legube B, Karpel Vel Leitner N. Catalytic ozonation: a promising advanced oxidation technology for water treatment. Catal Today 1999; 53:61–72.
236. Kang JW, Hoffman MR. Kinetics and mechanism of the sonolytic destruction of methyl *tert*-butyl ether by ultrasonic irradiation in the presence of ozone. Environ Sci Technol 1998; 32:3194–3199.
237. Fung PC, Huang Q, Tsui SM, Poon CS. Treatability study of organic and colour removal in desizing/dyeing wastewater by UV/US system combined with hydrogen peroxide. Water Sci Technol 1999; 40:153–160.
238. Kang JW, Hung HM, Lin A, Hoffman MR. Sonolytic destruction of methyl tert-butyl ether by ultrasonic irradiation: the role of O_3, H_2O_2, frequency and power density. Environ Sci Technol 1999; 33:3199–3205.
239. Lang PS, Ching WK, Willberg DM, Hoffman MR. Oxidative degradation of 2,4,6 trinitrotoluene by ozone in an electrohydraulic discharge reactor. Environ Sci Technol 1998; 32:3142–3148.

2

Photocatalytic Degradation of Pollutants in Water and Air: Basic Concepts and Applications

Pierre Pichat
Ecole Centrale de Lyon, Ecully, France

I. INTRODUCTION

Several of the advanced oxidation processes described in this book involve photons used to generate oxidizing species, directly or indirectly, from H_2O, H_2O_2, and O_3. Heterogeneous photocatalysis is the only one of these processes that is based on the photonic excitation of a solid, which renders it more complex. Considering the very high number of papers and patents in this domain, the yearly publication of a bibliography [1], which includes organized references, the existence of several review articles (e.g., see Refs. 2–6, published since 1997), and the publication of a recent book [7], this chapter cites only some of the studies as a starting point in order to cover the principal issues. The choice of the particular references cited here is somewhat arbitrary and is influenced by the author's knowledge of the individual topics. Certainly, excellent reports have not been included, but they can be found in Ref. 1, which is a rich source of information.

II. BACKGROUND AND FUNDAMENTALS OF THE TECHNIQUE

A. General Description

1. Role of Photonic Excitation, Electron Transfer, and Adsorption

The term photocatalysis may designate several phenomena that involve photons and a catalyst [8]. In this chapter, the terms heterogeneous photocatalysis and photocatalysis refer only to cases where the photosensitizer is a semiconductor. Moreover, only TiO_2 is considered among the semiconductors, except for the general concepts.

Electrons pertaining to an isolated atom occupy discrete energy levels. In a crystal, each of these energy levels is split into as many energy levels as there are atoms. Consequently, the resulting energy levels are very close and can be regarded as forming a continuous band of energies. For a metal (or conductor), the highest energy band is half-filled and the corresponding electrons need only a small amount of energy to be raised into the empty part of the band, which is the origin of electrical conductivity at room temperature. In contrast, in insulators and semiconductors, valence electrons completely fill a band, thus called the valence band, whereas the next higher-energy band, termed the conduction band, is empty, at least at 0 K. In a perfect crystal, the energy band separating the highest level of the valence band from the lowest level of the conduction band is forbidden. Its width is referred to as the band gap. It is smaller for semiconductors (viz., ca. < 4 eV) than for insulators, in accordance with the names of these materials.

The absorption of exciting photons, most often in the ultraviolet spectral range, by a semiconductor promotes electrons from the filled valence band (where electron vacancies, electron deficiencies, or holes are thus formally created) to the vacant conduction band (Fig. 1). The electron–hole pairs can recombine either directly (band-to-band recombination) or, most

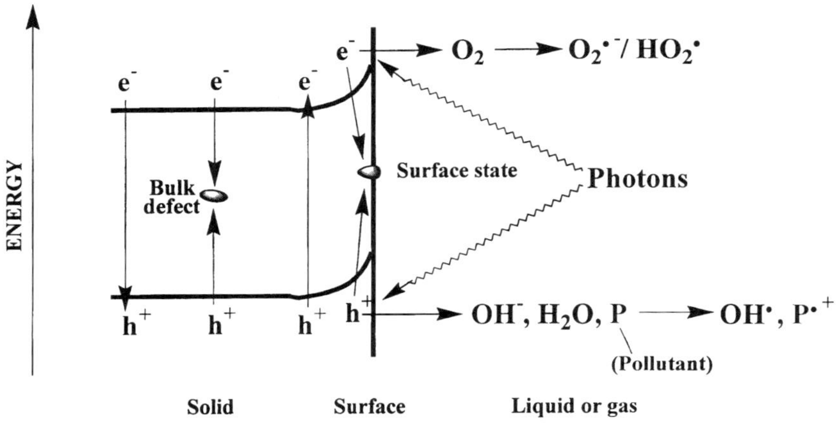

Figure 1 Simplified scheme illustrating in space-energy coordinates, the photogeneration, the bulk and surface recombination, the reaction with dioxygen, hydroxide ions, water and electron-donor pollutants, of charge carriers in an n-type semiconductor such as TiO_2.

often, indirectly (e.g., via bulk or surface defects) by radiative and non-radiative processes. If the charges are localized by trapping at surface states, their mean lifetime can be long enough to allow their transfer to adsorbed electron donors or acceptors. Provided that the resulting intermediates are transformed before backelectron transfer occurs, a photocatalytic redox reaction is produced. For colloidal TiO_2 samples, electrons can be trapped within about 30 psec after their excitation to the conduction band, and holes can be trapped within a period shorter than 250 nsec (9). Interfacial charge transfers take place within nanoseconds to milliseconds (10).

In the presence of dioxygen, adsorbed oxygen species are the most probable electron acceptors. Undissociated oxygen leads to the superoxide radical ion $O_2{\bullet}^-$ (Fig. 1), or its protonated form, the hydroperoxyl radical $HO_2{\bullet}$ ($pK_a =$ ca. 4.7). In liquid water, two $HO_2{\bullet}$ radicals can combine if their concentrations allow them to react significantly yielding H_2O_2 and O_2 (disproportionation reaction). In turn, H_2O_2 can scavenge an electron from the conduction band or from the superoxide, and accordingly be reduced to a hydroxyl radical $OH{\bullet}$ and a hydroxide ion OH^-. Because these reactions are known to take place in homogeneous aqueous phases, they are believed to occur at the TiO_2 surface as well. In other words, the very oxidizing hydroxyl radical might be produced, in principle, by the three-electron reduction of O_2:

$$O_2 + e^- \rightarrow O_2{\bullet}^- \qquad (1)$$

$$O_2{\bullet}^- + H^+ \rightarrow HO_2{\bullet} \qquad (2)$$

$$2\ HO_2{\bullet} \rightarrow H_2O_2 + O_2 \qquad (3)$$

$$H_2O_2 + e^- \rightarrow OH{\bullet} + OH^- \qquad (4)$$

$$H_2O_2 + O_2{\bullet}^- \rightarrow OH{\bullet} + OH^- + O_2 \qquad (5)$$

This series of chemical equations is equivalent to:

$$O_2 + 2\ H^+ + 3e^- \rightarrow OH{\bullet} + OH^- \qquad (6)$$

The formation of H_2O_2 by the reaction:

$$HO_2{\bullet} + R - H \rightarrow H_2O_2 + R{\bullet} \qquad (7)$$

where R-H is an organic species with a labile H atom, has also been envisaged, but this reaction would compete with H-atom abstraction from R-H by the $OH{\bullet}$ radical.

A much more direct way of forming the $OH{\bullet}$ radical is through the oxidation of an adsorbed water molecule or an OH^- ion by a valence band hole (h^+), (i.e., by an electron transfer from these entities to the photo-excited semiconductor) (Fig. 1).

Electron spin resonance (ESR) has been used to show the formation of $HO_2{\bullet}$ radicals on UV-irradiated TiO_2 at 77 K (11). Spin-trapping molecules have been added in the reaction medium to follow the reaction of OH• radicals with various organic pollutants (12–15). However, these experiments present the disadvantages either of being carried out under conditions that are quite distinct from those under which photocatalytic reactions are usually performed, or of relying on ESR signals whose origin is perhaps ambiguous as in the case of the DMPO–OH adduct (11,16).

On the other hand, many organic compounds have a redox potential at a higher energy than the valence band edge of common semiconductor oxides and, therefore, they can act as electron donors and thus yield a radical cation (Fig. 1), which can further react, for example, with H_2O, $O_2^{\bullet -}$, or O_2.

To summarize, the chemistry occurring at the surface of a photoexcited semiconductor is based on the radicals formed from O_2, H_2O, and electron-rich organic compounds. Also note that cations in aqueous solution can be directly reduced by conduction band electrons provided that the redox potentials of these cations are adequate (i.e., lying below the conduction band energy) (6).

This model, generally called the collective electron model of semiconductors, refers to thermodynamics because it considers the energy levels that can be occupied by electrons in the solid, the energy levels of the so-called surface states (Fig. 1), and, finally, the redox potentials of the species present in the external medium. The surface states can be intrinsic; this latter term designates defects due to the termination of the crystal lattice. The extrinsic surface states include impurities, various surface defects such as ion vacancies, surface groups, and adsorbed species. Whereas the energy levels of the valence band and the conduction band, as well as those of redox compounds in a solution, are generally known, data yielding the positions of defined surface states on the energy scale are rare. In addition, this collective electron model, however useful in indicating whether a given type of electron transfer is possible or not, has the disadvantage of considering adsorbed species and surface features from the energetic viewpoint only.

The localized model, which is founded on the concept of surface sites, allows one, at least on a qualitative basis, to consider other factors. It not only refers to the nature of the semiconductor, which provides the energy levels of its bands, but also tries to take into account the identity of the particular sample used. For powder samples, the preparation determines the texture (i.e., mean grain size), the porosity (and therefore the surface area), the morphology (spheres, polyhedra, needles, etc.), and the degree of crystallinity. As a result, the exposed crystal planes differ, and the number of surface irregularities, such as steps, kinks, etc., as well as the density of surface hydroxyl groups, vary for a given powder semiconductor. These irregularities correspond to electron energy levels that differ from the energy levels of the bulk.

Because the active species that can affect chemical transformations are those created at the photocatalyst surface or those reaching it, the photocatalytic reaction occurs, at least principally, in the adsorbed phase, and the overall process can be formally divided into five steps:

1. Transfer of the reactants from the fluid phase to the surface;
2. Adsorption of at least one of the reactants;
3. Reaction in the adsorbed phase;
4. Desorption of the product(s); and
5. Removal of the product(s) from the interfacial region.

As the adsorption and desorption rates are temperature-dependent, temperature can have an effect on the photocatalytic reaction rates. Increased rates on raising the temperature above the ambient temperature have been reported for the gas-phase removal of some pollutants (17,18) and, above all, for their mineralization rate (18).

2. Photocatalytic Character of a Reaction

From the abovedescribed principle of heterogeneous photocatalysis, it follows that photocatalytic reaction rates depend upon the characteristics of the irradiation, the mass of the photocatalyst, and the concentration (or partial pressure) of the reactants.

Irradiation Wavelength Dependence. Clearly, efficient photons are those that can be potentially absorbed by the semiconductor. Action spectra are most often determined by employing a series of optical cut-off filters or filtering solutions because the reaction rates are generally too weak to allow the use of a monochromator.

Blank experiments should be carried out to evaluate the photochemical transformations that can occur, and optical filters can be selected to cancel or to minimize these transformations, if desired. Even if the initial organic reactant(s) do(es) not absorb the photons that are used, some of the intermediate products may absorb the photons because, as a result of the gradual oxidation, they contain chromophore groups such as carbonyl and carboxyl groups.

Radiant Flux Dependence. Radiant flux can be measured by utilizing calibrated metallic grids and neutral density filters, or by filtering solutions of various absorbances without changing the geometry of the irradiating beam. For low radiant fluxes ϕ, a linear relation between the reaction rate and ϕ is observed. For higher ϕ values, the rate becomes proportional to $\phi^{1/2}$. This square root dependence arises from the predominance of electron–hole recombination [i.e., the rate of hole or electron capture by species involved in the chemical reaction(s) is small compared to that of the recombination of charges at high electron–hole generation rates] (19). In principle, there should be a ϕ value above which the reaction rate does not increase further; in other words, the rate becomes photon-independent (20), and rate limitations result from other causes.

Photocatalyst Mass Dependence. The maximum penetration depth of photons into the semiconductor is only a fraction of a micron, with light attenuation following the Beer–Lambert law. Consequently, for slurries in a given reactor, the photocatalytic reaction rate r increases linearly with increasing photocatalyst mass m up to a critical mass corresponding to the complete absorption of the photons—at the beginning of a plateau in the curve $r = f(m)$. For much higher m values, r can decrease because the coverage of the reactant on the irradiated particles is diminished because of reactant adsorption on nonirradiated particles, at least when the reactant concentration is rate-limiting.

Finally, to ascertain the photocatalytic character of a reaction, the reaction should be carried out over a period long enough to ensure conversions many folds greater than those expected from stoichiometric reactions involving preadsorbed or preexisting nonrenewable species (21).

3. Chemical Kinetics and Information on Reaction Mechanisms

From simple measurements of the rate of a photocatalytic reaction as a function of the concentration of a given reactant or product, valuable information can be derived. For example, these measurements should allow one to know whether the active species of an adsorbed reactant are dissociated or not (22), whether the various reactants are adsorbed on the same surface sites or on different sites (23), and whether a given product inhibits the reaction by adsorbing on the same sites as those of the reactants. Referring to kinetic models is therefore necessary. The Langmuir–Hinshelwood model, which indicates that the reaction takes place between both reactants at their equilibrium of adsorption, has often been used to interpret kinetic results of photocatalytic reactions in gaseous or liquid phase. A contribution of the Eley–Rideal mechanism (the reaction between one nonadsorbed reactant and one adsorbed reactant) has sometimes been proposed.

However, conclusions from the kinetic results should be drawn with caution. For example, assuming that OH• radicals formed at the surface of the solid are the active species in an aqueous-phase photocatalytic reaction, the question arises as to whether these radicals predominantly react in the adsorbed phase, or in the solution at a very short distance from the solid surface (i.e., in the double layer) (24–26). Four possibilities can be considered: the Langmuir–Hinshelwood model, the Eley–Rideal model with either the pollutant or the OH• radicals reacting when adsorbed, and, finally, the case where the organic compound and the active species react in the homogeneous phase. In each case, the expression of the rate r is:

$$r = k_{obs}\kappa C/(1 + \kappa C + \kappa_i C_i) \quad (8)$$

where k_{obs} is the observed reaction rate constant, C is the concentration of the organic compound, and the subscript i indicates an intermediate. This equation has the form expected for the Langmuir–Hinshelwood mechanism if κ and κ_i are adsorption constants. But if κ and κ_i are regarded as mere kinetic parameters, the other mechanisms can be considered as valid (27). In other words, the kinetic experiments, by themselves, do not allow one to discriminate between the kinetic models. Moreover, values of the adsorption constant derived from the Langmuir–Hinshelwood kinetics have been found to depend on both the radiant flux and the time interval used to measure the initial rate (28).

4. General Advantages and Disadvantages of Treatments by TiO_2 Photocatalysis

Photocatalytic treatments of gases and solutions offer several advantages:

1. No chemicals are used.
2. TiO_2 is a nontoxic compound used as an additive in the food and pharmaceutical industries. Its cost is on the order of a few dollars per kilogram (ca. \$2 kg^{-1} for the pigment grade; presumably 5–10 times more for a photocatalytic grade, depending on the future development of TiO_2 photocatalysis). It is synthesized in very large quantities for other purposes, so that its preparation is well mastered. It is stable and, in principle, self-regenerated when used appropriately (i.e., when its amount is in accordance with the pollutants' concentrations). The products of the initial organic pollutants, which may transitorily accumulate at its surface, are ultimately mineralized.
3. Owing to adsorption, which concentrates dilute pollutants at the TiO_2 surface where the active species are produced and/or can interact, photocatalysis is very appropriate in purifying/deodorizing indoor air, as well as gaseous and aqueous effluents containing only traces of toxic and/or malodorous pollutants.
4. In contrast to technologies that are exclusively based on adsorption or absorption phenomena and result in pollutant transfer with the need for supplementary treatments, photocatalysis completely mineralizes the organic pollutants or, at least, enables

one to reach low-enough concentrations of both the initial pollutants and their products.

However, the rates of the photocatalytic chemical transformations are limited by the rates of electron–hole recombination in the bulk of TiO_2 or at the surface (Fig. 1). These latter rates depend particularly on structural defects and on foreign cations in substitutional and interstitial positions. They are not easily controlled and consequently limit the application fields of photocatalysis.

In addition, in the case of water treatment, TiO_2 surface coverage is dominated by water molecules, which are linked to the surface hydroxyl groups by hydrogen bonds. This surface organization renders the approach of the organics to the surface difficult, especially for those compounds that are very soluble. However, studies have shown that even poorly adsorbed pollutants can be degraded, presumably because the organic molecules degraded in these cases are not limited to those located in the surface monolayer (26,29–31).

B. In-Depth Treatment of the Technique

1. Roles of O_2 and Effects of H_2O_2 and O_3

Roles of O_2. Dioxygen is believed to play several roles in the photocatalytic degradation of pollutants. First, as is illustrated in Fig. 1, it is able to scavenge electrons at the surface of UV-irradiated TiO_2, thereby allowing the separation of the photogenerated charges. This process is essentially equivalent to decreasing the recombination rate of electrons and holes. Second, O_2 can react with alkyl radicals or, more generally, organic radicals, yielding peroxyl radicals en route to the mineralization of the organic precursor. Third, the reduced form of O_2, viz. $O_2^{\bullet -}$, the radical anion superoxide, can react with an organic radical cation (32)—resulting from the reaction of holes with electron-rich organic pollutants (Fig. 1)—which is one of the primary steps in the chemical degradation events.

One of the consequences of this multiple involvement of dioxygen is that the photocatalytic reactors used for treating water should allow O_2 (air) to easily access the TiO_2 surface. In other words, the rates of gas-to-liquid and liquid-to-solid transfers should be maximized (33). This condition can be achieved by one or several of the following means: (1) bubbling air in the water; (2) producing a turbulent flow of the water in contact with air; and (3) limiting the water film thickness at the TiO_2 surface [the use of TiO_2-coated rotating disks or of TiO_2-coated beads (or small tubes) floating on water, etc.] (see Sec. IV.B, "Water Treatment").

Effect of H_2O_2. Adding H_2O_2 to the water to be photocatalytically treated can also be viewed as a way of increasing dioxygen concentration at the TiO_2 surface because H_2O_2 is disproportionated to H_2O and O_2 over UV-irradiated TiO_2 (34). However, hydroxyl radicals can also react with the added H_2O_2 instead of reacting with the organic pollutants:

$$OH\bullet + H_2O_2 \rightarrow H_2O + HO_2\bullet \tag{9}$$

Therefore, the net effect depends on the type of water, the TiO_2 specimen, and other experimental conditions. Reported beneficial effects are less than one order of magnitude (35–38).

Effect of O_3. Adding ozone in dioxygen or air is a very efficient means of enhancing the photocatalytic rates of the removal and, above all, the mineralization of organic pollutants both in air and in water, even if the wavelengths are intentionally selected so as not to excite ozone (39–41). This substantial effect is attributed to the difference in electron affinity between O_3 (2.1 eV) and O_2 (0.44 eV). Consequently, in the presence of ozone, the electrons photopromoted to the TiO_2 conduction band can be captured more easily, either directly:

$$e^- + O_3 \rightarrow O\bullet^- + O_2 \tag{10}$$

or indirectly:

$$O_2\bullet^- + O_3 \rightarrow O_2 + O_3\bullet^- \tag{11}$$

The radical anion $O_3\bullet^-$ is more unstable than O_3 and can presumably split easily at the surface of TiO_2:

$$O_3\bullet^- \underset{TiO_2}{\rightarrow} O\bullet^- + O_2 \tag{12}$$

Alternatively, it might react with adsorbed water:

$$O_3\bullet^- + H_2O \underset{TiO_2}{\rightarrow} OH\bullet + OH^- + O_2 \tag{13}$$

Furthermore, the increase in the scavenging rate of photoproduced electrons resulting from the presence of ozone should decrease the recombination rate of electrons and holes, and thereby augment the formation rate of hydroxyl radicals from basic OH surface groups and adsorbed water molecules (Fig. 1).

Irrespective of the mechanism, very oxidizing species, viz. $O\bullet^-$ and $OH\bullet$, would thus be generated. However, similar to H_2O_2, O_3 can act as a scavenger of hydroxyl radicals:

$$O_3 + OH\bullet \rightarrow O_2 + HO_2\bullet \tag{14}$$

Therefore, there is a limit to the favorable effect (40).

As ozone is employed in various industrial processes, such as paper bleaching, TiO_2 photocatalysis could be of interest in exploiting the presence of ozone to mineralize pollutants at higher rates while removing excess ozone. When O_3 is not used, the cost of its generation can be prohibitive; the interest of adding O_3 will then obviously be subordinate to the particular case and the regulations.

2. Properties Influencing the TiO_2 Photocatalytic Activity

Allotropic Form. In some papers, it is claimed that anatase TiO_2 is more photocatalytically active than rutile TiO_2. For those who are familiar with heterogeneous, thermally activated catalysis, this assertion cannot be valid because for every type of single-component catalyst, samples whose catalytic activities differ substantially exist. Indeed, the photocatalytic activities of various anatase and rutile samples overlap. The allegation about the superiority of anatase per se is based on the fact that, at least until now, the most active TiO_2 samples are anatase specimens. Also, some studies have shown that an increase in the percentage of rutile results in a lower photocatalytic activity (42); however, as other parameters (e.g., surface area, porosity, etc.) vary simultaneously, these results do not demonstrate that anatase is intrinsically more active. Conversely, it is the author's feelings that the relatively high photocatalytic activity of TiO_2 Degussa P-25, which is commonly used in laboratory and pilot plant studies as a reference sample, is not due to the supposedly optimum percentage of rutile (ca. 20%) with anatase. For example, this commercialized sample has an activity that corresponds to the expected value when measuring the activities of a series of anatase samples (with very low contents of rutile) prepared in the laboratory by the same method (i.e., in a flame reactor), utilizing the test reaction of the removal of 3-chlorophenol in water (43).

Several reasons have been proposed to explain that the most photocatalytically active samples have been found within the series of anatase samples. Insofar as electron capture by dioxygen (Fig. 1; Eq. 1) can be a limiting factor of the activity as was mentioned above, the higher energy position of the anatase conduction band could be the reason because it increases the driving force for the electron transfer to O_2. A higher mobility of the charge carriers, possibly caused by the less dense structure of anatase, might be another reason (44). However, the relationship between charge mobility and photocatalytic activity is not straightforward because a higher mobility may generate both a higher recombination rate of the photoproduced charges and a faster surface trapping of these charges to produce active species and/or faster reaction rates with adsorbed compounds (Fig. 1).

Nevertheless, this interpretation offers the advantage of being directly linked to the crystal structure of TiO_2.

Particle Size. Although a comparison of various TiO_2 samples is useful from the practical viewpoint, drawing definitive conclusions concerning the role of a given parameter ideally requires being able to restrict the number of parameters that vary simultaneously, which is extremely difficult (45).

A flame reactor is an excellent means for preparing pure or mixed oxides in the form of nonporous particles. The absence of porosity allows an average homogeneous irradiation of the particles. The morphologies (spheres, polyhedra) and the particle sizes (usually with a narrow distribution) can be mastered by adjusting the temperature, the flow rates (H_2 and O_2), and the concentration(s) of the compound(s) employed to generate the oxide. In the particular case of TiO_2, anatase samples whose rutile content is very low can be produced. Consequently, the effect of the surface area S on the photocatalytic activity can be determined, in principle.

For example, in the case of the degradation of 3-chlorophenol in TiO_2 aqueous suspensions, the initial degradation rate r_0 was found to increase linearly with increasing S, with the exception of the samples having the highest S (43). Consequently, the corresponding rate per unit of surface area r_s was constant within the experimental error for S, between approximately 10 and 150 $m^2\ g^{-1}$ (i.e., for particle diameters between about 170 and 10 nm); a significant decrease in r_s was observed for the samples with the highest S (up to ca. 210 $m^2\ g^{-1}$). It is tempting to attribute this decrease to an increase in the density of defects that can act as recombination centers of the photoproduced charge carriers. It must be kept in mind that the combined effects of S and the density of defects can vary with the pollutants and the medium (water or air).

Crystallinity. It seems that the degree of crystallinity of TiO_2 is an important factor in obtaining active TiO_2 (46). Indeed, amorphous titania is poorly active. An optimum calcination temperature of amorphous samples presumably corresponds to a compromise between an enhanced crystallinity, together with a decreased density of lattice defects, and limited decreases in surface area and coverage by OH groups.

Surface Coverage by Hydroxyl Groups. The surface of solid oxides necessarily carries OH groups, which can be free or linked by hydrogen bonds, depending on the hydroxylation/hydration state. These groups arise from water dissociation in the course of the solid oxide preparation. In addition, water molecules create a three-dimensional network by forming hydrogen bonds with the surface OH groups. The organizing effect of the surface is believed to extend to several molecular layers.

Surface OH groups can trap holes, and adsorbed water molecules can act as electron donors, creating hydroxyl radicals (Fig. 1). Also, water can be involved in the transformation of organic radical cations, e.g.,

$$R - C_6H_5 + h^+ \rightarrow R - C_6H_5{\bullet}^+ \tag{15}$$

$$R - C_6H_5{\bullet}^+ + H_2O \rightarrow R - C_6H_4OH + H^+ \tag{16}$$

and hydrolysis reactions, e.g.,

$$R - COCl + H_2O \rightarrow R - COOH + Cl^- + H^+ \tag{17}$$

Accordingly, it is tempting to conclude that the higher is the surface hydroxylation, the more active is the semiconductor oxide. However, the well-ordered adsorbed water molecules can hinder the adsorption of non-polar organic compounds because breaking hydrogen bonds between the water molecules close to the surface is more difficult than in bulk water. Therefore, an optimum surface hydroxylation may exist and vary according to the pollutant. Correlatively, an important factor could be the reversibility of the surface coverage by OH groups when relatively dry air is purified by photocatalysis.

Optical Properties. It should not be forgotten that the absorption of photons by the semiconductor is the initial step of heterogeneous photocatalysis. Therefore, the utilization of incident photons must be carefully considered.

For instance, a high porosity can increase the extent of adsorption of certain molecules, but at the same time, the internal surface of the pores is not fully irradiated so that the density of photoproduced active species inside the pores can be lower than on the external surface. Photons are not only absorbed but also reflected and scattered by the semiconducting particles, whether they are in the form of powders or films. Consequently, the texture, surface rugosity, and agglomeration of particles affect the fraction of photons that are absorbed and therefore are potentially useful for photocatalytic chemical transformations. In addition, scattering depends on the refractive index of the medium and is therefore very different depending on whether TiO_2 is exposed to air or liquid water.

Both the energy per photon and the light penetration depth vary with wavelength. Therefore, although TiO_2 absorbs about 60% of incident photons at 365 nm and absorbs nearly 100% at 254 nm, the energy required to create one electron–hole pair was found to be almost constant between 250 and 370 nm for a given TiO_2 sample dispersed in water (47).

Calculations and measurements have been made to assess the fraction of absorbed vs. incident photons in a number of cases, in particular for suspensions in water (48–51) and also for TiO_2 coated on glass fiber tissues.

The important message is that the optical properties should not be dismissed when trying to evaluate the effects of various factors on the photocatalytic activity. Obviously, taking into account these properties makes clear conclusions even more difficult to reach, but should help avoid false deductions.

3. Modifications of TiO_2

Doping. Studies (52,53) have shown that if substitutional cationic doping at low levels is homogeneous, which can be achieved by the use of a flame reactor (see Sec. II.B.2, "Particle Size"), it has a detrimental effect on the photocatalytic activity under UV irradiation. Furthermore, no activity is observed under irradiation in the visible spectral range in spite of an absorption by these samples. These observations regarding various reactions in different media have been attributed to electron–hole recombination at the site of the foreign cations (52,54).

However, increases in the photocatalytic activity have been reported for TiO_2 "doped" by lanthanides, tin, and iron (III) (55). It may be questioned whether these increases could, in fact, arise from the photo–Fenton reaction between the cations located at the surface and the hydrogen peroxide formed in situ, or even possibly because of a partial dissolution of these cations in the case of aqueous-phase reactions.

In contrast, a deep cation implantation in TiO_2, followed by calcination in O_2, has been shown to produce samples whose activity in the UV spectral region is unchanged and which are active when irradiated by visible radiation (56). The coordination of chromium and other transition metal cations differs from that obtained by aqueous-phase doping. However, why these cations do not act as recombination centers of photoproduced charges is not clear. The concentration of the implanted cation increases by a factor of ca. 50 from the TiO_2 surface to a depth of ca. 200 nm (56) so that the electron–hole pairs resulting from visible light irradiation could be created principally in deep layers. On this basis, we suggest here that the photocatalytic activity in the visible spectral range observed for implanted TiO_2 might be due to the fact that the charge carriers, which move from these deep layers to the surface layers, have a lower recombination rate because of a smaller density of foreign cations near the surface. If this interpretation is valid, samples comprising a shell of undoped TiO_2 around a core of doped TiO_2 absorbing in the visible spectral range might also be effective.

A recent study (57) has shown that the anionic substitutional doping of TiO_2 by N with an optimum of ca. 2.25 atom% can displace the spectral absorption limit from ca. 400 to 500 nm. This shift has been attributed to a change in the valence band energy resulting presumably from the mixing of N orbitals with O orbitals. However, TiON bonds were created in addition to TiN bonds. In the case of acetaldehyde mineralization rate, a threefold increase was observed for N-doped TiO_2 using 436 nm excitation when compared to TiO_2 before N-doping. For supported TiO_2, doping was effected by the sputtering of N_2; the doped films had a yellowish color. In a much simpler way, TiO_2 powder was doped by mere heating in an ammonia atmosphere, its surface area decreased by a factor of 4.

Irreversible oxygen vacancies can be generated by a plasma treatment of TiO_2 (58,59). These vacancies introduce an electron energy level in the band gap so that sensitization in the visible spectral range becomes possible. This visible-range photocatalytic activity corresponds to the excitation of electrons from the valence band to the additional energy level.

All of these studies are of great interest in extending the range of solar applications of TiO_2 photocatalysis (see Sec. IV.B.1).

Combining TiO_2 with Another Adsorbent. To increase the photocatalytic degradation activity of trace pollutants, attempts have been made to add an adsorbent, especially activated carbon (60–63) and zeolites (62,64), with a surface area higher than that of TiO_2. The problem is how to avoid the mere coexistence of two phenomena regarding the pollutants: destruction on TiO_2 and adsorption on the other solid. Also, photon losses must be minimized. What was postulated was the migration of pollutants from the nonphotocatalytic solid to TiO_2 or, alternatively, a transfer of active species from TiO_2 to the pollutants adsorbed on the added solid. An increased rate of degradation was observed from experiments carried out in the gas phase or in the aqueous phase, with various adsorbents and pollutants, and different means of depositing or mixing TiO_2. For each case, the ratio of the amounts of the two solids should be adjusted to maximize the area of the interfaces. In parallel, from experiments using well-defined structures consisting of alternate microstripes or islands of TiO_2 on Si, it has been inferred that pollutants can diffuse over distances as long as 20 μm from Si to TiO_2 (65).

4. Fixing TiO_2: Supporting Materials and Depositing Methods

The material chosen to support TiO_2 must withstand photocatalytic degradation. This condition excludes polymers, either synthetic (except those containing only C–F bonds) or natural, unless these materials can be used as replaceable, interchangeable parts. If cost and use allows it polymers can be coated with a layer of a substance, such as silica or alumina, which is inert with respect to photocatalytically produced active species. Another condition is that, during the coating process, the support must not release into TiO_2 chemical elements that can decrease the photocatalytic activity. Such unfavorable migrations have been found for sodium from glass (7,66), and for chromium and iron from stainless steel (67). Again, an intermediate layer is the remedy as, for example, in the case of SiO_2-coated glass (68). The abovementioned conditions being fulfilled, the choice of the material supporting TiO_2 depends on the use, mechanical properties, cost, etc. Glass, fused silica, ceramics, tiles, concrete, metals, polymers, paper, and textile materials have been tested. These materials can come in the form of plates, pellets, beads, thin sheets, honeycomb structures, etc.

The coating method must both preserve the TiO_2 photocatalytic activity and make TiO_2 solidly fixed on the support. This second requirement can be met by heating the coating, by adding an anchoring substance to TiO_2, or by combining both procedures. Achieving a better adhesion is clearly opposite to obtaining a better photocatalytic activity. Thermal treatment can induce the sintering of the TiO_2 particles, thus decreasing the surface area accessible to the pollutants. The addition of another substance can embed the TiO_2 particles and also restrict the mobility of the charge carriers if this substance is an insulator (e.g., silica). The choice of the procedure depends on the objective, type of support, cost, etc. Every type of coating technique can be used to spread the coating mixture over the support, such as dipcoating, spincoating, and spraycoating. Equipment employed for other types of coatings in the glass (68), paper (69), and printing industries can be utilized (7).

III. DEGRADATION OF POLLUTANTS

A. Laboratory-Scale Experimental Design

The choice of the light source—form, emitted wavelengths, radiant power—depends inter alia on whether TiO_2 is unsupported or supported, and on the type and shape of the supporting material. For example, a TiO_2-coated flexible material can be wrapped around a cylindrical lamp placed inside the reactor. A plate covered by TiO_2 can be installed perpendicularly to the beam of a lamp located outside the reactor. Obviously, the choice of the lamp, especially of the emission characteristics, also depends on the objective. For instance, in view of solar photocatalytic applications, lamps mimicking solar irradiation at the Earth's level or a solar box can be used in the laboratory.

As in photochemistry, filter solutions, solid optical filters, and grids can be used to monitor the effects of the wavelength and of the radiant power. Controlling and measuring these parameters are important, and they should be taken into account when designing the reactor. To avoid heating effects due to the infrared fraction of the emission, a water jacket can be placed in the light beam.

Both batch reactors and continuous-flow reactors have been used. Because TiO_2 photocatalysis is generally considered of interest in purifying air or water with low concentrations of pollutants, the absorption of the incident photons by the pollutants is most often insignificant. If it is not the case, a falling film annular reactor (49) can be used as in photochemistry.

For the particular case where the self-cleaning properties of TiO_2-coated plates have to be evaluated, a photoreactor allowing a comparison of several plates under simulated solar irradiation has been designed (70).

B. Examples

1. Mechanisms

Active Species. The respective importance, in the chemical steps that transform organic pollutants, of the various species that can exist at the surface (or near the surface) of photoexcited TiO_2 (Fig. 1) is being debated on.

Most of the experiments conceived to clarify this topic are based on the chemical analysis of the intermediate products of a given pollutant. For water treatment, comparisons have been made between the product distributions obtained with and without the addition of an OH• scavenger (e.g., Ref. 71) or an enzyme (72) that dismutates either $O_2^{\bullet -}$ or H_2O_2. The product distribution of the photocatalytic degradation can also be compared to that yielded by processes generating OH• radicals in homogeneous phase, such as the Fenton and photo–Fenton reactions (32), the H_2O_2–UV system, and radiolysis (with added substances to scavenge the other species formed in this latter case). These procedures can be criticized because (1) some include the addition of chemicals; (2) they generally do not discriminate between the effects originating from the existence of species other than OH• radicals in photocatalysis and the effects related to adsorption phenomena; and (3) controversies exist about whether the Fenton or photo–Fenton processes act uniquely through OH• radicals or involve Fe (IV) ions (73).

Nevertheless, these efforts to unravel the role of the various active species have brought about interesting conclusions even if they are not definitive. For example, the photocatalytic disappearance rate of 1,2-dimethoxybenzene (1,2-DMB) (74) and of quinoline (32) in aqueous TiO_2 suspensions was decreased in the presence of superoxide dismutase (SOD), which catalyzes the overall reaction:

$$2\ O_2{\bullet}^- + 2\ H^+ \rightarrow O_2 + H_2O_2 \qquad (18)$$

Both the inhibition of this effect by CN^- ions, which form complexes with the enzyme metal cations, and the change in the distribution of the degradation intermediates at equivalent transformation rates of the organic pollutant, with or without an enzyme, suggest that the observed phenomenon really stemmed from the catalytic activity of SOD. These results emphasize the importance of superoxide.

To address the question of the respective importance of pathways in heterogeneous photocatalysis, a molecule susceptible to yield different primary products, depending on the initial attack by an OH radical or a hole, has been used. Six-membered aromatic carbon cycles do not fulfil this condition. For example, from a substituted benzene, the monohydroxycyclohexadienyl radical can be formed either from addition of the hydroxyl radical or from the capture of a hole, followed by the hydration and deprotonation of the resulting radical cation.

In contrast, the degradation of quinoline:

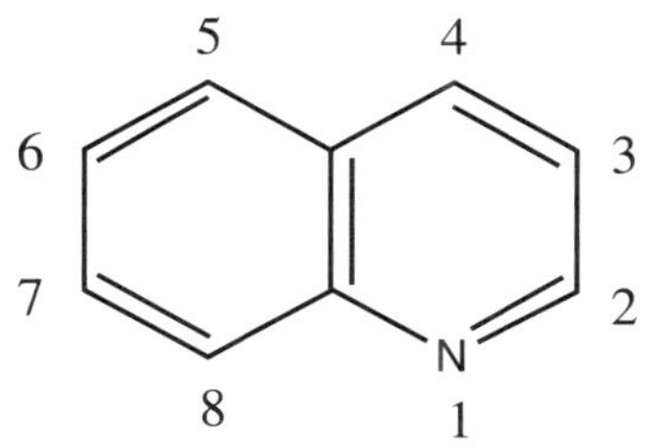

by the electrophilic hydroxyl radical should yield primary products reflecting a dominant attack onto the benzene ring, whose electron density is higher than that of the pyridine ring because of electron attraction by the more negative N atom, whereas the formation of primary products in which the pyridine ring is altered will reveal the existence of other pathways (32).

The photo–Fenton reaction was used to produce OH• radicals in the absence of TiO_2. H_2O_2 and Fe^{3+} ion concentrations were chosen so as to have a quinoline disappearance rate similar to that obtained by TiO_2 photocatalysis degradation in order to make comparisons more sound.

The three intermediate products that reached the highest concentrations in the case of quinoline degradation by the photo–Fenton process were in the order:

$$5 - \text{hydroxyquinoline} > 8 - \text{hydroxyquinoline} \geq \text{quinoline} - 5,\ 8 - \text{dione}$$

The dione was a secondary product. These three products show a preferential attack of the hydroxyl radical on the benzene ring at positions 5 and 8, as is expected from literature results (see Fig. 2 in the case where the OH• radical attack is initiated at position 5). The amount of 5-hydroxyquinoline was diminished by a factor of ca. 2 for the photocatalytic degradation. Although 8-hydroxyquinoline was not entirely extracted from the TiO_2 surface, it was clear that a lower amount was formed by the heterogeneous

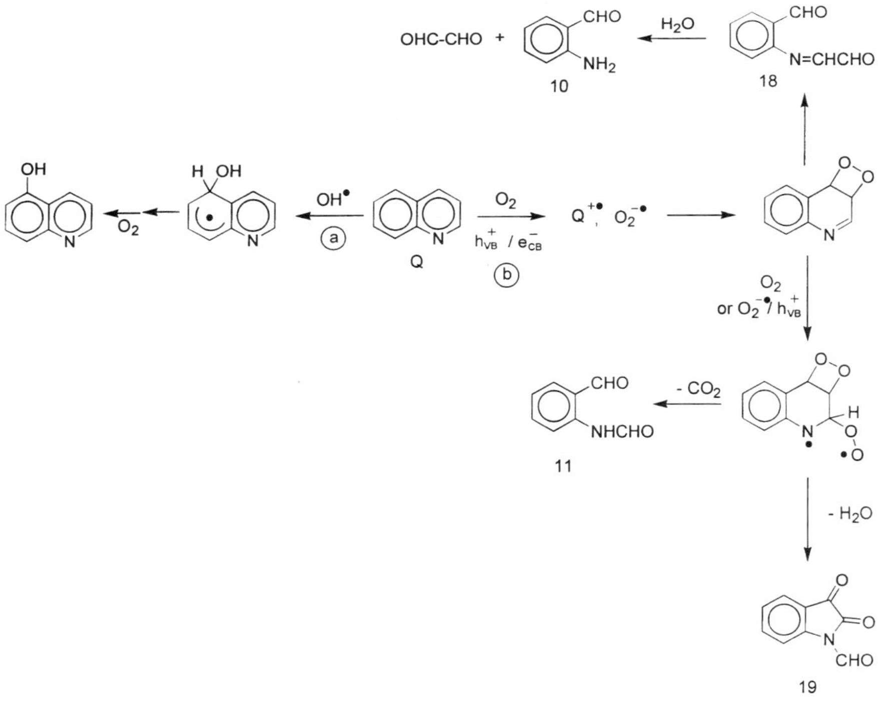

Figure 2 Scheme showing degradation pathways of quinoline depending on whether the initial attack occurs through a hydroxyl radical (a) or through direct electron transfer to TiO_2 and subsequent reaction with superoxide (b).

photocatalysis rather than by the photo–Fenton reaction. Furthermore, quinoline-5,6-dione was absent in the degradation over TiO_2.

Regarding the main products corresponding to the opening of the pyridine moiety of quinoline, 2-aminobenzaldehyde and, to a lesser extent, its *N*-formyl derivative were formed by photocatalysis, whereas only traces of this latter product were detected when the photo–Fenton process was employed. Also, (2-formyl)phenyliminoethanol was detected only in the case of the degradation over TiO_2.

It may be questioned whether a solid of large surface area onto which the quinoline molecules are adsorbed can change per se the distribution of the products. To test that, the photo–Fenton degradation process was performed in the presence of silica powder, which was not photocatalytically active. The distribution of intermediate products was not significantly influenced.

Insofar as the OH• radicals are considered to be the only species intervening in the photo–Fenton process at the concentrations used, these results (32) mean that other active species are also involved in TiO_2 photocatalysis. From the aforementioned results obtained with SOD, it was inferred that the $O_2{\bullet}^-$ radical anion was involved in the degradation of TiO_2. As quinoline does not react significantly with KO_2, which is a source of superoxide, the activation of quinoline appears to be a prerequisite. Quinoline activation can result from hole capture. The oppositely charged radical ions, $Q{\bullet}^+$ and $O_2{\bullet}^-$, likely react at a diffusion-controlled rate to form a dioxetane (Fig. 2). The regioselectivity is expected to depend on the spin density on $Q{\bullet}^+$, which is reasonably larger at positions 2 and 4.

To substantiate this mechanism, haloquinolines (75) were used. The strategy was to hinder sterically the addition of superoxide. In the case of 6-chloroquinoline, the products were the same as those formed from quinoline, except that they were chlorinated, which was expected because position 6 is not involved in either mechanism. Halogen substitution on the pyridine moiety in part directed oxygen addition to the benzene moiety, which was consistent with superoxide addition onto the more accessible positions on the benzene ring of the halogenated radical cation. This result supports the fact that a cycloaddition mechanism can take place in the photocatalytic degradation of quinoline. This mechanism has been proposed in the case of other aromatics, such as 4-chlorophenol (76) and 4-chlorocatechol (77).

The primary event may be contingent not only upon the pollutant but also upon its concentration (78) because of a competition for holes between the adsorbed pollutant and basic surface OH groups (Fig. 1) (79).

Whether the hole trapped as a surface-bound OH• radical transfers a hydroxyl group or only a positive charge to the pollutant has also been discussed (78–82) and probably depends on the pollutant. Another method of investigating the surface mechanisms is to replace surface OH groups by fluoride ions (83,84). Phenol being the test pollutant, that method has allowed the delineation of the roles of hydroxyl radicals either bound to TiO_2 or in the solution. It was inferred that, in the absence of fluoride, phenol is predominantly attacked by bound OH• radicals, although this compound is poorly adsorbed.

Diffuse reflectance flash photolysis has also been employed to distinguish products of direct electron transfer between the pollutant and TiO_2 in aqueous suspensions from those resulting from the reaction of the pollutant with hydroxyl radicals, provided that the absorption spectra of the radical cation and the OH adduct are unambiguously assigned (85).

The possible formation of singlet dioxygen on UV-irradiated TiO_2 has been suggested on the basis of the observation, in some cases, of products that can arise from the oxidation by this species. However, these products could also be formed through other pathways and, above all, $^1O_2$ traps did not have the expected effect (86). For TiO_2 dispersed in ethanol, a recent paper (87) has presented evidence based on ESR spectroscopy. The question is whether the trapping agent used to obtain the ESR signal is really specific. The authors suggest that $^1O_2$ might be generated by the energy released by the electron–hole recombination.

Examples of Basic Degradation Pathways. On the basis of data provided by radiochemists, the following reactions have been proposed for the photocatalytic transformations of aliphatic alkyl radicals (e.g., formed by OH• radicals):

$$RH + {\bullet}OH \rightarrow R{\bullet} + H_2O \quad (19)$$

$$R{\bullet} + O_2 \rightarrow RO_2{\bullet} \quad (20)$$

$$RO_2{\bullet} + RH \rightarrow ROOH + R{\bullet} \quad (21)$$

$$ROOH \rightarrow RO{\bullet} + {\bullet}OH \quad (22)$$

$$RCH_2O{\bullet} + O_2 \rightarrow RCHO + HO_2{\bullet} \quad (23)$$

$$RCH_2O{\bullet} + RH \rightarrow RCH_2OH + R{\bullet} \quad (24)$$

The initial propagating reaction corresponds to the addition of dioxygen. Subsequent propagating reactions are (1) the abstraction of H atoms from the initial pollutants and their products by the alkylperoxyl and alkoxyl radicals; (2) the elimination of a hydroperoxyl radical following O_2 addition; and (3) the splitting of unstable alkylperoxides.

In the case of benzenic compounds, a hydroxyl radical is believed to be initially added to the ring, with the resulting cyclohexadienyl radical adding dioxygen and eliminating a hydroperoxyl radical, as is discussed above for quinoline (32).

To account for the formation of NH_3 or NH_4^+ from an amino group (i.e., without change in the N oxidation number), the following reactions could be suggested:

$$\gt C-NH_2 \rightarrow \gt C\bullet + \bullet NH_2 \tag{25}$$

$$\bullet NH_2 + RH(+H^+) \rightarrow NH_3(\text{or } NH_4^+) + R\bullet \tag{26}$$

2. Products

The following paragraphs provide examples of the chemistry involved in the degradation of aromatic and nonaromatic cyclic compounds.

Hydroxylation of Benzenic Rings. Intermediate products corresponding to the monohydroxylation of the aromatic ring have been identified in many cases (43,76,77,83,84,88–90). Electron-donating substituents exert the expected orientation effects to *para* and *ortho* positions. No orientation dominates for nitrobenzene (90). Dihydroxylated intermediates have also been identified, as well as the corresponding quinones.

Monohydroxylated and dihydroxylated aromatic intermediate products are very unstable in UV-irradiated TiO_2 aqueous suspensions, and they generally disappear within the same time period as the original aromatic pollutant. Trihydroxybenzenes are even more unstable and are therefore difficult to detect. They are suspected precursors to ring opening, although ring cleavage may compete with ring hydroxylation especially when *ortho* hydroxy (77) or methoxy (89) groups are present.

Case Study of Pyridine. For photocatalytic degradation of pyridine, the only monohydroxylated aromatic intermediate product was 2-hydroxypyridine (91). The –CH=CH– bonds in pyridine were also present in several aliphatic intermediate products, although only one of them contained the same number of C and N atoms as pyridine. All aliphatics contained one or several C=O groups; N was included in amido groups (i.e., its formal oxidation number was unchanged). Acetate and formate ions reached, by far, the highest concentrations among aliphatic products. For high initial pyridine concentrations, bipyridines and carbomoylpyridines were identified, illustrating the formation of coupling products.

In short, besides the expected formation of C=O and C–OH bonds and the anticipated cleavages of C–C and C–N bonds, the nature of the intermediate products of pyridine degradation shows the transfer of H atoms to N atoms to form NH and NH_2 groups, and finally NH_4^+ ions (see Sec. III.B.1, "Examples of Basic Degradation Pathways"). It also shows the transfer of H atoms onto C atoms (such as in acetate ions) and the probable formation of the pyridinyl radical, which leads to coupling products.

Case Study of 2,3-Dimethylpyrazine (1,2-DMB). The oxidation products of 2,3-dimethylpyrazine (Fig. 3) indicate attack at all the molecular sites: the methyl groups, the N atoms, and the CH groups of (41). Further transformations lead to a variety of aliphatic products with a decreasing number of atoms. Also, coupling reactions reflect the probable generation of monomethylpyrazinyl and dimethylpyrazinyl radicals.

Case Study of 1,2-Dimethoxybenzene. For 1,2-DMB degradation all of the intermediate aliphatic compounds, with the exception of methanol, contained one or two ester or acid functionalities (89). Methyl ester functionalities obviously stem from the methoxy groups in 1,2-DMB; C–C bonds were present in most of these intermediates. Some of the aliphatic intermediate products included a CHOH and/or a C–O group. The most interesting feature was the existence of aliphatic intermediate products

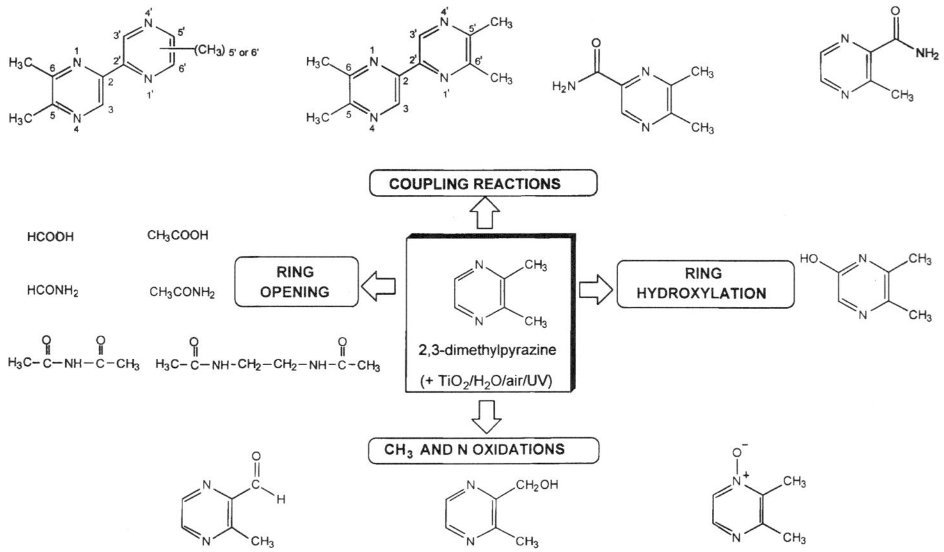

Figure 3 Intermediate products whose identification was derived from GC–MS analyses performed in the course of the photocatalytic degradation of 2,3-dimethyl-pyrazine.

resulting from the opening of the aromatic ring before its hydroxylation at position 3, 4, 5, or 6 in line with the case of 4-chlorocatechol (77).

Case Study of Tetrahydro-1,4-Oxazine (Morpholine). As in the case of aromatic pollutants, the oxidation of the N atom and the C atoms was observed with or without the destruction of the morpholine cycle (92). Also, the abstraction of H atoms to form methyl groups was illustrated by the formation of *N*-hydroxyamino-2-ethenyl ethyl ether, *N*-formylformamine, and acetate ions. When the starting morpholine concentration was increased, intermediate products with mass peaks much greater than the molecular mass of morpholine were detected, showing that coupling reactions occurred; the concentrations of these intermediates were very low.

Main Intermediate Aliphatic Products. Acetate and formate have been detected in the course of the degradation of various aromatic pollutants. The maximum amount of acetate formed from the degradation of the aromatic ring was higher than that of formate. These maxima were reached within approximately the same time period as the disappearance of all aromatics (43). The amounts of formate or acetate can be increased markedly by the degradation of substituents containing one or two carbon atoms, respectively. These carboxylic ions constitute the major fraction of the remaining total organic carbon in the last stages of the degradation. From the practical viewpoint, they could be removed advantageously by a subsequent, more economical technique without trying to achieve complete mineralization by photocatalysis alone.

Chlorine and Nitrogen Atom Mineralization. The release of chloride from various monochlorophenols, dichlorophenols, and trichlorophenols (and other chlorinated aromatics) was found to be relatively facile and to occur with an apparent first-order kinetic law (within a satisfactory approximation). In the case of monochlorophenols, the Cl atoms at *meta* positions were not as easily abstracted as those at the other positions, as was expected from the orientation due to the OH substituent for electrophilic attack (43,88).

Organic nitrogen is mineralized to both ammonium (see Sec. III.B.1, "Examples of Basic Degradation Pathways") and nitrate (90,93,94). Therefore, in some cases, it might be necessary to combine heterogeneous photocatalysis (as well as other advanced oxidation processes) with ion exchange or microbiological denitrification. The proportion of ammonium depends mainly on the oxidation state of nitrogen in the starting pollutant, as was expected. However, even in the case of nitro-substituted monocyclic aromatics, ammonium is detected (90). That indicates the existence of reduction steps of the nitro group. The formation of nitrate occurs quite slowly when nitrogen is at a low oxidation state as in amines, amides, and N-containing heterocycles because the conversion of ammonium to nitrate is very difficult, as can be predicted from the difference of eight units in the oxidation states. The presence of a N–N bond in the initial pollutant leads to the formation of N_2 as in the case of other oxidation methods.

3. Interferences

Effects of Inorganic Anions on Photocatalytic Water Treatment. Inorganic ions can significantly affect photocatalytic treatment only if they are located at the photocatalyst surface or in the Helmholtz layer. This location depends both on the chemical affinity of the ions for titania and on the point of zero charge (PZC) of this solid.

At pH greater than the TiO_2 PZC (between 6 and 6.5), the surface is negatively charged and anions are repelled. At pH values below the TiO_2 PZC, the excess anion concentration at the surface over that in the bulk of the solution can reach very substantial values because it is proportional to the square of the surface charge density. Nevertheless, even for $pH < 3$, nitrate anions do not seem to have unfavorable effects for concentrations up to 0.1 mol L^{-1}, at least in the case of monochlorophenols. In contrast, chloride, sulfate, and, above all, hydrogen phosphate anions decrease the disappearance rate of various organic pollutants, presumably because these anions penetrate into the TiO_2 inner coordination sphere (95,96). Indeed, adsorption measurements (97) showed the following order for the amounts adsorbed on TiO_2 at a pH less than the PZC:

$$H_2PO_4^- > HSO_4^- > Cl^-$$

Alternatively (98), the effect of some anions can be explained by their reaction with the photoproduced holes to give radicals (e.g., $HCO_3\bullet$) capable of oxidizing the pollutant, but at rates lower than those resulting either from the reaction of the pollutant with $OH\bullet$ radicals or from the direct interaction of the pollutant with photogenerated holes.

Effect of pH on Photocatalytic Water Treatment. For uncharged organic pollutants, photocatalytic treatment is unfavorable under very acidic conditions, whereas very basic conditions appear to be favorable. At near-neutral pH values, the variations in degradation with pH are modest or nonexistent (35,43). Because even at the extreme pH values the change in the photocatalytic degradation rate is generally less than one order of magnitude, the TiO_2-photocatalyzed water treatment has, from this respect, an advantage over biological H_2O_2–Fe^{2+}, H_2O_2–UV, O_3–UV, and O_3–H_2O_2 processes.

Competition Between Pollutants. Competition between several organic pollutants may affect the photocatalytic degradation rate of each species, depending on whether the process is limited by the irradiation or by the total organic matter. The factors intervening in the competition are the respective concentrations, the partition coefficients between the fluid phase and the adsorbed phase, and the relative reactivities with respect to the active species. Consequently, interference effects may or may not be observed.

Effect of Humic Acid on Photocatalytic Water Treatment. In the case of water treatment, it has been shown recently (31) that the inhibiting effect of humic acids at the concentrations found in natural waters is mainly due to UV absorption by these substances in competition with UV absorption by TiO_2. However, at low concentrations of humic acids, this negative effect can be more than counterbalanced by a positive effect tentatively attributed to the sequestration (99,100) of pollutant molecules in the micro-environment created in the vicinity of the TiO_2 surface by the adsorbed humic acid (99).

Effect of Water Vapor on Photocatalytic Air Treatment. Several studies have reported on the effects of water vapor on the photocatalytic treatment of air (101–108). The effect of water vapor very much depends on the type of pollutant and, obviously, on the partial pressure of water against that of the pollutant. On one hand, water can compete with the adsorption of organic pollutants, especially those that are structurally related, such as alcohols. On the other hand, water can behave as a reactant in some of the successive steps of the degradation of organics and, in particular, can limit the formation of products that inhibit the photocatalytic activity. Water can be at the origin of the formation of hydroxyl radicals; however, the importance of these radicals in gas-phase photocatalytic reactions is being debated on (109–111). The conclusion is that some humidity seems necessary for optimum photocatalytic activity.

Inactivation of Microorganisms. For the inactivation of a variety of microorganisms in water, UV-irradiated TiO_2 has been found to be more efficient than the same UV irradiation in the absence of TiO_2 (7,112–117). Cellular membranes and DNA appear to be damaged (112,115). This is not surprising because TiO_2 photocatalysis is able to degrade nearly all categories of organics, including polymers. Although the effects on a greater number of microorganism families should be tested, the method already appears promising both for water and air treatment (118). These antibacterial properties undoubtedly increase the interest of TiO_2 photocatalysis because, simultaneous to disinfection, organic pollutants can be degraded and certain toxic metal cations can be reduced or oxidized (6,90,119). Furthermore, the bactericidal activity can be enhanced by spraying an aqueous solution of silver ions on the TiO_2 coating during its preparation (7); under UV irradiation, these cations are reduced to Ag, which also presents anti-microbial properties.

4. Posttreatment

Certain chemical products generated in the course of a photocatalytic treatment en route to complete mineralization can be more toxic than the original organic pollutants. Consequently, the overall toxicity as monitored by some tests can be higher than the initial toxicity during some periods of the treatment (120,121). This situation is a problem common to every oxidation method, including those commonly in use today. Such issues necessitate appropriate tests according to the intended destination of the treated effluent. As a precaution, an adsorption stage on activated carbon can be incorporated after the photocatalytic treatment, just as might be applied after any other oxidation treatment. On the other hand, photocatalysis, used as a pretreatment, can eliminate pollutants that are biorecalcitrant; for instance, photocatalysis can effectively dechlorinate various compounds (35,43,88). The main problem is to determine the optimum duration of the photocatalytic treatment in order to reach the best overall efficiency of the combined treatments. Solving this problem requires kinetic studies (122,123).

Photocatalysis can be associated not only with a biological treatment, but also with other processes whose properties differ in some aspects from photocatalytic treatment. For instance, an initial ultrasonic treatment can allow one to destroy the CF_3 group (124,125), which withstands most oxidation reactions. Ultrasound can also be more appropriate to start the destruction of hydrophobic compounds with long hydrocarbon chains, which interact poorly with TiO_2 in water (125).

IV. SCALE-UP STUDIES AND ENVIRONMENTAL/INDUSTRIAL APPLICATIONS

A. Powder TiO_2 vs. Supported TiO_2

Regarding air treatment, to our knowledge, large-scale photocatalytic reactors based on fluidized beds are not used; TiO_2 is always supported. For water treatment, the system designed by Purifics Environment Technologies (126) uses a TiO_2 slurry, which is circulated in a narrow annulus between two tubes, one of which contains a UV lamp. TiO_2 is filtered by a ceramic membrane. High-pressure air is applied intermittently and very briefly to remove the TiO_2 particles that have collected in the membrane. Despite the added investment, the treatment is more economical, according to Purifics Environment Technologies, than when TiO_2 is supported on a fiber glass mesh, as the one used by Matrix Photocatalytic, Inc. (127) in an otherwise similar setting, because the efficacy of unsupported TiO_2 is higher.

Analogously, comparisons of plates coated by TiO_2 or plates onto which a TiO_2 aqueous suspension is flown showed a better efficiency for the suspension (128). Hence, the use of the suspension was recommended for a solar application; water was treated during the day, the TiO_2 slurry was allowed to settle at night, and the supernatant water was reused in the industrial process on the following day while another water batch was treated.

By use of the procedures mentioned in Sec. II.B.4, TiO_2 has been anchored on various materials. Fiber glass mesh—or organic fiber meshes previously coated with a photocatalytically inert material such as silica—presents the advantage of being flexible and, consequently, being shaped easily into filters. Pressure drops can be adjusted in these applications. This type of material has been used by Matrix Photocatalytic, Inc. (127), and is fabricated by Ahlstrom Research (69) and others. By analogy with automotive catalytic converters of exhaust gases, coated monoliths have been used for the photocatalytic treatment of air (129,130). The advantages are low-pressure drops and a good contact of the pollutants with TiO_2. Conversely, light penetration into the channels is restricted (130) so that honeycombs of only a few millimeters thick are used. TiO_2 has been coated on magnetite to permit a magnetic separation of the solid grains from the treated water (131).

TiO_2-coated glass, tiles, concrete, and polymers are being manufactured (7). The main purpose is to obtain self-cleaning materials (1,7,68). Some of them can also be used as decontaminating materials indoors when very weak ambient UV light needs to be counterbalanced by a large area covered by these materials (7).

In several cases, TiO_2 has been mixed with activated carbon with the same support (7,132). However, figures are lacking regarding the suggested improvements achieved by this approach.

B. Examples of Photocatalytic Reactors and Results

A major challenge in developing efficient photocatalytic reactors is to ensure a good contact between TiO_2 and the pollutants to be treated. Furthermore, for water treatment, a good contact with O_2 is also necessary. A second major challenge for photocatalytic treatment is the maximization of TiO_2 irradiation. Maximizing both pollutant/reagent contact with TiO_2 and achieving a certain degree of TiO_2 irradiation are arduous because they require different designs. Consequently, the design of a photocatalytic reactor is more difficult than that of a catalytic reactor.

Waveguides in the form of fibers or sheets coated with TiO_2 have been proposed to improve light distribution to TiO_2 through successive internal reflections. The problems with this design are: (1) difficulty in the efficient collection of lamp output; (2) occurrence of refraction of radiation at the exit of the waveguide; and (3) fragility of many waveguide materials. Silica-passivated polymer sheets coated with TiO_2 overcome these problems (133).

1. Water Treatment

In the systems commercialized by Matrix Photocatalytic, Inc. (127), water is circulated in a tube surrounding the UV lamp; this tube is enveloped by a TiO_2-coated fiber glass mesh. The system designed by Purifics Environment Technologies (126) circulates a TiO_2 slurry in a tube surrounding the UV lamp. A high degree of UV irradiation of TiO_2 is achieved in these systems, in which the water thickness is narrow enough to provide a turbulent flow rate. These systems are modular, and several tubes can be used in series or in parallel.

The treatment cost stems principally from the electrical power required to operate the systems. A major secondary expense is the cost of lamps.

An evaluation of ultraviolet oxidation methods has been carried out for the removal of 2,4,6-trinitrotoluene and 1,3,5-trinitrobenzene from groundwater (126). The methods used were powdered TiO_2–UV, O_3–UV, and H_2O_2 + additive–UV. Each firm involved in the evaluation determined the method of operation for its system. Independent consultants controlled the procedures. The total cost (capital, operation, and maintenance) per volume of treated water varied from a relative value of 1 to about 4. Heterogeneous photocatalysis was found to be the most economical. Two of the O_3–UV systems tested differed substantially in their cost, illustrating the importance of the specific system design in governing cost. Even though some criticisms can be made of this evaluation, heterogeneous photocatalysis appears to be a method that can compete economically with other UV oxidation processes for water treatment.

Purifies Environment Technologies (126) provides other examples of on-site treatments carried out continuously for several years, which demonstrates that TiO_2 photocatalysis is a cost-effective technique for the remediation of groundwater. Comparisons were made with H_2O_2–UV or with activated carbon; these comparisons are based on observed costs for photocatalysis and costs drawn from the proposals of competitors for the other techniques. The long-lasting operations have also demonstrated that TiO_2 photocatalysis is robust. Ambient temperatures of ca. 40°C did not affect the treatment units or their efficiency. The system was able to return to operation rapidly after interruptions caused by flooding or severe frost.

Mobile units for photocatalytic treatment have been constructed (126,127). The European Joint Research Center laboratory pilot plant, placed on a truck, includes TiO_2 loaded on membranes in UV-irradiated tubular reactors behind microfiltration and ultrafiltration modules. The waste water flow rate for this unit was typically 40 L hr^{-1}, and hydrogen peroxide was added to the photocatalytic process (134).

Solar reactors that do not concentrate the incident light have been shown to be preferable to concentrators (135–137). They do not need costly hardware to follow the position of the sun, and they take advantage of diffuse sunlight, which is an important contributor to total intensity. Furthermore, photon losses at reflecting surfaces are eliminated. A common design consists of plates or tubes facing the sun (Fig. 4). In some cases, water or the TiO_2 slurry is poured on top of the plate; the plate can be compartmentalized, and air can be injected in each compartment from the bottom of the plate. In the installation at "Plataforma Solar" (Spain), the TiO_2 suspension circulates into polyvinylidene fluoride (PVDF) tubes placed into ellipsoidal reflectors (137). In this latter case, the yearly capacity has been evaluated to be on the order of 2.6 kg m^{-2} C of reactor for water coming from the rinsing of barrels that had contained pesticides (137).

As mentioned in Sec. II.B.1, "Roles of O_2," oxygenation is crucial in some cases. Because of this fact, special designs have been proposed: a fountain-like reactor, a cascade reactor (138) (Fig. 5), reactors using floating TiO_2-coated glass beads (or small tubes) (139), and rotating disks (140), etc. In addition, these reactors can use either UV lamps or solar irradiation; they also minimize the amount of direct light absorption by the pollutants because only a thin film of liquid is irradiated before the light reaches the TiO_2.

2. Air Treatment

To our knowledge, all photocatalytic purifiers/deodorizers for indoor air comprise cylindrical UV lamps consuming a few tens of electrical watts and a TiO_2-coated material through which, or along which, the air passes (40,132). Because of size limitations, the geometrical surface of this material is on the order of 500–2000 cm^2 for an apparatus designed for an ordinary room (i.e., a volume of 20–40 m^3). An incorporated fan allows the air in the

Figure 4 View of the demonstration plant built in Spain and comprising 100 m^2 of collectors whose tubes contain 675L of TiO_2 aqueous supension exposed to sun light.

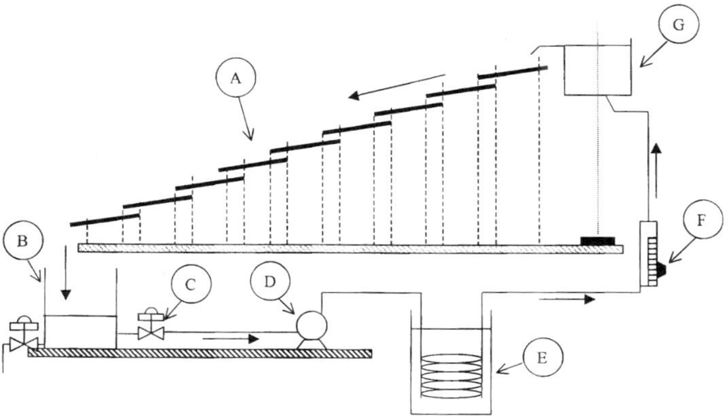

A) TiO_2-coated plates arranged stepwise
B) Liquid reservoir with drain valve
C) Valve
D) Pump
E) Cooling coil in water bath
F) Flowmeter
G) Reactor tank

Figure 5 Scheme of a pilot-scale cascade photocatalytic reactor. (From Ref. 138.)

room to be circulated in and out of the purifier at a flow rate of a few hundreds of cubic meters per hour. Numerous arrangements of the TiO_2-coated material, the UV lamps, and the fan have been patented (1).

The effect of an air purifier prototype on air quality in an ordinary (i.e., nonair-tight) room having a volume of 83 m^3 was quantitatively demonstrated for benzene, toluene, and xylenes (40). Average concentrations of these pollutants were decreased by a factor of ca. 2. Of course, as TiO_2 photocatalysis is a nonspecific method, similar results would be obtained for other pollutants having concentrations on the same order of magnitude. In contrast, CO and CH_4 concentrations in ambient air are too high (>1 ppmv) to be significantly diminished by the air purifier.

Quantitative measurements were also performed for formaldehyde, acetaldehyde, and acetone (141) because these compounds are likely to be formed by the oxidation of common organic pollutants found in air. Although these carbonyl compounds can easily be further oxidized, their high volatility allows them to be significantly desorbed from TiO_2, unlike the corresponding carboxylic acids with the same numbers of C atoms. For formaldehyde and a given series of measurements, the concentration (in μmol m^{-3}) range was 0.37–0.65 when the purifier lamps were off and became 0.52–0.76 when they were on; for acetaldehyde, the corresponding range was 0.06–0.13 with no irradiation, becoming 0.33–0.48 with irradiation; and for acetone, nonirradiated values were 0.19–0.78 and irradiated values were 0.33–1.1. Even though the ranges were shifted to higher values when photocatalysis occurred, the concentrations either overlapped (formaldehyde and acetone) or remained low (acetaldehyde). Note that the prototype purifier was not optimized, and that the values indicated above correspond to average concentrations over 2 hr with the lamps off or on, and with the fan always on (141).

For purifying air in road tunnels, reactors that consist of a series of tubes comprising black light lamps at the center and the TiO_2-activated carbon mixture fixed on the inner wall of the tubes have been proposed. Chicanes were used to enable a better transfer of the pollutants to the photocatalyst/adsorbent (142).

Large-scale installations of air purification by TiO_2 photocatalysis have been built and operated in North America by Trojan Technologies, Inc. (143) and United Technologies of Connecticut (144) (Fig. 6). The objective was to avoid polluting groundwater by chlorinated aliphatic

Figure 6 View of the United Technologies of Connecticut 1000 cfm (1700 $m^3\ h^{-1}$) photocatalytic unit (foreground) used to treat the air issued from stripping towers for groundwater purification at a former manufacturing plant.

hydrocarbons contained in the soil of abandoned industrial sites. These hydrocarbons were air-stripped from the soil and/or groundwater, and the contaminated air was then photocatalytically treated. For an electrical power of 19 kW and an air flow rate of ca. 1500 m^3 hr^{-1}, the elimination of trichloroethene at an initial concentration of ca. 1000 ppmv was found to be greater than 99% despite a relative humidity of 100%. Caustic scrubbers were installed to remove HCl and possibly traces of $COCl_2$. Irradiation can be automatically controlled in response to air flow changes. No information is available concerning the photocatalyst. The treatment and capital costs were claimed to be much lower than that for carbon adsorption. Note, however, that trichloroethene is a favorable case for photocatalysis because its removal rate is very high, and it can even enhance the elimination rates of other pollutants (110), presumably by yielding Cl• radicals at the surface of UV-irradiated TiO_2.

V. CONCLUSIONS

For the treatment of air and water, photocatalysis over TiO_2 is a technique that is simple, robust, easily automated, and flexible. It is versatile because organic pollutants, some inorganic pollutants, and microorganisms can be eliminated or inactivated, respectively. In water, some ions can also be transformed. Its operating cost arises mainly from the UV lamps and the corresponding electrical consumption. Provided that the needs and location of the treatment allow it, this cost can be reduced substantially by the use of solar irradiation instead of UV lamps. For these reasons, photocatalytic water purification appears well adapted to isolated communities, in particular if no competent labor force is available.

Regarding indoor air purification/deodorization, TiO_2 photocatalysis seems appropriate with the available UV lamps and TiO_2-coated materials, provided that the purifier design maximizes the retention of the intermediate products. Among common indoor pollutants, CH_4 and CO are the only compounds whose concentrations cannot be abated because they are too high. Also, concentrations of formaldehyde, acetaldehyde, and acetone—which are volatile, intermediate oxidation products—can decrease, remain the same, or even slightly increase, depending on the concentrations of pollutants present and on the relative destruction rates.

Concerning odors, photocatalysis can be an interesting technique if the target malodorous compound has an olfactory threshold on the order of 1 ppbv or less, and if the concentration is low enough. In that case, the pollutant quantity to be eliminated per time unit may fall within the range of removal rates treatable by photocatalysis, depending on the flow rate.

Because of the poor yield of photon utilization, the treatment of gaseous and aqueous industrial effluents may require reactor sizes that are prohibitive in some cases. Efficiency increases based on the synthesis of more highly active TiO_2 (and modified TiO_2 that can be activated by visible irradiation), on the addition of another solid to TiO_2, on better means of immobilizing TiO_2, and on the design of more adequate reactors appear to be limited to gains of one order of magnitude, at best. The addition of H_2O_2 in the case of water treatment is easy, but the expected enhancement is also lower than one order of magnitude. Very marked removal rate increases can be obtained by the use of ozonated air; therefore, TiO_2 photocatalysis is a technique to be kept in view when O_3-containing effluents must be treated; otherwise, the additional cost due to O_3 production can be prohibitive. Progress in UV lamp performance is useful but limited. Consequently, for each gaseous or aqueous effluent, calculations based on previous laboratory and field data should be performed to estimate whether photocatalysis might be a suitable method before if need be proceeding to a pilot study.

REFERENCES

1. Blake DM. Bibliography of Work on the Photocatalytic Removal of Hazardous Compounds from Water and Air. NREL/TP-510-31319, Golden, CO: National Renewable Energy Laboratory (htpp:/www.nrel.gov/chemistry bioenergy/chemistry.html).
2. Pichat P. Heterogeneous photocatalysis. In: Ertl G, Knözinger H, Weitkamp J, eds. Handbook of Heterogeneous Catalysis Vol. 4. Weinheim: Wiley-VCH, 1997:2111–2122.
3. Mills A, Le Hunte S. An overview of semiconductor photocatalysis. J Photochem Photobiol A Chem 1997; 108:1–35.
4. Ollis DF. Heterogeneous photocatalysis. Cat Tech 1998; 2:149–157.
5. Muskat LEA. Photochemical processes. In: Kearney P, Roberts T, eds. Pesticide Remediation in Soils and Water. New York: Wiley, 1998:307–337.
6. Litter MI. Heterogeneous photocatalyses transition metal ions in photocatalytic systems. Appl Catal B Environ 1999; 23:89–114.
7. Fujishima A, Hashimoto K, Watanabe T. TiO_2 Photocatalysis. Tokyo: BKC, 1999.
8. Kisch H. Preparative photoreactions catalyzed by semiconductor powders. J Prakt Chem 1994; 336:635–648.
9. Rothenberger G, Moser J, Graetzel M, Serpone N, Sharma DK. Charge carrier trapping and recombination dynamics in small semiconductor particles. J Am Chem Soc 1985; 107:8054.
10. Martin ST, Herrmann H, Choi W, Hoffmann MR. Time-resolved microwave

conductivity: Part 1. TiO_2 photoreactivity and size quantization. J Chem Soc Faraday Trans 1994; 90:3315–3322.
11. Soria J, Lopez-Munoz ML, Augugliaro V, Conesa JC. Electron spin resonance study of radicals formed during photooxidation of phenol on titania. Colloids Surf A Physicochem Eng Asp 1993; 78:73–83.
12. Jaeger CD, Bard AJ. Spin trapping and electron spin resonance detection of radical intermediates in the photodecomposition of water at TiO_2 particulate systems. J Phys Chem 1979; 83:3146–3152.
13. Brezova V, Stasko A, Biskupic S, Blazkova A, Havlinova B. Kinetics of hydroxyl radical spin trapping in photoactivated homogeneous (H_2O_2) and heterogeneous (TiO_2, O_2) aqueous systems. J Phys Chem 1994; 98:8977–8983.
14. Nada H, Oikawa K, Ohya-Nishiguchi H, Kamada H. Efficient hydroxyl radical production and their reactivity with ethanol in the presence of photoexcited semiconductors. Bull Chem Soc Jpn 1994; 67:2031–2037.
15. Riegel G, Bolton JR. Photocatalytic efficiency variability in TiO_2 particles. J Phys Chem 1995; 99:4215–4224.
16. Hanna PM, Mason RP. Direct evidence for inhibition of free radical formation from Cu(I) and hydrogen peroxide by glutathione and other potential ligands using the EPR spin-trapping technique. Arch Biochem Biophys 1992; 295:205–213.
17. Fu X, Zeltner WA, Anderson MA. The gas-phase photocatalytic mineralization of benzene on porous titania-based catalysts. Appl Catal B Environ 1995; 6:209–224.
18. Selvaggi A, David A, Zappelli P. Thermophotocatalytic degradation of gaseous organic pollutants. J Adv Oxid Technol 2002; 5:107–112.
19. Egerton TA, King CJ. The influence of light intensity on photoactivity in TiO_2 pigmented systems. J Oil Colour Chem Assoc 1979; 62:386–391.
20. Buechler KJ, Nam CH, Zawistowski TM, Noble RD, Koval CA. Design and evaluation of a novel-controlled periodic illumination reactor to study photocatalysis. Ind Eng Chem Res 1999; 38:1253–1258.
21. Childs LP, Ollis DF. Is photocatalysis catalytic? J Catal 1980; 66:383–390.
22. Pichat P, Herrmann JM, Disdier J, Courbon H, Mozzanega MN. Photocatalytic hydrogen production from aliphatic alcohols over a bifunctional Pt/TiO_2 catalyst. Nouv J Chim 1981; 5:627–636.
23. Pichat P, Mozzanega MN, Courbon H. Investigation of the mechanism of the photocatalytic alcohol dehydrogenation over Pt/TiO_2 by using poisons and labelled ethanol. J Chem Soc Faraday Trans I 1987; 83:697–704.
24. Cunningham J, Al-Sayyed G. Factors influencing efficiencies of TiO_2-sensitised photodegradation. J Chem Soc Faraday Trans 1990; 86:3935–3941.
25. Cunningham J, Sedlak P. Interrelationships between pollutant concentration, extent of adsorption, TiO_2-sensitized removal, photon flux and levels of electron or hole trapping additives. J Photochem Photobiol A Chem 1994; 77:255–263.
26. Cunningham J, Al-Sayyed G, Srijaranai S. Adsorption of model pollutants onto TiO_2 particles in relation to photoremediation of contaminated water. In:

Helz GR, Zepp RG, Crosby DG, eds. Aquatic and Surface Photochemistry. Boca Raton, FL: Lewis, 1994:317–348.
27. Turchi CS, Ollis DF. Photocatalytic degradation of organic water contaminants: mechanisms involving hydroxyl radical attack. J Catal 1990; 122:178–192.
28. Xu Y, Langford CH. Variation of Langmuir adsorption constant determined for TiO_2-photocatalyzed degradation. J Photochem Photobiol A Chem 2000; 133:67–71.
29. Peterson MW, Turner JA, Nozik AJ. Mechanistic studies of the photocatalytic behavior of TiO_2 particles in a photoelectrochemical slurry cell and the relevance to photodetoxification reactions. J Phys Chem 1991; 95:221–225.
30. Robert D, Parra S, Pulgarin C, Krzton A, Weber JV. Chemisorption of phenols and acids on TiO_2 surface. Appl Surf Sci 2000; 167:51–58.
31. Enriquez R, Pichat P. Interactions of humic acid, quinoline and TiO_2 in water in relation to quinoline photocatalytic removal. Langmuir 2001; 17: 6132–6137.
32. Cermenati L, Pichat P, Guillard C, Albini A. Probing the TiO_2 photocatalytic mechanisms in water purification by use of quinoline, photo–Fenton generated OH radicals and superoxide dismutase. J Phys Chem B 1997; 101:2650–2658.
33. Bideau M, Claudel B, Faure L, Kazouan H. Diffusional limitations in liquid phase photocatalysis. Prog React Kinet 1994; 19:195–209.
34. Jenny B, Pichat P. Determination of the actual photocatalytic rate of H_2O_2 decomposition over suspended TiO_2. Langmuir 1991; 7:947–954.
35. Bahnemann D, Cunningham J, Fox MA, Pelizzetti E, Pichat P, Serpone N. Photocatalytic treatment of waters. In: Helz GR, Zepp RG, Crosby DG, eds. Aquatic and Surface Photochemistry. Boca Raton, FL: Lewis, 1994:261–316.
36. Lindler M, Theurich J, Bahnemann DW. Photocatalytic degradation of organic compounds: accelerating the process efficiency. Water Sci Technol 1997; 35:79–86.
37. Cornish BJPA, Lawton LA, Robertson PKJ. Hydrogen peroxide enhanced photocatalytic oxidation of microcystin-LR using titanium dioxide. Appl Catal B Environ 2000; 25:59–67.
38. Dionysou DD, Suidan MT, Bekou E, Baudin I, Laîné JM. Effect of ionic strength and hydrogen peroxide on the photocatalytic degradation of 4-chlorobenzoic acid in water. Appl Catal B Environ 2000; 25:153–171.
39. Ohtani B, Zhang SW, Nishimoto SI. Catalytic and photocatalytic decomposition of ozone at room temperature over titanium (IV) oxide. J Chem Soc Faraday Trans 1992; 88:1049–1053.
40. Pichat P, Disdier J, Hoang-Van C, Mas D, Goutailler G, Gaysse C. Purification/deodorization of indoor air and gaseous effluents by TiO_2 photocatalysis. Catal Today 2000; 63:363–369.
41. Pichat P, Cermenati L, Albini A, Mas D, Delprat H, Guillard C. Degradation processes of organic compounds over UV-irradiated TiO_2. Res Chem Intermed 2000; 26:161–170.

42. Tanaka K, Capule MFV, Hisanaga T. Effect of crystallinity of TiO_2 on its photocatalytic action. Chem Phys Lett 1991; 187:73–76.
43. Pichat P, Guillard C, Maillard C, Amalric L, D'Oliveira JC. TiO_2 photocatalytic destruction of water aromatic pollutants: intermediates; properties–degradability correlation; effects of inorganic ions and TiO_2 surface area; comparisons with H_2O_2 processes. In: Ollis DF, Al-Ekabi H, eds. Photocatalytic Purification and Treatment of Water and Air. Amsterdam: Elsevier, 1993; 207–223.
44. Warman JM, de Haas MP, Pichat P, Koster TPM, Van der Zouwen-Assink EA, Mackor A, Cooper R. Electronic processes in semiconductor materials studied by nanosecond time-resolved microwave conductivity: Al_2O_3, MgO and TiO_2 powders. Radiat Phys Chem 1991; 37:433–442.
45. Tomkiewicz M. Scaling properties in photocatalysis. Catal Today 2000; 58:115–123.
46. Stone VF Jr, Davis RJ. Synthesis, characterization, and photocatalytic activity of titania and niobia mesoporous molecular sieves. Chem Mater 1998; 10:1468–1474.
47. Brucato A, Grifasi F. Economic evaluation of UV sources for photocatalytic applications. J Adv Oxid Technol 1999; 4:47–54.
48. Yue PL. Modelling of kinetics and reactor for water purification by photooxidation. Chem Eng Sci 1993; 48:1–11.
49. Yue PL. Oxidation reactors for water and wastewater treatment. Water Sci Technol 1997; 35:189–196.
50. Gimenez J, Curco D, Queral MA. Photocatalytic treatment of phenol and 2,4-dichlorophenol in a solar plant in the way to scaling-up. Catal Today 1999; 54:229–243.
51. Cassano AE, Alfano OM. Reaction engineering of suspended solid heterogeneous photocatalytic reactors. Catal Today 2000; 58:167–197.
52. Courbon H, Herrmann JM, Pichat P. Effect of platinum deposits on oxygen adsorption and oxygen isotope exchange over variously pretreated, UV-illuminated powder TiO_2. J Phys Chem 1984; 88:5210–5214.
53. Mu W, Herrmann JM, Pichat P. Room temperature photocatalytic oxidation of liquid cyclohexane into cyclohexanone over neat and modified TiO_2. Catal Lett 1989; 3:79–84.
54. Highfield JG, Pichat P. Photoacoustic study of the influence of platinum loading and bulk doping with chromium III ions on the reversible photochromic effect in titanium dioxide, correlation with photocatalytic properties. New J Chem 1989; 13:61–66.
55. Choi W, Termin A, Hoffmann MR. The role of metal ion dopants in quantum-sized TiO_2: correlation between photoreactivity and charge carrier recombination dynamics. J Phys Chem 1994; 98:13669–13679.
56. Anpo M, Ichihashi Y, Takeuchi M, Yamashita H. Design and development of unique titanium oxide photocatalysts capable of operating under visible light irradiation by an advanced metal ion-implantation method. Stud Surf Sci Catal 1999; 121:305–310.

57. Asahi R, Morikawa T, Ohwaki T, Aoki K, Taga Y. Visible-light photocatalysis in nitrogen-doped titanium oxides. Science 2001; 293:269–271.
58. Nakamura I, Negishi N, Kutsuna S, Ihara T, Sugihara S, Takeuchi K. Role of oxygen vacancy in the plasma-treated TiO_2 photocatalyst with visible light activity for NO removal. J Mol Catal A Chem 2000; 161:205–212.
59. Takeuchi K, Nakamura I, Matsumoto O, Sugihara S, Ando M, Ihara T. Preparation of visible-light-responsive titanium oxide photocatalysts by plasma treatment. Chem Lett 2000; 1354–1355.
60. Uchida H, Itoh S, Yoneyama H. Photocatalytic decomposition of propyzamide using TiO_2 supported on activated carbon. Chem Lett 1993; 1995-1998.
61. Ibusuki T, Takeuchi K. Removal of low concentration nitrogen oxides through photoassisted heterogeneous catalysis. J Mol Catal 1994; 88:93–102.
62. Takeda N, Ohtani M, Torimoto T, Kuwabata S, Yoneyama H. Evaluation of diffusibility of adsorbed propionaldehyde on titanium dioxide-loaded adsorbent photocatalyst films from its photodecomposition rate. J Phys Chem B 1997; 101:2644–2649.
63. Matos J, Laine J, Herrmann JM. Synergy effect in the photocatalytic degradation of phenol on a suspended mixture of titania and activated carbon. Appl Catal B Environ 1998; 18:281–291.
64. Sampath S, Uchida H, Yoneyama H. Photocatalytic degradation of gaseous pyridine over zeolite-supported titanium dioxide. J Catal 1994; 149:189–194.
65. Haick H, Paz Y. Remote photocatalytic activity as probed by measuring the degradation of self-assembled monolayers anchored near microdomains of titanium dioxide. J Phys Chem B 2001; 105:3045–3051.
66. Paz Y, Luo Z, Rabenberg L, Heller A. Photooxidative self-cleaning transparent titanium dioxide films on glass. J Mater Res 1995; 10:2842–2848.
67. Fernandez A, Lassaletta G, Jimenez VM, Justo A, Gonzalez-Elipe AR, Herrmann JM, Tahiri H, Aït-Ichou Y. Preparation and characterization of TiO_2 photocalysts supported on various rigid supports (glass, quartz and stainless steel). Comparative studies of photocatalytic activity in water purification. Appl Catal B Environ 1995; 7:49–63.
68. Patent WO 97/10185: Titanium dioxide-based photocatalytic coating substrate and titanium dioxide-based organic dispersions; Patent WO 97/10186: Photocatalytic coating substrate.
69. Patent WO 9951345: Procédé de dépôt de TiO_2. Ahlstrom Research, 38780 Pont-Evêque, France.
70. Roméas V, Pichat P, Guillard C, Chopin T, Lehaut C. Degradation of palmitic (hexadecanoic) acid deposited on TiO_2-coated self-cleaning glass: kinetics of disappearance, intermediate products and degradation pathways. New J Chem 1999; 23:365–373.
71. Richard C, Boule P. Photocatalytic oxidation of phenolic derivatives: influence of OH^0 and h^+ on the distribution of products. New J Chem 1994; 18:547–552.

72. Pichat P, Guillard C, Amalric L, Renard AC, Plaidy O. Assessment of the importance of the role of H_2O_2 and O_2^{0-} in the photocatalytic degradation of 1,2 dimethoxybenzene. Sol Energy Mater Sol Cells 1995; 38:391–399.
73. Pignatello JJ, Liu D, Huston P. Evidence for an additional oxidant in the photoassisted Fenton reaction. Environ Sci Technol 1999; 33:1832–1839.
74. Amalric L, Guillard C, Pichat P. Use of catalase and superoxide dismutase to assess the roles of hydrogen peroxide and superoxide in the TiO_2 or ZnO photocatalytic destruction of 1,2-dimethoxybenzene in water. Res Chem Intermed 1994; 20:579–594.
75. Cermenati L, Albini A, Pichat P, Guillard C. TiO_2 photocatalytic degradation of haloquinolines in water: aromatic products GM-MS identification, role of electron transfer and superoxide. Res Chem Intermed 2000; 26:221–234.
76. Li X, Cubbage JW, Tetzlaff TA, Jenks WS. Photocatalytic degradation of 4-chlorophenol: 1. The hydroquinone pathway. J Org Chem 1999; 64:8509–8524.
77. Li X, Cubbage JW, Jenks WS. Photocatalytic degradation of 4-chlorophenol: 2. The 4-chlorocatechol pathway. J Org Chem 1999; 64:8525–8536.
78. Goldstein S, Czapski G, Rabani J. Oxidation of phenol by radiolytically generated ^0OH and chemically generated SO_4^{0-}. A distinction between ^0OH transfer and hole oxidation in the photolysis of TiO_2 colloid solution. J Phys Chem 1994; 98:6586–6591.
79. Kesselman JM, Weres O, Lewis NS, Hoffmann MR. Electrochemical production of hydroxyl radical at polycrystalline Nb-doped TiO_2 electrodes and estimation of the partitioning between hydroxyl radical and direct hole oxidation pathways. J Phys Chem B 1997; 101:2637–2643.
80. Lawless D, Serpone N, Meisel D. Role of OH^0 radicals and trapped holes in photocatalysis. J Phys Chem 1991; 95:5166–5170.
81. Micic OI, Zhang Y, Cromack KR, Trifunac AD, Thurnauer MC. Trapped holes on TiO_2 colloids studied by electron paramagnetic resonance. J Phys Chem 1993; 97:7277–7283.
82. Soana F, Sturini M, Cermenati L, Albini A. Titanium dioxide photocatalyzed oxygenation of naphthalene and some of its derivatives. J Chem Soc Perkin Trans 2000; 2:699–704.
83. Minero C, Mariella G, Maurino V, Pelizzetti E. Photocatalytic transformation of organic compounds in the presence of inorganic anions: 1. Hydroxyl-mediated and direct electron-transfer reactions of phenol on a titanium dioxide–fluoride system. Langmuir 2000; 16:2632–2644.
84. Minero C, Mariella G, Maurino V, Vione D, Pelizzetti E. Photocatalytic transformation of organic compounds in the presence of inorganic ions: 2. Competitive reactions of phenol and alcohols on a titanium dioxide–fluoride system. Langmuir 2000; 16:8964–8972.
85. Draper RB, Fox MA. Titanium dioxide photosensitized reactions studied by diffuse reflectance flash photolysis in aqueous suspensions of TiO_2 powder. Langmuir 1990; 6:1396–1402.
86. Fox MA, Dulay MT. Heterogeneous photocatalysis. Chem Rev 1993; 93:341–357.

87. Konaka R, Kasahara E, Dunlap WC, Yamamoto Y, Chen KC, Inoue M. Irradiation of titanium dioxide generates both singlet oxygen and superoxide anion. Free Radic Biol Med 1999; 27:294–300.
88. Minero C, Pelizzetti E, Pichat P, Sega M, Vincenti M. Formation of condensation products in advanced oxidation technologies. The photocatalytic degradation of dichlorophenols over TiO_2. Environ Sci Technol 1995; 29:2226–2234.
89. Almaric L, Guillard C, Pichat P. The GC-MS identification of some aliphatic intermediates from the TiO_2 photocatalytic degradation of dimethoxybenzenes in water. Res Chem Intermed 1995; 21:33.
90. Maillard-Dupuy C, Guillard C, Pichat P. The degradation of nitrobenzene in water by photocatalysis over TiO_2: kinetics and products; simultaneous elimination of benzamide or phenol or Pb^{2+} cations. New J Chem 1994; 18:941–948.
91. Maillard-Dupuy C, Guillard C, Courbon H, Pichat P. Kinetics and products of the TiO_2 photocatalytic degradation of pyridine in water. Environ Sci Technol 1994; 28:2176–2183.
92. Doherty S, Guillard C, Pichat P. Kinetics and products of the photocatalytic degradation of morpholine. J Chem Soc Faraday Trans 1995; 91: 1853–1859.
93. Hidaka H, Nohara K, Pelizzetti E, Serpone N, Guillard C, Pichat P. Photooxidation processes in amphoteric surfactants catalyzed by irradiated TiO_2 suspensions. J Adv Technol 1996; 1:27–33.
94. Nohara K, Hidaka H, Pelizzetti E, Serpone N. Processes of formation of NH_4^+ and NO_3^- ions during the photocatalyzed oxidation of N-containing compounds at the titania/water interface. J Photochem Photobiol A Chem 1997; 102:265–272.
95. Augustynski J. Aspects of photo-electrochemical and surface behaviour of titanium (IV) oxide. Struct Bond 1988; 69:1–61.
96. O'Shea KE, Pernas E, Saiers J. The influence of mineralization products on the coagulation of TiO_2 photocatalyst. Langmuir 1999; 15:2071–2076.
97. Kazarinov VE, Andreev VN, Mayorov AP. Investigation of the adsorption properties of the TiO_2 electrode by the radioactive tracer method. J Electroanal Chem Interfacial Electrochem 1981; 130:277–285.
98. Low GKC, Mc Evoy SR, Matthews RW. Formation of nitrate and ammonium ions in titanium dioxide mediated photocatalytic degradation of organic compounds containing nitrogen atoms. Environ Sci Technol 1991; 25:460–467.
99. Puchaiski MM, Morra MJ. Fluorescence quenching of synthetic organic compounds by humic materials. Environ Sci Technol 1992; 26:1787–1792.
100. Lindsey ME, Tarr MA. Inhibition of hydroxyl radical reaction with aromatics by dissolved natural organic matter. Environ Sci Technol 2000; 34: 444–449.
101. Obee TN, Brown RT. TiO_2 photocatalysis for indoor air applications: effects of humidity and trace contaminant levels on the oxidation rates of

formaldehyde, toluene, and 1,3-butadiene. Environ Sci Technol 1995; 29: 1223–1231.

102. Fu X, Clark LA, Zeltner WA, Anderson MA. Effects of reaction temperature and water vapor content on the heterogeneous photocatalytic oxidation of ethylene. J Photochem Photobiol A Chem 1996; 97:181–186.
103. Obee TN. Photooxidation of sub-parts-per-million toluene and formaldehyde levels on titania using a glass-plate reactor. Environ Sci Technol 1996; 30:3578–3584.
104. Obee TN, Hay SO. Effects of moisture and temperature on the photo-oxidation of ethylene on titania. Environ Sci Technol 1997; 31:2034–2038.
105. Kim JS, Itoh K, Murabayashi M. Effect of humidity on the gas phase photocatalytic degradation of trichloroethylene over pretreated TiO_2. Denki Kagaku 1997; 65:966–968.
106. Obee TN, Satyapal S. Photocatalytic decomposition of DMMP on titania. J Photochem Photobiol A Chem 1998; 118:45–51.
107. Poncet-Vincent S. Desodorisation de l'air par photocatalyse au contact de TiO_2: dégradation de l'acide pentanoïque, du limonène et du sulfure d'hydrogène. Thesis. France: University of Lyon, 1999; 309–399.
108. Zorn ME, Tompkins DT, Zeltner WA, Anderson MA. Photocatalytic oxidation of acetone vapor on TiO_2/ZrO_2 thin films. Appl Catal B Environ 1999; 23:1–8.
109. Linsebigler AL, Lu G, Yates JT. Photocatalysis on TiO_2 surfaces: principles, mechanisms, and selected results. Chem Rev 1995; 95:735–758.
110. d'Hennezel O, Ollis DF. Photocatalytic oxidation of air contaminants by chlorine (Cl) or hydroxyl (OH) radicals or holes (h^+): mechanistic correlations. Stud Surf Sci Catal 1996; 101:435–442.
111. Benoît-Marquié F, Wilkenhöner U, Simon V, Braun A, Oliveros E, Maurette MT. VOC photodegradation at the gas–solid interface of a TiO_2 photocatalyst: Part I. 1-Butanol and 1-butylamine. J Photochem Photobiol A Chem 2000; 132:225–232.
112. Saito T, Iwase T, Horie J, Morioka T. Mode of photocatalytic bactericidal action of powdered semiconductor TiO_2 on mutant streptococci. J Photochem Photobiol B Biol 1992; 14:369–379.
113. Ireland JC, Klosterman P, Rice EW, Clark RM. Inactivation of *Escherichia coli* by titanium dioxide photocatalytic oxidation. Appl Environ Microbiol 1993; 59:1668–1670.
114. Wei C, Lin WY, Zainal Z, Williams NE, Zhu K, Kruzic AP, Smith RL, Rajeshwar K. Bactericidal activity of TiO_2 photocatalyst in aqueous media: toward a solar-assisted water disinfection system. Environ Sci Technol 1994; 28:934–938.
115. Xu M, Ma J, Gu J, Lu Z. Photocatalytic TiO_2 nanoparticles damage to cellular membranes and genetic supramolecules. Supramol Sci 1998; 5:511–513.
116. Horie Y, Taya M, Tone S. Evaluation of photocatalytic sterilization rates of *Escherichia coli* cells in titanium dioxide slurry irradiated with various light sources. J Chem Eng Jpn 1998; 31:577–584.

117. Armon R, Laot N, Narkis N. Photocatalytic inactivation of different bacteria and bacteriophages in drinking water at different TiO_2 concentration with or without exposure to O_2. J Adv Oxid Technol 1998; 3:145–150.
118. Blake DM, Maness PC, Huang Z, Wolfrum EJ, Huang J, Jacoby WA. Application of the photocatalytic chemistry of titanium dioxide to disinfection and the killing of cancer cells. Sep Purif Methods 1999; 28:1–57.
119. Tanaka K, Harada K, Murata S. Photocatalytic deposition of metal ions onto TiO_2 powder. Sol Energy 1986; 36:159–161.
120. Manilal VB, Haridas A, Alexander R, Surrender GD. Photocatalytic treatment of toxic organics in wastewater: toxicity of photodegradation products. Water Res 1992; 26:1035–1038.
121. Lu MC, Roam GD, Chen JN, Huang CP. Microtox bioassay of photodegradation products from photocatalytic oxidation of pesticides. Chemosphere 1993; 27:1637–1647.
122. Updaya S, Ollis DF. Simple photocatalysis model for photoefficiency enhancement via controlled periodic illumination. J Phys Chem B 1997; 101:2625–2631.
123. Parra S, Sarria V, Malato S, Péringer P, Pulgarin C. Photochemical versus coupled photochemical–biological flow system for the treatment of two biorecalcitrant herbicides: metobromuron and isoproturon. Appl Catal B Environ 2000; 27:153–168.
124. Théron P, Pichat P, Guillard C, Guillard C, Chopin T. Degradation of phenyltrifluoromethylketone in water by separate or simultaneous use of TiO_2 photocatalysis and 30 or 515 kHz ultrasound. Phys Chem 1999; 1:4663–4668.
125. Théron P, Pichat P, Pétrier C, Guillard C. Water treatment by TiO_2 photocatalysis and/or ultrasound: degradation of phenyltrifluoromethylketone, a trifluoroacetic acid-forming pollutant, and octan-1-ol, a very hydrophobic pollutant. Water Sci Technol 2001; 44:263–270.
126. Purifics Environment Technologies (www.purifics.com).
127. Matrix Photocatalytic, Inc., 22 Pegler Street, London, Ontario, Canada.
128. Golnish R, Dillert R, Bahnemann DW. Solar water treatment principles and reactors. Water Sci Technol 1997; 35:137–148.
129. Avila P, Bahamonde A, Blanco J, Sanchez B, Cardona AI, Romero M. Gas-phase photo-assisted mineralization of volatile organic compounds by monolithic titania catalysts. Appl Catal B Environ 1998; 17:75–88.
130. Hossain MdM, Raupp G, Hay SO, Obee TN. Three-dimensional developing flow model for photocatalytic monolith reactors. AIChE J 1999; 45:1309–1321.
131. Beydoun D, Amal R, Low GKC, McEvoy S. Novel photocatalyst: titania-coated magnetite. Activity and photodissolution. J Phys Chem B 2000; 104:4387–4396.
132. www.daikin.com
133. Miller LW, Tejedor-Tejedor MI, Perez Moya M, Johnson R, Anderson MA. Photocatalyst-coated acrylic waveguides for oxidation of organic compounds. Stud Surf Sci Catal 2000; 130:1925–1930.

134. Vacca S, et al. The application of alternative techniques for the recovery of water from effluents for reuse. EUR 19645 EN, European Communities, 2000.
135. Feitz AJ, Waite TD, Boyden BH, Jones GJ. Solar pilot-scale photocatalytic degradation of microcystin-LR. J Adv Oxid Technol 2002; 5:22–26.
136. Blake DM. Solar processes for the destruction of hazardous chemicals. In: Sterrett F, eds. Alternative Fuels and the Environment. Boca Raton, FL: Lewis, 1994:175–186.
137. Blanco J, et al. Compound parabolic concentrator technology development to commercial solar detoxification applications. Sol Energy 1999; 67:317–330.
138. Chan AHC, Porter JF, Barford JP, Chan CK. Photocatalytic thin film cascade reactor for treatment of organic compounds in wastewater. Water Sci Technol 2001; 44:187–195.
139. Franke R, Franke C. Model reactor for photocatalytic degradation of persistent chemicals in ponds and waste water. Chemosphere 1999; 39:2651–2659.
140. Dionysou DD, Balasubramanian G, Suidan MT, Baudin I, Laîné JM. Thin film photocatalytic reactor for the destruction of organic contaminants in industrial wastewater and drinking water. In: Abraham MA, Hesketh RP, eds. Reaction Engineering for Pollution Prevention. Amsterdam: Elsevier, 2000:137–153.
141. Pichat P, Disdier J. Unpublished results.
142. Ibusuki T, Kutsuna S, Takeuchi K, Shin-Kai K, Samamato T, Miyamato M. Removal of low concentration air pollutants through photoassisted heterogeneous catalysis. In: Ollis DF, Al-Ekabi H, eds. Photocatalytic Purification and Treatment of Water and Air. Amsterdam: Elsevier, 1993:375–386.
143. Trojan Technologies, Inc. (www.trojanuv.com).
144. United Technologies of Connecticut (www.utc.com).

3

Supercritical Water Oxidation Technology

Indira Jayaweera
SRI International, Menlo Park, California, U.S.A.

I. INTRODUCTION

Supercritical water oxidation (SCWO) is a waste treatment technology that uses supercritical water as a medium for oxidizing organic material. SCWO can also be described as an extension of the subcritical oxidation process, wet air oxidation (WAO) process, or widely known Zimpro process [1].* Both processes (SCWO and WAO) involve bringing together organic waste, water, and an oxidant (such as air or oxygen) to elevated temperatures and pressures to bring about chemical oxidations. Generic operational conditions for the two processes are as follows:

- WAO: 150–300°C, 10–200 bar
- SCWO: > 374–675°C, > 220 bar.

Fig. 1 shows a graphical representation of these operational regions. WAO and SCWO processes are often referred to as hydrothermal oxidation technologies (HTOs). The major difference between the processes is that, in SCWO, organics are completely oxidized in a relatively short time (seconds to minutes), whereas in WAO, the reaction may require a longer time (minutes to hours). Furthermore, in WAO, some refractory organics are not completely oxidized because of the lower temperature of operation

* The pioneering work of Fred Zimmerman in the 1950s led to the creation of the Zimpro process [1].

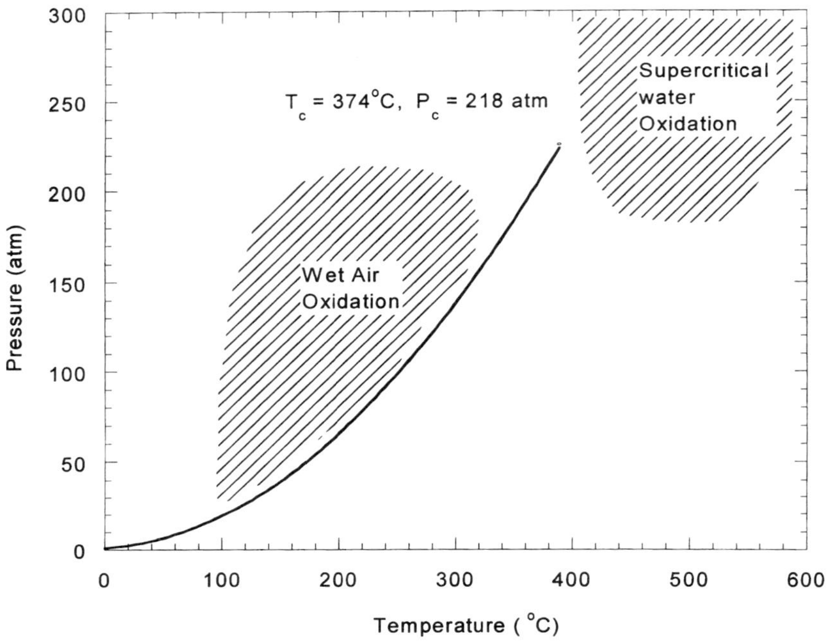

Figure 1 Phase diagram for pure water. Solid line: liquid–gas equilibrium.

($<350\,°C$), thus requiring a secondary treatment process. As information on WAO technology is readily available from other sources [2–5], the SCWO process is mainly discussed here.

The SCWO process is ideal for the disposal of many aqueous hazardous materials (e.g., EPA priority pollutants, industrial wastewater and sludges, municipal sludges, agricultural chemicals, and laboratory wastes), but has also been demonstrated to effectively destroy military wastes (e.g., ordnance, rocket propellants, and chemical agents) [6–18]. The effluent from the SCWO process, consisting primarily of water and carbon dioxide, is relatively benign. Therefore, the SCWO process can easily be designed as a full-scale containment process with no release of pollutants to the atmosphere. Compared with incineration and other high-temperature treatments, such as the plasma process, SCWO processes achieve high organic destruction efficiencies ($>99.99\%$) at much lower temperatures ($<700\,°C$) and without NO_x production.

Sanjay Amin, a student of Michael Modell at Massachusetts Institute of Technology, first discovered in the mid-1970s the effect of supercritical water for decomposition of organic compounds without forming tar [19].

This information, together with the information available at the time from Connolly's 1966 publication [20], which stated that organics can be solubilized in all proportions in high-temperature pressurized water, has led to the birth of the SCWO process. The breakthrough of the SCWO process seems to stem from the work of E. U. Frank, Karlsruhe, Germany, and Marshall and coworkers, Oak Ridge, TN, on the thermodynamics of binary mixtures of gases, organics, and inorganics in subcritical and supercritical water [21–23 and references therein]. Although the technology was invented in the late 1970s, much of the development work was conducted from 1980 to the early 1990s. During this period, researchers demonstrated the great utility of SCWO as a method for waste disposal without production of harmful products.

However, during the same period, the major technical obstacles to commercialization of the process had also been discovered. The two major technical challenges were reactor corrosion and reactor plugging. Reactor corrosion is caused by the formation of acids during the processing, especially when waste streams containing acid-forming components (e.g., chlorinated organics) are treated. Reactor plugging occurs when inorganic salts present in the waste stream are precipitated during the processing. Thus, the major criteria for designing the process involve consideration of possible corrosion and reactor plugging, as most industrial waste streams contain inorganic solids or heteroatoms that form inorganic solids for a majority of the SCWO systems. In addition, the problems associated with salt plugging and corrosion vary with the SCWO operating conditions (or the type of SCWO system). In general, there are several different versions of SCWO systems (low- and high-temperature SCWO, moderate and very high pressure SCWO, catalytic and -noncatalytic SCWO, etc.). Most of these different SCWO systems have been developed to overcome problems and to improve the performance of the process. However, only a few of those SCWO processes are commercially available and commonly practiced SCWO systems are discussed in this chapter.*

II. BACKGROUND AND FUNDAMENTALS OF SCWO

A. General Description

The SCWO process involves bringing together an aqueous waste stream and oxygen in a heated pressurized reactor operating above the critical point of

* Only the aboveground SCWO systems are discussed here. There is an underground SCWO system that uses hydrostatic pressure to avoid the need for high-pressure pumps. This system was designed by Oxydyne Corporation, Dallas, TX.

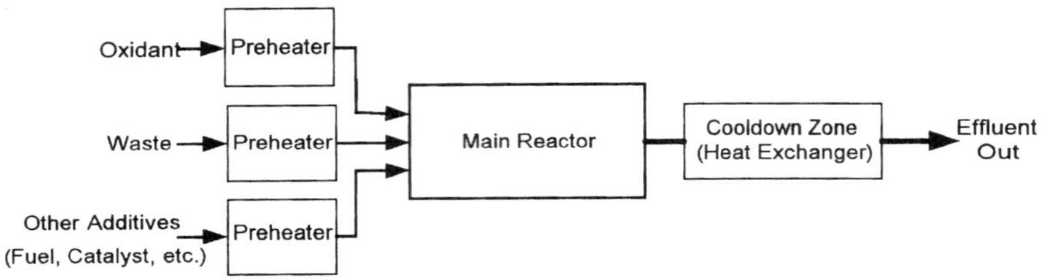

Figure 2 A generic hydrothermal oxidation (WAO, SCWO) process flow diagram.

water (374°C, 22.1 MPa or 218 atm). Under these conditions, the solubility properties of water are reversed (i.e., increased organic solubility and decreased inorganic solubility*), and the viscosity of the media is decreased to a value similar to -gaslike values, thus enhancing the mass transfer properties. These unique properties of hot pressurized water allow oxygen and organics to be contacted in a single phase in which oxidation of organics proceed rapidly. At 400–650°C and 3750 psi, SCWO can be used to achieve complete oxidation of many organic compounds with destruction rate efficiencies of 99.99% or higher.

A generic flow diagram for the SCWO process is given in Fig. 2. The aqueous waste is brought to system pressure using one or more high-pressure pumps. Compressed air or oxygen is added to the pressurized waste, and the waste–air mixture is initially heated to about 300°C using a preheater. The preheated mixture is directed to the main reactor operated at the desired temperature (above 374°C), where the complete oxidation occurs. The effluent from the reactor then travels through a heat exchanger, a pressure letdown valve, and a solid/liquid/gas separator.

The preheater section of the system mimics a miniature WAO system because the reaction conditions in the preheater are similar to those of a WAO system except that WAO systems need longer reaction times. In the heat-recovery mode of operation, the SCWO uses the heat from the reaction to preheat the influent. As a rule of thumb, if the aqueous waste stream contains about 4 wt.% of organics, the SCWO can be processed under self-sufficient heat conditions. However, for dilute aqueous waste streams, the SCWO process may not be cost-effective because of the additional thermal energy required to maintain the reactor temperature in the 400–650°C range.

* The details of the inorganic solubility are given in Sec. B.2., Phase Separations.

B. In-Depth Treatment of SCWO

1. Fluid Characteristics

The basic properties of water such as viscosity, dissociation constant, dielectric constant, compressibility, and the coefficient of thermal expansion play a major role in determining optimal reaction conditions for obtaining maximum benefits in both SCWO and WAO processes. The properties of water change dramatically with temperature, particularly near the critical point [24–26]. A well-known example, the variation of pK_w with temperature at the saturation pressure, is shown in Fig. 3. The dissociation constant of water goes through a maximum around 250°C (pK_w minimum), and then undergoes a sharp decline as the temperature approaches the critical point. The density and the dielectric constant of water also show sharp changes close to the critical point, as shown in Fig. 4.

The rate-limiting properties of any organic reaction that includes the mixing of several components are the solubility of the contaminant in the liquid phase or its equilibrium solubility, and the mass transfer step (i.e.,

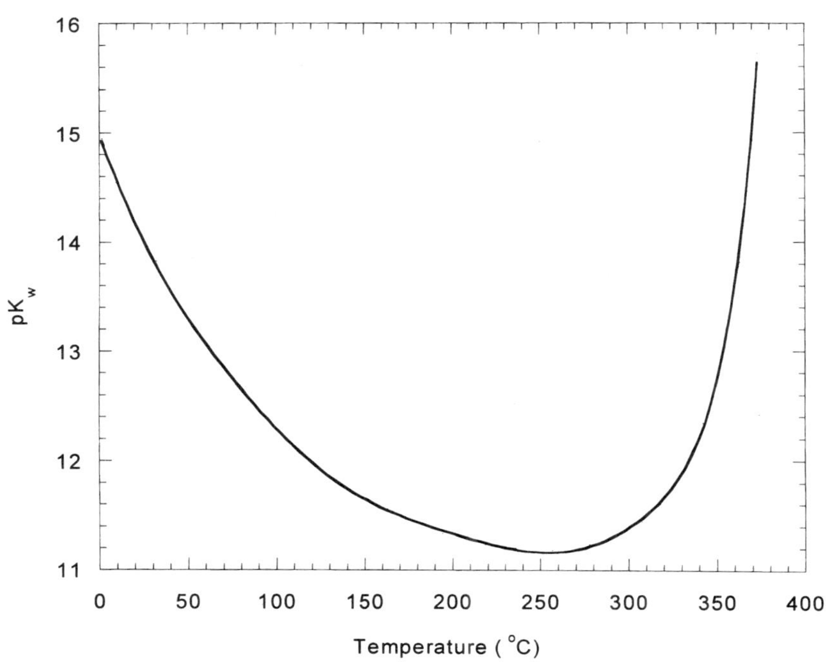

Figure 3 Variation of pK_w with temperature at the saturation pressure.

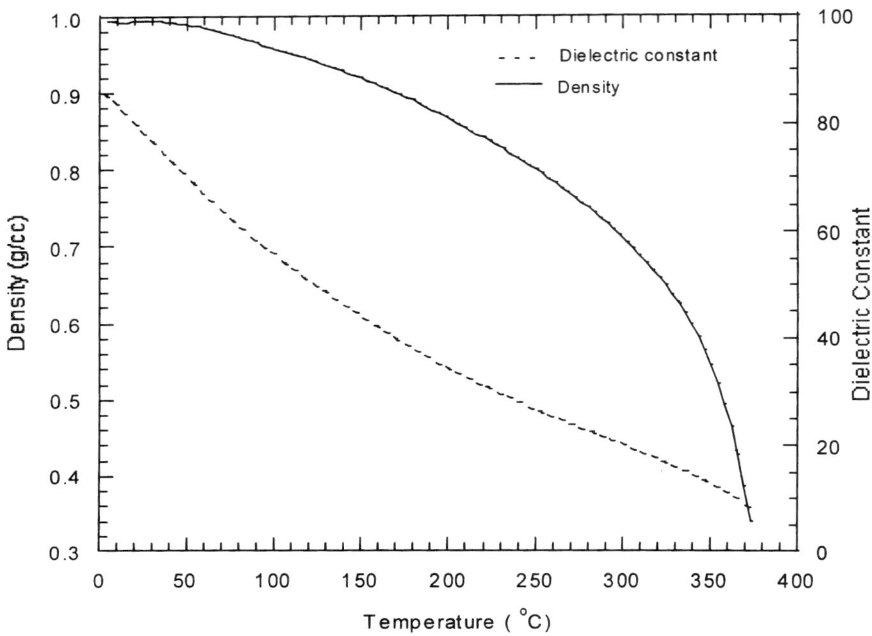

Figure 4 Variation of density and dielectric constant with temperature at the saturation pressure.

dissolution into the aqueous phase). Therefore, the transport properties of the reaction media are very important for efficient waste processing. The viscosity of water decreases with temperature, thus providing rapid diffusion. The conductance of heated water remains high in spite of the decrease in the dielectric constant because of the increased ion mobility brought about by the decreased viscosity. However, as the dielectric constant of water decreases with the increase in temperature, electrolytes that are completely dissociated at low temperature become much less dissociated at high temperature, particularly in the supercritical region. At the critical point (374°C, 218 atm pressure, dielectric constant, $\varepsilon = 5$), water acts as a mildly polar organic solvent, and thus supercritical water can readily solubilize nonpolar organic molecules. In fact, most hydrocarbons become soluble in water between 200° and 250°C [27], allowing opportunities to enhance reactivities of organics even under subcritical water conditions. The enhanced diffusivity and the decreased dielectric constant at elevated temperatures make water an excellent solvent for dissolving organic materials that are tightly bound to solid material (important for treatment of solid waste). As an example, hot pressurized water has been shown to break and separate "highly stable" water–oil–solid emulsions, generated in petroleum wastewater treatment and other industrial operations [28].

Compared with ambient values, the specific heat capacity of water approaches infinity at the critical point and remains an order of magnitude higher in the critical region [26], making supercritical water an excellent thermal energy carrier. As an example, direct measurements of the heat capacity of dilute solutions of argon in water from room temperature to 300°C have shown that the heat capacities are fairly constant up to about 175–200°C, but begin to increase rapidly at around 225°C and appear to reach infinity at the critical temperature of water [29].

The static dielectric constant is a measure of hydrogen bonding and reflects the characteristics of the polar molecules in water. However, very little is known about the degree of hydrogen bonding under supercritical water oxidation conditions. The lack of data on the character of hydrogen bonding in water at high temperatures and pressures hinders the understanding of the structure and properties of supercritical water. The important question is: Up to what temperature can hydrogen bonding in water exist? It was initially believed that hydrogen bonds do not exist above 420 K. Later, Murchi and Eyring [30], using the approach of significant structures, showed that the hydrogen bonds disappear above 523 K and that water above this temperature consists of free monomers. Later, Luck [31], studying the IR absorption in liquid water, extended the limit of hydrogen bonding at least up to the critical temperature. Recently, a theoretical model developed by Gupta et al. [32] has shown that in supercritical water, significant amounts of hydrogen bonding are still present despite all the thermal energy and are highly pressure and temperature dependent. An interesting result has emerged from Sandia National Laboratories' theoretical estimation of hydrogen bonding of supercritical media by calculating the equilibrium population of water polymers (dimers, trimers, etc.) [33]; however, this contradicts the Murch and Eyring findings above [30]. Their calculations have shown that at 450–650°C and 240–350 bar, the water polymer concentration can be as high as 40%. It is also cited in later work by Kalinichev and Bass [34] that hydrogen bonding is still present in the form of dimers and trimers in the supercritical state. More details and new theoretical discussions can be found in Refs. 35, 36, and the references therein.

2. Phase Separations

It is important that the phase behavior of the influent at high temperature and pressure conditions be clearly understood for designing process components such as the main reactor. Under the operational conditions of SCWO, the conditions can be easily adjusted to attain a single phase when only organics are present. However, when inorganic salts are present (either as a reagent or as a by-product from the process) under SCWO conditions, it is challenging or even impossible to predict the phase behavior of the medium.

The presence of electrolytes changes the saturation–vapor boundary line for water. Liquid–vapor equilibria in a soluble salt–water system above the critical temperature are complex. However, the situation below the critical temperature of pure water is simpler, at least for solutes that are so involatile at this temperature that their concentrations in the vapor phase are negligible. Liquid solutions of these solutes have vapor pressures that are lower at a given temperature than that of pure water. Equivalently, they have boiling points that are higher at a given total pressure than that of pure water. Fig. 5 shows the relationship between vapor pressure and temperature for Na_2CO_3–H_2O and NaCl–H_2O systems [36]. It is clear that these two systems have different phase behaviors under SCWO conditions. Because of the complex nature of the phase diagrams for salt-water systems and the

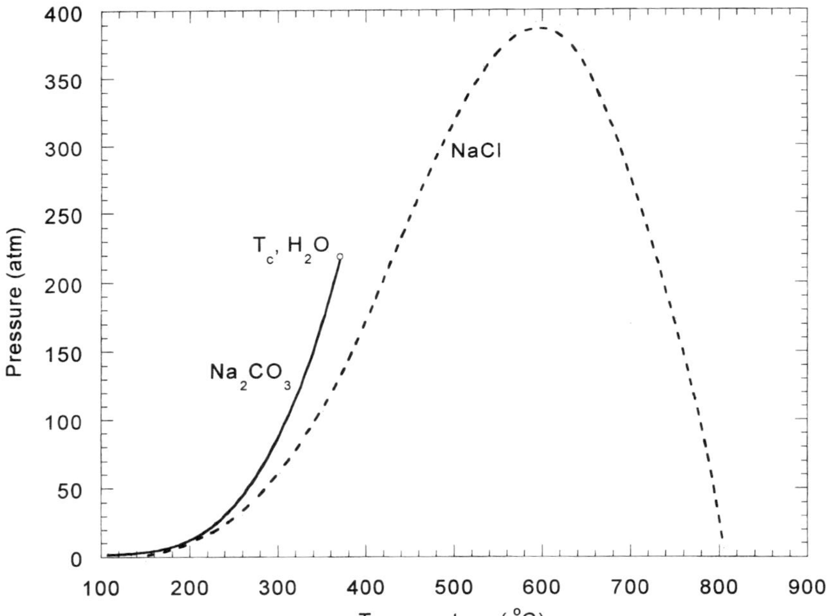

Figure 5 The relationship between vapor pressure and temperature for the Na_2CO_3–H_2O and NaCl–H_2O systems.

inconsistencies of the available literature data, only a brief discussion is given below with appropriate references.

In recent years, studies of the phase behavior of salt-water systems have primarily been carried out by Russian investigators (headed by Prof. Vladimir Valyashko) at the Kurnakov Institute in Moscow, particularly for fundamental understanding of the phase behavior of such systems. Valyashko [37,39,42,43], Ravich [38], Urosova and Valyashko [40], and Ravich et al. [41] have given a classification of the existence of two types of salts, depending on whether the critical behavior is observed in saturated solutions. Type 1 does not exhibit critical behavior in saturated solutions. The classic example of Type 1 is the NaCl–water system and has been studied by many authors [36,37,44–47]. The Type 2 systems exhibit critical behaviors in saturated solutions, and therefore have discontinuous solid–liquid–vapor equilibria. Table 1 shows the classification of binary mixtures of salt–water systems.

In brief, the salts that are classified as Type 1 have increasing solubility with increasing temperature, whereas Type 2 salts show an opposite trend. For example, sodium carbonate, a Type 2 salt, has a 30 wt.% solubility under ambient conditions and its solubility near the critical point approaches zero [36] whereas sodium chloride, a Type 1 salt, has a 37 wt.% solubility under subcritical conditions at 300°C and about 120 ppm at 550°C [46].

In real systems, organic–inorganic multicomponent phase systems are possible, and the information gathered from binary or ternary systems cannot be extended to these real situations. Currently, Valyashko from Kurnakov Institute and Jayaweera from SRI International are jointly study-

Table 1 Saltwater Binary Systems

Type 1 salts	Type 2 salts
KF, RbF, CsF	LiF, NaF
LiCl, LiBr, LiI	Li_2CO_3, Na_2CO_3
NaCl, NaBr, NaI	Li_2SO_4, Na_2SO_4, K_2SO_4
K_2CO_3, $RbCO_3$	Li_2SiO_3, Na_2SiO_3
Rb_2SO_4	Na_3PO_4
Na_2SeO_4	CaF_2, SrF_2, BAF_2
K_2SiO_3	SiO_2, Al_2O_3
K_3PO_4	
$CaCl_2$, $CaBr_2$, CaI_2	
$BaCl_2$, $BaBr_2$	
NaOH, KOH	

Source: Ref. 37.

ing both the phase behavior and the morphology changes of salts precipitated from a salt-water system containing Na_2CO_3, K_2CO_3, and NaCl [48].

Importance of Electrochemistry in SCWO Processing. In both WAO and SCWO processing, where the reactor surfaces experience extremes in pH and high inorganic salt concentrations under high temperature/high pressure conditions, enhanced electrochemical processes could cause corrosion and rapid metallurgical degradations of the reactor vessels. Therefore, materials should be evaluated to determine if they could withstand the SCWO conditions. In general, researchers have been mainly focused on understanding the corrosion processes such as pitting corrosion (disruption of the protective oxide surface layer followed by the heavily localized dissolution of the underlying alloy, forming holes or pits), crevice corrosion (localized form of corrosion associated with stagnant solutions in crevices), and stress corrosion cracking (cracking induced by the combined influence of the tensile stress and corrosive medium) under SCWO conditions [49]; a detailed description of metallurgical aspects, material properties, thermodynamics of the corrosion process, corrosion kinetics, and corrosion phenomena under hydrothermal conditions can be found in Refs. 51 and 52.

In predicting metal stability under aqueous environments, it is customary to use electrochemical potential–pH diagrams (E_h–pH or Pourbaix diagrams). Many workers have derived and published potential–pH diagrams for metal–water systems under varying temperature and pressure conditions [52–54]. Cr, Fe, and Ni systems are the most widely studied systems (the alloys currently used for WAO and SCWO studies contain Cr, Fe, and Ni, e.g., stainless steel 316 and Hastelloy C2-276). Under oxidative conditions, metal oxide films are formed on the reactor surfaces. Some metal oxides, such as Fe_3O_4, Cr_2O_3, etc., will passivate the metal surface reducing the corrosion, and thus both immunity and passivation regions, where a process can be operated with minimal corrosion, are possible. In the case of chromium, the shift in the equilibrium line for the oxidative dissolution of Cr_2O_3

$$Cr_2O_3 + 5H_2O \rightarrow 2CrO_4^- + 10H^+ + 8e^-$$

with increasing temperature is of practical importance for stainless steel, because it is the formation of chromic oxide (or at least a chromium-containing spinel, e.g., $FeCr_2O_4$) that confers passivity to the alloy. Researchers have tried to evaluate the effect of secondary metals on the primary metal in alloys by adding corresponding salts to the corrosion medium. For example, Fig. 6 shows the possible passivity, immunity, and corrosion areas for iron in the presence of CrO_4^{2-} under ambient conditions [52].

The potential–pH diagrams under ambient conditions cannot be used to predict the stability at higher temperatures. The passivation region for

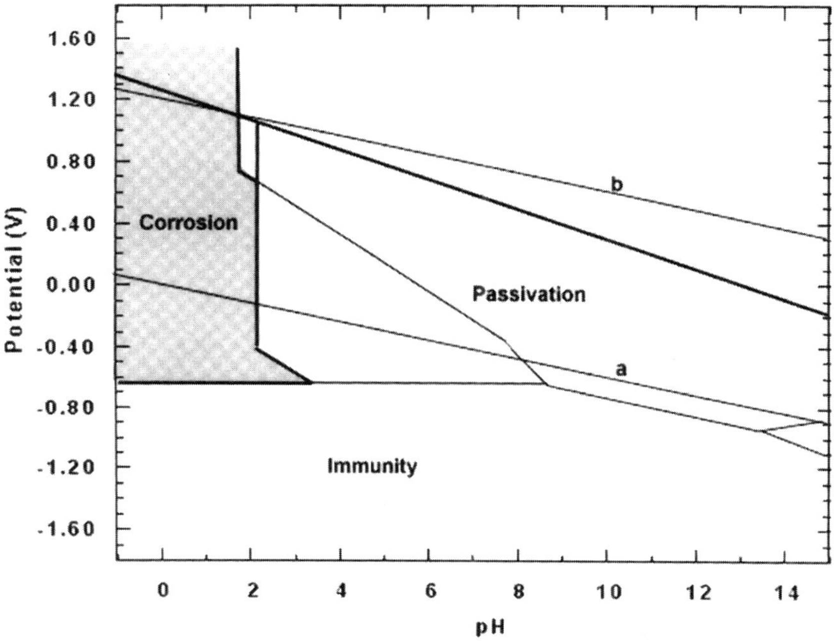

Passivation: possible passivity by stable Fe_2O_3 or metal stable Fe_3O_4
Immunity: no corrosion possible
Corrosion: no passivity posible

Figure 6 Potential–pH diagram: the possible passivity, immunity and corrosion areas for iron in the presence of CrO_4^{2-} under ambient conditions.

iron is different at supercritical conditions compared to ambient conditions. High-temperature - thermodynamic properties have to be properly incorporated when evaluating the diagrams for elevated temperatures. However, it should be noted that accurate determination of potential–pH diagrams is impossible because of uncertainties about existing equilibria of different species at elevated temperatures.

It is also important to note that the electrochemical potential of the system is dependent on the equilibrium between various ions present in the system and cannot be changed without application of an external potential (e.g., cathodic protection) or addition of a chemical (e.g., corrosion inhibitor). Emphasis should thus be placed on the interpretation of the data contained within the E_h–pH diagrams, with particular attention being paid to the limitations that are imposed by the fact that they are at equilibrium rather than the kinetic descriptions of a system. One must take precautions when using potential–pH diagrams for predicting possible corrosion in metal–water systems. They are used only as a guide, and the experimental data must be used for accurate prediction of corrosion rate. Further details can be found in the "Interference from Reactor Corrosion." section.

III. DEGRADATION OF POLLUTANTS

A. Laboratory-Scale Experimental Design

The laboratory-scale experimental setups are designed typically to conduct chemical reaction studies under a range of pressures, temperatures, densities, oxidant and organic concentrations, and residence times in several reactor configurations. In general, model compounds for simulating common pollutants in industrial waste streams are used in laboratory-scale experiments.

Selection of the reactor to achieve the required reaction time is one of the key aspects of designing the laboratory-scale experiments.* There are several types of reactors that can be used for this purpose: small batch reactors or bombs, tubular plug-flow reactors (PFRs), and stirred tank reactor systems (STRs) on either batch or continuous mode. Small batch reactor setups are the most convenient and ideal for initial scouting experiments to determine general conversion and suitable conditions for continuous flow operation. In addition, it is convenient when the change in surface-to-volume ratio is required for studying the surface effects on the reaction rates. Batch reactor setups are generally small bombs that can easily be custom-made using high-pressure stainless steel tubing. During testing, these reactors are loaded with predetermined amounts of water, oxidant, and the organic material, and the closed reactor is then heated in an isothermal oven. These systems are self-pressurized, and the reactor pressure can be changed by increasing the water loading, which in turn increases the density; steam tables provide the relationships between the water loading and the pressure [24,25].

Tubular reactor advantages include their well-defined residence time distributions, turbulent mixing reactants, ease of obtaining and applying kinetic data, efficient use of reactor volume, and mechanical simplicity. However, great care must be taken when applying the correct flow model (e.g., plug

---

* It is important to note that most large-scale SCWO reactors are designed to be turbulent flow. In addition, some of the reactors (e.g., transpiring wall reactor) are not possible to scale down for laboratory-scale experiments. The method given here is a generic approach for understanding the reaction kinetics of pollutants under SCWO conditions.

flow assumption) for evaluating reaction times, and numerous limitations accompany the use of the plug-flow treatment of tubular flow reactor data. A critical evaluation of the plug-flow idealization for supercritical water oxidation is reported by Cutler et al. [55]. More detailed criteria for evaluating the legitimacy of plug-flow idealization for general applications are given by Mulcahy and Pethard [56] and Furue and Pacey [57].

Stirred tank reactors are very useful when the reagents contain multiple components that could exist in separate phases. In addition, it is a convenient way to achieve long reaction times for the study of slow kinetic processes. The attainable reaction time depends on the size of the reactor (e.g., 100 to 2000 mL). Sample schematics of STR and PFR systems are given in Figs. 7 and 8, respectively.

In designing laboratory-scale experiments, it is important to use proper analytical methods for determining both organic and inorganic species to

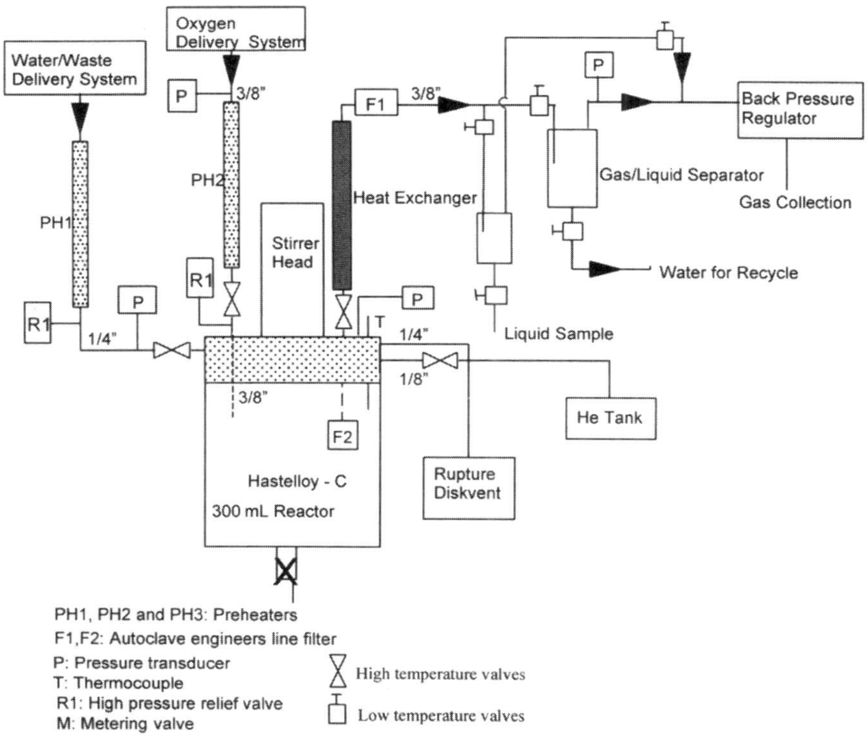

Figure 7 A bench-scale stirred tank reactor system for subcritical and supercritical studies. (From SRI International.)

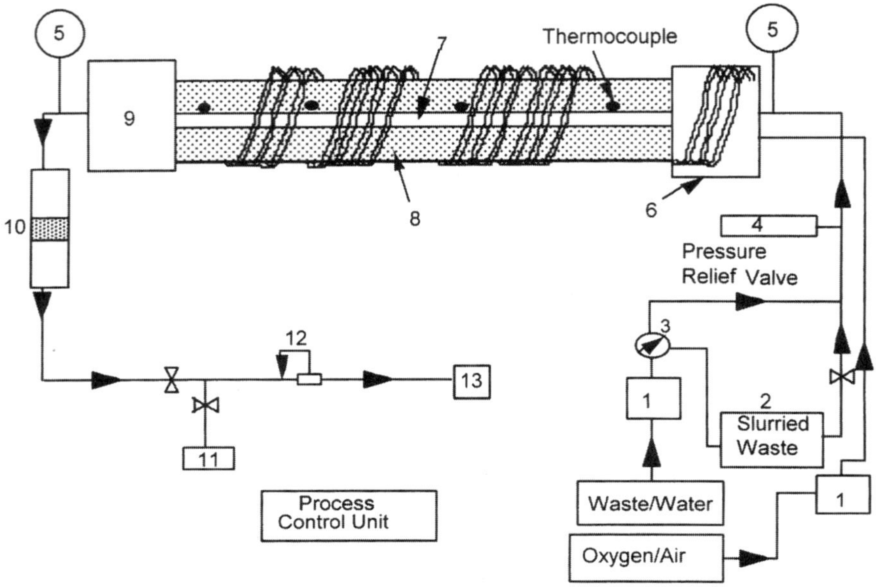

1. HPLC pump
2. Stirred autoclave for slurry introduction
3. Three-way valve for solvent bypass
4. Pressure relief valve
5. Pressure transducer
6. Preheater
7. Hastelloy C-276 reactor
8. Heating coil and copper jacket
9. Heat exchanger
10. Product accumulator
11. Gas cylinders
12. Back pressure regulator
13. Gas liquid separator

Figure 8 A continuous flow reactor system for subcritical and supercritical studies. (From SRI International.)

examine the kinetics of the oxidation of pollutants, because both species are present in the treatment of waste. In most cases, where only the destruction rate efficiency (DRE) is required, it is customary to use either the total organic content (TOC) or the chemical oxygen demand (COD) as the analytical parameter. However, when the studies are targeted for detailed understanding of the process, an in-depth analysis on all the phases is required. Commonly used analytical techniques for analyzing liquid and gas samples from the treated pollutants include gas chromatography (GC) (with flame ionization and thermal conductivity detectors), gas chromatography–mass spectrometry (GC-MS), ion chromatography for analysis of inorganic anions and cations, and inductively coupled plasma–atomic emission spectroscopy (ICP) for metal analysis. During the initial development of SCWO, some researchers have also used in situ optical methods for quantification of gaseous species in the supercritical media [58,59] (SF Rice, personal communication, 1998). Recently, a hot stage microscope or diamond cell observation cell has been used to visually observe the supercritical media [61]. Such visual observation reactors provide insight into the phase behavior of supercritical fluids containing inorganic salts.

B. Examples

To date, numerous model compounds simulating the pollutants in common waste streams have been studied under laboratory-scale conditions by many researchers to determine their reactivities and to understand the reaction mechanisms under supercritical water oxidation conditions. Among them, hydrogen, carbon monoxide, methanol, methylene chloride, phenol, and chlorophenol have been extensively studied, including global rate expressions with reaction orders and activation energies [58–70] (SF Rice, personal communication, 1998).

1. Mechanisms

Because the reactions of organic compounds with oxygen are very complex and because it is not essential to understand the reaction mechanisms for engineering purposes, the oxidation mechanisms were not addressed in early SCWO studies [70,71]. In the late 1980s and early 1990s, several researchers [72–74] attempted to develop kinetic models for SCWO oxidation of methanol, methane, carbon dioxide, and ethanol based on combustion theory. In the combustion reactions (at high temperatures and low pressures), the OH radical plays an important role. Because it has a large electron affinity, it oxidizes all organic compounds containing hydrogen.

As more applications of SCWO started to emerge, researchers began to work on understanding the stable, common reaction intermediates and their reaction pathways. By the early 1990s, it was clear that one of the main stable intermediate forms of oxidation of most organic compounds under hydrothermal conditions (especially at temperatures $<450\,^{\circ}\mathrm{C}$) is acetic acid [75]. In most mechanistic studies, researchers have always used phenol and chlorophenol because they are the most common pollutants in commercial waste streams (e.g., paper industry) and they are readily soluble in water and are easily studied (when flow reactors are used). Although several mechanisms have been proposed [76], the following mechanism involving a complex set of competing free radicals in which organic structures are oxidized and cleaved via carbon, peroxy, and oxyradicals [77,78] can be given as an acceptable mechanism in the absence of a reactive ionic species. Each initiating reaction

(1), creating a pair of radicals, is matched by a terminating reaction (6), destroying a pair of radicals.

$$RH + O_2 \rightarrow R^{\bullet} + HO_2^{\bullet} \quad \text{initiation} \tag{1}$$

$$R^{\bullet} + O_2 \rightarrow RO_2^{\bullet} \tag{2}$$

$$RO_2^{\bullet} + RH \rightarrow RO_2H + R^{\bullet} \quad \text{propagation} \tag{3}$$

$$2RO_2^{\bullet} \rightarrow 2RO^{\bullet} + O_2 \tag{4}$$

$$RO^{\bullet} \rightarrow C\text{–}C \quad \text{cleavage} \tag{5}$$

$$2RO_2^{\bullet} \rightarrow \text{Termination products} \tag{6}$$

Here R is an organic functional group

The organic radical ($R^{\bullet}$) can readily react with oxygen [Eq. (2)] to form a peroxyradical, which then abstracts hydrogen from the organic compound, producing a hydroperoxide (ROOH) and another organic radical. The formed organic hydroperoxides are relatively unstable, and decomposition of such intermediates leads to the formation of subsequent intermediates containing lower carbon numbers until acetic and formic acids are finally formed. These acids will eventually be converted to carbon dioxide. When hydrogen peroxide is used as the oxidant, the thermal decomposition of hydrogen peroxide is very rapid, and the reaction proceeds as $H_2O_2 \rightarrow H_2O + 1/2O_2$ [79].

Although it is not easy to evaluate the exact mechanism for each organic molecule under a wide temperature range, a remarkable agreement was found by two laboratories [81,82] that had studied the phenol oxidation in both subcritical and supercritical conditions. The rate constant for phenol oxidation in the temperature range of 100–420°C is presented in the Arrhenius form in Fig. 9. The work by Thornton and Savage [80] was done using a flow reactor system (temperatures between 300° and 420°C; pressures from 188 to 278 atm; varying oxygen concentration). The data from Mill et al. [81] were collected from a continuously stirred tank reactor system (temperatures between −100° and 250°C; saturated pressures; varying oxygen concentration). The rate constant is evaluated assuming an overall second-order rate relationship: reaction rate $= k[O_2][\text{phenol}]$.

The Arrhenius plot shows an apparent overall activation energy of about 8 kcal/mol, well below the initiation by a hydrogen abstraction [(Eq. (7)] and more consistent with an electron transfer model for initiation reaction [(Eq. (8)].

$$C_6H_5OH + O_2 \rightarrow C_6H_5O^{\bullet} + HO_2^{\bullet} \tag{7}$$

$$C_6H_5OH + O_2 \rightarrow C_6H_5OH^{\bullet +} + O_2^{\bullet -} \tag{8}$$

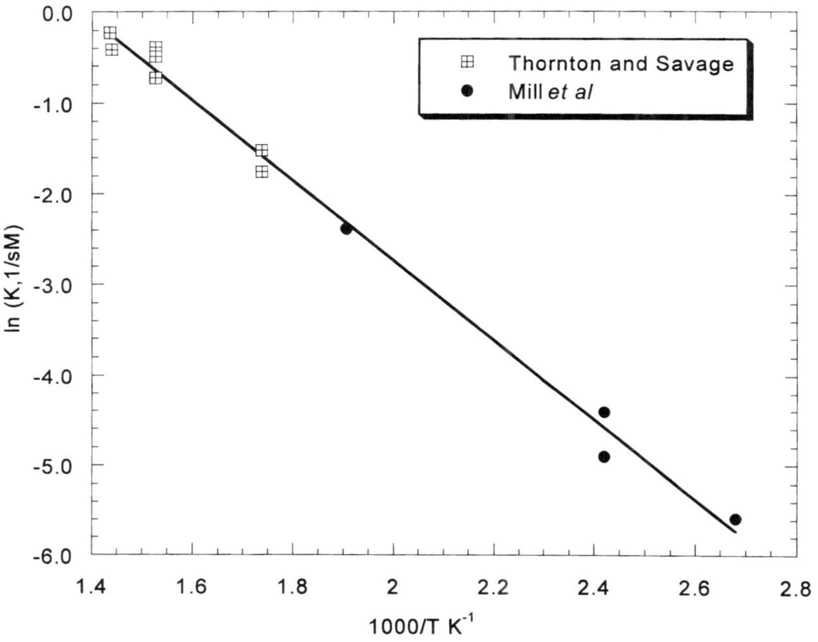

Figure 9 Arrhenius plot for hydrothermal oxidation of phenol between 100° and 420°C. (From Ref. 81.)

The linear dependence of log k on $1/T$ is contrary to the expectation that a curved Arrhenius plot should result from the change in properties of water that undergoes transition from subcritical to supercritical. However, the above data suggest that the kinetic features of the process are similar in the entire temperature range. It should be noted, however, that the changing temperatures and pressures will affect both reaction rates and pathways. This is one example that shows the importance of both radical and ionic pathways, depending on the organic species, the density of the water [10], and the temperature of the operation. However, the majority view is that only a -free radical reaction mechanism is accountable for SCWO of organic compounds. The complexity of the understanding of the pathways for single components demonstrates that it is impossible to extrapolate the reaction rates from single-component systems to predict the reaction rates involved in real-world samples, which contain mixtures of organic and inorganic components. However, it is clear that cooxidation of mixtures of phenols with other alkyl aromatics could lead to significant enhancement in reaction rates of the alkyl aromatics. This is simply because of the fast electron-transfer reaction of phenol and oxygen to form radicals capable of oxidizing the alkyl aromatics; radicals are not formed by reaction of oxygen directly with alkyl aromatics.

Because of the complexity of the pathways involved, most of the researchers have turned to global rate laws to determine the overall reaction rates for organic oxidation [82–84]. The objective of such global kinetic analysis was to determine the Arrhenius parameters (A and E_a) and the reaction orders (a, b, c) with respect to organic compound, oxygen, and water, respectively, for the rate expression for the disappearance of the starting organic during SCWO as given in Eq. (9).

$$\text{Rate} = A\exp(-E_a/RT)\,[\text{organic}]^a[\text{O}_2]^b[\text{H}_2\text{O}]^c \tag{9}$$

In this method, the rate of disappearance of the organic compound is measured with varying temperatures, oxygen concentrations, pressures, and organic concentrations. The optimal values for A, E_a, a, b, and c are then evaluated from the best fit. This type of global rate formula typically captures the general trends in the data, but it cannot provide the details of the oxidation chemistry. One example for the best fit for chlorophenol (CP) oxidation is given in Li et al. [84] [Eq. (10)], where $E_a = 11$ kcal/mol and the preexponential factor is 10^2 s^{-1}.

$$\text{Rate} = 10^{2.0}\exp(-11,000/\text{RT})\,[2\text{CP}]^{0.88}[\text{O}_2]^{0.41}[\text{H}_2\text{O}]^{0.34} \tag{10}$$

One has to be careful when using these global rate laws, as there may be more than one solution. The overall rate expression may provide some information on the kinetics; however, individual parameters cannot be used separately to predict their effect on the rate. Kinetic lumping is another method that is often used by scientists to derive a simple rate formula avoiding the use of elementary reactions [85].

2. Products from SCWO

Identifying the products (both intermediates and final products) from the SCWO process is an essential prerequisite for evaluating the environmental impact of the technology. Additionally, identification of products is key to optimizing the process parameters to obtain the desired conversion for the destruction of the pollutant. The intermediate products and their composition depend on the temperature, water density (or pressure), oxidant concentration, concentrations of other additives, if present, reactor surface, and the extent of the conversion.

To date, the most extensive efforts have been on the identification of intermediate products of phenol and substituted phenols [83,86,87]. However, most of the studies have been carried out at temperatures only slightly above the critical temperature but far from the actual operational temperatures of the conventional SCWO operational temperatures (450–650°C). Nonetheless, these data provide information for mechanistic development and process optimization purposes. These studies also provide information for developing SCWO processing under lower operating temperatures (e.g., <450°C). One group has identified 16–40 different intermediate products during phenol oxidation, including carboxylic acids, dihydroxybenzenes, phenol dimers (phenoxy phenol and biphenyls), and the related products dibenzofuran and dibenzo-*p*-dioxin [82,83,87]. Li et al. [84] studied the intermediates from the oxidation of 2-chlorophenol and noted the production of chlorinated dibenzofurans and chlorinated dibenzo-*p*-dioxins, which are potentially more hazardous than 2-chlorophenol, the starting pollutant. It is worth noting that these intermediates are formed during the very early stages of the reaction, and both these compounds would be ultimately converted to the intended product, CO_2.

Ross et al. [10,88] conducted an extensive study on the conversion of several model compounds (e.g., parachlorophenol, dichlorobenzene, hexachlorobenzene, and tetrachlorobiphenyl) to simulate the waste streams containing PCBs, under supercritical conditions at 400°C and 3700 psi with sodium carbonate added as a promoter. In their study, no formation of dibenzofurans or dibenzo-*p*-dioxins was noted during the decomposition of the starting material, even at conversions as low as 50%. These results were confirmed by Mitsubishi Heavy Industries (MHI) in their laboratory-scale testing.

Ammonia and acetic acid have been identified as the slowly oxidizing intermediates of degraded organics [74,89]. However, far fewer studies have been done on the oxidation of ammonia than on the oxidation of acetic acid [74]. Because acetic acid is resistant to oxidation under WAO conditions, it has been identified as the main refractory product from that process. Consequently, it is not surprising that numerous data are available on the oxidation of acetic acid [90–96]. The recent data from the treatment of waste simulants for some of the most hazardous waste streams have shown that acetic acid is the key component that determines the required process operational temperature when the complete elimination of organic carbon is required [97]. Because acetic acid has been singled out as the key organic intermediate by many authors, it deserves special attention here. It is not clear whether the researchers' interests in understanding the reactivity of acetic acid in water comes from its implications for the origin of natural gas (a process called hydrous pyrolysis) or environmental impacts from the waste streams containing low molecular weight carboxylic acid (e.g., textile and leather industries). Whatever the reason, there is plenty of literature data available from natural gas studies to allow an understanding of the reactivity of acetic acid in water under high-pressure conditions [98,99]. There is also a fair amount of literature available on SCWO of acetic acid. However, a quick analysis of the data available on acetic acid oxidation from SCWO shows that there is very poor agreement between the data from different laboratories. Some of the available data on the oxidation of acetic acid under SCWO are given in Table 2 [11,75,96]. This table provides Arrhenius coefficients, A (Arrhenius factor), and E_a (activation energy), which can be related to the rate constant, k, for the oxidation of acetic acid by $k = A\ \exp(E_a/RT)$. Later, it was found that these differences are due to the different types of reactors (e.g., surface effects are important for acetic acid oxidation). As an example, the data from Lee [94] indicate that the main contribution to the acetate decomposition comes from heterogeneous processes. The effects of surface area are discussed in the next section.

Heterogeneous and Homogeneous Catalysis Under SCWO. In view of slow oxidation rates for certain intermediates (e.g., acetic acid), the use of catalysts to enhance the rate of oxidation has received great attention. A large body of data is available on both homogeneous and heterogeneous catalysis under SCWO conditions. A homogeneous catalyst is superior to a heterogeneous catalyst because with the former the reaction is in a single phase, which eliminates diffusion and mass transfer problems. Copper salts are the most active catalysts when hydrogen peroxide is used as the oxidant [90,104]. Manganese chloride and manganese acetate have also been tested as homogeneous catalysts [86]. Homogeneous catalysts have the disadvantage of requiring -posttreatment recovery to minimize their toxicity in the effluent.

Heterogeneous catalysts, either as metals or as metal oxides, are easier to separate from the effluent stream and when coated onto porous carriers are more active than homogeneous catalysts in promoting oxidation. Some examples of heterogeneous catalyzed systems operating at subcritical temperatures (WAO conditions) include the following: ruthenium supported on cerium (IV) oxide, the most active metal catalyst among precious metals

Table 2 Arrhenius Parameters for Acetic Acid Oxidation

Log A	E_a (kJ/mol)	Conditions	Reference
4.91	106	400°–500°C, batch reactor	75
9.73	165	380°–470°C, batch reactor	96
18.0	231	400–445 bar, 338°–445°C	96
13.4	205	441°–532°C, 269–276 bar	11
9.9	168	246 bar, 425°–600°C	96

Source: Ref. 96 and references therein.

tested for the wet air oxidation of 1-propanol, 1-butanol, phenol, acetamide, poly(propylene glycol), and acetic acid [98]; CuO and Mn_2O_3 on Al_2O_3 for phenol oxidation [101]; Co-Bi complex oxides [100] and ferric oxide [102] for oxidation of acetate; cerium-based composite oxides [103] for oxidation of ammonia; and Co_2O_3 oxide for the oxidation of a number of compounds containing oxygen and nitrogen [104].

Many heterogeneous reaction studies under supercritical water conditions have been reported, including the use of MnO_2, V_2O_5, or Cr_2O_3 [104–110]. Some studies used catalysts to increase the rates of oxidation of organics, whereas others attempted to assess the role of a heterogeneous reaction in a nominally homogeneous system. Several researchers have observed increases in rates of oxidation by increasing the surface area of the reactor. For example, 1) the rate of oxidation of *p*-chlorophenol in supercritical water was enhanced more by increasing the surface-to-volume ratio of the reactor than by adding copper (II) tetrafluoroborate [86]; 2) Webley et al. [89] showed that SCWO of ammonia in a packed bed reactor made from Inconel 25 was more rapid than oxidation in an unpacked tubular reactor made of Inconel 625; and (3) Lee [109] observed an increased rate of oxidation of acetic acid with increased surface-to-volume ratio of the reactor.

It is clear from these studies that although higher rates of oxidation of organics can be achieved by the addition of selected heterogeneous catalysts they have some limitations. Because of surface contamination, heterogeneous catalysts can effectively treat only homogeneous waste streams. Other significant issues to be considered are catalyst stability, poisoning [110], recovery and regeneration, toxicity, and costs. Therefore, there is a great need for the development of catalysts that not only speed the destruction of organic compounds below 450°C to make the process economic but also satisfy the other concerns.

Recently it was demonstrated that the rate of oxidation can be increased by the introduction of surface under basic conditions [111]. This work has introduced a new catalyst concept that meets the above criteria for use under moderate SCWO conditions in a continuous tubular flow reactor. The concept involves -in situ precipitation of the catalyst (e.g., sodium carbonate) under SCWO conditions, but the catalyst is otherwise soluble under ambient conditions. -In situ precipitation is a unique way to generate a high-surface-area catalyst in the reaction zone, thereby ensuring maximum surface contact with the medium while minimizing catalyst poisoning.

In situ precipitation also provides a method for preparing surfaces with uniform stoichiometry and purity, a small range of grain and particle diameters, and minimum excess surface energy. These properties should maximize catalyzed rates and minimize differences between experiments caused by nonuniformity of mixing and contact between solution and surface [112].

3. Interferences

During the laboratory-scale testing of pollutant oxidation under SCWO, the main interference would come from reactor corrosion and reactor plugging. These issues are discussed in detail in the "Scale-Up Studies" section.

4. Posttreatment

In theory, the SCWO process can be operated under closed-system conditions with minimal exhaust release to the atmosphere. Therefore, during the laboratory-scale testing, post treatments are not required if the waste stream is treated under optimized conditions to completely oxidize the organic carbon to carbon dioxide. However, the effluent from the reactor should be treated under EPA guidelines for the waste (e.g., waste model compounds).

IV. SCALE-UP STUDIES AND ENVIRONMENTAL/ INDUSTRIAL APPLICATIONS

A. Experimental Design

The selection requirements for each of the components of the SCWO system for treating a variety of waste types comes from environmental regulations, waste characteristics, and cost and safety criteria. Similar to the bench-scale experimental design, the major components to be included in the SCWO design involve three main subsystems (influent introduction, reactors, and effluent removal systems). Other auxiliary systems such as heat exchangers and effluent exhaust systems must also be designed. In addition, for scale-up operations, the waste pretreatment and handling systems have to be considered. Fig. 10 shows a schematic of a complete system.

For scale-up operations, the selection of the reactor is considered to be the key element in designing SCWO systems. Environmental regulations set the requirement for the destruction efficiency, which in turn sets requirements on the temperature and residence time in the reactor (as an example, the required DRE is 99.99% for principal hazardous components and 99.9999% for polychlorinated biphenyls, PCBs). The reactor parameters for the scale-up designs can be extrapolated from the available bench-scale data. A detailed discussion on available reactor types is given below.

1. Reactor Selection

Early Reactor Designs for Scale-Up Operations. A review on different types of reactors considered during the early stages of SCWO development is given in a report on the evaluation of the suitability of SCWO for

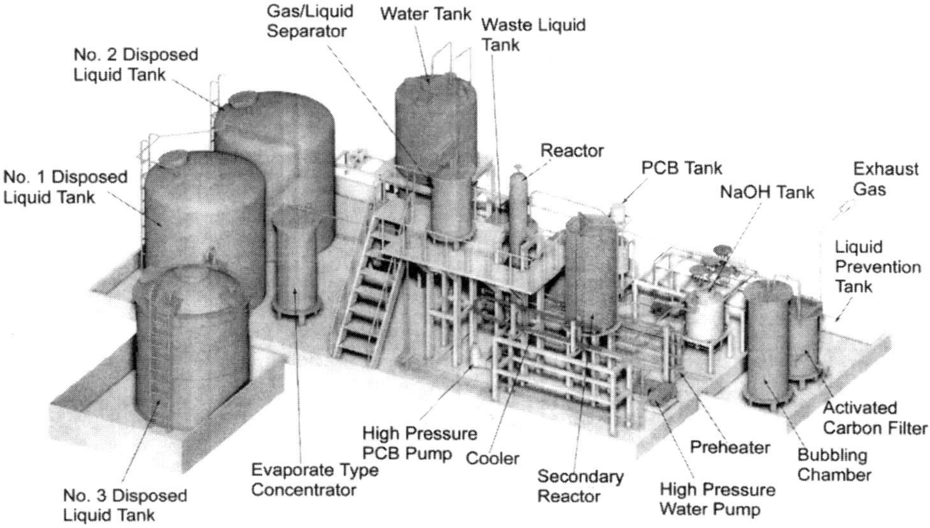

Figure 10 A schematic of an SCWO processing plant for PCB disposal. (Courtesy of Mitsubishi Heavy Industries, Japan.)

treatment of U.S. Department of Energy mixed waste [112]. These reactors vary in shape from tubular to cylindrical. Several types of tubular reactors are described in the published literature for SCWO. A double pipe or annular reactor is described in a patent issued to Welch and Slegwarth [113]. A patent issued to Dickenson contains designs of both annular reactors and a U-tube configuration [114]. An annular reactor using sintered separators has been presented by Li and Gloyna [115]. Tubular reactors with substantially constant diameter of several configurations are discussed in a patent issued to Modell [116].

Many tubular reactor designs depend on high velocities to avoid deposition of particles in the reactor [112,117,118]. However, when the high velocities are applied by use of a small diameter, a reactor length of hundreds of feet is required to achieve the required residence time. Therefore, the principles of keeping solids from depositing by high velocities have not been demonstrated at any acceptable scale.

Selected Novel Reactor Designs for Scale-Up Operations. Novel reactor designs currently being tested at pilot scale for treating challenging waste include Foster Wheeler, General Atomic, and MHI reactor designs. The Foster Wheeler technology uses a transpiring wall reactor design that is intended to protect the liner of the pressure vessel from salt deposition and

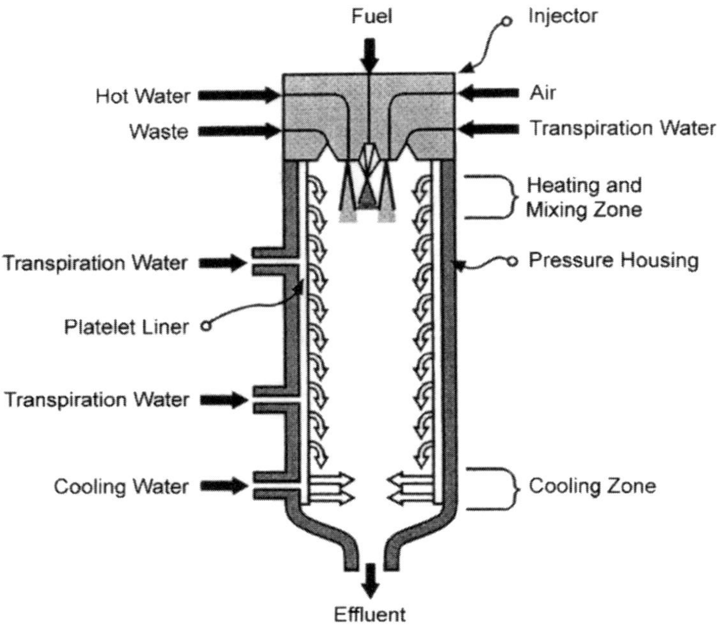

Figure 11 Transpiring wall reactor by Foster Wheeler Corporation. (From Ref. 119.)

corrosion and provide a thermal and corrosion barrier for the pressure vessel. The General Atomic technology uses a solid wall design with a liner to protect the main reactor (down-flow design) from corrosion. The MHI reactor uses a conventional cylindrical reactor combined with a tubular reactor. These three reactor designs are described in detail here.*

A schematic of the Foster Wheeler reactor design is given in Fig. 11 [119]. This reactor uses a transpiring wall platelet technology by GenCorp/Aerojet. The transpiring wall is based on the use of platelet devices, which provide an intricate circuitry that meters and repeatedly divides a flow steam into thousands of small injection pores that form a protective boundary layer to inhibit salt deposition and corrosion. Platelet devices are made by diffusion bonding a stack of thin plates (or platelets), each of which is etched with flow-control passages. Platelets differ from porous liners by providing for precise flow controls. This reactor operates as a down-flow reactor.

* Current reactor designs are provided in http://www.aro.army.mil/chemb/people/shaw/hague1.htm.

General Atomic's down-flow reactor system (described in U.S. Patent 6,0504, 057) is designed to operate in the temperature range 374–800°C and at pressures above 250 bar. It includes a substantially cylindrically shaped vessel that forms a reactor chamber. The vessel is oriented to assist the down-flow of material through the vessel by gravity. A jet assembly mounted on the top end of the vessel directs the stream of pressurized feed material into the reactor chamber in a direction so that the stream does not directly impinge on the walls of the reactor chamber. The velocity at which the stream is introduced into the reactor chamber causes a back-mixing action to be established in the upper back-mixing section of the reactor chamber. This back-mixing is said to promote reactions within the reactor chamber. Below the back-mixing section is a plug flow section, which in comparison with the back-mixing section is characterized by minimal back-mixing, and is added to accomplish additional reaction, if necessary. This reactor has dimensions that allow for effective flow through the reactor without causing a buildup of sticky solids (e.g., the length-to-diameter ratio of the reactor is typically between 1:1 and 50:1).

The MHI (Japan) reactor, which uses the SRI International technology, operates at temperatures from 374° to 450°C and 250 bar. It is also designed to attain >99.99999% conversions, as the regulatory requirements in Japan require the amount of chlorinated compounds in the liquid discharge to be less than 0.5 ppb. This system contains a cylindrical reactor (up-flow) with a jet mixing system at the lower end of the reactor. Their reactor is specially designed for handling inorganic salts and avoids any precipitation of solids on the reactor wall. The cylindrical reactor provides 99.9% conversion of organics, and the tubular coiled reactor is used to achieve the additional conversion [116,120,121].

2. System Operation Procedure

The basic operation of an SCWO system involves influent pressurization and introduction to the heated reactor and removal of the effluent from the reactor for discharge. However, the details of the operation of each subsystem of the SCWO processing system differ from one system to another, and are specific to both the type of waste and the reactor. A detailed description of the reactor heating method is given below, as the selection of the proper reactor heating method is important for the cost-effective operation of the SCWO process.

There are two methods of raising the reactant temperatures: indirect and direct heating. Indirect feed heating, the most common method, involves transferring energy across a solid boundary to the waste feed stream. Regenerative heat exchange (e.g., feed/effluent heat exchangers, bayonet reactors, and separate heat transfer liquid loops) and external heating (e.g., gas-fired, resistive, or radiant heating) are examples of indirect heating. Direct feed heating occurs when a cold (or partially preheated to < 250 °C) aqueous waste stream is mixed with a hot (500° to 650 °C) supercritical water/oxidant stream to achieve an unreacted "mix" temperature above 374 °C. This supercritical water for direct heating is from either recycled reactor effluent or separate, indirectly heated water stream. Regenerative heat exchange is especially amenable to treating low-heating-value feed streams, but it does have a number of drawbacks. Because the feed/effluent heat exchange surfaces experience the most corrosive environment in SCWO systems, the heat exchanger can cost several times more than the reactor itself. The feed heat exchanger is also susceptible to organic char and oxide/salt fouling. By keeping the waste stream cold until it is injected into the reactor, direct feed heating avoids the highly corrosive 250–400 °C temperature range. Organic polymerization and charring are also minimized. Initiating oxidation by recycling the reactor effluent does, however, require either a temperature-canned pump or an educator. Likewise, an indirectly heated supercritical stream requires an additional water pump and heater. This clean SCW stream, however, does not have the processing problems associated with preheating waste feeds.

B. Examples of Scale-Up, Environmental Application, and Industrial Application

Over a dozen pilot demonstrations of SCWO treatment for a variety of waste streams ranging from industrial waste to military waste (e.g., chemical weapons) are currently in operation or planned for near-future operation. Among them, the most visible are government-funded demonstrations such as those at Los Alamos, Tyndall Air Force Base, U.S. Army facilities, and U.S. Navy facilities. Currently, SCWO is being evaluated under the U.S. government's Assembled Chemical Weapons Assessment (ACWA) program [17,126]. Table 3 provides information on the waste being treated or considered to be treated under government installations.

Unlike in the United States, the pilot installations in Europe are geared toward treating industrial and municipal waste types. Some of the examples include the following: 1) tube reactor (15 kg/hr, 30 MPa, 15 m long, 8 mm i.d., air as oxidant) for treatment of pharmaceutical, chemical, paper, and municipal wastes, in Forschungszentrum Karlsruhe, Germany; 2) tube reactor (4 kg/hr, 400–600 °C, 20–60 MPa, 5–15% organic feed) to treat organic components of radio active waste, at Commissariat a l'Energie Atomiqui, Pierelatte, France; 3) tube reactor (250 kg/hr, 1.5–2.0 wt.% organic, 25 MPa at 600 °C) for treatment of deinking sludge and spent cutting liquids containing amines, at Chemature Engineering,* Sweden; and 4) tube reac-

Table 3 Supercritical Water Oxidation Demonstrations at U.S. Government Locations

Agency	Reactor type	Waste feed	Miscellaneous information
Los Alamos Laboratory	Down-flow reactor	Actinide-contaminated waste	540°C, 46 MPa
	Vertical coil reactor	Explosives	450°C, 110 MPa
Tyndall Air Force Base	Tubular reactor	Gray waste, sewage	
U.S. Army	Transpiring wall reactor	Colored smokes and dyes	600°C, 10 g/s
U.S. Navy	Combination of down-flow and transpiring wall reactors	Black water, gray water, paint, solvent, oil, etc.	40 g/s

Source: Ref. 17 and references therein.

tors—stationary and mobile (30–35 MPa, 600°C, 12–20 L/hr), for liquid or suspended wastes including electronic scrap, at Fraunhofer Institute, Germany. This is only a partial list of installations in Europe [17].

Japan can be considered one of the countries where very active programs in fundamental research of systems under supercritical water conditions are currently in operation for various applications. However, there are only a few pilot-scale installments of SCWO in Japan because of stringent regulations. Among them are the following: 1) the first pilot-scale SCWO reactor installed by Organo Corporation for treatment of municipal waste and sludge, electronic wastes, and polychlorinated biphenyls [17] and (2) the very recently installed MHI pilot plant to provide a complete solution for destroying waste streams containing 100% PCBs and other PCB-contaminated material (capacitors, transistors, transformer containers, etc.). The MHI system is based on SRI International's sodium carbonate technology [123,124].

1. Efficiency

In this section, three important examples are selected for discussion: 1) treatment of aqueous streams containing EPA priority pollutants [17] by

* Chematur Engineering AB, Karlskoga, Sweden, has acquired the exclusive worldwide rights to Eco Waste SCWO technologies and the process is demonstrated under the trade name Aqua Critox.

MODAR, Co; (2) shipboard waste treatment using a novel reactor design [123,126] by Foster Wheeler and General Atomics, and (3) treatment of PCB waste streams containing very low levels to 100% concentration [125] by MHI.

Efficiency of SCWO for Aqueous Waste Streams Containing EPA Priority Pollutants. During the early development of the process, MODAR conducted two field demonstrations, one in 1985 in Niagara Falls, New York, and one in 1986 at a chemical company site in Pennsylvania. The objectives of these demonstrations were to field test with real-world hazardous waste streams and to demonstrate the achievability of steady-state DRE of 99.99% of waste streams with and without particulate solids. One of the waste streams treated at Niagara Falls was an aqueous waste stream contaminated with EPA priority pollutants, and because the heating value of this stream was below 4200 kJ/kg, isopropanol was used as a supplemental fuel during the waste destruction tests. During this testing, the reactor temperature was in the range of 615–635°C. The waste streams tested in Pennsylvania contained significant amounts of organic salts, organic chlorine, and sulfur and organic particulates, and it was reported that the MODAR design was modified to handle these components. Table 4 shows the efficiency of the process for the contaminants selected.

Efficiency of SCWO for Treatment of Shipboard Waste. Foster Wheeler has tested the treatability of U.S. Navy excess hazardous materials (EHMs), which represent the organic materials found aboard Navy ships [119,126,127], using an SCWO plant that utilizes a transpiring wall reactor design. The test plant is reported to have had a nominal waste feed rate of 45 kg/hr with compressed air as the oxidant. In this testing, corrosive wastes, including chlorinated solvents (diluted with kerosene) and lube oils, and a salt-producing photographic solution simulant have been processed at feed

Table 4 Selected Results of MODAR Pilot Demonstration in Niagara Falls

Contaminant	Feed concentration (mg/kg)	Feed rate (g/min)	DRE (%)
2-Chlorophenol	1,200	0.49	>99.997
1,1,2-Trichloroethane	190	0.077	>99.98
Chloroform	1	0.0001	>98.83
Carbon tetrachloride	0.59	0.00024	>96.53
PCB	1600	0.091	>99.995

Source: Ref. 13.

rates between 45 and 95 kg/hr. The operating pressures of 24.1 MPa and reactor temperatures between 594° and 816°C have been applied in this testing (this high-temperature use is not common for SCWO, and it may reflect the very high Btu value of the treated waste).

The efficiency of the SCWO process for the destruction of EHMs was greater than 99.98%. Table 5 summarizes the data for EHM feed constituents, kerosene, polychlorotrifluoroethylene (PCTFE), trichloroethane (TCA), and photographic simulant.

General Atomics has also conducted a treatability study to evaluate the applicability of SCWO to treat EHM. Details of this testing taken from EPA 542-R-00-004 [119] are given in Table 6.

Efficiency of the MHI Process for Treatment of PCBs. The MHI process was developed for the treatment of chlorinated organics varying from neat PCBs to PCB-contaminated liquids and solid material. The MHI processing system provides a total treatment of PCB and PCB-contaminated material. The process involves the removal of PCBs from contaminated containers, recovery of the usable material, and treatment of recovered PCB using an advanced SCWO process. This process operates at 400°C and 250 bar with sodium carbonate as a catalyst. The MHI system has a capacity of 4 gal/min and could treat a wide range of PCB concentrations (100 ppm–100 wt.%) [125]. The process efficiency is given in Table 7.

2. Products

The products from SCWO processing of pollutants are specific to the constituents present in the feed waste stream including actual waste components and other additives such as catalyst, fuel etc.

The usual gaseous effluent of SCWO treatment of waste containing carbon, hydrogen, oxygen, and nitrogen consists of the oxidation products

Table 5 Treatability Study Results for Navy Excess Hazardous Waste Material (EHM)

Contaminant	Outlet temperature (°C)	Feed rate (kg/hr)	DRE (%)
Kerosene	307	36.3	>99.98
PCTFE	310	36.3	>99.99
TCA	316	46.4	>99.99
Photographic simulant	307	46.9	>99.99

Source: Refs. 119, 128.

Table 6 Summary of SCWO Process Operating Conditions and Efficiencies of Some Selected Waste Types

System parameters or results	Fluorochlorocarbon	Chlorinated solvent	Motor oil	Gray water	Paint waste	Lube oil	Photographic solution
Steady-state time (hr)	30	30	30	30	27	30	30
Waste feed rate (lb/hr)	102	107	101	136	136	102	105
Reactor temperature (C	638	638	648	648	648	657	658
Reactor pressure (psig)	3250	3430	3430	3430	3430	3430	3430
Cooling water (gpm)	95	91	84	96	95	93	91
Reactor residence time (sec)	10	11	10	15	13	11	15
Airflow rate (lb/hr)	1690	1660	1670	1650	1690	1710	1650
Feed water flow rate (lb/hr)	721	703	744	70	322	622	70
Quench water flow rate (lb/hr)	2240	2290	2020	1710	2780	3020	2430
Kerosene—startup (lb)	11	9	8	15	6	11	15
Kerosene—steady state (lb/hr)	0	0	0	41	0	0	40
Kerosene—shutdown (lb)	8	—	—	5	—	6	10
Startup time (hr)	1.03	0.95	0.6	1.03	0.431.25	1	
Organic carbon DRE (%)	> 99.998	> 99.998	> 99.998	> 99.997	> 99.997	> 99.997	> 99.994
Mineral acids produced (%)	0.6	8.2	0	0	0	5.5	7.7
CO (ppm)	< 2	< 2	< 2	< 2	< 2	< 2	< 2
Phenols (ppm)	< 0.03	< 0.03	< 0.2	< 0.2	< 0.2	< 0.1	< 0.1
Chromium (ppm)	< 0.003	< 0.003	< 0.001	< 0.002	0.007	0.003	0.004

Source: Ref. 119.

Table 7 Efficiency of MHI Process for Processing of PCBs

Contaminant	Feed concentration	Feed rate (kg/day)	DRE (%)
PCBs	100% PCB	2	>99.999995 (<0.5 ppb)

Source: MHI/SRI, private communication, 2001. Ref. 125.

CO_2, N_2, and some N_2O and residual O_2 and N_2 (from the air when used as the oxidant). However, for nitrogen containing organic compounds, nitrogen gas is confirmed to be the predominant SCWO end product [121]. The formation of ammonia and nitrous oxide has been reported in SCWO of various nitrogen-containing organic compounds and wastewaters. Ammonia (more favorable product under WAO conditions) is usually a hydrolysis product of nitrogen-containing organic compounds, and nitrous oxide is a partial oxidation product of ammonia. For waste streams containing sulfur, phosphorous, and halides, if the gas stream is not neutralized, the gas effluent may also contain acids of halides and traces of P_2O_5 and SO_3.

The aqueous liquid products will contain the acidic components or corresponding salts. For example, if the waste stream includes chlorides, sulfur, and phosphorous-containing materials, the liquid effluent stream would contain HCl, H_2SO_4, H_3PO_4, or their respective salts if neutralized using a base. In addition, if NaOH is used as base it would be converted to sodium carbonate reacting with CO_2.

In Foster Wheeler EHM testing, the samples collected included gas and liquid effluents (see Sec. b, "Efficiency of SCWO for Treatment of Shipboard Waste"). Foster Wheeler reported that the priority air pollutants NO_x and CO were below 25 and 100 ppm, respectively, for their testing of EHM [126].

3. Interferences

The potential application of SCWO to a number of wastes is hindered by interferences arising from inorganic chemistry (i.e., corrosion) and phase separations, particularly deposition of solids on reactor system components. Other notable interferences come from excess heat generated by the treatment of high-fuel-content waste and formation of tar when the system is oxygen starved.

Interference from Reactor Corrosion. Pilot SCWO treatment systems constructed for disposal of both civilian and military wastes perform extremely well in destroying hydrocarbons. However, most hazardous wastes contain additional components that pose direct or indirect barriers to the proper operation of the SCWO process as explained below [17,127]. The direct barrier comes from the formation of corrosive acids generated during the destruction of hazardous waste containing chlorine, phosphorus, nitrogen, and sulfur, which in turn degrade (corrode) the reactor. One approach to avoid this corrosion is by neutralizing reactor acids using NaOH. However, in an effort to neutralize these acids with NaOH, for example, the second barrier arises from the precipitation of neutralization products such as sodium chloride, phosphates, nitrates, and sulfates, which lead to reactor plugging. These -interrelated problems of acidic corrosion and salt plugging are the two most serious impediments to more widespread civilian and military use of SCWO.

At present, several different reactor types and several approaches for handling corrosive waste have been tested by different organizations; some of which were discussed in Sec. III.B ("Reactor Types"). Although the complex reactor types have proven their usefulness for treating selected waste types, their complex designs may be cost prohibitive. These complex reactor systems may have very long downtime during reactor replacement, which in turn would add to the cost of operation. Therefore, vigorous corrosion estimation/analysis studies are required before installation of the plant. The experimental measurements to determine the metal stability or corrosion are straightforward (e.g., weight loss methods, scanning electron microscopy, and other standard mechanical tests). Corrosion rates of various metallic and -nonmetallic materials in SCWO environments have been carried out by many workers and a review can be found in Gloyna and Li [12] and later articles by Kim et al. and Mitton et al. [Refs. 50, 51, and references therein]. In addition, SRI International is involved in long-term corrosion measurement under -subcritical and supercritical conditions not only for SCWO applications but also for power boiler and supercritical water nuclear power plant applications. The conclusion from all the published work is that corrosion is severe under near-critical conditions and then decreases above the critical temperature. Therefore, corrosion mitigation is important in SCWO systems in areas where near-critical conditions exist (e.g., preheater and cool-down sections). In addition, reactor corrosion becomes a major issue when treating feeds containing acid-forming compounds (e.g., chlorinated compounds). Gloyna and Li [12] report that in one comprehensive study related to the destruction of chemical warfare agents, platinum showed a corrosion rate of 10 to 30 mil/yr when exposed to supercritical water at 450–550°C and high concentrations of chlorides [12].

Interference from Phase Separations. The potential phase separations that occur during treatment of highly concentrated waste streams containing both organic and inorganic material on the preheater, reactor, or cool-down section cause major interference to the operation of the SCWO process. The extent of the interference depends on the operational conditions at each individual section of the process. In the preheater section in which the pressurized process fluid is typically passed through heat exchanges to raise the fluid temperature to 300–375°C range, inorganic salts may exhibit decreasing solubility (e.g., Type 2 salts such as sodium sulfate and sodium carbonate) and may lead to scaling of heat transfer surfaces and cause reactor plugging. In this section, it is typical to have components in all three phases: gas, liquid, and solid. The phase separation becomes even more complex when the fluid approaches the main reactor where temperatures in the range 375° to 650°C are encountered. For SCWO processes, which operate at higher temperatures (e.g., 650°), the phase separation arises from both Type 2 and Type 1 salts (e.g., NaCl would separate as a brine, Na_2SO_4 would separate as a solid). However, for SCWO operations in which the maximum reactor temperature is maintained at less than 450°C, the situation is less complex because below 450°C and at pressures $>$250 bar most of the Type 1 salts (e.g., NaCl) will stay in the aqueous phase and only the Type 2 salts will be in a separate solid phase. For special SCWO operations, very high pressures (e.g., 1000 bar) have been applied to solubilize the inorganic salts [9]; however, these high pressures are not recommended for common use.

Finally, in the cool-down section, in which the process stream temperatures are reduced below the critical temperature of water for product separation and recovery, precipitated inorganics may redissolve, but gas phase immiscibility will rise.

It is important to note that the reaction conditions (temperature and pressure) under which the various input streams are introduced into the SCWO system are of key importance in controlling phase separations. For waste streams that are highly complex and require upper-limit operation of the SCWO system, appropriate flow velocities can be used to minimize the solid deposition on the reactor components. The critical flow velocities required for solids are found in Refs. 117 and 118.

Interferences from Excess Heat Generation. Another obstacle to the smooth operation of an SCWO process consists of limitations imposed by the high Btu value of wastes, which would require additional heat exchangers to remove the excess heat generated from the process. Current SCWO designs are capable of accommodating 10–20 wt.% of organic waste in the feed stream. For feed material with sufficient heat value, it is recommended that cooling water be introduced to the reactor to avoid runaway reactions. The coolant can be introduced directly into the main reactor, bypassing the preheating (see Sec. II.A.2, "Phase Separations"). Barnes [112] reports that for DOE waste containing high caloric value a dilution of the initial feed waste is required, thus adding cost to the process.

4. Posttreatment

In general, there is no need for posttreatment of the effluent from a properly operated SCWO process other than the handling of permanent gases, the liquid water, and the recovered solids, if any. However, the discharge of the effluent from the process may require further treatment, depending on the waste classification and the regulatory requirements of the location of the plant. For example, treatment of organics contaminated with radioactive inorganics may require further handling of the recovered radioactive material (only the organic contaminants will be destroyed by the SCWO process).

5. Other Practical Considerations and Limitations

Because water can be considered as one of the most benign solvents available, together with its solvation powers at high temperature, its applicability for a variety of wastes is tremendous. However, based on the current developments of the process, its applicability remains limited to pumpable waste streams. In addition, it is generally understood that the waste streams with organic content in the 1–20 wt.% range is most suitable for this technology. Waste streams with 100 wt.% can be treated with appropriate dilution. However, the cost for the dilution may add prohibitively to the total cost.

Hydrothermal oxidation (especially SCWO) occurs in a pressurized vessel, thus requiring stringent safety and maintenance operations. In addition, SCWO is better suited for processing solutions and slurries, whereas solid wastes require sizing and pretreatment.

V. CONCLUSIONS

SCWO is among the best technologies available today for efficient and environmentally sound disposal of both civilian and military waste. Scale-up demonstrations for development of this process have been explored around the world for a variety of waste-disposal applications (e.g., municipal waste).

Civilian applications are numerous, but most funding of SCWO technology has stemmed from the military's need to find a safe and effective alternative to incineration of their wastes, as well as the need to clean up mixed wastes (radioactive and hazardous organic materials) at DOE weapons facilities. For better utilization of SCWO for its application to a wide range of waste types, a better fundamental understanding of reaction media, including reaction rates, reaction mechanisms, and phase behavior of multicomponent systems is required. Such an understanding would help optimize the process conditions to minimize reactor corrosion and salt deposition, the two major obstacles for development of the technology, and to devise a SCWO process that operates at much more favorable temperature and pressure conditions (< 450 °C and < 300 bar).

Currently, SCWO and WAO processes are considered as two separate technologies. However, integrated systems would be more beneficial and would provide opportunities for cost-effective waste treatment because an integrated single process could be used for various applications. Special, complex reactor designs (e.g., transpiring wall reactor) will remain important for treating special types of wastes. However, for common waste types (e.g., industrial waste streams), more generic, simpler reactor designs can be used. The cost numbers available for the process thus far have come from estimations and extrapolations of data from bench- and pilot-scale operations [112,129], because no industrial-scale SCWO plant exists to date. However, the estimated numbers compare very well with the competing technologies, and the technology is expected to become more cost effective with time. For example, the cost for the treatment of one wet ton of organic content of 10 wt.% is estimated to be US$300 [127].

Finally, as a result of an increase in industrial activities to support the expected increase in population growth (the global population is expected to grow to 10 billion by 2050), the industrial waste (e.g., electronic waste, wastes from energy-related operations) generation would increase at a considerable rate. Considering its young stage of maturity, SCWO has already shown great promise for its use as an alternative to incineration for those industrial waste types that are most difficult to destroy (e.g., chlorinated compounds). Therefore, SCWO processes (and other hydrothermal oxidation processes such as WAO) are potential technologies that would play a major role in managing and controlling industrial waste generated from future operations.

REFERENCES

1. Zimmermann FJ. Waste disposal. U.S. Patent 2,665,249, Jan. 5, 1954.
2. Copa WM, Gitchel WB. Wet oxidation. In: Freeman HM, ed. Standard Handbook of Hazardous Waste Treatment and Disposal. New York: McGraw-Hill, 1989, Section 8.8.
3. Joshi JB, Shah YT, Parulekar SJ. Engineering aspects of the treatment of aqueous waste streams. Indian Chem Eng 1985; 37:3–37.
4. Laughlin RGW, Gallow T, Robey H. Wet air oxidation for hazardous waste control. J Hazard Mater 1983; 8:1–9.
5. Mishra VS, Mahajani VV, Joshi JB. Wet air oxidation. Indian Eng Chem Res 1995; 34:2–48.

6. Foy BR, Waldthausen K, Sedillo MA, Buelow SJ. Hydrothermal processing of chlorinated hydrocarbons in a titanium reactor. Environ Sci Technol 1996; 30(9):2790–2799.
7. Timberlake SH, Hong GT, Simson M, Modell M. Supercritical water oxidation for wastewater treatment: preliminary study of urea destruction. SAE Tech, Paper 820872, Society of Automotive Engineers, Inc., Warrendale, PA, 1982.
8. Harradine DM, Buelow SJ, Dell'orco PC, Dyer RB, Foy BR, Robinson JM, Sanchez JA, Sportarelli T, Wander JD. Oxidation chemistry of energetic materials supercritical water. Hazard Waste Hazard Mater 1993; 10(2):233–246.
9. Rofer CK, Buelow SJ, Dyer RB, Wander JD. Conversion of hazardous materials using supercritical water oxidation. U.S. Patent 5,133,877, 1992; Dell'orco PC, Foy BR, Robinson JM, Buelow SJ. Hydrothermal treatment of Hanford waste constituents. Hazard Waste Hazard Mater 1993; 10:221.
10. Ross DS, Jayaweera IS, Leif R. Improved method for hydrothermal oxidation of halogenated organic compounds with addition of specific reactants. U.S. Patent 5,746,926, 1998.
11. Rice SF, Steeper La Jenunesse RR. Destruction of representative Navy wastes using supercritical water oxidation. Sandia National Laboratory Report SAND4-88203, 1993.
12. Gloyna EF, Li L. Supercritical oxidation research and development update. Environ Prog 1995; 14(3):182–192.
13. Thomson B, Hong GT, Swallow KC, Killilea WR. The MODAR supercritical oxidation process. In: Freeman HM, ed. Innovative Hazardous Waste Treatment Technology Series, 1:31–42.
14. Modell M, Gaudet GG, Simson M, Hong GT, Biemann K. Supercritical water testing reveals new process hold promise. Solid Wastes Manage 1982; 25:26.
15. Rofer CK, Buelow SJ, Dyer RB, Wander JD. Conversion of hazardous materials using supercritical water oxidation. U.S. Patent 5,133,877, 1992.
16. Shaw RW, Brill TB, Clifford AA, Eckart CA, Franck EU. Supercritical water—a medium for chemistry. Chem Eng News 1991; 69(51):26–39.
17. Shaw RW, Dahmen N. Destruction of toxic organic materials using supercritical waste oxidation. J Supercrit Fluids 2000; 17:425–437.
18. Thomson TB, Model M. Supercritical water destruction of aqueous wastes. Hazard Waste 1984; 1(4):453–467.
19. Amin SI. Reforming and decomposition of organics in water. Ph.D thesis, Massachusetts Institute of Technology, Cambridge, MA, 1975.
20. Connolly JF. Solubility of hydrocarbons in water near the critical solution temperatures. J Chem Eng Data 1996; 11(1):13–16.
21. Marshall WL, Messer RE. Pressure–density relationships and ionization equilibria in aqueous solutions. J Sol Chem 1984; 13:383–391.
22. Frank EU, Weingartner H. Supercritical water. In: Letcher TM, ed. Chemical Thermodynamics (Chemistry for the 21st Century Monograph). Oxford: Blackwell Science, 1999.

23. Marshall WL, Franck AU. J Phys Chem Ref Data. 1981; 10:295; M Uematsu, Franck EU. J Phys Chem Ref Data 1980; 9:1291.
24. Keenan JH, Keys FG. Thermodynamic properties of steam. 1st ed. New York: John Wiley & Sons, Inc., 1936:28–73.
25. Keenan JH, Keys FG, Hill PG, Moore JG. Steam Tables—Thermodynamic Properties of Water Including Vapor, Liquid, and Solid Phases New York: John Wiley & Sons, Inc., 1969.
26. Linsay WT Jr. Chemistry of steam cycle solutions: principles. The ASME Handbook on Water Technology for Thermal Power Systems. In: Cohen Paul, ed. New York: The American Society of Chemical Engineers, 1989:347–544.
27. Jayaweera IS. Hot water extraction technology (HWE): on site remedial technology for contaminated soils and sediments. Eye Environ 2001; 6(2):1–4.
28. Baker EG, Sealock LJ. Treatment and disposal of petroleum sludges. Presented at the Third Annual Symposium on Environmental Protection in the Energy Industry, Tulsa, Oklahoma, 1991.
29. Biggerstaff DR, White DE, Wood RH. Heat capacities of aqueous argon from 306 to 578 K. J Phys Chem 1985; 89:4378–4381.
30. Murchi RP, Eyring H. J Phys Chem 1992; 68(suppl 1):221.
31. Luck WAP. A model of hydrogen-bonded liquid. Angew Chem Int Ed Eng 1980; 19:28.
32. Gupta RB, Panayiotou CG, Sanchez IC, Johnston KP. Theory of hydrogen bonding in supercritical fluids. AIChE J 1992; 38:1243–1253.
33. Johnston SC. Supercritical water oxidation of hazardous wastes. White Paper for Public Distribution, Sandia National Laboratories, Menlo Park, CA, 1990.
34. Kalinichev AG, Bass JD. Hydrogen bonding in supercritical water. 2. Computer simulations. J Phys Chem A 1997; 101:9720–9727.
35. Kalinichev AG, Gorbaty YE, Bondarenko GV, Okhulkov AV. The effect of pressure on hydrogen bonding in water. Steam, water, and hydrothermal systems. Physics and Chemistry Meeting the Needs of Industry Ottawa, Canada: NRC-CNRC, 2000:538–545.
36. Keevil NB. Vapor pressures of aqueous solutions at high temperature. J Am Chem Soc 1942; 64:841–850.
37. Valyashko VM. Metastable and stable equilibria in water-salt systems at elevated temperatures and pressures. Zh Neorgan Khimii 1973; 18:1114.
38. Ravich MI. Water-salt systems at elevated temperatures and pressures. Moscow: Nauka Publ., 1974.
39. Valyashko VM. Phase Equilibria and Properties of Hydrothermal Systems Moscow: Nauka, Publ., 1990.
40. Urusova MA, Valyashko VM. Tricritical phenomena in the $NaCl\text{-}Na_2B4O_7\text{-}H_2O$ system and the transformation laws of complete phase diagrams. Russ J Inorg Chem 1998; 43:948–955.
41. Ravich MI, Borovaya FE, Ketkovich VYa. Solubility and vapor pressure of saturated solutions in the $KCl\text{-}K_2SO_4$ H_2O system. Izv Sekt Fiz-Him Anal Inst Obs Neorg Him Im NS Kurnakova 1953; 22:225.

42. Valyashko VM. Metastable and stable equilibria in water-salt systems at elevated temperatures and pressures. Zh Neorgan Khimii 1973; 18:1114.
43. Valyashko VM. Phase equilibria and solution properties of aqueous systems at high temperatures and pressures. Proceedings of the 13th International Conference on the Properties of Water and Steam. Steam, Water, and Hydrothermal Systems: Physics and Chemistry Meeting the Needs of Industry, Toronto, Canada, Sept 12–16, 1999. Ottawa, Canada: NRC Research Press, 2000, pp. 727–735.
44. Morey GW, Chen WT. Pressure temperature curves in some systems containing water and a salt. J Am Chem Soc 1956; 78:4249–4252.
45. Marshall WL. Water and its solutions at high temperatures and high pressures. Chemistry 1975; 48:2.
46. Sourirajan S, Kennedy GC. The system H_2O-NaCl at elevated temperatures and pressures. Am J Sci 1962; 260:115–141.
47. Pitzer KS, Pabalan RT. Thermodynamics of NaCl in steam. Geochim Cosmochim Acta 1986; 50(7):1445–1454.
48. Jayaweera IS, Valyashko VM. Supercritical water oxidation: solubility of mixed salt systems. Manuscript in preparation, 2003.
49. Macdonald DD, Cragnolino GA. Corrosion of steam cycle material. In: Cohen P, ed. The ASME handbook on Water Technology for Thermal Power System. New York: The American Chemical Society of Mechanical Engineers, 1989:653–1031.
50. Kim YS, Mitton DB, Latanision RM. Corrosion resistance of stainless steel in chloride containing SCWO systems. Korean J Chem Eng 2000; 17:58.
51. Mitton DB, Yoon JH, Cline JA, Kim HS, Eliaz N, Latanision RM. The corrosion behavior of Nickel base alloys in SCWO systems. Ind Eng Chem Res 2000; 39:4689.
52. Macdonald et al. Potential-pH diagrams for Iron and Nickel in high salinity geothermal brine containing low levels of hydrogen Sulfide. Corrosion 1979; 35:471. Macdonald et al. "The use of potential-pH diagrams..." Corrosion 1979; 35:1. Macdonald et al. "The corrosion of copper-nickel alloys 706 and 715 in flowing sea water", Corrosion 1979; 35:367.
53. Jayaweera P, et al. Solid state reference electrode for high temperature electrochemical measurements. U.S. Patent 5,425,871, June 1995.
54. Kriksunov LB, Macdonald DD. Corrosion in supercritical water oxidation systems—a phenomenological analysis. J Electrochemical Sci 1995; 142(12): 4069, 4073.
55. Cutler AH, Antal MJ Jr, Jones M Jr. A critical evaluation of plug-flow idealization of tubular-flow reactor data. Ind Eng Chem Res 1988; 27:691–697.
56. Mulcahy MF, Pethard MR. Errors in determining the rate constant of a first-order gaseous reaction by the flow method. Aust J Chem 1963; 16:527–543.
57. Furue H, Pacey PD. Performance of a cylindrical flow reactor in a kinetic study of the isomerization of cyclopropane. J Phys Chem 1980; 84:3139–3143.
58. Gorbaty YE, Bondarenka GV. High-pressure high-temperature Raman cell for corrosive liquids. Rev Sci Instrum 1995; 66(8):4347–4349.

59. Sphon PD, Brill TB. A Raman-spectroscopy cell for aqueous solutions to 500° C and 34.5 MPa. Applied spectrosc 1987; 41(7):1151–1156; Sphon PD, Brill TB. Raman spectroscopy of the species in concentrated aqueous solutions of $Zn(NO_3)_2$, $Ca(NO_3)_2$, $Cd(NO_3)_2$, $LiNO_3$, and $NaNO_3$ up to 450°C and 30 MPa. J Phys Chem 1989; 93(16):6224–6231.
60. Jayaweera IS, Ross DS, Mill T, Penwell P. Hydrothermolysis of energetic materials: safety and continuous process parameters. Final Report, DOD Contract F08637-94-C6050, SRI International, Menlo Park, CA, 1998.
61. Holgate HR, Webley PA, Tester JW. Carbon monoxide oxidation in supercritical water: the effects of heat transfer and the water-gas shift reaction on observed kinetics. Energy & Fuels 1992; 6:586–597.
62. Holgate HR, Tester JW. Oxidation of hydrogen and carbon monoxide in sub- and supercritical water: reaction kinetics pathways, and water-density effects: 1. Experimental results. J Phys Chem 1994; 98:800–809.
63. Holgate HR, Tester JW. Oxidation of hydrogen and carbon monoxide in sub- and supercritical water: reaction kinetics pathways, and water-density effects: 2. Elementary reaction modeling. J Phys Chem 1994; 98:810–822.
64. Tester JW, Welby PA, Holgate HR. Revised global kinetic measurements of methanol oxidation in supercritical water. Ind Eng Chem Res 1993; 32:236–239.
65. Webley PA, Tester JW, Holgate HR. Oxidation kinetics of ammonia and ammonia-methanol mixtures in supercritical water in the temperature range 530–700°C at 246 bar. Ind Eng Chem Res 1991; 30:1745–1754.
66. Chang CJ, Li S, Ko C. Catalytic wet oxidations of Phenol and p-Chlorophenol-contaminated waters. J Chem Technol Biotechnol 1995; 64:245–252.
67. Webley PA, Tester JW. Fundamental kinetics of methane oxidation in supercritical water. Energy Fuels 1991; 5:411–419.
68. Tester JW, Webley PA, Holgate HR. Revised global kinetic measurements of methanol oxidation in supercritical water. Ind Eng Ch R 1993; 32(1):236–239; Helling RK, Tester JW. Fundamental kinetics and mechanisms of hydrogen oxidation in supercritical water. Combust Sci Technol 1993; 88(5–6):369–397.
69. Salvatierra D, Taylor JD, Marrone PA, Tester JW. Kinetic study of hydrolysis of methylene chloride at 246 bar from 100° to 500°C. Ind Eng Chem Research 1999; 38(11):4169–4174.
70. Savage PE, Gopalan S, Mizan TI, Martini CJ, Brook EE. Reactions at supercritical conditions: applications and fundamentals. AIChE J 1995; 41(7): 1723–1778.
71. Model M. Processing methods for the oxidation of organics in supercritical water. US Patent 4,338,199, 1982.
72. Helling RK. Sc.D dissertation, Massachusetts Institute of Technology, Cambridge, MA, 1986.
73. Tester JW, Webley PA, Holgate HR. Revised global kinetic measurements of methanol oxidation in supercritical water. Ind Eng Chem Res 1993; 32(1):236–239. Webly and Tester, 1988.

74. Rofer CK, Streit GE. Phase II final report: oxidation of hydrocarbons and organics in supercritical water. Los Alamos National Laboratory Rep. LA-11700MS, 1989.
75. Lee DS, Gloyna EF, Li L. Efficiency of H_2O_2 and O_2 in supercritical water oxidation of 2,4-dichlorophenol and acetic acid. J Supercrit Fluids 1990; 3:249–255.
76. Yang HH, Eckert CA. Homogeneous catalysis in the oxidation of p-chlorophenol in supercritical water. Ind Eng Chem Res 1988; 27:2009–2014.
77. Walling C. Free Radical in Solution. New York: John Wiley and Sons, 1957.
78. Mill T, Hendry DG. Kinetics and mechanisms of free radical oxidation of alkanes and olefins in the liquid phase. In: Tipper C, ed. Comprehensive Chemical Kinetics. Vol. 16. Amsterdam: Elsevier, 1–83.
79. Schumb WC, Satterfield CN, Wenthorth RL. Hydrogen Peroxide. New York: Reinhold Publishing, 1955.
80. Thornton TD, Savage PE. Phenol oxidation in supercritical water. J Supercrit Fluids 1990; 3:240–248.
81. Mill T, Jayaweera IS, Yao D, Ross DS. Hydrothermal oxidation of phenol. Physical chemistry of aqueous systems. Meeting the Needs of Industry, Proceedings of the 12th International. In: White HJ Jr, ed. New York: Begel House, 1995: 589–593.
82. Thornton TD, Savage PE. Phenol oxidation in supercritical water. J Supercrit Fluids 1990; 3:240–248.
83. Thornton TD, Savage PE. Kinetics of phenol oxidation in supercritical water. AIChE J 1992; 38(3):321–327. Phenol oxidation pathways in supercritical water. Ind Eng Chem Res 1992; 31(suppl 11):2451–2456.
84. Li R, Savage PE, Szmukler D. 2-Chlorophenol oxidation in supercritical water: global kinetics and reaction products. AIChE J 1993; 39(1):178–187.
85. Li L, Crain N, Gloyna EF. Kinetic lumping applied to wastewater treatment. Water Environ Res 1996; 68(5):841–854.
86. Yang HH, Eckert CA. Homogeneous catalysis in the oxidation of *p*-chlorophenol in supercritical water. Ind Eng Chem Res 1988; 27(11):2009–2014.
87. Thornton TD, Ladue DE, Savage PE. Phenol oxidation in supercritical water: formation of dibenzofuran, dibenzo-*p*-dioxin, and related compounds. Environ Sci Technol 1991; 25(8):1507–1510.
88. Ross DS, Jayaweera IS, Bomberger DC, Lief R. Hydrothermal oxidation of organic compounds with heterogeneous neutralizing reagent. U.S. Patent No. 6,010,632, 2000.
89. Webley PA, Tester JW, Holgate HR. Oxidation kinetics of ammonia and ammonia-methanol mixtures in supercritical water in the range 530–700°C at 2466 bar. Ind Eng Chem Res 1991; 30(8):1745.
90. Lee DS. Heterogeneous oxidation kinetics of acetic acid in supercritical water. Environ Sci Technol 1996; 30(12):3487–3492.
91. Levec J, Smith JM. Oxidation of acetic acid solutions in a trickle-bed reactor. AIChE J 1976; 22(suppl 1):159–1668.
92. Webley PA, Tester JW, Holgate HR. Oxidation kinetics of ammonia and am-

monia-methanol mixtures in supercritical water in the temperature range 530–700°C at 246 bar. Ind Eng Chem Res 1991; 30(8):1745–1754.
93. Savage PE, Smith MA. Kinetics of acetic acid oxidation in supercritical water. Environ Sci Technol 1995; 29:216–221.
94. Lee DS. Heterogeneous oxidation kinetics of acetic acid in supercritical water. Environ Sci Technol 1996; 30:3487–3492.
95. Savage PE, Smith MA. Kinetics of acetic oxidation in supercritical water. Environ Sci Technol 1995; 29:216–221.
96. Meyer JC, Marrone PA, Tester JW. Acetic acid oxidation and hydrolysis in supercritical water. AIChE J 1995; 41(suppl 9):2108–2121.
97. Downey KW, Snow RH, Hazlebeck DA, Roberts AJ. Corrosion and chemical agent destruction. research on supercritical waste oxidation of hazardous military waste. Innovations in Supercritical Fluids, Chapter 21. Washington, D.C.: American Chemical Society, 1995.
98. Barth T, Borgund AE, Hopland AL. Generation of organic compounds by hydrous pyrolysis of Kimmerridge oil shale-bulk results and activation energy calculations. Org Geochem 1889; 14:69–76.
99. Borgund AE, Barth T. Generation of short-chain organic acids for crude oil by hydrous pyrolysis. Org Geochem 1994; 21(8–9):943–952.
100. Imamura S, Hirano A, Kawabata N. Wet oxidation of acetic acid catalyzed by Co-Bi complex oxides. Ind Eng Chem Prod Dev 1982; 21(suppl 4):570–575; Imamura S, Fukuda I, Ishida S. Wet oxidation catalyzed by ruthenium supported on cerium(IV) oxides. Ind Ing Chem Res 1988; 27(suppl 4):718–721.
101. Sadana A, Katzer JR. Catalytic oxidation of phenol in aqueous solution over copper oxide. Ind Eng Chem Fundam 1974; 13:127–134.
102. Levec J, Smith JM. Oxidation of acetic acid solutions in a Trickle-bed reactor. AIChE J 1976; 22(suppl 1):159–1668.
103. Imamura S, Doi A. Wet oxidation of ammonia catalyzed by cerium-based composite oxides. Ind Eng Chem Prod Res Dev 1985; 2491:75–80.
104. Ito MM, Akita K, Inoue H. Wet oxidation of oxygen and nitrogen containing organic compounds catalyzed by cobalt(III) oxide. Ind Eng Chem Rev 1989; 28(suppl 7):894–899.
105. Ding ZY, Aki NVSK, Abraham MA. Catalytic supercritical water oxidation: an approach for complete destruction of aromatic compounds. Presented at the Symposium on Supercritical Fluids, AIChE Meeting, San Francisco, CA, 1994.
106. Jin L, Abraham MA. Low temperature catalytic oxidation of 1,4-dichlorobenzene. Ind Eng Chem Res 1991; 30:89–95; Aki S, Abraham MA. Catalytic supercritical water oxidation of Pyridine: comparison of catalysts. Ind Eng Chem Res 1999; 38:358–367.
107. Jin L, Ding Z, Abraham MA. Catalytic supercritical water oxidation of 1,4-dichlorobenzene. Chem Eng Sci 1992; 47(9–11):2659–2664.
108. Aki SNVK, Ding Z, Abraham MA. Catalytic supercritical water oxidation: stability of Cr_2O_3 catalyst. AIChE J 1996; 42(7):1995–2004.

109. Lee DS. Heterogeneous oxidation kinetics of acetic acid in supercritical water. Environ Sci Technol 1996; 30:3487–3492.
110. Jin L, Ding Z, Abraham MA. Catalytic supercritical water oxidation of 1,4-dichlorobenzene. Chem Eng Sci 1992; 47(9–11):2659–2664.
111. Jayaweera IS. Montserrat Marti-Perez. SCWO treatment of dye waste. SRI Internal project Report, SRI International, Menlo Park, CA, 2002.
112. Barnes CM. Evaluation of tubular reactor designs for supercritical water oxidation of U.S. Department of Energy mixed waste. INEL-94/0223, Lockheed Idaho Technologies Co., Idaho Falls, ID, 1994.
113. Welch JF, Slegwarth JD. Method for the processing of organic compounds. U.S. Patent 4,861,497, 1989.
114. Dickinson NL. Pollution free low temperature slurry combustion process utilizing the supercritical state. U.S. Patent 4,380,960, 1983.
115. Li L, Gloyna EF. High temperature wet oxidation using sintered separators. International patent Application PCT/US92/06459, 1993.
116. Modell M. Supercritical water oxidation process of organics with inorganics. U.S. Patent 5,252,224, 1993.
117. Davies JT. Calculation of critical velocities to maintain solids in suspension in horizontal pipes. Chem Eng Sci 1987; 42(7):1667–1670.
118. Orosker AR, Turian RM. The critical velocity in pipeline flow of slurries. AIChE J 1980; 26:550–558.
119. EPA 542-R-00-004. Potential applicability of chemical weapons assessment technologies to RCRA waste streams and contaminated media. U.S. Environmental Protection Agency, Technology Innovation Office, Washington, D.C., August 2000; 2:21–31.
120. Tateishi; Masakazu (Nagasaki, JP); Tsuchiyama; Yoshihiko (Nagasaki, JP); Yamauchi; Yasuhiro (Nagasaki, JP); Fukuzumi. PCB decomposition process. U.S. Patent 6,162,958, Dec 19, 2000.
121. Yamauchi; Yasuhiro (Nagasaki, JP); Ogata; Kan (Nagasaki, JP); Moribe; Takashi (Nagasaki, JP); Tateishi; Masakazu (Nagasaki, JP); Fukuzumi; Tadatsugu (Nagasaki, JP); Iwao; Mitsuji (Nagasaki, JP); Ikeda; Nobuyuki (Nagasaki, JP); Shindo; Naoki (Nagasaki, JP); Hokao; Nobuhiro (Nagasaki, JP). PCB decomposition reactor. U.S. Patent 6,322,761, Nov 27, 2001.
122. Supercritical Water Oxidation Meeting. Achievements and Challengers in Commercial Applications. Aug 14–15, 2001, Arlington, VA.
123. Ross DS, Jayaweera IS, Nguyen L, Hum GP, Haag WR. Environmentally acceptable waste disposal by conversion of hydrothermally labile compounds. U.S. Patent No. 5, 409, 617, April 25, 1995.
124. Ross DS, Jayaweera IS, Lief R. Method for hot and supercritical water oxidation of material with addition of specific reactants. U.S. Patent 5, 837,149, Nov 17, 1998.
125. Mitsubishi Heavy Industries. PCB hydrothermal decomposition. Marketing Brochure, 2001.
126. Crooker PJ, Ahuluwalia KS, Fan Z. Operating results from supercritical water oxidation. Ind Eng Chem Res 2000; 39:4865–4870.

127. Kritzer P, Dinjus E. An assessment of supercritical water oxidation (SCWO) existing problems, possible solutions and new reactor concepts. Chem Eng J 2001; 83:207–214.
128. Schmieder H, Abeln J. Supercritical water oxidation state of the art. Chem Eng Technol 1999; 22:903–908.
129. Aki SNVK, Abraham MA. An economical evaluation of catalytic supercritical water oxidation: comparison with alternative waste treatment technologies. Environ Prog 1998; 17(4):246–255.

4

Fenton and Modified Fenton Methods for Pollutant Degradation

Matthew A. Tarr
University of New Orleans, New Orleans, Louisiana, U.S.A.

I. INTRODUCTION

The biorefractory nature of many persistent organic pollutants has resulted in efforts to devise chemical and physical methods of degrading pollutants in waste streams or in contaminated sites such as soils, sediments, and groundwater. Fenton reagent has been applied to the degradation of a wide range of contaminants, predominantly persistent organic pollutants. The primary benefits of the Fenton reagent are its ability to convert a broad range of pollutants to harmless or biodegradable products, its benign nature (residual reagents do not pose an environmental threat), and the relatively low cost of the reagents. The major drawbacks to utilization of Fenton reagent are interferences from nonpollutant species, difficulty in application to the subsurface, generation of excessive or explosive heat under aggressive conditions, and wasted reagent costs due to inefficient application or inefficient pollutant degradation in the subsurface. This chapter will detail the fundamental chemistry of Fenton reagent, discuss the kinetics and mechanisms of pollutant degradation, illustrate the advantages and disadvantages of its application to subsurface remediation, and evaluate recent advances that may improve the utility and efficiency of Fenton reagent for pollutant degradation and subsurface remediation.

The term Fenton reagent refers to aqueous mixtures of Fe(II) and hydrogen peroxide. The Fenton reagent was first reported by Fenton [1,2] in 1876. Although Fenton did not elucidate the mechanism of the reaction named after him, subsequent research has indicated the following net reaction as the predominant process:

$$Fe^{2+} + H_2O_2 \rightarrow Fe^{3+} + HO^{\bullet} + OH^- \qquad (1)$$

where Fe^{2+} and Fe^{3+} represent the hydrated species, $Fe(H_2O)_6{}^{2+}$ and $Fe(H_2O)_6{}^{3+}$, respectively.* Reaction (1) is often referred to as the Fenton reaction, although many other reactions occur in Fenton systems. The primary utility of the Fenton reagent in the degradation of pollutants is the formation of hydroxyl radical. Hydroxyl radical is a very strong, nonselective oxidant capable of degrading a wide array of pollutants. Numerous studies have addressed the applicability of Fenton reagent to pollutant degradation and remediation.

Although formation of hydroxyl radical is a key step in the Fenton reagent, other important reactions also occur. In fact, the overall process is dramatically affected by the conditions under which the reaction occurs. Additional important reactions occurring in aqueous mixtures of iron and hydrogen peroxide include the following:

$$Fe^{3+} + H_2O_2 \rightarrow Fe^{2+} + H^+ + HO_2^{\bullet} \qquad (2)$$

$$Fe^{3+} + HO_2^{\bullet} \rightarrow Fe^{2+} + H^+ + O_2 \qquad (3)$$

$$Fe^{2+} + HO_2 \cong 6Fe^{3+} + HO_2^- \qquad (4)$$

$$Fe^{2+} + HO^{\bullet} \rightarrow Fe^{3+} + HO^- \qquad (5)$$

$$H_2O_2 + HO^{\bullet} \rightarrow H_2O + HO_2^{\bullet} \qquad (6)$$

Reactions (2) and (3) indicate processes that regenerate Fe^{2+} in the catalytic cycle. As long as peroxide is available in the system, the iron species continually cycle between Fe^{2+} and Fe^{3+}, unless additional reactions result in formation of insoluble iron oxides and hydroxides. The rate of formation of hydroxyl radical can be expressed as:

$$\text{Rate} = k_1 [Fe^{2+}][H_2O_2] \qquad (7)$$

The concentration of Fe^{2+}, however, is governed by all of the processes involving iron (reactions (1) through (6)) as well as other reactions that may control the concentration of free radical species in the system. Furthermore, the second-order rate constant k_1 is dependent on the coordination chemistry

---

* Throughout this chapter, waters of coordination are omitted for clarity. However, the reader should be aware that any open coordination site on iron will be occupied by water. Iron typically has six coordination sites.

of the iron. For example, Fe^{2+} in pure aqueous solution may exist as the hexaaqua species, which has a rate constant different from that of iron bound by other ligands such as phosphate. Therefore, the presence of both inorganic and organic iron ligands in natural systems or waste streams can have a dramatic influence on the Fenton reaction. Not only can the rate of reaction (1) be influenced by other species present, but the lifetime of free radicals can also be altered, resulting in changes in the Fe^{2+} concentration, again influencing hydroxyl radical formation rate. Additional species can also affect the efficiency of hydroxyl radical formation by providing competing sinks for hydrogen peroxide. For example, manganese reacts with hydrogen peroxide to form products that do not include hydroxyl radical. Such reactions represent a waste of reagent, ultimately resulting in higher costs or ineffective pollutant degradation.

Perhaps the biggest limitation of Fenton reagent for subsurface remediation is the difficulty of delivering the reagent to the pollutant. In addition to physical limitations in reagent injection efficiency, there are also limitations on a molecular scale. Nonpollutant species act as hydroxyl radical scavengers, consuming the radical before it can reach the pollutant. Furthermore, hydrophobic pollutants typically sorb to nonpolar sites within the soil. These sites are not easily accessed by polar reagents such as iron and peroxide and, therefore, pollutant molecules trapped in inaccessible hydrophobic sites are likely to remain undegraded even under aggressive remediation conditions. A thorough understanding of how hydroxyl radical is formed, its reactions with nonpollutant species, and the interactions between pollutants and nonpollutant species is necessary in order to effectively design pollutant degradation protocols using Fenton reagent.

II. FENTON REAGENT MECHANISMS AND KINETICS

A. Fenton Reaction

The second-order rate constant k_1 for reaction (1) has been reported as 76 $M^{-1}\ sec^{-1}$ [3] (no detailed conditions provided) and as 41.4 $M^{-1}\ sec^{-1}$ [4] in the presence of 0.1 M $HClO_4$. The fundamental mechanism for reaction (1) has been proposed as [4]:

$$Fe^{2+} + HOOH \rightarrow Fe(OH)^{2+} + HO\cdot \tag{8}$$

$$Fe(OH)^{2+} \rightarrow Fe^{3+} + OH^- \tag{9}$$

with reaction (8) being rate-limiting. If excess Fe^{2+} is present in acidic solution, an additional reaction (reaction (11)) results in stoichiometric conversion of hydrogen peroxide to water, representing a two-electron reduction of the peroxide:

$$2Fe^{2+} + H_2O_2 + 2H^+ \rightarrow 2Fe^{3+} + 2H_2O \quad (10)$$

with mechanistic steps of

$$Fe^{2+} + HOOH \rightarrow Fe(OH)^{2+} + HO^{\bullet}$$

$$Fe^{2+} + HO^{\bullet} \rightarrow Fe(OH)^{2+} \quad (11)$$

$$Fe(OH)^{2+} + H^+ \rightarrow Fe^{3+} + H_2O \quad (12)$$

If hydrogen peroxide is available in excess in a Fenton system, or if additional Fe^{3+} is present, the overall reaction yields greater O_2 formation [4] by favoring reaction (3).

B. Scavengers

The presence of hydroxyl radical scavengers also plays a key role in the overall fate of the peroxide. Scavenging of hydroxyl radical minimizes the occurrence of reactions (5), (6), and (11). Organic scavengers generally react with hydroxyl radical in one of the two following types of reactions:

$$HO^{\bullet} + RH \rightarrow H_2O + R^{\bullet} \quad (13)$$

$$HO^{\bullet} + R \rightarrow ROH^{\bullet} \quad (14)$$

Reaction (13) is typical of aliphatics and alcohols, whereas reaction (14) is common for double bonds, especially in conjugated and aromatic systems. To form the final products, the radicals R· and ROH· undergo additional reactions. Free radical scavengers are a very important component of Fenton systems, and their importance in pollutant degradation will be discussed further in Secs. III and IV.

C. Haber–Weiss Reaction

Although Fenton did not observe hydroxyl radical-mediated reactions for mixtures of Fe^{3+} and hydrogen peroxide, more recent work has illustrated that such systems can produce hydroxyl radical. Haber and Weiss [5] originally proposed a free radical mechanism for the Fe^{3+}-catalyzed decomposition of hydrogen peroxide. These reactions include [3]:

$$Fe^{3+} + H_2O_2 \rightarrow Fe^{2+} + H^+ + HO_2^{\bullet}$$
$$Fe^{2+} + H_2O_2 \rightarrow Fe^{3+} + HO^{\bullet} + OH^-$$
$$H_2O_2 + HO^{\bullet} \rightarrow H_2O + HO_2^{\bullet}$$

$$Fe^{3+} + HO_2\cdot \rightarrow Fe^{2+} + H^+ + O_2$$
$$Fe^{2+} + HO_2\cdot \rightarrow Fe^{3+} + HO_2^-$$

The overall sum of these reactions, in the absence of radical scavengers, is the Fe^{3+}-catalyzed decomposition of 2 mol of hydrogen peroxide to 2 mol of water and 1 mol of diatomic oxygen. Although the treatment of pollutants with Fe^{3+}/H_2O_2 mixtures has been referred to as the Haber–Weiss process, it is clear that the same set of reactions is involved in both the Haber–Weiss process and the Fenton process. The hydroperoxyl radical ($HO_2\cdot$) formed in reactions (2) and (6) is a good reducing agent and, under some circumstances, it may reduce pollutant species. Studies suggesting this process will be discussed in Sec. V.

Not only are the series of reactions in iron/peroxide systems complex, but there are additional equilibrium processes that may effect the overall reaction kinetics and mechanisms. Some pertinent equilibrium steps are given below:

$$Fe^{3+} + H_2O_2 \rightleftharpoons FeOOH^{2+} + H^+ \qquad (15)$$
$$FeOOH^{2+} \rightarrow Fe^{2+} + HO_2\cdot \qquad (16)$$
$$HO_2\cdot \rightleftharpoons H^+ + O_2^{-}\cdot \qquad (17)$$
$$H_2O_2 \rightleftharpoons H^+ + HOO^- \qquad (18)$$
$$Fe^{2+} + nL^{-m} \rightleftharpoons (FeL_n)^{2-m\times n} \qquad (19)$$
$$Fe^{3+} + nL^{-m} \rightleftharpoons (FeL_n)^{3-m\times n} \qquad (20)$$

where L represents a ligand.

D. Iron Ligands and Coordination

Reactions (19) and (20) represent the coordination of Fe^{2+} and Fe^{3+} species. Common ligands may include OH^-, Cl^-, HCO_3^-, CO_3^{2-}, PO_4^{3-}, NO_2^-, NO_3^-, and SO_4^{2-}. Coordination by anionic ligands makes oxidation of Fe^{2+} easier, yielding an increased rate constant for reaction (1) [6]. Conversely, Fe^{3+}-coordinated by anions may be more difficult to reduce to the divalent species. Such changes in oxidation/reduction potentials for Fe^{2+} and Fe^{3+} not only alter the rate constant for reaction (1), but also change the relative concentrations of the two iron species, consequently changing the rates of several reactions in iron/peroxide systems. Some anions form insoluble salts with iron. In these cases, the precipitation of iron may severely decrease or eliminate its participation in Fenton reactions. Iron hydroxides and oxyhydroxides are likely to precipitate under higher pH conditions. As a result, Fenton processes generally must be carried out at pH values well below 7, unless iron-solubilizing agents (chelators) are added.

Although the ligands in reactions (19) and (20) are represented as negative ions, they could also be neutrals such as water. Furthermore, multidentate ligands are likely in natural systems or in industrial waste streams in which metal chelators are present. The presence of organic ligands has several important implications for Fenton chemistry: (a) the chelated iron species will have a kinetic behavior different from that of pure aqueous iron; (b) the distribution and cycling between Fe^{2+} and Fe^{3+} states will vary with different ligands; (c) the solubility of iron, especially at high pH values, can be dramatically increased through chelation; (d) oxidizable ligands bound directly to iron (the site of hydroxyl radical formation) may be more likely to react with the radicals than other species, including pollutants; and (e) some iron complexes form alternate oxidants other than hydroxyl radical.

A number of reports have evaluated the Fenton reaction as a function of iron ligands. Among ligands that substantially or completely inhibited Fenton reactions are phosphate [7,8], desferal, diethylenetriamine pentaacetic acid (DTPA), ethylenediamine di(*o*-hydroxyphenylacetic acid) (EHPG), and phytate [9]. These are all strongly coordinating ligands or chelators. Graf et al. [9] have proposed that both reactions (1) and (3) require either a free iron coordination site or an easily displaced ligand (e.g., H_2O) at one iron coordination site. Further evidence to support this hypothesis has been presented [10].

The activity of iron in the Fenton process varies as a function of its speciation. As mentioned above, some ligands inhibit the Fenton process, whereas others may enhance it. Some confusion exists in the literature regarding the enhancement of the Fenton process by iron chelators. For example, several studies have evaluated the effect of iron chelators on the Fenton reaction in the presence of phosphate buffers [11–13]. Phosphate is a strong iron-binding agent and results in the formation of insoluble iron phosphates. The addition of iron chelators resolubilizes iron, resulting in an increase in the occurrence of reaction (1) and other reactions necessary in the catalytic Fenton process. However, such results do not necessarily indicate that the iron–chelator complex is more efficient than pure aqueous iron, $Fe(H_2O)_6^{2+}$ [9,14]. Another complicating factor is the hydroxyl radical-scavenging ability of the chelator. A good scavenger may appear to have a lower production rate of hydroxyl radical due to rapid trapping of the radical by the chelator.

Croft et al. [15] reported significant increases in k_1 when iron was chelated by DTPA, ethylenediaminetetraacetic acid (EDTA), nitrilotriacetic acid (NTA), and several aminophosphonic acids. Compared to aqueous Fe^{2+}, increases in k_1 from 1000-fold to 50,000-fold were indicated. They also note that DTPA inhibits reaction (3). Graf et al. [10] suggest that reaction (3) is completely inhibited by DTPA, EHPG, phytate, and desferal. These two studies are not entirely consistent in their findings, possibly due to the use of aliphatic radical scavengers by Croft et al. These scavengers can form R· upon reaction with hydroxyl radical, possibly resulting in regeneration of Fe^{2+} by:

$$Fe^{3+} + R\cdot \rightarrow Fe^{2+} + R^{+} \tag{21}$$

This point indicates the important role that additional reducing species, whether free radicals or otherwise, have on the overall Fenton process.

Tannic acid has been reported to inhibit the formation of hydroxyl radical by chelation of Fe^{2+} [16]. Because tannic acid is a plant-derived material, it and other natural polyphenols may be important for subsurface applications of Fenton chemistry. A study of 50 different iron chelators assessed the affect of each chelator on the Fenton process initiated with Fe^{3+} and hydrogen peroxide [17]. Among the nine classes of chelators tested, results indicated that the chelators ranged from inactive to highly active in terms of hydroxyl radical formation.

E. Photo–Fenton

In the presence of light, additional reactions that produce hydroxyl radical or increase the production rate of hydroxyl radical can occur [18]:

$$H_2O_2 + h\nu \rightarrow 2HO\cdot \tag{22}$$

$$Fe(OH)^{2+} + h\nu \rightarrow Fe^{2+} + HO\cdot \tag{23}$$

$$Fe^{3+}(L^{-}) + h\nu \rightarrow Fe^{2+} + L\cdot \tag{24}$$

where L is an organic ligand. Often called the photo–Fenton system, irradiation of iron-peroxide solutions can result in more effective pollutant degradation than dark Fenton systems. Although reactions (22) and (23) can be important, the photochemical cycling of Fe^{3+} back to Fe^{2+} is perhaps the most important aspect of the photo–Fenton process. Recalling that the rate of hydroxyl radical formation is governed by the Fe^{2+} concentration (Eq. (7)), it is clear that any additional route for reducing iron to the 2+ state will enhance the production rate of hydroxyl radical.

Photochemical reduction of Fe^{3+} has been observed in many systems, including raindrops [19], water and wastewater [20–24], and natural surface waters [25]. Concentrations of Fe^{3+} lower than 10^{-8} M are reportedly capable of catalyzing the photo–Fenton reaction [26].

F. Ferryl Ion or Other High-Valent Iron Species

Although applications of Fenton chemistry to pollutant degradation most likely involve hydroxyl radical as the primary oxidant, an alternative mechanism that does not involve hydroxyl radical has also been proposed [4]. In this mechanism, instead of reaction (8), the following rate-limiting step occurs:

$$HOOH + Fe^{2+} \rightarrow FeO^{2+} + H_2O \qquad (25)$$

The importance of FeO^{2+} (the ferryl ion) as an alternate oxidant, instead of hydroxyl radical, has been extensively debated. Much of the debate exists due to difficulty in definitively detecting the ferryl ion and the hydroxyl radical. Because of their low concentrations and short lifetimes, these species are detected through indirect methods.

Fe(IV) has been observed for iron porphyrins reacting with *tert*-butyl hydroperoxide in toluene solution [27], and Pignatello et al. [18] have reported the existence of ferryl species in photo–Fenton systems under certain conditions. Additional evidence using ^{17}O-labeled H_2O_2 and water indicates that all of the hydroxyl radical oxygen molecules originate from H_2O_2, indicating that any ferryl ion formed must have its oxygen derived from H_2O_2 [28]. Shen et al. [29] observed chemiluminescence upon the addition of H_2O_2 to aqueous solutions of Fe(II). They postulated that an excited ferryl species may be the source of the chemiluminescence, but also suggested that the ferryl species resulted from the reaction of initially formed HO· with Fe^{3+}. Bossmann et al. [30] observed degradation products of 2,4-dimethylaniline in Fenton systems, photo–Fenton systems, and with hydroxyl radical generated by the photolysis of H_2O_2 in the absence of iron. Because different products were observed in the absence and presence of iron, the authors concluded that a different mechanism must occur in the iron-containing system. An Fe^{4+} species was suggested to be the electron transfer agent that oxidized 2,4-dimethylaniline to 2,4-dimethylphenol, whereas hydroxyl radical is not expected to form this product. Although this report provides evidence for a nonhydroxyl radical mechanism for the oxidation of 2,4-dimethylaniline, the nature of the oxidant was still not directly elucidated. One important factor not considered in this study is the possible importance of direct coordination of the amine to the iron. Such coordination would likely alter the kinetics and mechanisms of oxidation.

In contrast to the results of Bossmann et al., Lindsey and Tarr [31,32] observed equivalent rate constants for polycyclic aromatic hydrocarbon (PAH) degradation with Fenton systems as had been previously observed for PAH reaction with hydroxyl radicals as generated by pulse radiolysis techniques [33]. Such kinetic agreement suggests, but does not confirm, equivalent mechanisms. PAHs, unlike 2,4-dimethylaniline, are not expected to directly coordinate iron in aqueous solutions.

Whereas only a few examples have been discussed here, it is obvious that the exact kinetics and mechanisms for Fenton oxidations are highly dependent on conditions ($[Fe^{2+}]$, $[Fe^{3+}]$, $[H_2O_2]$, iron chelators, nature, and concentration of the species being oxidized, etc.) Consequently, generalized kinetic and mechanistic predictions cannot be made to Fenton systems; rather, a strong knowledge of the system under study is needed before kinetic models can be applied. Nevertheless, in most practical applications, hydroxyl radical appears to be the major oxidative species [18].

G. Other Metals

A number of other transition metal ions can also participate in Fenton-type cycles to produce hydroxyl radical. Examples include Cu^+, VO^{2+}, Ti^{3+}, Cr^{2+}, and Co^{2+} [4], although other reducing metals are also active in the formation of hydroxyl radical from peroxide.

III. HYDROXYL RADICAL REACTIONS WITH ORGANIC COMPOUNDS

Although there is still debate as to whether hydroxyl radicals or ferryl species are the key oxidants in Fenton systems, most literature reports on the mechanisms of degradation of organic compounds invoke the hydroxyl radical. Based on the reports discussed above, it seems likely that hydroxyl radical is a major oxidant during Fenton degradations. Although ferryl ions or other highly oxidized forms of iron may occur, either to a limited extent or more abundantly under specific conditions, this section will deal with documented reaction pathways and kinetics for hydroxyl radical or species assumed to be hydroxyl radical. The reader should keep in mind that ferryl pathways may need to be considered under certain conditions.

A. General Mechanisms

For the reaction of hydroxyl radical with organic species, there are three common reaction pathways: (a) hydroxyl radical addition to unsaturated systems (e.g., double bonds), (b) hydrogen abstraction (typically from alkyl or hydroxyl groups), and (c) direct electron transfer. These generic mechanisms are illustrated in Figure 1. For nonradical reactants, all three mechanisms result in initial products that are radicals. Subsequent reactions follow to yield nonradical products. Additional reactants are necessary to complete these subsequent reactions. Common reactants include Fe^{2+}, Fe^{3+}, O_2, H_2O, H^+, $HO\cdot$, other metals, other organics, and other radicals present in the system. Dimerization can also occur if the initially formed radical species reacts with

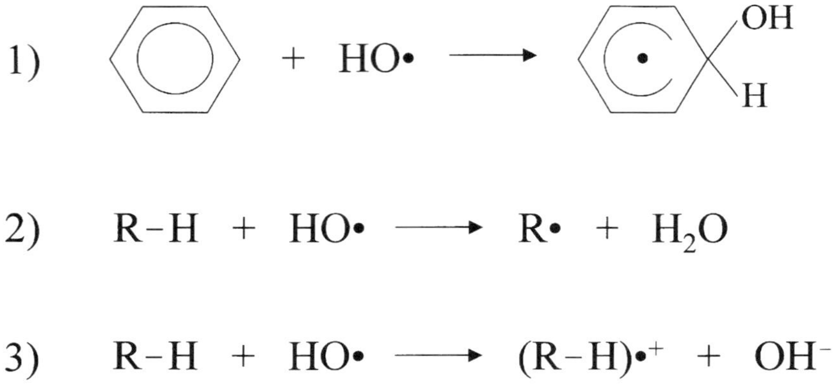

Figure 1 Generic mechanisms for hydroxyl radical reaction with organic compounds.

another identical radical. Typical reaction products include oxygenated products (alcohols, aldehydes, oxyacids, ketones, etc.), ring opening products, and dimers. In some cases, complete mineralization yields carbon dioxide as a final product.

B. Mechanistic Examples

The reaction of hydroxyl radical with unsaturated compounds often results in the addition of the radical to the double bond to produce an unsaturated alcohol. This reaction is quite common for aromatic systems. A number of proposed reaction pathways are depicted in Figure 2. These reactions can yield several products including hydroxylated species, dimers, and polyoxygenated products. Stable products, such as phenols and dihydroxybenzenes, can react further to yield additional degradation products. Ring opening products are the typical result of subsequent oxidation. The further oxidation of dihydroxybenzenes by multiple hydroxyl radical attack is illustrated in Figure 3. Whereas Figures 2 and 3 illustrate some of the possible reactions occurring in Fenton systems, other pathways are also possible.

The reactivity of initial products will have a dramatic effect on their buildup during oxidation. Those that react rapidly (e.g., benzoquinones) will be seen only in minor amounts, whereas less reactive initial or subsequent products will tend to accumulate. One weakness of in situ remediation applications, using Fenton or other biochemical or chemical degradation methods, is a failure to monitor potentially harmful degradation products.

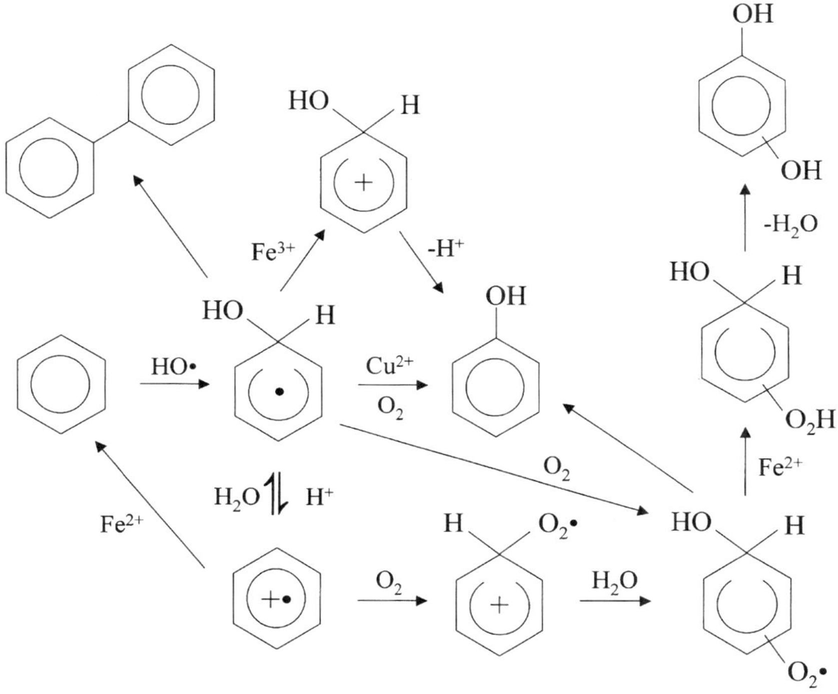

Figure 2 Proposed reaction pathways for hydroxyl radical reaction with benzene. (From Refs. 3, 4, 105, and 106.)

Substituted aromatics can react either by addition of hydroxyl radical to the aromatic ring structure or by reaction with the ring substituents. For example, phenol can react through the addition of hydroxyl radical as depicted for aromatics in Figure 2, or it can undergo hydrogen abstraction from the alcohol, followed by subsequent steps. The relative importance of these pathways will depend on the rate constants for each. Because the aromatic structure generally has a higher rate constant than alcohols, attack by HO· addition to the ring is more prevalent. Similar comparisons can be made for other substituents.

The final products of Fenton oxidation are dependent on several factors. The amount of hydroxyl radical is clearly important. In addition, the amount of hydroxyl radical scavengers and the amount of pollutant present will affect the array of final products. The concentrations of Fe^{2+} and Fe^{3+}, as well as those of other metals, also affect product distributions because these species are involved in some degradation pathways. The presence of

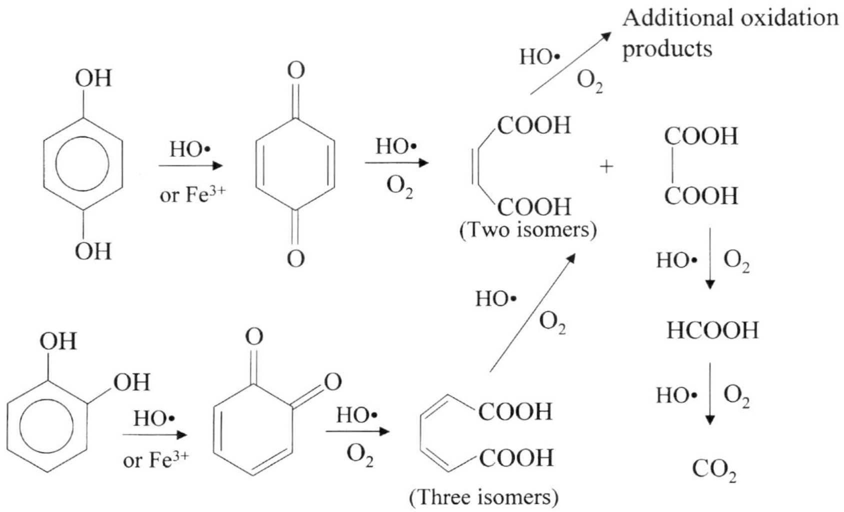

Figure 3 Proposed reaction pathways for hydroxyl radical reaction with hydroxylated aromatics. (From Refs. 3, 4, and 107.)

other radicals, including those produced from the Fenton cycle (e.g., $HO_2\cdot$), can provide alternate reaction pathways and thereby affect the distribution of products. Under certain conditions (e.g., high peroxide concentration), substantial amounts of reducing species ($HO_2\cdot$, $O_2^{-}\cdot$) are present in high-enough quantity that reductive pathways become important [34,35].

If hydroxyl radical availability is high enough, the pollutants present will be degraded to CO_2 and other mineralized products. However, if a species of low reactivity is produced, this species may become a major degradation product, and complete mineralization will not be observed. For example, oxalic acid has a relatively low rate constant for reaction with hydroxyl radical ($1.4 \times 10^6\ M^{-1}\ sec^{-1}$, about three to four orders of magnitude lower than aromatics). Such highly oxygenated compounds, however, are likely to be biodegradable. Consequently, such species can be acceptable end points for remediation or waste stream treatment.

C. Kinetics

Hydroxyl radical reacts rapidly with a wide array of species. Typically, reactions follow second-order kinetics. In many cases, aqueous reactions with hydroxyl radical are diffusion-limited. Two comprehensive sources of second-order rate constants for reactions with hydroxyl radical are available

[33,36]. Most compounds that have been studied have rate constants between 10^7 and 10^{10} M^{-1} sec^{-1}, but a few compounds with lower rate constants have been observed. These few examples include compounds with strong electron-withdrawing groups or without C-H bonds [4]. Electron-donating groups are believed to increase reactivity with hydroxyl radical [4]. In general, alkenes and aromatics react rapidly with hydroxyl radical ($k = 10^9$–10^{10} M^{-1} sec^{-1}), whereas alkanes react somewhat more slowly ($k = 10^7$–10^9 M^{-1} sec^{-1}). A few examples of rate constants are given in Table 1.

D. Other Important Factors

As has already been made clear, there are many factors that affect the kinetics and mechanisms of Fenton-based degradation. Rate constants, hydroxyl radical concentration, pollutant concentration, and the presence of other species are all among the factors previously discussed. Additional important factors include those related to solubility and homogeneity.

Iron solubility is an obvious factor in Fenton oxidation because the rate of hydroxyl radical formation is directly proportional to $[Fe^{2+}]$. At elevated pH values, iron hydroxides and oxides form and precipitate, causing a dramatic decrease in hydroxyl radical formation rate. Iron chelators can be used to offset this factor. A related issue is the rate of Fe^{3+} reduction to Fe^{2+}, which, if insufficient, can result in Fe^{2+} concentrations that are too low to

Table 1 Rate Constants for Reaction with Hydroxyl Radical for Selected Compounds

Compound	k (M^{-1} sec^{-1})	Reference
Urea	7.9×10^5	36
Oxalic acid	1.4×10^6	36
Benzene	7.8×10^9	33
Phenanthrene	1.9×10^{10}	32
Pyrene	1.5×10^{10}	32
2-Methyl-2-(methylthio)propanal, *O*-[(methylamino)carbonyl]oxime (Aldicarb)	8.1×10^9	36
Aniline	1.4×10^{10}	33
Hexachlorocyclohexane (Lindane)	9.1×10^8	36
Tetrachloroethylene (PCE)	2.6×10^9	33
Dichloromethane	6×10^7	33
Carbon tetrachloride	Unreactive	104
Ethanol	2.2×10^9	36

produce hydroxyl radical at sufficient rates. The presence of reducing species (e.g., $HO_2\cdot$ and citrate) and the state of the Fe^{3+} are both important. Iron(III) bound in very stable complexes may be resistant to reduction.

Homogeneity is a very important factor that has not been sufficiently discussed in the literature. Even in aqueous-based systems, hydrophobic pollutants may be isolated from iron that is coordinated by hydrophilic ligands. Because hydroxyl radical is formed at the site of the iron, any factor that causes the separation of iron and the pollutant will decrease the efficiency of pollutant degradation. Hydroxyl radical concentration will be highest near iron species, and its concentration will decrease with increasing distance from the iron site as the radical is scavenged. Consequently, any species that has a reduced concentration near the iron ions will experience a lower hydroxyl radical concentration than species that are randomly distributed throughout the solution. Furthermore, species that associate with iron will be more likely to scavenge hydroxyl radical, preventing its reaction with pollutants.

This effect has been illustrated for dissolved natural organic matter (NOM) [31,32]. Because iron is chelated in hydrophilic regions and nonpolar pollutants are often bound to hydrophobic regions of the NOM, the likelihood of pollutant oxidation is substantially reduced.

Soil systems present even greater challenges. In these systems, hydrophobic pollutants can be sorbed to soil in sites that have low accessibility for the hydroxyl radical. Dissolved iron is not likely to enter into hydrophobic soil pores and, therefore, pollutants in these sites can be extremely difficult to degrade.

Studies on the degradation of polychlorinated biphenyls (PCBs) have indicated that dissolved PCBs are much more readily degraded by hydroxyl radical than are PCBs sorbed to sand [37]. PCBs sorbed to glass were also difficult to degrade [38]. These studies suggest that hydroxyl radical is unlikely to cross a phase boundary in order to react with the solid or sorbed pollutant.

The inability of hydroxyl radical to reach sorbed or otherwise sequestered pollutants is one of the major drawbacks to the in situ application of Fenton degradation methods. Several reports have suggested that aggressive conditions (i.e., high H_2O_2 concentration) can allow for direct degradation of sorbed species [39–41].

IV. APPLICATIONS

A. Typical In Situ Applications

Fenton oxidation has been applied at many sites for in situ remediation of contaminated soils. Typical applications have involved the degradation of light nonaqueous phase liquids (LNAPLs) and dense nonaqueous phase liquids (DNAPLs), although some work has been done on larger molecules (e.g., PAHs, PCBs, and energetic compounds). Application in soil systems requires care and should only be carried out by experienced individuals. Fenton treatment can liberate both heat and gas, and therefore can present hazards when carried out in the subsurface, especially when high concentrations of peroxide are used.

Research conducted at Washington State University, as well as in situ applications by commercial entities, has indicated that stabilization of hydrogen peroxide is necessary for effective subsurface injection [39]. Without stabilization, added peroxide decomposes rapidly through interaction with iron oxyhydroxides, manganese oxyhydroxides, dissolved metals, and enzymes (e.g., peroxidase and catalase). Some of these peroxide decay pathways involve nonhydroxyl radical-forming mechanisms, and therefore are especially detrimental to Fenton oxidation systems.

Monobasic potassium phosphate was found to be an effective stabilizer to prevent premature decomposition of hydrogen peroxide in soil systems [39]. In this laboratory study, a disturbed silt loam was collected from a diesel-contaminated maintenance and equipment storage yard. Unstabilized 15 M hydrogen peroxide decomposed completely within the first 2–4 cm of a column packed with this soil. With 150 mM KH_2PO_4 added to the peroxide solution, the penetration depth of the peroxide was increased threefold. The protective effect of the stabilizer was diminished through the removal of phosphate from the solution as it passed through the soil column. Removal was believed to be via ion exchange or another fixation process. Hexadecane oxidation in this soil was improved by the addition of >30 mM KH_2PO_4. Furthermore, degradation of hexadecane was spread over a longer column depth when the stabilizer was added, whereas only the hexadecane in the top 2 cm was degraded in the absence of the stabilizer. These results further demonstrate that observation of peroxide decomposition rates is not necessarily a good indicator of pollutant degradation rate because nonhydroxyl radical-forming pathways may contribute substantially to peroxide decomposition.

In addition to in situ soil applications, Fenton reagent has also been studied for treatment of industrial waste streams. For example, the treatment of residual Kraft black liquor from the pulp-and-paper industry has been shown to be effective [42]. Near-complete degradation of lignin (95–100%) and decolorization were achieved under optimized conditions. Basic oxygen furnace slag was evaluated as a source of iron for degradation of 2-chlorophenol in industrial wastewater [43], and favorable results were achieved.

Treatment of olive mill wastewater has also been carried out with Fenton reagent [44]. This study also used Fenton treatment in conjunction with subsequent aerobic biotreatment. The Fenton treatment moderately reduced chemical oxygen demand (COD) and had a significant impact on the polyphenolic content of the wastewater.

Reports of industrial applications to waste stream treatment are less abundant than in situ soil applications because many of these applications are developed in commercial applications and are not disclosed in the literature.

B. Iron Minerals as Fenton Catalysts

Aqueous Fe^{2+} and many of its coordination complexes serve as excellent catalysts for the formation of hydroxyl radical from hydrogen peroxide. Iron oxyhydroxides have also been found to catalyze the formation of hydroxyl radical [45], although at a much slower rate than dissolved iron. Consequently, a number of researchers have investigated the potential for using soil minerals as catalyst to avoid the need for the addition of soluble iron to the system.

In laboratory studies, Ravikumar and Gurol [46] monitored the Fenton degradation of pentachlorophenol (PCP) and trichloroethylene (TCE) from sand. In both column and batch studies, they observed the degradation of PCP and TCE with the addition of hydrogen peroxide only. They concluded that iron naturally present in the sand was an effective catalyst for the formation of hydroxyl radical from the added peroxide. However, addition of soluble ferrous salts caused a more rapid degradation of the pollutants, indicating that either insufficient iron was present in the sand or the nature of the iron in the sand made it a poor catalyst.

Without added iron, 7 mol of hydrogen peroxide were needed per mole of PCP oxidized, whereas the peroxide demand for TCE degradation was 3.7 mol of peroxide per mole of TCE oxidized. With Fe^{2+}, these peroxide demands decreased to 4.3 and 2.1 for PCP and TCE, respectively. Chloride release was ~80–90% of the theoretical value for each compound and was only slightly higher with added iron. This early study demonstrated the potential to perform in situ Fenton oxidation without added iron. However, the added complexity of most real systems increases the need for added soluble iron.

Although the authors in this sand study reported the iron content of the sand, there remains a possibility that noniron species were acting as catalysts. In fact, a study of coal (anthracite) oxidation by hydrogen peroxide indicates that formation of hydroxyl radical occurs even after the removal of iron from the coal [47]. This study suggested that surface sites on the coal may have acted as catalytic centers for hydroxyl radical formation. Whereas coal is quite distinct from the sand system that Ravikumar and Gurol used, the work on coal raises the possibility that noniron soil minerals or organic matter could also serve as catalysts for hydroxyl radical formation.

In other laboratory studies, Watts et al. [48] found that iron oxyhydroxides yielded a better stoichiometric efficiency (mole of H_2O_2 consumed per mole of pollutant degraded) than soluble iron. This observation may be related to inefficiencies in hydroxyl radical utilization if it is formed too rapidly, as in the system with dissolved Fe^{2+}.

Watts et al. [49] have also studied the hydrogen peroxide-induced degradation of several chlorinated aromatics sorbed to hematite (α-Fe_2O_3) with no other source of iron. 1,3,5-Trichlorobenzene, 1,2,3,4-tetrachlorobenzene, pentachlorobenzene, and hexachlorobenzene were all degraded by sufficient concentrations of added peroxide (0.1–5%). Hydroxyl radical concentration was found to be independent of hydrogen peroxide concentration, indicating that the catalytic sites were peroxide-saturated at the concentrations used. As with other reports by Watts et al., at high peroxide concentration, degradation rates exceeded desorption rates. This result has been interpreted as evidence of direct oxidation of sorbed species, although alternate explanations are possible.

Goethite (α-FeOOH) has also been studied as a potential Fenton-like catalyst [50]. Hydrogen peroxide decay over goethite exhibited second-order kinetics according to:

$$-\mathrm{d}[H_2O_2]/\mathrm{d}t = k[\mathrm{FeOOH}][H_2O_2]; \quad k = 0.0031\,\mathrm{M}^{-1}\mathrm{sec}^{-1} \tag{26}$$

at pH = 7. The kinetics were determined to be dependent on the intrinsic reaction rates at the oxide surface. In comparison, the rate constant for Fe^{2+} reaction with H_2O_2 (Eq. (1)) is in the range of 40–80 M^{-1} sec^{-1} [3,4]. In additional studies, it was suggested that hydroxyl radical reacted with the pollutant sorbed on the mineral surface [51,52]. Although several reports make this assertion, such conclusions should be scrutinized carefully because some evidence may be indirect. In a separate study, goethite was observed to be a better catalyst than ferrihydrite or semicrystalline iron oxides for the Fenton-based degradation of quinoline [53].

Lepidocrocite (γ-FeOOH) has also been used as a catalyst for Fenton-like reactions [54]. First-order decomposition of hydrogen peroxide was observed in the presence of this catalyst. Peroxide decay at 20 g/L catalyst was found to be pseudo-first-order and pH-dependent, with rate constant values reported from 0.102 hr^{-1} at pH 3.3 to 0.326 hr^{-1} at pH 8.9. In this system benzoic acid degradation was fastest at the low pH value. Under these conditions, acid dissolution of the lepidocrocite was observed to produce

Fe^{2+}. Because peroxide decomposition was slowest under the same conditions at which benzoic acid decomposition was highest, it is important to consider the efficiency of hydroxyl radical formation from peroxide decomposition. With the surface catalyst, either hydroxyl radical is not readily available to benzoic acid and is scavenged by other species, or the mineral-catalyzed decomposition of hydrogen peroxide involves additional, non-hydroxyl radical-forming pathways for peroxide decomposition.

This last point emphasizes a weakness in the Fenton literature pointed out previously. Often, the rate of peroxide decomposition is assumed to be proportional to the hydroxyl radical formation rate. However, nonhydroxyl radical-forming pathways and the formation of radicals that are not accessible to pollutants may significantly decrease the yield of useable hydroxyl radical.

Hematite has been studied as a Fenton catalyst [55,56]. Catalytic decomposition of hydrogen peroxide by hematite was found to be slower than that for ferrihydrite or goethite, although hematite had the highest activity for 2-chlorophenol degradation [55]. The degradation efficiency of 2-chlorophenol was correlated with the inverse sequence of specific area and point-of-zero charge of the Fe oxides. Furthermore, particle size and porosity characteristics played important roles in governing the degradation kinetics. Unlike magnetite, hematite was not observed to catalyze the disproportionation of hydrogen peroxide, which, if it occurs, is a source of inefficiency [56]. The release of free iron was observed in these systems, but the use of ferrozine and desferrioxamine, strong iron chelators, prevented the activity of dissolved iron with respect to peroxide.

Ferrihydrite catalysis of hydroxyl radical formation from peroxide has also shown experimental results consistent with a surface reaction [57]. The yield of hydroxyl radical formation was lower for ferrihydrite than for dissolved iron, resulting in a higher peroxide demand to degrade a given amount of pollutant. As mentioned above, although ferrihydrite exhibited a faster rate of peroxide decomposition than goethite or hematite, the rate of 2-chlorophenol degradation with these catalysts was fastest for hematite [55]. In other studies, quinoline oxidation by peroxide was not observed when ferrihydrite was used as catalyst [53].

Mixed element oxides such as Fe–Si–O, Fe–Mg–O, and Fe–Mg–Si–O complex oxides have also been used as catalysts for Fenton-like oxidations [58]. Some of these catalysts exhibited selectivity for phenol hydroxylation. Whereas these catalysts are unlikely to be useful for in situ applications, they may be valuable for waste stream treatment, especially when a well-defined waste stream with specific pollutants exists.

Other iron species have also been evaluated as potential Fenton catalysts. These include magnetite [59], pyrite [60], and basic oxygen furnace slag [43]. As with other studies, these solids were able to catalyze pollutant degradation in the presence of hydrogen peroxide, but at a slower rate than for dissolved iron. Furthermore, the presence of Fe^{2+} (aq) dissolved from the solids may be the key catalyst in some systems.

Because of the slower kinetics for hydroxyl radical formation with mineral catalysts, most applications of Fenton oxidation for in situ remediation or waste stream treatment involve the addition of dissolved iron(II).

C. Iron Chelators

Iron speciation is a major factor in Fenton chemistry. As previously discussed, iron solubility, redox potentials, and concentrations of Fe^{2+} and Fe^{3+} are all dependent on the ligands that coordinate iron. In order to produce hydroxyl radical, there must be a readily accessible coordination site for H_2O_2 to bind to [9,10]. Very strong iron chelators, therefore, inhibit the formation of hydroxyl radical. Iron ligands can also act as hydroxyl radical scavengers. Because the radical is always formed in close proximity to these ligands, they are more likely to react with hydroxyl radical than pollutants that are not in close proximity to the iron.

In many in situ applications, phosphate or other peroxide stabilizers are added. These species are likely the dominant iron ligands; however, natural ligands and chelating agents may coordinate some of the iron. Such coordination will alter the kinetics of hydroxyl radical formation as well as the dynamics of hydroxyl radical interaction with pollutants.

Several studies have investigated the effect of chelators on the Fenton reaction. Addition of chelators to Fe(III)–H_2O_2 systems allows for effective degradation at near-neutral pH values [17,23,61]. Much of the influence of the iron chelators stems from increased solubility of iron species at higher pH values. These chelators were also effective for combined Fenton bioremediation treatment because they eliminate the need for low pH, which can harm useful bacteria.

Many phosphates have been observed to inactivate iron as a catalyst for the formation of hydroxyl radical [17]. Included in those studied are pyrophosphate, phytate, adenosine 5′-triphosphate, and others. Of the 50 compounds tested in one study, those having the highest activity in terms of 2, 4-dichlorophenoxyacetic acid degradation were: nitrilotriacetic acid, hydroxyetheliminodiacetic acid, picolinic acid, gallic acid, rhodizonic acid, tetrahydroxy-1,4-quinone, and hexaketocyclohexane [17]. Additional studies indicated that several of these iron chelators were effective in Fenton degradation of other pollutants at near-neutral pH, including 2,4,5-trichlorophenoxyacetic acid, atrazine, picloram, baygon, and carbaryl [23].

These iron chelators improved the Fenton oxidation of pollutants by increasing iron solubility, and perhaps also by increasing the rate constant for hydroxyl radical formation from peroxide. However, despite their positive effects, these chelators also acts as hydroxyl radical scavengers. Not only does this scavenging remove hydroxyl radical from potential interaction with pollutants, but it also degrades the chelator. Over time, loss of chelator results in poor iron solubility and slow rates of hydroxyl radical formation. Earlier studies indicated that at pH 7.3, each EDTA–Fe complex was able to produce more than 50 hydroxyl radicals before being degraded [13].

Flavenoids, common plant-derived compounds that are widespread in the environment, have been tested for their effect on the Fenton reaction [62]. These compounds are known hydroxyl radical scavengers. Myricetin, quercetin, catechin, morin, and kaempferol were all found to enhance the formation of hydroxyl radical in Fe(III)–EDTA systems. No change in radical formation was observed with added flavone. In the presence of ascorbate, a good reducing agent that converts Fe^{3+} to Fe^{2+}, the flavenoids had little effect on hydroxyl radical formation. This observation suggests that the flavenoids enhanced hydroxyl radical formation by reducing Fe^{3+} to Fe^{2+}. However, this study utilized a phosphate buffer and a strong iron chelator (EDTA). Therefore, the flavenoids may not have been able to effectively coordinate the iron under these conditions. In the absence of EDTA and phosphate, compounds such as flavenoids could act both as iron chelators and as iron-reducing agents. Unfortunately, insufficient data are available to determine the effects of these and other natural chelators in systems without other additives.

In the presence of adenosine triphosphate, flavenoids were observed to inhibit hydroxyl radical formation [63]. These contrasting results further emphasize that the effect of iron chelators is highly dependent on the other species present. The relative efficiencies of the chelators for hydroxyl radical formation will determine whether the added chelator will have a positive or negative effect on radical formation. Furthermore, increased radical formation is not always correlated with improved pollutant degradation. Catechol groups and 4-keto,5-hydroxyl benzene groups are believed to be important iron coordination sites for flavenoids [63].

Other catechol-containing natural compounds have been implicated as Fenton-enhancing compounds in white rot fungi, brown rot fungi, and mold-mediated degradation of lignin and wood [64,65]. In addition to their potential effect on in situ remediation, these compounds may be of potential benefit for the pulp-and-paper industry.

Tannic acid, a plant polyphenol, has also been studied with respect to its effect on Fenton reaction [16]. Formation of a Fenton-inactive Fe(II)–tannic acid was believed to be the mechanism by which tannic acid inhibits Fenton reaction. Although tannic acid is a good hydroxyl radical scavenger, its scavenging effects appeared to be secondary in this study. Tannic acid is also capable of reducing Fe^{3+} to Fe^{2+}, but because the Fe(II)–tannic acid is Fenton-inactive, the reduction of tannic-bound iron does not contribute to the catalytic cycle. Tannic acid is able to compete with EDTA for complexation of iron. Therefore, in soil systems with significant levels of tannic acid, addition of beneficial iron chelators may not be completely effective.

Numerous studies have evaluated the effects of chelators [8,12,15,66–71]. However, the effects of natural chelators have not been thoroughly studied, especially as related to in situ remediation. Consequently, in situ applications may suffer from decreased efficiency caused by interferences from natural chelators present in the site.

D. Photo–Fenton

As previously discussed, the concentration of Fe^{2+} is an important factor in the rate of hydroxyl radical formation from hydrogen peroxide. Consequently, any process that can speed the reduction of Fe^{3+} to Fe^{2+} will increase the formation rate of hydroxyl radical. UV or visible radiation can play this role by photoreducing iron. However, the photo–Fenton process involves three additional mechanisms that can contribute to pollutant degradation: (a) direct photolysis of H_2O_2 to yield two hydroxyl radicals (Eq. (22)); (b) photolysis of $Fe(OH)^{2+}$ to form hydroxyl radical (Eq. (23)); and (c) degradation of pollutants by direct photolysis (i.e., absorption of a photon by the pollutant molecule followed by decomposition of the photoexcited pollutant molecule).

Photo–Fenton systems have been studied for the degradation of a wide range of pollutants. Chlorophenoxy herbicide degradation by Fenton processes was found to be significantly enhanced by UV radiation [23,72]. Additional compounds that were effectively degraded using photo–Fenton systems include metolachlor and methyl parathion [73], organophosphorous compounds [74], PCBs [75], polychlorinated dibenzo-*p*-dioxins, and dibenzofurans [76]. Dyes found in textile wastewater have also been degraded using photo–Fenton techniques; Orange II (an azo dye) [77] and the reactive dyestuff, R94H [78], were degraded. Whereas Orange II was effectively mineralized, R94H was decolorized, but not mineralized. Interference from polyvinyl alcohol, a common component of textile wastewaters, was cited as interfering with R94H degradation. As discussed before, the presence of hydroxyl radical scavengers can be a major impediment to the degradation of targeted pollutants. In contrast, the photo–Fenton degradation of pentachlorophenol was found to be enhanced by the presence of humic acid [79]. Although humic acid is a good scavenger of hydroxyl radical, the phenolic acid groups of humic acid were believed to play a major role in Fe^{3+} photoreduction, thereby significantly enhancing the PCP degradation.

Lignin, a major structural component of wood, was also degraded by a photo–Fenton technique [80]. Although this report demonstrated the ability of the method to degrade lignin, the study did not elaborate on potential applications in the pulp-and-paper industry. The treatment of kraft mill bleaching effluent by Fenton and photo–Fenton methods has been studied, and was reportedly more effective and more economical than several other techniques [81].

Attempts have been made to immobilize iron in photo–Fenton systems. Iron immobilized in Nafion® was successfully used to degrade 4-chlorophenol [82] and Orange II [83]. In both studies, the authors indicated that the Nafion®-bound iron was resistant to aging or fouling.

Efforts to develop large-scale or prototype photo–Fenton reactors have also been reported [84–86]. As with any photocatalytic method, the ability of the photons to reach the catalyst is a key design issue.

E. Combined Fenton Biodegradation

Biodegradation is often a desirable approach to in situ remediation because it is of relatively low cost and low intervention. However, many microbial systems are limited by the toxicity of pollutants, especially when high concentrations or mixtures of compounds are present. Furthermore, microbial systems may be limited to a single compound or to a group of structurally related compounds. Consequently, diverse mixtures of contaminants can be problematical for biodegradation systems. Finally, biodegradation techniques are relatively slow and, therefore, cannot be used in sites where immediate action is required. Chemical systems do not generally suffer from the above listed limitations, although they may be more costly, require more extensive activity at the site, or may have other limitations. In order to capitalize on the benefits of both biological and chemical systems, it is often beneficial to utilize a chemical pretreatment step, followed by a longer-term biodegradation phase of treatment.

Tetramethyl ammonium hydroxide (TMAH) toxicity, as measured by a Microtox® assay, was dramatically reduced after pretreatment with Fenton reagent. Subsequent biodegradation was successful. Without the Fenton pretreatment, biodegradation was slow and required a substantial acclamation period [35]. The biodegradation of anthracene [87] and benz(*a*)anthracene [88] were improved with Fenton pretreatment. For anthracene biodegradation over 30 days, 30% of the compound was degraded without pretreatment, whereas 90% was degraded with Fenton pretreatment. Benz-

(*a*)anthracene was only 12% biodegraded in 63 days without pretreatment, but was 98% degraded with Fenton pretreatment. Both compounds were converted to intermediate products by Fenton oxidation: anthracene was mainly converted to 9,10-anthraquinone and benz(*a*)anthracene was substantially converted to benz(*a*)-7,12-dione. These intermediate products were readily biodegraded. Tetrachlordibenzo-*p*-dioxin (TCDD) pretreated with Fenton reagent yielded chlorophenols and chlorobenzenes that were readily biodegraded [89].

Although there are reports of improved biodegradation with Fenton pretreatment, there are also reports indicating limitations of Fenton pretreatment. Acid conditions often used for Fenton treatment are usually incompatible with microbial activity, and measures to overcome this incompatibility must be taken. For example, the use of iron chelators and higher pH values is one approach that has been taken [61].

In the case of a biorecalcitrant textile effluent from a site in southern France, photo–Fenton pretreatment was not successful in improving biodegradability even when 70% of the effluent was mineralized by photo–Fenton treatment [90]. Aromatic intermediates were believed to be responsible for the recalcitrance of the pretreated effluent.

F. Examples of In Situ Applications

In situ treatment of contaminated soil or groundwater requires specific design parameters that are site-dependent. Treatment of a site requires several steps, including site assessment (contaminant identification and distribution as well as site geology and hydrology), bench-scale and/or pilot field-scale testing, design of remediation strategy, implementation of remediation, and site monitoring (before, during, and after remediation). Reports detailing treatment on government properties are publicly available. Consequently, a large amount of information is available on the details of these remediation projects, either through government agencies or directly from the remediation firm that performed the work. An overview of two examples is given below, representing only a small sampling of the many contaminated sites that have been remediated with Fenton technologies.

Volatile organic compounds (VOCs) were degraded in groundwater at the Naval Air Station, Pensacola, FL (NAS Pensacola) using Fenton-based in situ chemical oxidation [91]. Trichloroethene (TCE) at 2440 μg/L and its degradation products *cis*-1,2-dichloroethene (403 μg/L) and vinyl chloride (976 μg/L) were the major contaminants at this site. These contaminants were focused predominantly in a region from 35 to 45 ft below the surface. The soil at this site consisted of marine terrace sands of fine to medium quartz. An impermeable clay layer existed at a depth of 35–45 ft. Dense nonaqueous phase liquids tend to settle to the lowest depth, and often reside just above such impermeable clay layers.

For this site, pressurized injection of concentrated hydrogen peroxide and ferrous ion was used to carry out in situ Fenton oxidation. Fourteen injectors were installed at various depths in order to maximize the effectiveness of the degradation approach. These injectors were placed either in the contaminant source zone or downgradient from the source in the groundwater flow. Injectors and wells were designed and constructed to withstand elevated temperatures and pressures expected during Fenton treatment. Reagents (hydrogen peroxide and ferrous ion catalyst solution) were mixed at the head of the injector just prior to injection into the subsurface. The design of the injectors used at this site is presented in Figure 4.

In the first phase of treatment, approximately 4089 gal of 50% H_2O_2 and an equal volume of ferrous ion catalyst solution were injected over a

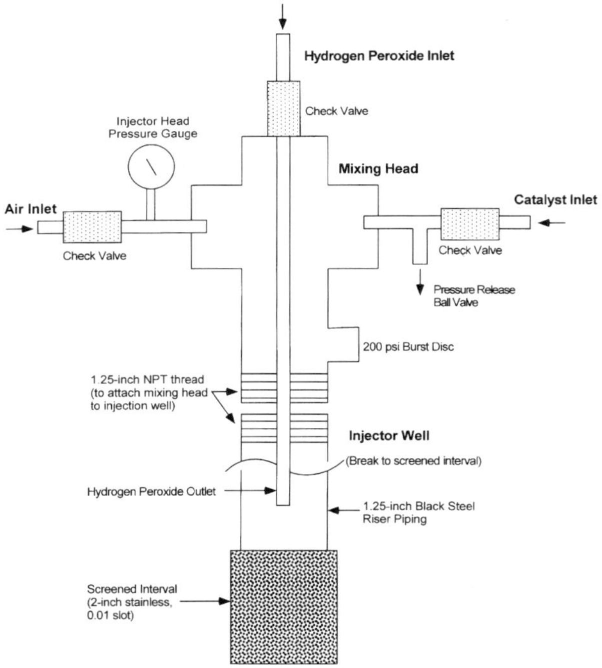

Figure 4 Design of hydrogen peroxide and iron injector used for remediation of contaminated sites. (From GeoCleanse International, Inc., U.S. Patents 5,525,008 and 5,611,642.)

5-day period with injection rates of each reagent of between 0.2 and 1.4 gal/min per injector. Air was also injected at rates of up to 3 ft^3/min to facilitate reagent distribution. Groundwater pH, alkalinity, total iron, hydrogen peroxide concentration, chloride concentration, headspace VOCs, oxidation–reduction potential, and conductivity were monitored. pH values of less than 6 were needed, and total iron levels of $\geq$5 mg/L in groundwater were optimal.

Offgas production was monitored in order to follow the progress of the degradation. Carbon dioxide was indicative of pollutant (or at least organic compound) degradation. Oxygen was taken as an indicator of Fenton reaction occurring in the absence of organic compounds (see reaction (3)).

After the first phase of treatment, the site was recharacterized to determine the extent of degradation and the new distribution of pollutants. A second phase of reagent injection was carried out (6038 gal of each reagent) over a 5-day period approximately 5 months after the first phase of injections. This phase of treatment was also accompanied by site monitoring.

After the first phase of treatment, VOC concentration decreased in one monitoring well, but remained relatively unchanged in the remaining wells. After the second phase of treatment, which utilized phosphoric acid as a peroxide stabilizer to increase its radius of influence, VOC concentrations were substantially decreased, on the order of 96–100%. In a 30-day posttreatment sampling, levels of VOCs were found to be the same as immediately after treatment, indicating successful treatment of the site. Chromium(III) present in the site was not oxidized to Cr(VI) to any detectable level during the treatment. The formation of Cr(VI) in oxidizing treatment technologies is always a concern in any site that has substantial levels of Cr(III).

In another study, bench-scale and field pilot tests were conducted to assess the applicability of in situ Fenton treatment of explosives (nitroaromatic compounds) dissolved in groundwater at the Pueblo Chemical Depot (PCD), CO [92]. The soil in the site consisted of eolian, alluvial, and colluvial deposits over a layer of shale bedrock. Groundwater concentrations of 1,3,5-trinitrobenzene (TNB) and 2,4,6-trinitrotoluene (TNT) were 18,000 and 3000 μg/L, respectively. Additional nitroaromatic compounds were present at lower concentrations. The pilot program was designed to test the ability of Fenton reagent to degrade these contaminants under the site geological and hydrological conditions, as well as to determine the optimum design parameters for delivery of the reagents (e.g., injector spacing, reagent composition, and injection rate). The effect of treatment on nitrate production, soil pH, and heavy metal mobility was also assessed.

Bench-scale tests revealed several important conclusions. First, the groundwater buffering capacity was found to be relatively high. This is a key issue because Fenton treatment is often ineffective at high pH due to iron precipitation. Bench tests also indicated that from some field-collected soil samples, explosives were leached from soil by application of iron and lowered pH. Other sites did not exhibit increased mobility of the pollutants. Increased mobility of pollutants may be beneficial because they are more easily degraded in the desorbed state. However, the mobility of undegraded pollutant must be carefully monitored. Lack of pollutant desorption from soils indicates the potential that some of the pollutants may be difficult to degrade if sorbed in inaccessible pores. Bench-scale oxidation trials did indicate some degradation of the pollutants upon Fenton treatment. However, interferences were observed when using the EPA Method 8330, and an alternate analysis method was required (Method 8321A). No significant increase in nitrate concentration was observed in these trials. Hexavalent chromium, total chromium, cadmium, lead, silver, and mercury showed no changes after treatment, whereas arsenic and barium levels increased slightly. The concentration of these latter two species remained below regulatory levels. The concentration of manganese showed a dramatic increase from 160 μg/L before treatment to 6400 μg/L after treatment.

The pilot-scale study involved the injection of 1975 gal of 50% H_2O_2 and 7964 gal of acidic ferrous solution into the site over a 7-day period. Injection wells were found to have a radius of influence of 20–24 ft. The groundwater concentrations of the minor contaminants were substantially reduced after treatment, although the major contaminant (TNB and TNT) groundwater concentrations increased. These increases were believed to be due to desorption. The observed increase of nitrate, over 100 times beyond that expected from dissolved pollutants, suggested that sorbed explosives were degraded in this treatment. Microbial activity was reduced but not eliminated after treatment. Metal desorption was noted during treatment, but subsided as ambient conditions returned. No oxidation of chromium to Cr(VI) was observed. Based on the results of this pilot study, Fenton treatment was considered to be a viable method for treating "hot spots" (i.e., high concentration areas) of explosives-contaminated soils.

V. NEW DEVELOPMENTS

A. Complexing Agents to Improve Fenton Selectivity and Efficiency

Recent work at the University of New Orleans has focused on methods of bringing pollutants and hydroxyl radical together to improve selectivity and to enhance the rate and efficiency of pollutant degradation. As previously discussed, sorption of pollutants into hydrophobic sites substantially inhibits their degradation because hydroxyl radicals are less likely to penetrate into these sites. Because the catalyst for hydroxyl radical formation (Fe^{2+}) is hydrophilic, it is unlikely that the pollutant will be near the formation site of the radical in most systems. In order to alleviate these problems, we have utilized natural compounds that are capable of simultaneously complexing with both a hydrophobic pollutant and a hydrophilic iron ion. In this manner, several improvements can be achieved: (a) the pollutant is more likely to encounter a hydroxyl radical because they are formed nearby; (b) the pollutant is removed from its isolated, hydrophobic site that is not accessible to dissolved hydroxyl radicals; and (c) the importance of scavengers is reduced because the pollutant is, on average, closer to the formation site of the radical than is a dissolved scavenger.

The natural molecules that have been found to be successful in improving Fenton degradation of hydrophobic pollutants are cyclodextrins (CDs). Cyclodextrins are cyclical oligosaccharides with six, seven, or eight glucose rings fused to form α-cyclodextrins, β-cyclodextrins, or γ-cyclodextrins, respectively. The interior portion of these torus-shaped molecules is relatively hydrophobic. The upper and lower rims contain primary and secondary hydroxyl groups that extend outward and give the molecules relatively high water solubility [93–95]. Cyclodextrins have been used for a number of applications, including chromatography [93,94]; introduction of catalysts that stabilize the transition state of the substrate [96–98]; and pharmaceutical, agricultural, and food applications. Because of their hydrophobic interior and ability to include nonpolar molecules, CDs enhance the aqueous solubility of hydrophobic compounds [99,100].

Wang and Brusseau [95] and Brusseau et al. [101] studied a derivitized CD, carboxymethyl-β-cyclodextrin (CMCD), for simultaneous soil washing of metals and organic contaminants. Metals such as Cd^{2+} are chelated directly by the oxygen atoms on the rim of the CMCD, and hydrophobic organic pollutants can also be included within the nonpolar cavity [95–101]. Consequently, both classes of pollutants can be mobilized with CDs.

In order to further profit from the dual complexing ability of CDs, we have studied Fenton-type processes in the presence of CDs. Several classes of compounds showed enhanced degradation rates in the presence of CDs in aqueous solution: PCBs, PAHs, TNT, and chlorinated phenoxyacetic acids [38,102]. Dissolved natural organic matter typically inhibits Fenton degradation by sequestering the iron away from the pollutant [31,32]. However, addition of cyclodextrins overcame the inhibitory effect of the NOM and resulted in enhanced degradation rates [38].

Fluorescence spectroscopy and mass spectrometry have suggested that a ternary complex of iron–cyclodextrin–pollutant exists in solution. Such a complex is believed to play a key role in increasing pollutant degradation rates. Studies using added scavengers also support this theory [38]. A schematic illustration of such a theorized ternary complex is depicted in Figure 5.

Table 2 presents representative enhancements in observed degradation rates for several aqueous pollutants upon the addition of CMCD. As can be

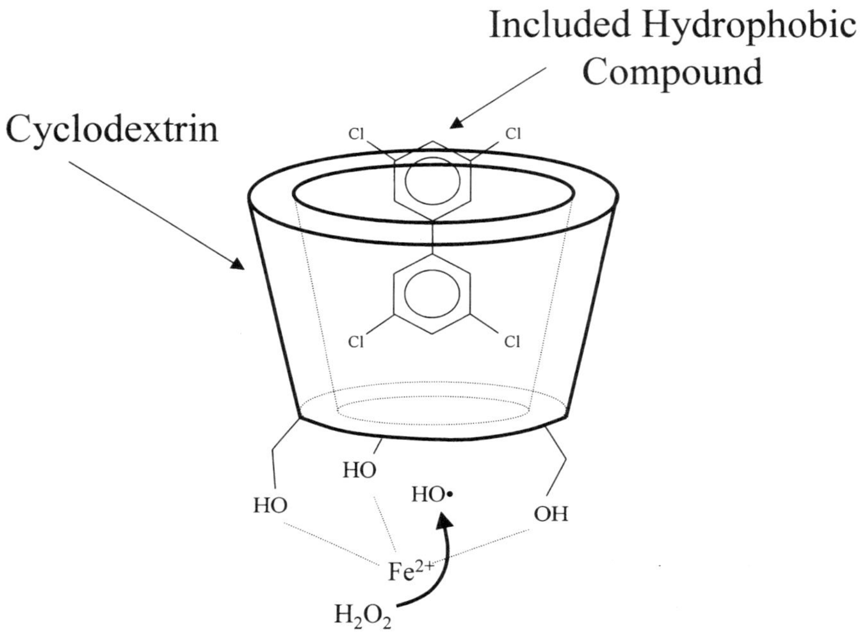

Figure 5 Preliminary schematic representation of a ternary iron–cyclodextrin–pollutant complex believed to be involved in improving the efficiency and selectivity of Fenton degradation of hydrophobic organic pollutants.

seen, up to 10-fold rate enhancements have been observed. In general, these studies used millimolar concentrations of iron and cyclodextrin to degrade pollutants present in low-micromolar to near-millimolar concentrations. The optimum concentrations of iron, peroxide, and cyclodextrin are dependent on the specific pollutant, its concentration, and the nature and concentration of other species present. Another advantage of the CMCD system is the ability to work at near-neutral pH values in many cases. Just as for other iron chelators discussed above, the cyclodextrin improves the water solubility of iron and helps prevent the precipitation of iron hydroxides. Although these results are promising, the demonstration of this approach in soil-containing systems has not yet been achieved.

B. Reductive Degradation

As can be seen from reactions (2), (6), (16), and (17), the reducing species $HO_2\cdot$ (hydroperoxyl radical) and $O_2^{-}\cdot$ (superoxide) are formed in Fenton

Table 2 Observed Enhancements in Degradation Rate Constants with CMCD Added

	CMCD added		No CMCD
Pollutant	k/k_o[a] (no NOM)	k/k_o (with NOM)	k/k_o (with NOM)
Naphthalene	2.3		
Phenanthrene	1.9		0.44
Pyrene	1.3	1.3	0.8
TNT	7.3 (pH = 3)	2.5 (pH = 3) 9.3 (pH ~ 7)	
2,4,5-T	1.2		0.5
PCB-54 (sorbed to glass)	13[b]		

[a] k/k_o = ratio of observed rate constant with additives (k) to observed rate constant without additives (k_o). Additives were either CMCD or NOM.

[b] Ratio of total amount of PCB degraded with CMCD to total amount of PCB degraded without CMCD.

systems. Whereas the concentrations of these species are generally low, under high peroxide concentrations, substantial amounts of these reducing species can be produced. Under such conditions, reductive degradation can become an important pathway for the removal of pollutants. Furthermore, reductive pathways have been suggested as potential mechanisms for the desorption of pollutants from soil sites.

Enhanced desorption of chlorinated aliphatic compounds from a sandy loam soil was observed in Fenton systems with peroxide concentrations greater than 0.3 M [34]. Observed degradation of hexachloroethane, which does not react with hydroxyl radical, was provided as evidence of alternate degradation pathways other than hydroxyl radical or other strongly oxidizing species. Addition of a reductant scavenger (chloroform) resulted in a complete inhibition of hexachloroethane degradation, again providing evidence for a reductive pathway. Furthermore, the appearance of a reduced product, pentachloroethane, also suggested a reductive pathway. A report written in Korean also detailed the degradation of CCl_4, presumably through reaction with superoxide and hydroperoxide [103].

Studies of CCl_4-containing systems treated with Fenton reagents with high peroxide concentrations have led to the conclusion that a reductive pathway is involved [104]. Carbon tetrachloride was unreactive with hydroxyl radical, but was degraded in high-peroxide-concentration Fenton systems to yield free chloride. Hexachloroethane, bromotrichloromethane, and tetranitromethane, additional compounds that are difficult to degrade with hydroxyl radical, were also degraded using high-peroxide Fenton systems. Superoxide was suggested as the key reductant responsible for the degradation of these compounds. The observation of a reductive pathway in Fenton systems may expand the applicability of Fenton treatment to pollutants that can only be degraded by reductive pathways. Initial reduction products may be further degraded by reaction with hydroxyl radical. Of even greater importance may be the ability of the reducing species to desorb hydrophobic pollutants [104], which can then be further degraded by reductive or oxidative pathways. Although the high concentration of hydrogen peroxide needed to produce significant levels of reducing species might seem cost-prohibitive, such conditions are typically used in current site remediation protocols. Consequently, it is likely that current remediation strategies are capable of desorbing bound pollutants. Additional research is necessary to further understand and optimize the reductive and desorbing properties of these systems.

REFERENCES

1. Fenton HJH. On a new reaction of tartaric acid. Chem News 1876; 33:190.
2. Fenton HJH. Oxidation of tartaric acid in presence of iron. J Chem Soc 1894; 65:899–910.
3. Walling C. Fenton's reagent revisited. Acc Chem Res 1975; 8:125–131.
4. Edwards JO, Curci R. Fenton type activation and chemistry of hydroxyl radical. In: Strukul G, ed. Catalytic Oxidations with Hydrogen Peroxide as Oxidant. Dordrecht, Netherlands: Kluwer Academic Publishers, 1992:97–151.
5. Haber F, Weiss J. The catalytic decomposition of hydrogen peroxide by iron salts. Proc R Soc Ser A. Math Phys Sci 1934; 147:332–351.
6. Wells CF, Salam MA. Hydrolysis of ferrous ions: a kinetic method for the determination of the Fe(II) species. Nature 1965; 205(4972):690–692.
7. Vella PA, Munder JA. Toxic pollutant destruction. In: Tedder DW, Pohland FG, eds. Emerging Technologies in Hazardous Waste Management III, ACS Symposium Series No. 518. New York: Oxford University Press, 1993:85–105.
8. Yoshimura Y, Matsuaki Y, Watanabe T, Uchiyama K, Ohsawa K, Imaeda K. Effects of buffer solutions and chelators on the generation of hydroxyl radical and the lipid peroxidation in the Fenton reaction system. J Clin Biochem Nutr 1992; 13:147–154.
9. Graf E, Mahoney JR, Bryant RG, Eaton JW. Iron-catalyzed hydroxyl radical formation: stringent requirement for free iron coordination site. J Biol Chem 1984; 259:3620–3624.
10. Szulbinski WS. Fenton reaction of iron chelates involving polyazacyclononane. The ligand structure effect. Pol J Chem 2000; 74:109–124.
11. Smith JB, Cusumano JC, Babbs CF. Quantitative effects of iron chelators on hydroxyl radical production by the superoxide-driven Fenton reaction. Free Rad Res Commun 1990; 8:101–106.
12. Gutteridge JMC. Superoxide-dependent formation of hydroxyl radicals from

ferric-complexes and hydrogen peroxide: an evaluation of fourteen iron chelators. Free Radic Res Commun 1990; 9:119–125.
13. Sutton HC. Efficiency of chelated iron compounds as catalysts for the Haber–Weiss reaction. J Free Radic Biol Med 1985; 1:195–202.
14. Van Dyke BR, Clopton DA, Saltman P. Buffer-induced anomalies in the Fenton chemistry of iron and copper. Inorg Chim Acta 1996; 242:57–61.
15. Croft S, Gilbert BC, Smith JRL, Stell JL, Sanderson WR. Mechanisms of peroxide stabilization. An investigation of some reactions of hydrogen peroxide in the presence of aminophosphonic acids. J Chem Soc Perkin Trans 1992; 2:153–160.
16. Lopes GKB, Schulman HM, Hermes-Lima M. Polyphenol tannic acid inhibits hydroxyl radical formation from Fenton reaction by complexing ferrous ions. Biochim Biophys Acta 1999; 1472:142–152.
17. Sun Y, Pignatello JJ. Chemical treatment of pesticide wastes. Evaluation of Fe(III) chelates for catalytic hydrogen peroxide oxidation of 2,4-D at circumneutral pH. J Agric Food Chem 1992; 40:322–327.
18. Pignatello JJ, Liu D, Huston P. Evidence for an additional oxidant in the photoassisted Fenton reaction. Environ Sci Technol 1999; 33:1832–1839.
19. Graedel TE, Mandich ML, Weschler CJ. Kinetic model studies of atmospheric droplet chemistry: 2. Homogeneous transition metal chemistry in raindrops. J Geophys Res, D Atmos, 1986; 91(D4):5205–5221.
20. Larson RA. Sensitized photodecomposition of organic compounds found in Illinois wastewaters (78 pp.). 1990 Report (HWRIC-RR-045; order no. PB91-132456). Champaign, IL, USA: Institute of Environmental Studies, University of Illinois, Urbana.
21. Larson RA, Schlauch MB, Marley KA. Ferric ion promoted photodecomposition of triazines. J Agric Food Chem 1991; 39:2057–2062.
22. Sun Y, Pignatello JJ. Activation of hydrogen peroxide by iron(III) chelates for abiotic degradation of herbicides and insecticides in water. J Agric Food Chem 1993; 41:308–312.
23. Sun Y, Pignatello JJ. Photochemical reactions involved in the total mineralization of 2,4-D by iron(3+)/hydrogen peroxide/UV. Environ Sci Technol 1993; 27:304–310.
24. Lipczynska-Kochany E. Degradation of nitrobenzene and nitrophenols in homogeneous aqueous solution. Direct photolysis versus photolysis in the presence of hydrogen peroxide and the Fenton reagent. Water Pollut Res J Can 1992; 27:97–122.
25. Zepp RG, Faust BC, Hoigne J. Hydroxyl radical formation in aqueous reactions (pH 3–8) of iron(II) with hydrogen peroxide: the photo–Fenton reaction. Environ Sci Technol 1992; 26:313–319.
26. Lunak S, Sedlak P, Lederer P. Methylene-blue-sensitized photochemical hydroxylation of 2-hydroxybenzoic acid by hydrogen peroxide: photocatalytic effect of ferric chloride. J Photochem 1987; 39:239–250.
27. Arasingham RD, Cornman CR, Balch AL. Detection of alkylperoxo and ferryl, $(Fe^{IV}=O)^{2+}$, intermediates during the reaction of *tert*-butyl hydroperoxide with iron porphyrins in toluene solution. J Am Chem Soc 1989; 111:7800–7805.

28. Lloyd RV, Hanna PM, Mason RP. The origin of the hydroxyl radical oxygen in the Fenton reaction. Free Radic Biol Med 1997; 22:885–888.
29. Shen X, Tian J, Li J, Li X, Chen Y. Formation of the excited ferryl species following Fenton reaction. Free Radic Biol Med 1992; 13:585–592.
30. Bossmann SH, Oliveros E, Göb S, Siegwart S, Dahlen EP, Payawan L, Straub M, Wörner M, Braun AM. New evidence against hydroxyl radicals as reactive intermediates in the thermal and photochemically enhanced Fenton reactions. J Phys Chem 1998; 102:5542–5550.
31. Lindsey ME, Tarr MA. Inhibited hydroxyl radical degradation of aromatic hydrocarbons in the presence of dissolved fulvic acid. Water Res 2000; 34:2385–2389.
32. Lindsey ME, Tarr MA. Inhibition of hydroxyl radical reaction with aromatics by dissolved natural organic matter. Environ Sci Technol 2000; 34:444–449.
33. Buxton GV, Greenstock CL, Helman WP, Ross AB. Critical review of rate constants for reactions of hydrated electrons, hydrogen atoms and hydroxyl radicals ($\cdot OH/\cdot O^-$) in aqueous solution. J Phys Chem Ref Data 1988; 17:513–886.
34. Watts RJ, Bottenberg BC, Hess TF, Jensen MD, Teel AL. Role of reductants in the enhanced desorption and transformation of chloroaliphatic compounds by modified Fenton's reactions. Environ Sci Technol 1999; 33:3432–3437.
35. Kim Y-S, Kong S-H, Bae S-Y, Hwang G-C. The mechanisms of oxidation and reduction under various pH regimes in Fenton's reaction. Kongop Hwahak 2001; 12:755–760.
36. Radiation Chemistry Data Center, Notre Dame Radiation Laboratory (http://www.rcdc.nd.edu).
37. Sedlak DL, Andren AW. The effect of sorption on the oxidation of polychlorinated-biphenyls (PCBs) by hydroxyl radical. Water Res 1994; 28:1207–1215.
38. Lindsey ME, Xu G, Lu J, Tarr MA. Enhanced Fenton degradation of hydrophobics by simultaneous iron and pollutant complexation with cyclodextrin. Sci Total Environ. 2003; 307:215–229.
39. Kakarla PKC, Watts RJ. Depth of Fenton-like oxidation in remediation of surface soil. J Environ Eng 1997; 123:11–17.
40. Watts RJ, Stanton PC. Process conditions for the total oxidation of hydrocarbons in the catalyzed hydrogen peroxide treatment of contaminated soils. WA-RD 337.1. Olympic, WA: Washington State Department of Transportation, 1994.
41. Watts RJ, Kong S, Dippre M, Barnes WT. Oxidation of sorbed hexachlorobenzene in soils using catalyzed hydrogen peroxide. J Hazard Mater 1994; 39:33–47.
42. Araujo E, Rodriguez-Malaver AJ, Gonzalez AM, Rojas OJ, Penaloza N, Bullon J, Lara MA, Dmitrieva N. Fenton's reagent-mediated degradation of residual kraft black liquor. Appl Biochem Biotechnol 2002; 97:91–103.
43. Li Y-S, Liu C-C, Chern B-R. Oxidation of 2-chlorophenol wastewater by hydrogen peroxide in the presence of basic oxygen furnace slag. Bull Environ Contam Toxicol 1999; 62:336–343.

44. Beltran-Heredia J, Torregrossa J, Garcia J, Dominguez JR, Tierno JC. Degradation of olive mill wastewater by the combination of Fenton's reagent and ozonation processes with an aerobic biological treatment. Water Sci Technol 2001; 44:103–108.
45. Kitajima N, Matsumura T, Fukuzumi SI, Ono YJ. Formation of superoxide ion during the decomposition of hydrogen peroxide on supported metals. J Phys Chem 1977; 81:1307–1311.
46. Ravikumar JX, Gurol MD. Chemical oxidation of chlorinated organics by hydrogen peroxide in the presence of sand. Environ Sci Technol 1994; 28:394–400.
47. Heard I, Senftle FE. Chemical oxidation of anthracite with hydrogen peroxide via the Fenton reaction. Fuel 1984; 63:221–226.
48. Watts RJ, Udell MD, Monsen RM. Use of iron minerals in optimizing the peroxide treatment of contaminated soils. Water Environ Res 1993; 65:839–844.
49. Watts RJ, Jones AP, Chen P-H, Kenny A. Mineral-catalyzed Fenton-like oxidation of sorbed chlorobenzenes. Water Environ Res 1997; 69:269–275.
50. Lin S-S, Gurol MD. Catalytic decomposition of hydrogen peroxide on iron oxide: kinetics, mechanisms, and implications. Environ Sci Technol 1998; 32: 1417–1423.
51. Gurol MD, Lin S-S. Hydrogen peroxide/iron oxide-induced catalytic oxidation of organic compounds. Water Sci Technol: Water Supply 2001; 1:131–138.
52. Gurol MD, Lin S-S. Hydrogen peroxide/iron oxide-induced catalytic oxidation of organic compounds. J Adv Oxid Technol 2002; 5:147–154.
53. Valentine RL, Wang HCA. Iron oxide surface catalyzed oxidation of quinoline by hydrogen peroxide. J Environ Eng 1998; 124:31–38.
54. Chou S, Huang C. Applications of supported iron oxyhydroxide catalyst in oxidation of benzoic acid by hydrogen peroxide. Chemosphere 1999; 38:2719–2731.
55. Huang H-H, Lu M-C, Chen J-N. Catalytic decomposition of hydrogen peroxide and 2-chlorophenol with iron oxides. Water Res 2001; 35:2291–2299.
56. Fubini B, Mollo L, Giamello E. Free radical generation at the solid/liquid interface in iron containing minerals. Free Radic Res 1995; 23:593–614.
57. Kwan WP, Voelker BM. Decomposition of hydrogen peroxide and organic compounds in the presence of dissolved iron and ferrihydrite. Environ Sci Technol 2002; 36:1467–1476.
58. Xiong C, Chen Q, Lu W, Gao H, Lu W, Gao Z. Novel Fe-based complex oxide catalysts for hydroxylation of phenol. Catal Lett 2000; 69:231–236.
59. Kong S-H, Watts RJ, Choi J-H. Treatment of petroleum-contaminated soils using iron–mineral-catalyzed hydrogen peroxide. Chemosphere 1998; 37:1473–1482.
60. Arienzo M. Oxidizing 2,4,6-trinitrotoluene with pyrite–H_2O_2 suspensions. Chemosphere 1999; 39:1629–1683.
61. Nam K, Rodriguez W, Kukor JJ. Enhanced degradation of polycyclic aromatic hydrocarbons by biodegradation combined with a modified Fenton reaction. Chemosphere 2001; 45:11–20.

62. Puppo A. Effect of flavenoids on hydroxyl radical formation by Fenton-type reactions: influence of the iron chelator. Phytochemistry 1992; 31:85–88.
63. Cheng I, Breen K. On the ability of four flavonoids, baicilein, luteolin, naringenin, and quercetin, to suppress the Fenton reaction of the iron–ATP complex. BioMetals 2000; 13:77–82.
64. Goodell B, Qian Y, Jellison J, Richard M, Qi W. Lignocellulose oxidation by low molecular weight metal-binding compounds isolated from wood degrading fungi: a comparison of brown rot and white rot systems and the potential application of chelator-mediated Fenton reactions. Prog Biotechnol 2002; 21: 37–47.
65. Qian Y, Goodell B, Felix CC. The effect of low molecular weight chelators on iron chelation and free radical generation as studied by ESR measurement. Chemosphere 2002; 48:21–28.
66. Rush JD, Koppenol WH. Oxidizing intermediates in the reaction of ferrous EDTA with hydrogen peroxide. J Biol Chem 1986; 261:6730–6733.
67. Egan TJ, Barthakur SR, Aisen P. Catalysis of the Haber–Weiss reaction by iron–diethylenetriaminepentaacetate. J Inorg Biochem 1992; 48:241–249.
68. Rahhal S, Richter HW. Reduction of hydrogen peroxide by the ferrous iron chelate of diethylenetriamine-*N*,*N*,*N*′,*N*″,*N*″-pentaacetate. J Am Chem Soc 1988; 110:3126–3133.
69. Smith JB, Cusumano JC, Babbs CF. Quantitative effects of ion chelators on hydroxyl radical production by the superoxide-driven Fenton reaction. Free Radic Res Commun 1990; 8:101–106.
70. Iwahashi H, Kawamori H, Fukushima K. Quinolinic acid, α-picolinic acid, fusaric acid, and 2,6-pyridinedicarboxylic acid enhance the Fenton reaction in phosphate buffer. Chem-Biol Interact 1999; 118:201–215.
71. Nakagawa H, Wachi M, Woo J-T, Kato M, Kasai S, Takahashi F, Lee I-S, Naga K. Fenton reaction is primarily involved in a mechanism of (−)-epigallocatechin-3-gallate to induce osteoclastic cell death. Biochem Biophys Res Commun 2002; 292:94–101.
72. Pignatello JJ. Dark and photoassisted Fe^{3+}-catalyzed degradation of chlorophenoxy herbicides by hydrogen peroxide. Environ Sci Technol 1992; 26:944–951.
73. Pignatello JJ, Sun Y. Complete oxidation of metolachlor and methyl parathion in water by the photoassisted Fenton reaction. Water Res 1995; 29:1837–1844.
74. Doong R-A, Chang W-H. Photoassisted iron compound catalytic degradation of organophosphorous pesticides with hydrogen peroxide. Chemosphere 1998; 37:2563–2572.
75. Pignatello JJ, Chapa G. Degradation of PCBs by ferric ion, hydrogen peroxide and UV light. Environ Toxicol Chem 1994; 13:423–427.
76. Pignatello JJ, Huang LQ. Degradation of polychlorinated dibenzo-*p*-dioxin and dibenzofuran contaminants in 2,4,5-T by photoassisted iron-catalyzed hydrogen peroxide. Water Res 1993; 27:1731–1736.
77. Bandara J, Morrison C, Kiwi J, Pulgarin C, Peringer P. Degradation/decomposition of concentrated solutions of Orange II. Kinetics and quantum yield for

sunlight induced reactions via Fenton type reagents. J Photochem Photobiol, A Chem 1996; 99:57–66.

78. Kang S-F, Liao C-H, Po S-T. Decolorization of textile wastewater by photo–Fenton oxidation technology. Chemosphere 2000; 41:1287–1294.
79. Fukushima M, Tatsumi K, Morimoto K. Effect of phenolic acids in humic acid on the degradation of pentachlorophenol by the photo–Fenton reaction. Toxicol Environ Chem 2001; 79:9–21.
80. Gold MH, Kutsuki H, Morgan MA. Oxidative degradation of lignin by photochemical and chemical radical generating systems. Photochem Photobiol 1983; 38:647–651.
81. Perez M, Torrades F, Domenech X, Peral J. Removal of organic contaminants in paper pulp effluents by AOPs: an economic study. J Chem Technol Biotechnol 2002; 77:525–532.
82. Maletzky P, Bauer R. Immobilisation of iron ions on Nafion® and its applicability to the photo–Fenton method. Chemosphere 1999; 38:2315–2325.
83. Dhananjeyan MR, Kiwi J, Albers P, Enea O. Photo-assisted immobilized Fenton degradation up to pH 8 of azo dye Orange II mediated by Fe^{3+}/Nafion®/glass fibers. Helv Chim Acta 2001; 84:3433–3445.
84. Malato S, Blanco J, Vidal A, Richter C. Photocatalysis with solar energy at a pilot-plant scale: an overview. Appl Catal, B Environ 2002; 37:1–15.
85. Krutzler T, Bauer R. Optimization of a photo–Fenton prototype reactor. Chemosphere 1999; 38:2517–2532.
86. Oliveros E, Legrini O, Hohl M, Müller T, Braun AM. Large scale development of a light-enhanced Fenton reaction by optimal experimental design. Water Sci Technol 1997; 35:223–230.
87. Lee B-D, Hosomi M, Murakami A. Fenton oxidation with ethanol to degrade anthracene into biodegradable 9,10-anthraquinone. A pretreatment method for anthracene-contaminated soil. Water Sci Technol 1998; 38:91–97.
88. Lee B-D, Hosomi M. A hybrid Fenton oxidation–microbial treatment for soil highly contaminated with benz(*a*)anthracene. Chemosphere 2001; 43:1127–1132.
89. Kao CM, Wu MJ. Enhanced TCDD degradation by Fenton's reagent peroxidation. J Hazard Mater 2000; 74:197–211.
90. Rodriguez M, Sarria V, Esplugas S, Pulgarin C. Photo–Fenton treatment of a biorecalcitrant wastewater generated in textile activities: biodegradability of the photo-treated solution. J Photochem Photobiol, A Chem 2002; 151:129–135.
91. Geo-Cleanse International, Inc. Effectiveness evaluation report. Geo-Cleanse® Treatment Program. Pensacola, FL: Naval Air Station Pensacola, September 24, 1999.
92. Geo-Cleanse International, Inc. Effectiveness evaluation report. Geo-Cleanse® Pilot Test Program. Pueblo, CO: Pueblo Chemical Depot, July 25, 2001.
93. Mohseni RM, Hurtubise RJ. Retention characteristics of several compound classes in reversed-phase liquid chromatography with β-cyclodextrin as a mobile phase modifier. J Chromatogr 1990; 499:395–410.

94. Lamparczyk H, Zarzycki P, Ochocka RJ, Sybilska D. Structure-retention studies on the inclusion complex formation of some polycyclic aromatic hydrocarbons with β-cyclodextrin. Chromatographia 1990; 30:91–94.
95. Wang X, Brusseau ML. Simultaneous complexation of organic compounds and heavy metals by a modified cyclodextrin. Environ Sci Technol 1995; 29:2632–2635.
96. Granados A, de Rossi RH. Effect of cyclodextrins on the hydrolysis of amides. J Org Chem 1993; 58:1771–1777.
97. Breslow R, Dong SD. Biomimetic reactions catalyzed by cyclodextrins and their derivatives. Chem Rev 1998; 98:1997–2011.
98. Monflier E, Tilloy S, Castanet Y, Martreux A. Chemically modified β-cyclodextrins—efficient supramolecular carriers for the biphasic hydrogenation of water-insoluble aldehydes. Tetrahedron Lett 1998; 39:2959–2960.
99. Tachibana M, Kiba N. Correlation between inclusion formation constant and distribution coefficient in a liquid–liquid extraction system consisting of hydrocarbon solvents and aqueous dimethyl sulfoxide solutions of β-cyclodextrin. Analyst 1997; 122:903–909.
100. Szente L, Szejtli J. Non-chromatographic analytical uses of cyclodextrins. Analyst 1998; 123:735–741.
101. Brusseau ML, Wang X, Wang W-Z. Simultaneous elution of heavy metals and organic compounds from soil by cyclodextrin. Environ Sci Technol 1997; 31: 1087–1092.
102. Lindsey ME, Tarr MA. Directed pollutant oxidation using simultaneous catalytic metal chelation and organic pollutant complexation. US Patent 6,459,011, October 1, 2002.
103. Kim CG, Yoon TI, Seo HJ, Yu YH. Hybrid treatment of tetramethyl ammonium hydroxide occurring from electronic materials industry. Kor J Chem Eng 2002; 19:445–450.
104. Teel AL, Watts RJ. Degradation of carbon tetrachloride by modified Fenton's reagent. J Hazard Mater 2002; 94:179–189.
105. Ito S, Mitarai A, Hikino K, Hirama M, Sasaki K. Deactivation reaction in the hydroxylation of benzene with Fenton's reagent. J Org Chem 1992; 57:6937–6941.
106. Kunai A, Hata S, Ito S, Sasaki K. The role of oxygen in the hydroxylation reaction of benzene with Fenton's reagent. ^{18}O tracer study. J Am Chem Soc 1986; 108:6012–6016.
107. Scheck CK, Frimmel FH. Degradation of phenol and salicylic acid by ultraviolet radiation/hydrogen peroxide/oxygen. Water Res 1995; 29:2346–2352.

5
Sonochemical Degradation of Pollutants

Hugo Destaillats and Michael R. Hoffmann
California Institute of Technology, Pasadena, California, U.S.A.

Henry C. Wallace
Ultrasonic Energy Systems Co., Panama City, Florida, U.S.A.

I. INTRODUCTION

The term *sonochemistry* describes all chemical processes in which ultrasound irradiation is involved. The interaction of an acoustical field with the irradiated fluid provides new reaction pathways and alters existing chemical processes in the system, usually yielding an enhancement of reaction rates. A distinction should be pointed out between the effects of ultrasound in homogeneous and in heterogeneous media. In the first case, sonochemical reactions are related to new chemical species produced during acoustical cavitation, whereas the enhancement of heterogeneous reactions can also be related to mechanical effects induced in the fluid system by sonication. These effects include an increase in the surface area between the reactants, a faster renovation of catalyst surfaces, and accelerated dissolution and mixing. Although this chapter focuses only on the uses of ultrasound related to the degradation of organic pollutants, we should mention that many other sonochemical processes exist, such as organic and organometallic synthesis, and polymer synthesis and modifications [1]. Also a variety of industrial applications of power ultrasound (20–100 kHz) cover a vast range of industrial processes: welding of thermoplastics; cleaning and degreasing of surfaces; dispersion of solids in paint, inks, and resins; filtration; crystallization; homogenization; and defoaming and degassing. In medicine, high-frequency ultrasound (2–10 MHz) is employed for diagnostical imaging, and lower-frequency ultrasound (20–50 kHz) is used in physiotherapy and kidney stone treatments.

The attractiveness of sonochemistry in environmental engineering seems to stem from three major facts. Firstly, sonochemistry can cause real chemical changes to a solution without the necessity of adding any other compounds. Secondly, sonochemistry is often conducted at low or ambient temperatures and pressures; thus, no heating or pressurization is required. These two features simplify enormously the design and operation of reactors. Thirdly, in many cases, the peculiar nature of sonochemical reactions offers alternative pathways, providing a faster or environmentally safer degradation of contaminants. The last 20 years witnessed a dramatical increase in the amount of scientific work in the field of sonochemistry, joined by the development of new and more powerful instrumentation. Some sonochemical syntheses have also been successfully scaled up to plant size, providing convenient advantages such as lower operation costs and shorter times of operation compared to traditional techniques [2]. The present challenge for sonochemists and acoustical physicists in the field of environmental remediation is to provide cost-effective sonochemical solutions to large-scale problems.

Sonochemistry proceeds because the passing of acoustical waves of large amplitude, called finite amplitude waves, through solutions causes cavitation. Cavitation can be generated when large pressure differentials are applied in a flowing liquid (*hydrodynamical* cavitation), or by means of an electromechanical transducer, piezoelectrical or magnetostrictive, in contact with the fluid (*acoustical* cavitation). Cavitation consists of the formation of bubbles in a solution. These bubbles generally start very small, and grow in diameter by joining together until they become buoyant enough to escape the solution. At some point in this growth, they become resonant, which means that their diameter is such that their bubble wall motion is completely determined by the acoustical wave. The coupling of the bubble wall movements with the acoustical field is a complex physical and chemical phenomenon on the microscale (lengths are on the order of micrometers and times of oscillation are in the microsecond range), which triggers a series of processes that yield chemical reactions and light emission (sonoluminescence) [3,4]. In a sonochemical reactor, the energy input is focused on small hot spots of the solution (i.e., the cavitation bubbles), instead of heating the whole system. This provides localized, highly extreme reaction conditions.

II. PHYSICAL AND CHEMICAL PRINCIPLES

A. Acoustical Cavitation and Bubble Dynamics

Ultrasound occurs at frequencies above 16 kHz (i.e., the human audible frequency threshold). The range 16–1000 kHz is usually named *power* ultrasound, being the historical choice for most chemical and industrial applications. Frequencies lying above 1 MHz are used for evaluation in medical diagnosis because they are nondestructive when operating at moderate intensities. Sound waves are longitudinal perturbations transmitted through an elastic medium. Thus, in each point of a fluid under acoustical irradiation, consecutive compression and rarefaction cycles take place. Fig. 1 illustrates the harmonic fluctuation of local pressure in the liquid under the action of the applied acoustical field. In this simplified one-dimensional case, the circles represent unit volumes of the fluid, which are removed from their equilibrium position, thus generating cycles of rarefaction and compression. The fluctuations of the displacement from the equilibrium position and the particle velocity are also shown in Fig. 1. The acoustical pressure P_a is determined by the frequency (ν) and the driving pressure (P_0, proportional to the square root of the ultrasonic intensity):

$$P_a = P_0 \sin(2\pi \nu t) \quad (1)$$

When an acoustical field is applied to a fluid, the acoustical pressure is superimposed on the hydrostatic ambient pressure P, the total pressure being:

$$P_{total} = P + P_a \quad (2)$$

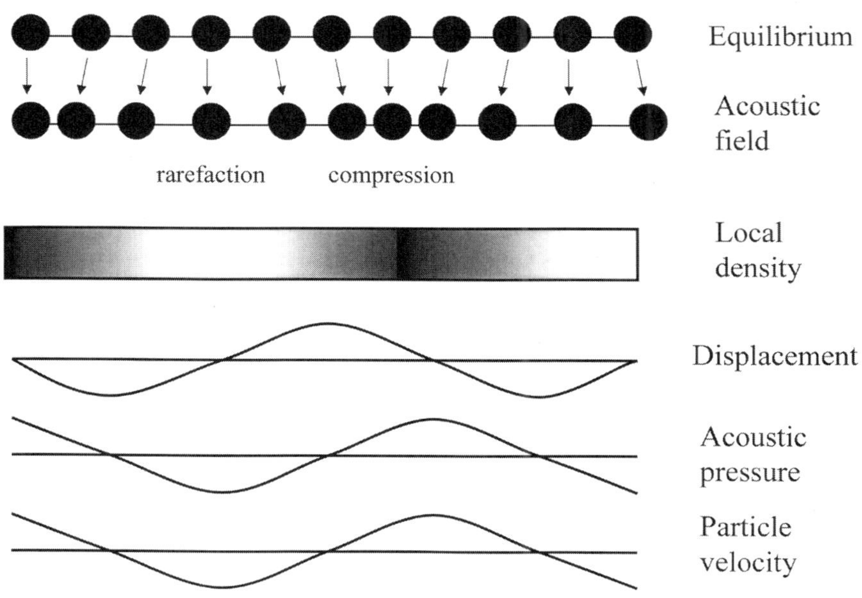

Figure 1 Propagation of a one-dimensional ultrasound wave. (From Ref. 4.)

The velocity of sound (c) in water, and in many aqueous solutions, is 1500 m/sec. For the usual frequencies employed in high-power, near-megahertz sonochemistry (100–1000 kHz), the wavelengths corresponding to the acoustical field range from $\lambda = c/v = 1.5$ cm to 1.5 mm. These lengths are about a thousand times larger than the radii of cavitation bubbles and more than a million times higher than molecular dimensions. It is important to keep these different magnitudes in mind: a macroscopical mechanical process (the sound wave) is able to produce changes in the molecular domain by means of cavitation bubbles, which oscillate in an intermediate micrometer range.

Acoustical cavitation is produced during the rarefaction cycle of the sound wave, when the liquid is pulled apart, generating a void. These cavitation bubbles vanish upon the next compression cycle if the radius is smaller than the critical value. If the radius is larger than the critical value, the bubbles undergo oscillatory growth and shrinking during the subsequent successive acoustical cycles. In an ideal degassed and pure liquid, the minimum acoustical pressure required to produce stable cavities is about 1500 bar [1]. However, different experiments have shown that a much lower sound intensity (equivalent to less than 20 bar) is enough to cause cavitation in real liquids. This lower threshold for cavitation is due to the existence of the so-called weak spots in the liquid, such as rugosities in the containers walls, which lower the liquid's tensile strength. A threshold for cavitation is determined then not only by the irradiation parameters (frequency and intensity of the ultrasound) but also by experimental factors such as gas nucleation site density, gas content and solubility, temperature (mainly through its effect on the vapor pressure), and hydrostatic pressure of the liquid.

Once a bubble is produced in the liquid, two different kinds of cavitation phenomena can take place: *stable* or *transient* cavitation. In the first case, the bubble wall couples with the acoustical field oscillating about the equilibrium radius for several acoustical cycles. This process is of little significance in terms of chemical effects because the size changes are not very dramatical, and the evaporation and condensation of the solvent inside the bubble are quasi-reversible phenomena in this regime. When the rates of mass transfer across the bubble interface are not equal, some vapor can be retained in the cavity, resulting in its growth. This mechanism is usually referred to as *rectified diffusion*, and is also applicable to the dissolved gases and volatile solutes, which can be concentrated in stable cavitation bubbles during many cycles. A slow-growing cavitation bubble will become unstable after a number of cycles, and then transient cavitation is observed. During transient cavitation, the bubble size suffers a dramatical increase, from tens to hundreds of times the equilibrium radius, followed by a fast collapse. It is upon this final collapse that the major chemical effects are produced. The Rayleigh–Plesset equation (Eq. (3)) describes the dynamics of the bubble radius R, as a function of the principal parameters affecting its behavior: the acoustical pressure (P_a), the initial radius (R_0), the vapor pressure of the solution (P_v), the surface tension (σ), the viscosity (η), the liquid density (ρ), the hydrostatical pressure (P), and the polytropical index of the gas inside the cavity (K) [3]:

$$R\overset{**}{R} + \frac{3}{2}\overset{*}{R}^2 = \frac{1}{\rho}\left[\left(P + \frac{2\sigma}{R_0} - P_v\right)\left(\frac{R_0}{R}\right)^{3K} - \frac{2\sigma}{R} - 4\eta\frac{\overset{*}{R}}{R} - (P - P_a)\right] \quad (3)$$

The first and second derivatives of R with respect to time are represented by $\overset{*}{R}$ and $\overset{**}{R}$, respectively. Solving this equation for different values of R_0 can be quite illustrative on the complex nonlinear dynamics of cavitation bubbles. Fig. 2 shows two different cases when a frequency of 20 kHz and an intensity equivalent to $P_0 = 2.7$ bar are used. In the first case (Fig. 2a), a relatively large ($R_0 = 2$ mm) bubble couples with the sonic field through small-amplitude growth and compression cycles (*stable* cavitation). In contrast, a smaller bubble ($R_0 = 20$ μm) experiences resonant coupling, which results

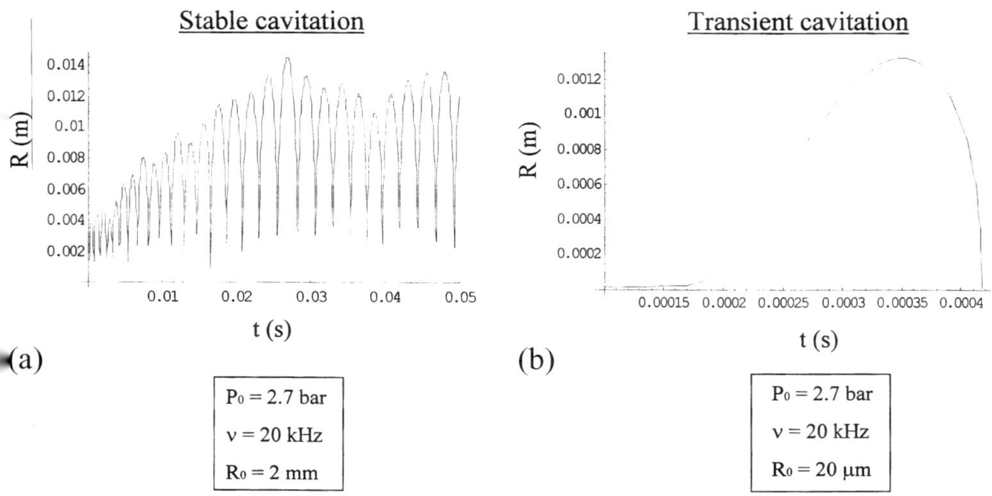

Figure 2 (a) Stable and (b) transient cavitation regimes for different initial bubble radii.

in an expansion of 50–100 times the initial bubble radius, followed by a drastic implosion of the bubble during the compression part of the final cycle (*transient* cavitation). It is in this last stage of the cycle that sonochemical effects take place.

B. Cavitation Chemistry as a Series of Processes

Although not all authors are in agreement [5], it is widely accepted that sonolysis, the breaking apart of molecules in a solution by the action of a sound field, is produced by the extreme conditions reached upon transient cavitation collapse. The fast implosion of the bubble produces a quasi-adiabatic heating of the vapor phase inside the cavity that yields localized transient high temperatures and pressures, in the range of thousands of degrees Kelvin and hundreds of bars, respectively. Didenko et al. [6] estimated a collapse temperature in water near 4000 K by using a spectroscopic technique. Some controversies exist because much higher temperatures have also been reported. Discussions extend also to the physical description of the transient hot spot. Some authors postulate the existence of shock waves in the final stages of the collapse, when the bubble wall velocity becomes near supersonic, whereas others consider a homogeneous temperature distribution and pressure distribution in the gas phase during the bubble implosion [7,8]. Skipping the details of this academical controversy, a general agreement that water molecules under such extreme conditions undergo thermal dissociation to yield •OH and •H radicals exists:

$$H_2O(g) \rightarrow \bullet H(g) + \bullet OH(g) \tag{4}$$

These active radicals have been observed in electron paramagnetic resonance (EPR) spin-trapping studies during the sonication of water and aqueous solutions [9,10]. Other molecules present in the gaseous pool, such as dissolved O_2, N_2, and eventually volatile solutes, also undergo thermal decomposition. Both volatile and nonvolatile solutes react in the liquid bulk with the free radicals released by the collapsing bubbles, particularly with •OH. Upon stationary ultrasound irradiation, the sonochemical output of radicals in the solution reaches a steady state. Recombination reactions occur both in the gaseous cavity and in the solution, resulting mainly in the production of H_2 and H_2O_2, the latter being related to many secondary reactions in the liquid phase. It is important to keep in mind that H_2 and H_2O_2 are, by far, the main products of water sonolysis and the main sonochemical products when aqueous solutions are irradiated with ultrasound. The chemical process of interest (e.g., the reaction of a pollutant with •OH radicals in the bulk solution) is, then, always a secondary reaction from an energy balance point of view.

Eventually, if H_2O_2 does not react with other solutes, it disproportionates, yielding H_2O and O_2 as follows:

$$H_2O_2(g) \rightarrow H_2O(g) + 1/2\ O_2(g) \tag{5}$$

For all these reasons, most of the acoustical energy involved in generating the cavities and in their collapse is ultimately spent in decomposing water into H_2 and O_2. This is the main factor affecting sonochemical efficiency (i.e., the ratio between the rate of the reaction of interest and the applied power density, W/L). In order to improve the efficiency of a sonochemical process, chemical or physical modifications can be introduced into the system, which may reduce this loss (see Sec. IV.G). The efficiency can also be affected by the presence of other chemicals in the solution, which may react with the radicals, thus reducing the number of reactive species available to the target molecules. A "preprocess" might be conceived to separate some of these unwanted chemicals from the solution prior to sonochemical treatment.

The implosive nature of the bubble collapse triggers many chemical and physical processes on a short time scale, driven by the exponential increase in temperature. Together with the water molecule splitting (Eq. (4)), other side reactions take place inside the bubbles involving other radicals generated from water, gases present such as O_2 and N_2, and volatile solutes [11]. Radical recombination is one of the sources related to the emission of light during the sonication of water, known as *sonoluminescence*. This light radiation induced by the ultrasonic irradiation of the liquid consists of a wide polychromatic emission in the visible and UV region of the spectrum.

C. Effects of Experimental Parameters on Cavitation

As was already outlined, the experimental conditions for sonochemistry must be carefully considered when a process is designed, and these conditions must be carefully controlled during operation. Here is a brief account of the main parameters influencing cavitation chemistry.

1. Ultrasonic Frequency

The frequency of the ultrasound determines the critical size of the cavitation bubbles. The reaction rate dependence on the ultrasonic frequency has been observed in many cases [12–14]. Usually, an optimum intermediate value of frequency exists, lying in the range of hundreds of kilohertz. For volatile solutes reacting inside the cavity, this effect can be understood as a balance between increasing numbers of excited bubbles and a lower solute content because the maximum radius of transient cavitation diminishes with the frequency [12]. Early sonochemical works employing higher ultrasonic frequencies include reports by Petrier et al. [15,16] and Kruus and Entezari [17]. Petrier et al. showed a fivefold enhancement in the reaction rate by simply raising the ultrasonic frequency from 20 to 487 kHz. Kruus and Entezari found a 30-fold enhancement by raising the frequency from 20 to 900 kHz.

2. Intensity of the Acoustical Wave

Many authors report a fairly linear enhancement of reaction rates with increasing power density (power applied/irradiated volume ratio, W/L) [14,18–20]. A saturation power was reached in some cases [21], probably related to the formation of clouds of cavitation bubbles near the transducer, which block the energy transmitted from the probe to the fluid, at high intensities.

3. Nature of the Background Gas

The choice of the saturation gas is critical. When Ar and Kr were sparged in water irradiated at 513 kHz, an enhancement in the production of •OH radicals of between 10% and 20%, respectively, was observed, compared with O_2-saturated solutions [22]. The higher temperatures achieved with the noble gases upon bubble collapse under quasi-adiabatic conditions account for the observed difference. Because the rate of sonochemical degradation is directly linked to the steady state concentration of •OH radicals, the acceleration of those reactions is expected in the presence of such background gases. The use of ozone as saturation gas (in mixtures with O_2) provided new reaction pathways in the gas phase inside the bubbles, which also increase the measured reaction rates (see Sect. IV.G.1).

4. Temperature

The concentration of volatile compounds in the cavitation bubbles increases with temperature; thus, faster degradation rates are observed at higher temperatures for those compounds [23]. Conversely, in the case of nonvolatile substrates (that react through radicals reactions in solution), the effect of temperature is somehow opposed to the chemical "common sense." In these cases, an increase in the ambient reaction temperature results in an overall decrease in the sonochemical reaction rates [24]. The major effect of temperature on the cavitation phenomenon is achieved through the vapor pressure of the solvent. The presence of water vapor inside the cavity, although essential to the sonochemical phenomenon, reduces the amount of energy available upon collapse because it cushions the implosion through a recondensation on the bubble walls and an enhancement of thermal transport.

5. Nature of the Solvent

For the same reason as above, excess solvent molecules in the cavitation bubble also seriously limit the applicability of many volatile organic solvents as a medium for sonochemical reactions [2,25,26]. In fact, water becomes a unique solvent in many cases, combining its low vapor pressure, high surface tension, and viscosity with a high yield of active radical output in solution. Its higher cavitation threshold results in subsequently higher final temperatures and pressures upon bubble collapse. Most environmental remediation problems deal with aqueous solutions, whereas organic solvents are mostly used in synthesis and polymer modifications processes.

6. Reactor Geometry and Operation Conditions

Most sonochemical reactors are based on acoustical cavitation, which is generated by one or more transducers. The most common transducer configurations consist of a planar source (usually for frequencies in the kilohertz range) or a probe immersed in the solution, provided with a replaceable tip (operating in the 20–50 kHz range) [1,2]. The control of temperature, stirring, inlet and outlet ports, and circulation flow rates between different units of the sonoreactor should optimize the reaction yields. In flow reactors, the reaction rate dependence upon the residence time of the solution in the reactor should be evaluated in detail, and the flow rate can become a critical operational parameter. In both flow and batch reactors, proper focusing and proper reflection of ultrasound increase their utilization. Gupta and Wallace [27] showed a twofold enhancement in the oxidation rate of KI (aq) by modifying the configuration of an efficient reactor vessel operating at 660 kHz. The scale-up of ultrasonic reactors to pilot plant size has been reported for chemical synthesis [28] and water treatment applications [29].

7. Sample Features

Matrix effects can be complex and difficult to predict because most co-solutes may compete with the reactant of interest for reactions with radicals, becoming effective scavengers that reduce the sonochemical efficiency. Taylor et al. [30] have observed a significant inhibition of the sonolysis of polycyclic aromatic hydrocarbons (PAHs) in the presence of dissolved organic matter. Substrate concentration effects on the rate constants have also been reported. When the target molecules are volatile, they partition between the bulk solution and the vapor phase inside the cavitating bubbles [12,31], where reactions are faster. Another case where concentration effects are remarkable is surfactant degradation. The amphiphilic molecules distribute between the liquid and the gas–liquid interfaces of the cavities, where sonochemical reactivity is also enhanced [32]. In both cases (volatile and surface active compounds), a lower concentration of the substrate molecules increases the sonochemical efficiency because the fraction of these compounds in the bubble or in the interface is higher.

III. ULTRASONIC INSTRUMENTATION

A. Some Ultrasonic Basics: Near-Megahertz Sonochemistry

In the past, it was argued that 20–50 kHz was the optimal frequency range for sonochemistry [1,33]. More recently, however, a large number of experiments have shown that this is often not the case. Generally, sonochemistry progresses as well, if not better, at the near-megahertz frequencies (i.e., 100–1000 kHz; see Sec. II.C.1). It is expected that near-megahertz ultrasonic frequencies will soon become as important to sonochemistry as 20–50 kHz has been in the past. Generally, higher-frequency ultrasound behaves differently because the structures involved are large in terms of the ultrasonic wavelength. For this reason, we will briefly introduce a few equations, which should give the sonochemist a passing knowledge of the behavior of beams of sound at the near-megahertz frequencies [33,34]. These dimensional relationships have been summarized in Fig. 3.

Sonochemistry is conducted at such high sound amplitudes that water can no longer be considered a linear material [35]. This means, among other things, that the local velocity of the ultrasound in the solution is dependent on the local sound amplitude. It is thus difficult to mathematically predict the resulting sound pattern at points removed from the transducer, even if the sound pattern is well known near the transducer. On the other hand, sound propagation through a linear material is well known and can be computed very accurately. Thus, most attempts at mathematically predicting nonlinear sound fields will only go so far as to predict the field in a linear material, and then to assert that the nonlinear field is similar. This is the approach followed here, and nonlinear effects will be discussed where possible. The acoustical wave velocity c in liquids is approximately defined by the equation $c^2 = B/\rho$, where B is the bulk modulus, which has the units of pressure required to compress (or shrink in volume) the bulk fluid by a known amount, and ρ is the mass density (mass per unit volume) of the liquid [34]. Note that the presence of any bubbles in the solution will

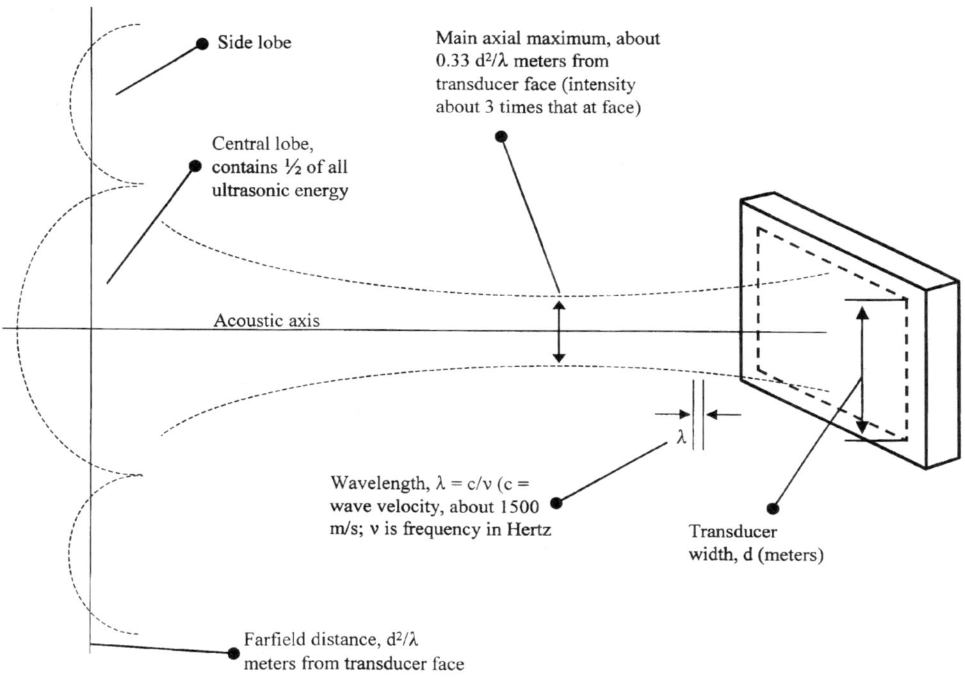

Figure 3 Beam geometry of linear ultrasound.

decrease both B and ρ, generally by unequal amounts, with a net effect of reducing c. For instance, B for pure water is 316,000 psi, and because of this large bulk modulus, water is often considered incompressible. It is obvious that the addition of even a few bubbles to the water would dramatically increase this compressibility and dramatically decrease B, while only affecting ρ by a small amount. Such bubbles are easily generated by the passing of large waves through the solution, as was described in Sec. II.A.

The wavelength of linear or low-amplitude sound in water [m] is: $\lambda = c/v$. Here, the wave velocity c is approximately 1500 m/sec, and v is the frequency in Hertz. Sound from a flat source (called a piston source if round, or a plate source if rectangular) does not retain the shape of the source as it propagates away, but spreads out fairly predictably. The rate of this spreading is often stated as the beamwidth, which means the width of a cone [°], that contains one-half the power of the beam. This width of this cone follows the relationship $\sin(\theta) = \lambda/d$, where d is the diameter or width of the source [m]. However, this equation loses its utility for d smaller than one wavelength because there are no angles θ with $\sin(\theta)$ greater than one. When this condition is encountered, it simply means that the flat source is tending to become a spherical radiator and to project its sound equally in all directions. Let us consider a numerical example: the wavelength of a 20-kHz beam in water is 7.5 cm, or 2.95 in. The wavelength of 500 kHz sound is 3 mm, or 0.118 in. Comparing a 1-in.-diameter plate source at 20 and 500 kHz, we find that the 20-kHz beam is approaching a spherical radiator, whereas the 500-kHz beam is concentrated into a fairly narrow beam only 7° wide.

Sound does not begin to spread exactly at the transducer's surface. The closest distance from the transducer where one will find a well-formed beam is called the farfield distance, and is approximately d^2/λ m from the transducer. The farfield distance is sometimes called the "d^2 over λ distance." From this point to infinity (or until the beam encounters a reflector or refractor), the beam retains the same general shape, spreading as predicted by its beamwidth. The farfield distance of our 20-kHz example is about 9 mm, and about 21.5 cm for the 500-kHz source. By contrast, the nearfield of a source is the region near the transducer. In this region, the sound field basically looks like the source, and the nearfield is sometimes called the shadow zone. As one moves away from the transducer, there is a point, at about one-third of the way to the farfield distance, where the sound field's width has shrunk to just over half of the plate's width. There is a commensurate increase in intensity to about three times the level at the transducer. This point in the beam is sometimes called the main axial maximum of the beam. Many high-frequency sonochemistry sources will cavitate the water at this point. However, this point is always moved closer to the transducer by the nonlinear, self-focusing action of the intense sound field [36]. At higher frequencies, it is sometimes possible to conduct sonochemistry at this point in the beam, without further focusing.

Between the main axial maximum and the farfield distance, the beam widens and takes on the familiar beam pattern. This pattern consists of a main lobe surrounded by many sidelobes of lower intensity. It is helpful to know that, at the farfield distance, the main lobe is again about the width of the source, and the intensity is about 80% of the intensity at the transducer face (not including losses due to absorption). This information has been condensed into Figs. 3 and 4. Fig. 4 illustrates a computer simulation from a three-dimensional solution of the wave equation considering a 1-in. square plate transducer operating at 500 kHz and utilizing only the propagating terms. Ignoring the evanescent wave contribution results in a small error, as can be seen in Fig. 4, where the intensity at the transducer face is incorrectly predicted to be slightly less than one. The x-axis has been normalized by dividing distance by the farfield distance; the maximum on the x-axis (one)

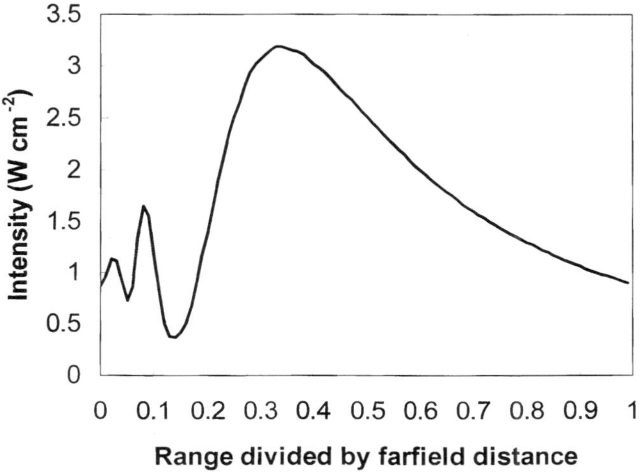

Figure 4 Normalized intensity for a plate transducer, 1×1 in., 500 kHz.

corresponds to the farfield distance (21.5 cm). The main axial maxima can be seen at a distance of about 0.33 d^2/l. From this point outward, the intensity is seen to decrease in a way that approaches the familiar $1/r^2$ pattern. In summary, the ultrasound originating from the transducer is "choked down" to three times the intensity at one-third the farfield distance, then reaches about 80% of the original size and intensity at the farfield. By conducting sonochemistry anywhere between the transducer and the farfield, one is assured that the intensity is somewhere between 0.8 and three times the intensity at the source.

Cavitation begins at much smaller intensities when low sound frequencies are applied. Fig. 5 describes how the threshold intensity increases with increasing frequency. Drawing a vertical line at approximately 20 kHz, as one moves up this vertical line, wave intensity increases [W/cm^2]. The first thing one encounters as the intensity is increased is the curve for "aerated water," or water saturated with air. The intensity at this point is sufficient to produce cavitation as desorbed air contributes to bubble nucleation. As one continues to increase intensity, one will encounter the curve for degassed cavitation. This intensity is the absolute maximum intensity allowed (at standard conditions) for sound traveling in water at this frequency. Most of sonochemistry are performed at intensity levels between these two values.

The discussion above of the main axial maximum is a good example of a nonlinear effect. As the sound reaches the main axial maximum, the intensity can become sufficient to approach the lower curve on Fig. 5, and

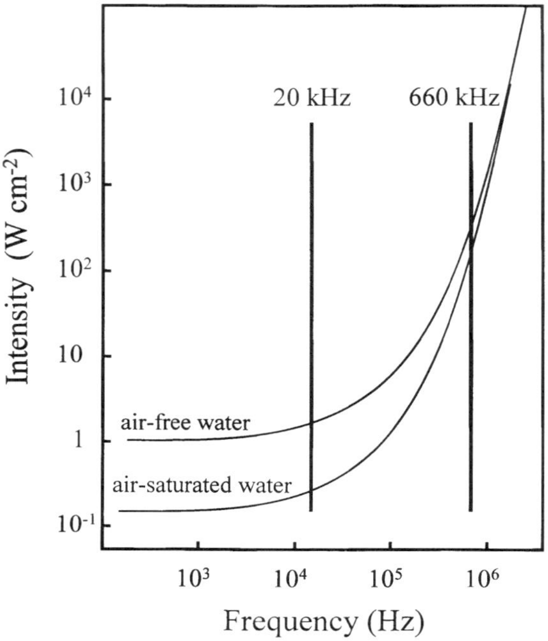

Figure 5 Variation in threshold intensity with frequency. (From Ref. 1.)

gaseous cavitation can begin. This creates local bubbles in the water, which reduce the bulk modulus of the water, thus reducing the local sound velocity. Because there are more bubbles in the center of the beam where the intensity is highest, the sound waves travel slowest at the beam's center and "self-focusing" occurs, which steers more sound towards the center of the beam, and so on. The result of these effects is that the apparent main axial maxima are moved closer to the transducer source.

B. Role of Higher Ultrasonic Frequencies in the "Scale-Up" of Sonochemical Processes

A scale-up of a sonochemical process is usually required in order to treat commercially viable quantities of a solution. It is now becoming apparent that higher ultrasonic frequencies present perhaps the best way to scale up a process. The energy required to cavitate water is provided by the transducer in the form of mechanical waves. How does this energy ultimately dissociate the water molecule and produce chemical reactions? Fig. 5 shows that for the lower ultrasonic frequencies (i.e., 20 kHz), the water itself cannot support compressional wave energy in excess of a few watts per square centimeter. Yet, there are many 20-kHz sonochemical experiments that have been performed at claimed energies in excess of tens of watts per square centimeter. How has this been accomplished? How does the energy get to the cavitating bubble if it cannot be transmitted through the surrounding water from the tip of the ultrasonic horn transducer? It would appear that all high-intensity sonochemical experiments at low frequencies are performed at, or very near, the surface of the horn's tip. To this extent, low-frequency sonochemistry would appear to be a surface effect: it is confined to a thin zone in intimate contact with the tip of the horn. An undesired consequence of this is tip erosion due to the destructive effects of cavitation [1,37]. Cavitation erosion is not a phenomenon unique to sonochemistry. It has long been recognized as a limit in mechanical pumps and ship propellers [38–42]. In order to scale up a low-frequency sonochemical process, the horn tip must be increased in area. Instead, higher frequencies can be used to cavitate the liquid far from the surface of the transducer. Referring back to Fig. 5, note that at just above 1 MHz, water can carry intensities on the order of 1000 W/cm^2. Thus, at higher ultrasonic frequencies, cavitation is not limited to a thin zone near the surface of the transducer and can extend into the bulk of the solution, where it can be focused and harnessed for use in sonochemistry.

C. Incidental Destructive Effects of Cavitation

Cavitation generally has a destructive nature, as was previously mentioned in the case of ship propellers and pump impellers. Sonochemistry depends on harnessing this destructive nature of high-intensity sound fields into desired processes, and it is necessary to manage the cavitation in such a way that the equipment producing the cavitation is not also destroyed. As we have discussed, low-frequency ultrasound tends to destroy the titanium tips of the applicators used to deliver it [1]. There is a rule of ultrasound production that has been adhered to by the sonar community for many years, "To prolong transducer life, never allow the water to cavitate at the face of the transducer." Cavitation at the transducer face will ultimately lead to tiny microcracks in the materials of the transducer. These tiny cracks will eventually cause transducer failure. Sonochemists need reliable, dependable sources of cavitation that will operate hour after hour, even year after year, to successfully bring a sonochemical process from the laboratory to the plant. Thus, low-frequency cavitation is limited to the transducer's face, in violation of the rule above. But high-frequency ultrasound can be generated at low intensities (at the transducer face) and then brought up in intensity to the cavitation threshold by focusing at a remote point, preserving the integrity of the transducer.

It can be argued that proper low-frequency transducers would become too large and expensive if designed similarly to the near-megahertz transducers, to appeal to the marketplace. To deliver 1 kW of acoustical power with an excitation limit (at the face) of 1/4 W would require a transducer face or horn tip of around 4000 cm^2, or a diameter on the order of 0.71 m.

Focusing involves converting expanding wave fronts into collapsing wave fronts. A planar transducer face or horn tip being driven uniformly sinusoidally over its entire extent generates plane acoustical wave fronts in the working medium. The resulting family of wave fronts in the working medium is spaced one acoustical wavelength apart, when measured along their direction of propagation, perpendicular to the transducer face or horn tip. At short distances from the transducer face, these wave fronts have basically the same shape as the transducer face or horn tip. As this distance in increased, the wave fronts begin to take on shapes other than that of the transducer face, as dictated by the laws of wave propagation. At large distances from the transducer face or horn tip, these wave fronts have acquired an expanding geometry, wherein they are no longer flat (or planar) but have acquired the shape of spherical sections, with the origin of the spheres being near the transducer face or horn tip.

A characteristic of these expanding wave fronts is that the acoustical energy of subsequent wave fronts is decreasing because each subsequent wave front is slightly larger than its predecessor. Thus, the "surface energy density" [W/cm^2] of such expanding wave fronts becomes lower as the distance is increased from the transducer face or horn tip.

Expanding wave fronts can be converted into collapsing wave fronts by reflection from concave acoustical reflectors. A collapsing, or focused, wave front actually increases watts per square centimeter as it travels. Because each successive wave front is reduced in area from its predecessor, it must carry a higher watts per square centimeter than its predecessor.

An expanding wave front looks like the surface of a balloon that is being inflated. The amount of energy it carries is constant (except for absorption), and so the larger is the wave front, the lower is the intensity in watts per square centimeter. It helps to visualize the intensity as the thickness of the balloon. A collapsing wave front looks like a balloon that is being deflated. The smaller the wave front (balloon) becomes, the thicker is the balloon and the higher is the intensity. This continues until cavitation begins. Cavitation produces bubbles, and sound in a solution cannot pass unaffected through a region of bubbles. The bubbles scatter the otherwise perfect collapsing wave fronts, and generate a new set of expanding wave fronts originating in the region of the bubbles. Whether or not these newly created expanding wave fronts are focused again, to produce more cavitation, is a subject of proper sonochemical vessel design.

Focusing is accomplished by three concentrating methods: (1) mirrors, (2) lenses, and (3) conformal or mechanically focused transducers. The low-frequency probes, consisting of an ultrasonic actuator coupled to the solution by a tapered titanium horn, fall between (2) and (3). For high-intensity work, lenses can prove unreliable because they are composed of materials, often polystyrene, which themselves can degrade over time at the intensities required. Mirrors are often used and can be easily made from thin, concave, air-backed glass (watch glass). The major disadvantage of mirrors is that the path length of the ultrasound from the transducer to the focal zone is about twice that of the conformal transducer's path length. Two examples of conformal transducers are shown in Fig. 6.

D. Absorption of Ultrasound

Absorption is an unavoidable consequence of passing ultrasonic waves through matter. In particular, to reach the cavitation zone in a solution, large-amplitude ultrasound must pass through the solution. Generally, the higher is the viscosity of a solution, the higher is the ultrasonic absorption. There is an additional loss associated with finely divided particles, known as

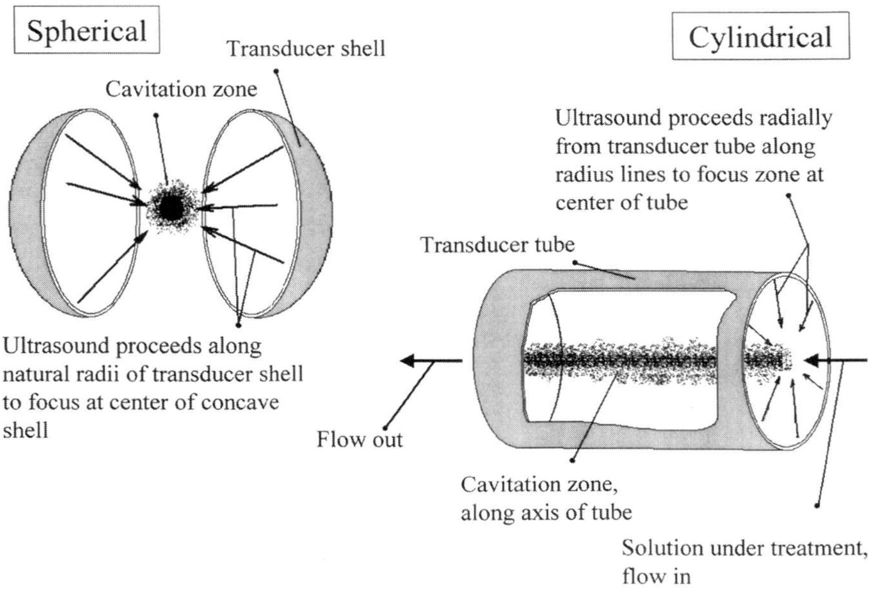

Figure 6 Conformal transducers.

fines, suspended in the solution [34]. The attenuation of sound in some solutions is reported in the section, "Acoustic Properties of Liquids" of the *American Institute of Physics Handbook* [43]. Attenuation and ultrasonic velocity are also subjects of active research [44–46].

Ultrasonic energy, generated at great expense in the transducer, should not be lost to absorption while in transit to the cavitation zone. For this reason, conformal transducers are usually the best choice for process transducers, in that they allow for the shortest distance from the transducer to the cavitation zone. On the other hand, for laboratory work, mirrors offer easy separation of the waste heat generated in the transducer from the heat generated by ultrasound in the solution under test. Thus during process development, when accurate determinations of acoustical output is desired, judicious use of mirrors and windows is often called for.

All ultrasonic transducers operate at something less than 100% efficiency. Of the electrical energy delivered to a transducer, some fraction is converted directly into acoustical energy. This fraction defines the transducer efficiency, and the ultimate fate of this acoustical energy is to be absorbed as heat. The other fraction of the electrical energy never produces acoustical energy: it is converted directly into heat in the process of providing the ultrasonic output. Thus, all the electrical energy delivered to the transducer ultimately becomes heat energy. Some of it goes directly into heat, and some of it goes through a phase as acoustical energy before being converted into heat. It is a small fraction of the energy in this acoustical phase which is utilized by sonochemical reactions.

The desire of the sonochemist is to know how much ultrasonic energy is being provided and absorbed in the acoustical phase, and not to be confused by the waste heat associated with the transducer's inefficiency. In many calorimetric calibration configurations, these two sources of heat are not suitably separated. Consider, for instance, a titanium probe immersed into a water bath of known volume, run for a known time at a known electrical power input, and whose temperature rise is noted. Such a calibration cannot distinguish between the waste heat and the acoustic-phase heat. Indeed, if the transducer were replaced by an electrical heater, with no ultrasonic output, the results would be the same as for the transducer.

Consider the configuration of Fig. 7. Here the sonochemical vessel is immersed in a tank packed in ice. The transducer is mounted several inches from the open end of the vessel, which is covered by an acoustical window. This window allows the ultrasonic energy to enter the vessel, while preventing the mixing of the tank water (which contains the waste heat of the transducer) with the solution under test. Thus, the transducer waste heat is confined to the tank water, and has no opportunity to enter the treatment

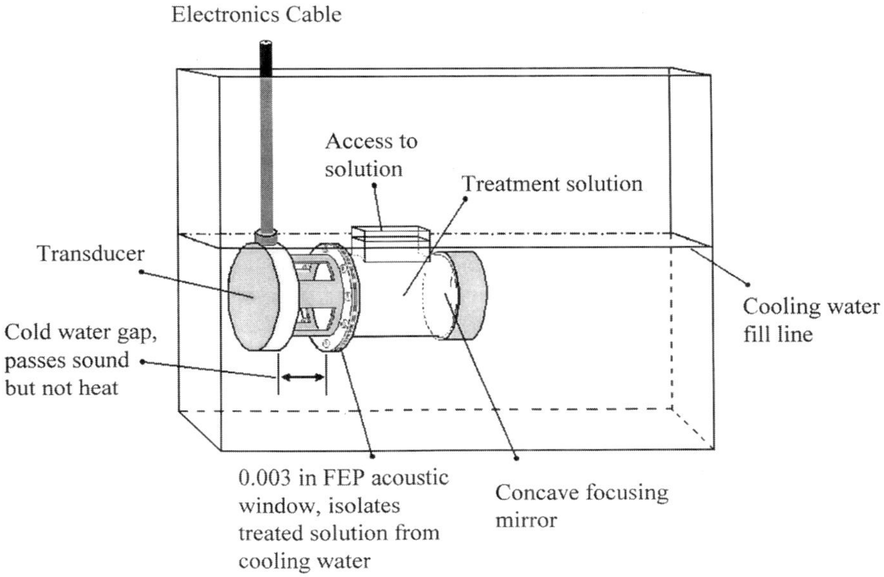

Figure 7 Ice bath-cooled transducer.

vessel. As a practical matter, only energy in the acoustical phase will make the transit across the several-inch space of cold water.

Thin sheets of polyethylene or FEP Teflon® (approximately 0.003 in. thick) can be used as acoustical windows. However, these materials must be placed close to the transducer face, in the region of low ultrasonic intensity, for them not to degrade during an experiment. Furthermore, they should be placed far enough from the transducer face so as not to interfere with the cooling of the transducer. Also, if the window is placed too close to the cavitation zone, it will experience the degrading effects of the ultrasound. Fig. 7 shows how a piston source can be coupled to a focusing mirror through an acoustical window. Fig. 7 also shows how an accurate calibration can be made by measuring the temperature rise in the vessel (accounting for heat loss to the cooling water).

Note that such a calibration only measures the ultrasonic energy that actually enters the vessel and is converted into heat in the vessel. The waste heat generated in the transducer as it produces the ultrasonic output does not introduce a calibration error. Nor does ultrasonic energy, which is lost in the transit from the transducer to the window (or reflected back into the cooling water from the window), introduce any calibration error.

IV. SONOLYSIS OF ENVIRONMENTAL POLLUTANTS

In the last decade, sonochemistry was successfully employed to eliminate a variety of organic pollutants from water on the laboratory scale. Here we summarize these results, which illustrate the promising applicability of this advanced oxidation technique in solving environmental problems.

A. Chlorinated Organic Chemicals

One of the most significant achievements in the sonochemical treatment of contaminants has been the degradation of CCl_4 in water [31,47,48]. Carbon tetrachloride is widespread in the environment due to its refractory nature and its large-scale production. It is considered as one of the most harmful volatile organic compounds (VOCs) due to its short-term and chronic human health effects and because it is a suspected human carcinogen. Under ambient conditions, it does not degrade easily and is unreactive towards hydroxyl radical (half-life > 330 years). Sonochemistry provides a unique, high-temperature gaseous environment inside the cavitation bubbles where the thermolysis of CCl_4 molecules takes place at fast rates, yielding •Cl and $\bullet CCl_3$ radicals as primary intermediate products. These radicals further react with •OH radicals or O_2, yielding stable HCl, HOCl, Cl_2, and CO_2 as final degradation products. The following gas phase reactions have been proposed by Hua and Hoffmann [31] for the sonolysis of CCl_4 inside the cavitation bubbles:

$$CCl_4(g) \rightarrow \bullet Cl(g) + \bullet CCl_3(g) \tag{6}$$

$$\bullet CCl_3(g) \rightarrow \bullet Cl(g) + :CCl_2(g) \tag{7}$$

$$2\bullet Cl(g) \rightarrow Cl_2(g) \tag{8}$$

$$CCl_4(g) \rightarrow :CCl_2(g) + Cl_2(g) \tag{9}$$

$$\bullet CCl_3(g) + \bullet OH(g) \rightarrow HOCCl_3(g) \tag{10}$$

$$HOCCl_3(g) \rightarrow HCl(g) + COCl_2(g) \tag{11}$$

$$\bullet CCl_3(g) + O_2(g) \rightarrow \bullet O - OCCl_3(g) \tag{12}$$

$$\bullet O - OCCl_3(g) \rightarrow HOCl(g) + COCl_2(g) \tag{13}$$

The hydrolysis of phosgene proceeds quickly once $COCl_2$ dissolves in the liquid phase, yielding CO_2 and H_2O as stable products. Part of the released Cl_2 hydrolyzes, yielding HCl and HOCl. The production of transient dichlorocarbene (:CCl_2) was also verified through its reaction with the scavenger, tetramethyl ethylene. The final products of dichlorocarbene hydrolysis are CO and HCl. Hexachloroethane, which is formed as the primary intermediate in the recombination of two •CCl_3 radicals, degrades at the same rate as CCl_4 [47]. Working at 20 kHz and 135 W output power, 90% of the initial CCl_4 ($C_0 = 2\times10^{-4}$ mol dm^{-3} and $C_0 = 2\times 10^{-5}$ mol dm^{-3}) was destroyed in only 12 min. Nearly 70% of the chlorine atoms remained in the solution as HCl and HOCl. Another important finding related with CCl_4 is that when multicomponent samples were treated [i.e., mixtures of CCl_4 with other contaminants such as *p*-nitrophenol (PNP)], the degradation rate of the concomitant reagent was enhanced [31]. This may result from the fact that CCl_4 degradation releases a residual oxidant, HOCl, which can continue to attack other refractory molecules a in solution. This result is relevant because actual environmental samples likely contain a mixture of contaminants rather than CCl_4 alone.

The sonochemical degradation of many other chloromethanes and chloroethanes was reported by Bhatnagar and Cheung [49]. Aqueous solutions of CCl_2H_2, CCl_3H, CCl_4, 1,2-dichloroethane, 1,1,1-trichloroethane, trichloroethylene, and perchloroethylene in concentrations ranging from 50 to 350 ppm were treated with 20-kHz ultrasound at a power density of 100 W/L. In less than 30 min, a depletion of the target molecules was observed, ranging from 72% to 100%. HCl is the main product reported in each case. The degradation rate dependence on the frequency and on the Henry's law constants for some of these compounds was analyzed by Colussi et al. [12]. Working in the near-megahertz range, an optimum frequency was found at 600 kHz for all species. The degradation rate constant increased for higher Henry's constants (H), as can be expected for solutes of increasing volatility, but the nonlinear dependence observed ($k \sim H^{0.30}$) reveals that the rate constants (k) are not solely determined by equilibrium parameters.

Chlorophenols are other chlorinated organic compounds of environmental interest. Petrier et al. [50] performed the total degradation of pentachlorophenate (PCP) in aqueous solutions saturated with O_2 and Ar in less than 1 hr of irradiation with 530 kHz at 200 W/L. An almost quantitative conversion to HCl was observed under O_2, whereas a 70–80% dechlorination was measured when Ar was the background gas. The mineralization of the sample, measured as the volume of CO_2 released, was accomplished with only 30% of the theoretical yield. The sonolysis of 2-chlorophenol, 3-chlorophenol, and 4-chlorophenol with 20 kHz and 500 W/L was reported by Serpone et al. [51]. These molecules experienced a much slower reaction than PCPs, degrading within 700–800 min. Dechlorination was also nearly quantitative, and it was shown to occur soon after the initiation of the disappearance of the initial substrate. Weavers et al. [52] compared the degradation rate for 4-chlorophenol at 20 and 500 kHz, finding the latter to be one order of magnitude faster, even if the power applied was 2.5 times higher for the lower frequency. This is a good example of how the near-megahertz ultrasonic frequency, also employed by Petrier et al. in the degradation of PCPs, resulted in a more effective sonolysis compared with lower frequencies.

B. Aromatic Compounds

The sonochemical degradation rates for several aromatic compounds of environmental interest such as phenol [15], nitrobenzene [52], and nitrophenol [18,52,53] have been determined recently. In the three cases, two frequencies were analyzed (20 and ~ 500 kHz), with the treatment being more efficient at the higher frequency. The same tendency as chlorophenol was observed: a

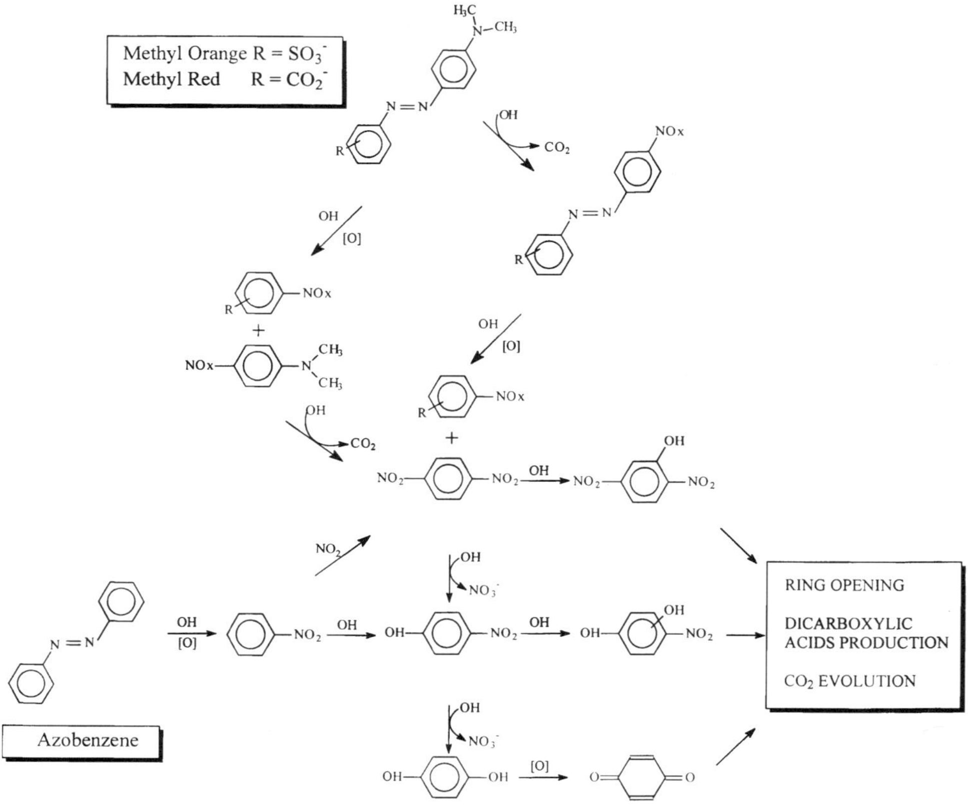

Figure 8 Sonochemical degradation of azobenzene, methyl orange, and methyl red. (From Ref. 58.)

degradation that was an order of magnitude faster in the near-megahertz range, operating at the same sound power, or even at a lower sound power. The dependence of the degradation rate on the initial concentration of phenol [15] showed limiting values for initial substrate concentrations higher than 5 mM. Above those levels, only the available •OH radicals produced in the cavitation bubbles limit the reaction rate. A parallel reduction in the rate of production of H_2O_2 was observed with increasing initial concentration of substrate, showing the competition of the degradation reaction with the recombination of •OH radicals to yield H_2O_2. The mechanism of the reaction of PNP was studied in detail by Kotronarou et al. [53], revealing that denitration is the primary process occurring in the gas phase inside the cavitation bubbles:

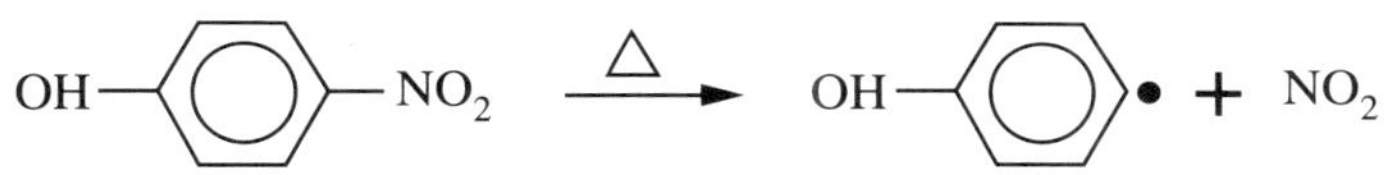

(14)

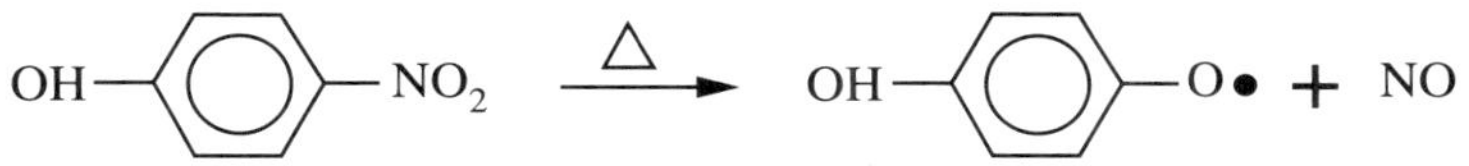

(15)

These thermolytic reactions yield the stable NO_2^- and NO_3^- anions (from a combination of NO and NO_2 with •OH radicals) and hydroquinone and benzoquinone as major by-products. Further oxidative ring-opening reactions yield organic acids such as formic and oxalic acids. The addition of •OH radicals to the aromatic ring is another reaction observed, yielding nitrocatechol and the same products after the oxidative ring opening. These reactions are included in Fig. 8 to illustrate the degradation of azo dyes (see Sec. IV.F), from which nitrophenol is an important by-product. The degradation of PNP in a nearfield acoustical processor (NAP) operating at 16 and 20 kHz was also investigated by Hua et al. [18]. A linear increase in the reaction rate with the acoustical intensity was registered in the range 1000–7000 W/L. The effect of employing different background gases was investigated, and Ar was found to be more efficient than O_2. A 4:1 Ar/O_2 mixture yielded the highest degradation rate.

C. Pesticides

The degradation of parathion (*O,O*-diethyl *O-p*-nitrophenyl thiophosphate) was reported by Kotronarou et al. [54], operating a 20-kHz probe reactor at 75 W with 25 mL of a saturated aqueous solution of the pesticide ($C_0 = 82$ μM) at 30 °C. Under these mild irradiation conditions, the substrate was completely degraded in less than 2 hr, yielding inorganic products such as SO_4^{2-}, PO_4^{3-}, NO_3^-, and CO_2, together with ethanol and light organic acids, as well as intermediate aromatic compounds like *p*-nitrophenol and benzoquinone. Fig. 9 describes the observed reaction. Most products were detected and quantified using ion chromatography (IC). The concentration of PNP was determined spectrophotometrically.

Parathion, having a low vapor pressure, is not expected to partition into the gas phase of the cavitating bubbles, but most of the organic byproducts do so and follow the same processes mentioned in Sec. IV.B for PNP. The initial reaction of the substrate with •OH radicals in the bulk solution is slower than those described previously for PNP in the gas phase of the bubbles. Consequently, the bulk aqueous phase reaction of parathion with hydroxyl radical is the rate-limiting step for the overall process.

D. Methyl *tert*-Butyl Ether (MTBE)

The contamination of groundwater with MTBE is a very recent environmental problem, which is related to MTBE use as a gasoline additive. Leakage from underground storage tanks has produced very negative

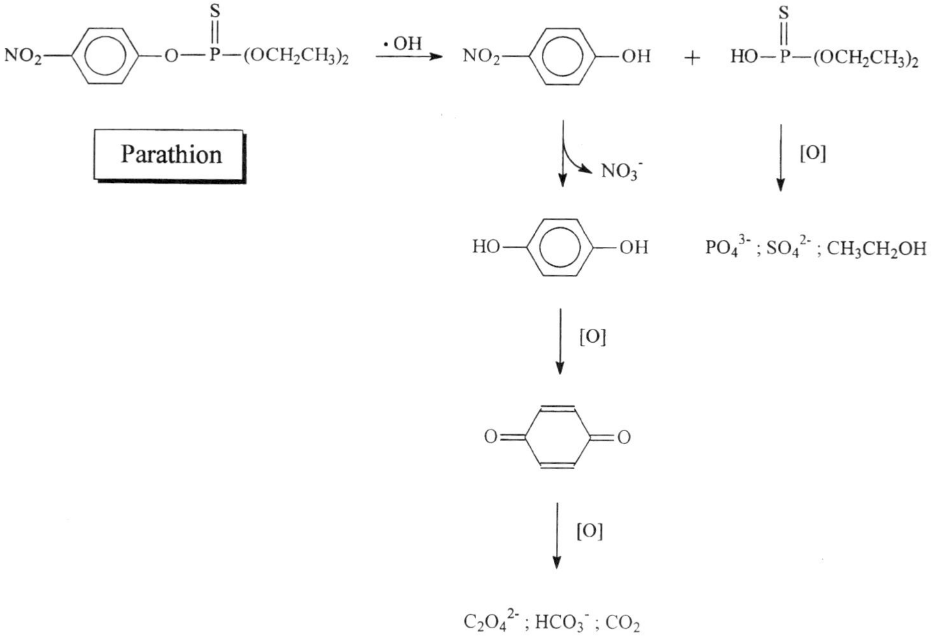

Figure 9 Sonochemical degradation of Parathion. (From Ref. 54.)

impacts on water quality. MTBE is highly soluble in water (0.35–0.71 M), is resistant to both aerobic and anaerobic microbial degradation, is poorly adsorbed on activated carbon, and has low volatility. The sonochemical degradation of aqueous MTBE was studied by Kang et al. [14] and Kang and Hoffmann [55] under O_2 saturation, working at 205 kHz and 200 W/L at 20 °C. An almost complete removal of MTBE was achieved in less than 1 hr for initial concentrations of the pollutant in the range $C_0 = 0.01$–1 mM. The observed first-order degradation rate constant increased from 4.1×10^{-4} to 8.5×10^{-4} sec^{-1} as the concentration of MTBE decreased. The reaction with •OH radicals in a solution proceeds via H atom abstraction from the methyl group, resulting in the formation of *tert*-butyl formate. The hydrolysis of this ester yields formic acid and *tert*-butyl alcohol, which further reacts to yield acetone, methanol, and formaldehyde. Because these last steps involve molecules that are relatively volatile, most of these reactions occur via pyrolysis in the cavitation bubbles. The sonolysis of MTBE and other related ethers was also reported by Ondruschka et al. [56].

E. Surfactants

Surface active compounds are discharged into domestic and industrial wastewater in large amounts, becoming ubiquitous in the environment. Alkylphenol ethoxylates (APEs), a group of nonionic surfactants, have been the subject of controversy because it was found recently that alkylphenols, produced during the biodegradation of APE, exhibit some estrogenic activities. Most techniques employed in the degradation of these molecules (including ozonation and photochemical degradation) are not able to destroy the toxic by-product. In a cavitated liquid, surfactant molecules tend to concentrate in the gas–liquid interface of the bubbles, with their hydrophobic tails pointing into the gas phase. Therefore, sonodegradation becomes a unique method where the active radicals are able to reach the alkyl chains and degrade them preferentially. Thus, the sonochemical attack is focused into the hydrophobic part of the molecule. Destaillats et al. [32] studied the degradation of Triton X-100 (*tert*-octylphenoxypolyethoxyethanol), a commercial APE, and *tert*-octylphenol, the corresponding alkylphenol, at 358 kHz and 50 W (83 W/L) at 15 °C under air saturation. Alkylphenols, or short-chain ethoxylated phenols, were not generated as by-products of Triton X-100 degradation. Instead, many products derived from the oxidation of the hydrophobic chain were identified. Similar results were observed when *tert*-octylphenol was treated. The reaction rate constant exhibited a drastic dependence on the concentration of the surfactant. When the initial concentration was above the critical micelle concentration (CMC), the degradation rate became very slow (with half-life times on the order of 700–1000 min). Below the CMC, the reaction proceeded much faster, with half-lives for concentrations below the CMC ranging from 85 min (for $C_0 = 0.11$ mM) to 32 min (for $C_0 = 0.03$ mM). These results indicate that micelle formation serves to isolate effectively the free surfactant monomers from the liquid–gas interface of the cavitation bubbles, decreasing the overall efficiency of the sonochemical reaction.

F. Dyes

Azo dyes constitute a major part of all commercial dyes employed in a wide range of processes in the textile, paper, food, cosmetics, and pharmaceutical industries. They are characterized by the presence of the azo group (NñN) attached to two substituents, mainly benzene or naphthalene derivatives, containing electron-withdrawing and/or electron-donating groups. The major sources of dyes in the environment are effluents from the textile industry. Due to their high solubility in water, these compounds can be transported over very large distances once they are discharged into streams and rivers. Vinodgopal et al. [57] reported the degradation of the reactive azo dye, Remazol Black B, using 640 kHz at 240 W. Although the dye was completely removed after 200 min of irradiation, an important organic load still remained in solution. After 6 hr of irradiation, only 65% mineralization was reached with this method. The analysis of the products performed with high-performance liquid chromatography (HPLC) and ion chromatography revealed the presence of small organic acids such as oxalic and formic acids, together with SO_4^{2-} and NO_3^- ions as final by-products. A similar trend was observed by Joseph et al. [58] for the degradation of azobenzene and the related monoazo dyes, methyl orange, *o*-methyl red, and *p*-methyl red. Working at 500 kHz and 100 W, sonochemical bleaching (i.e., total degradation of the parent molecule, yielding a loss of color of the solution) was achieved in 40 min. Nevertheless, based on a decrease of the total organic carbon (TOC), only 30–50% mineralization was achieved after 5 hr of irradiation of the treated dye solutions. Product analysis performed with HPLC electrospray mass spectrometry (ESMS) was used to develop the reaction scheme presented in Fig. 8, which was proposed as a general reaction mechanism for the degradation of this group of dyes. The addition of •OH radicals to the azo double bond is considered to be the first step of a sequence of oxidative bond cleavages leading to the production of carboxylic acids, quinones, CO_2, and NO_3^- as the main degradation products. Thus, even when total mineralization of the dye is not reached, the environmental hazard is substantially eliminated because the substrate is ultimately converted into nontoxic organic acids and inorganic compounds. Similar observations have also been reported for the sonodegradation of the dyes naphthol blue black [59], alizarin, and Procion Blue [19].

G. Combined Techniques

The examples mentioned above reveal that sonochemical methods can be useful tools in a variety of water systems and effluent treatments. Although, in many cases, the sonodegradation reactions are relatively fast, there is much work still ahead in order to improve the performance of this new technique. As was already mentioned (see Sec. II.B), most of the active radicals generated upon bubble collapse recombine before reaching the solution, or very soon after entering the liquid phase. Most of the mechanical energy input into the system, initially converted into cavitation, is later lost through trivial reactions such as the decomposition of water to yield H_2 and H_2O_2, which later disproportionates to yield H_2O and O_2. Together with heat release, nonproductive radical reactions are the major source of energy loss in sonochemical systems. A recent research in sonochemistry focuses on finding new mechanisms and reaction pathways in order to enhance the energy efficiency by two different means: (1) providing new sources of reactive radicals, and (2) exploiting the relatively large amounts of H_2O_2 produced in the process. Here we illustrate these two approaches where a combination of sonolysis with other advanced oxidation processes is applied.

1. Sonolysis/Ozonolysis

Ozone gas sparging has been widely used for water treatment. Two reactions of O_3 (aq) with dissolved organic substances can be distinguished in water. In the direct reaction, a highly selective attack of molecular ozone takes place on the double bonds of organic molecules, especially at low pH. In the reaction via radicals formation, free radicals from ozone decomposition react nonselectively with organic compounds. Hydroxide ions are the best known initiators of aqueous ozone decomposition, yielding •OH radicals via complex chain mechanisms. Other parameters influencing ozone decomposition are the chemical composition of the water (i.e., carbonate, phosphate, and dissolved organic carbon) and ionic strength. Ozonation, combined with ultrasonic irradiation, has been studied recently [14,52,55,60,61,63] and is a promising advanced oxidation process. The addition of ozone into sonolytic systems has been demonstrated to increase the net transformation rates of a wide range of chemical species. The thermal decomposition of O_3 (g) inside collapsing cavitation bubbles appears to be the main mechanism for the enhanced destruction of chemical contaminants [60]. In the extreme conditions reached upon cavitation, ozone decomposes into molecular and atomic oxygen as follows:

$$O_3(g) + M(g) \rightarrow O_2(g) + O(^3P)(g) + M(g) \tag{16}$$

$$O(^3P)(g) + H_2O(g) \rightarrow 2{\bullet}OH(g) \tag{17}$$

Thus, the presence of O_3 as a background gas provides new sources for •OH radical formation during sonolysis. The direct pyrolysis in the cavitation bubbles of volatile intermediates generated during ozonation also explains the improvement in the overall efficiency. In addition to these direct chemical effects, sonication also increases mass transfer coefficients of ozone from the bubbles to the solution, where the direct reaction of O_3 with the substrates, or further radical formation, takes place.

Among the latest reported studies, this combined technique was applied to the destruction of MTBE [14,55], aromatic compounds [62], azo dyes [63], and natural organic matter [61]. A threefold increase in the pseudo-first-order rate constant for the degradation of MTBE was observed when the ozone concentration was between 0.25 and 0.31 mM, compared to the sonolysis under O_2 with the same experimental conditions ($k = 31.3 \times 10^{-4}$ vs. 8.7×10^{-4} sec^{-1}). An increase of a factor of 50 was reported for the same reaction rate compared to ozonation alone under similar conditions ($k = 31.3 \times 10^{-4}$ vs. 0.6×10^{-4} sec^{-1}) [55]. The mineralization yields of aqueous solutions of azo dyes were improved by the combined sonolysis–ozonolysis treatment, reaching levels higher than 80% mineralization as compared to a maximum of 30–50% obtained with any of the treatments applied separately [63]. In the case of natural organic matter, under constant ultrasonic irradiation and continuous ozone gas application, TOC removal rates were enhanced by ultrasound (20 kHz, 55 W). After 60 min of treatment, 91% of the TOC in a 10-mg/L fulvic acid solution was removed. In the same period, only 40% of the TOC was removed with O_3 alone and merely 10% with ultrasound alone, operating under the same experimental conditions [61]. The combination of ultrasound and ozonation does not guarantee, however, an improved degradation of all types of substrates. Weavers et al. [62] reported the absence of a synergistic kinetic effect for the combined method as applied to aqueous pentachlorophenol solutions.

2. Sonolysis/Fenton's Reaction

The classical Fenton's reaction:

$$Fe^{2+}(aq) + H_2O_2(aq) \rightarrow Fe^{3+}(aq) + {\bullet}OH(aq) + OH^-(aq) \tag{18}$$

has been widely used as a chemical source of •OH radicals in aqueous solution. By adding $FeSO_4$ to the system before irradiation, this reaction becomes a secondary source of •OH, recovering part of the chemical activity otherwise lost in the production of relatively large amounts of H_2O_2 during sonication. Joseph et al. [58] reported an enhancement in the sonochemical bleaching of the azo dye, methyl orange, due to the presence of Fe(II) ions. As a function of the concentration of $FeSO_4$, in the range 10 μM–5 mM, a maximum effect was reached when [Fe(II)] was between 0.5 and 0.1 mM. The pseudo-first-order reaction rate constant determined under those optimum conditions reached a threefold increase compared with that measured in the absence of Fe(II) ions ($k = 0.12$ vs. 0.046 min^{-1}). This increase was due to the higher •OH concentration produced through the Fenton's reaction. Further increases in [Fe(II)] showed no further catalytic activity, due to the direct reduction of •OH radicals by the metal ions:

$$Fe^{2+}(aq) + \bullet OH(aq) \rightarrow Fe^{3+}(aq) + OH^{-}(aq) \qquad (19)$$

The direct scavenging of •OH radicals by Fe(II) imposes a limit to the applicability of this method. Furthermore, the limited solubility of the metal ions when alkaline media are employed is an additional constraint in these systems.

V. CONCLUSION

Sonochemical reactors can be used to degrade a variety of contaminants in aqueous solution, including some very harmful and recalcitrant compounds such as chlorinated hydrocarbons. The scale-up of these reactors in order to meet industrial needs (i.e., faster rates and high volume processes) is the present challenge in the development of the technique. The many advantages of the technique, described in this chapter, will surely encourage researchers and engineers to face some unsolved problems in the future, in order to provide alternatives to conventional waste treatment.

REFERENCES

1. Mason TJ, Lorimer JP. Sonochemistry: Theory, Applications and Uses of Ultrasound in Chemistry. Chichester: Ellis Horwood Limited, 1988.
2. Thompson LH, Doraiswamy LK. Sonochemistry: science and engineering. Ind Eng Chem Res 1999; 38:1215–1249.
3. Leighton TG. The Acoustic Bubble. London: Academic Press, 1992.

4. Brennen CE. Cloud cavitation: observations, calculations and shock waves. Multiph Sci Technol 1998; 10:303–321.
5. Margulis MA, Margulis IM. Theory of local electrification of cavitation bubbles: new approaches. Ultrason Sonochem 1999; 6:15–20.
6. Didenko YT, McNamaraIII WB, Suslick KS. Hot spot conditions during cavitation in water. J Am Chem Soc 1999; 121:5817–5818.
7. Hilgenfeldt S, Grossmann S, Lohse D. A simple explanation of light emission in sonoluminescence. Nature 1999; 398:402–405.
8. Colussi AJ, Hoffmann MR. Vapor supersaturation in collapsing bubbles. Relevance to the mechanisms of sonochemistry and sonoluminescence. J Phys Chem A 1999; 103:11336–11339.
9. Misik V, Miyoshi N, Riesz P. EPR spin trapping study of the sonolysis of H_2O/D_2O mixtures: probing the temperature of cavitation regions. J Phys Chem 1995; 99:3605–3611.
10. Misik V, Kirschenbaum LJ, Riesz P. Free radical production by sonolysis of aqueous mixture of *N,N*-dimethylformamide: an EPR spin trapping study. J Phys 1995; 99:5970–5976.
11. Colussi AJ, Weavers LK, Hoffmann MR. Chemical bubble dynamics and quantitative sonochemistry. J Phys Chem A 1998; 102:6927–6934.
12. Colussi AJ, Hung HM, Hoffmann MR. Sonochemical degradation rates of volatile solutes. J Phys Chem A 1999; 103:2696–2699.
13. Entezari MH, Kruus P. Effect of frequency on sonochemical reactions: I. Oxidation of iodide. Ultrason Sonochem 1994; 1:S75–S79.
14. Kang JW, Hung HM, Lin A, Hoffmann MR. Sonolytic destruction of methyl *tert*-butyl ether by ultrasonic irradiation: the role of O_3, H_2O_2, frequency and power density. Environ Sci Technol 1999; 33:3199–3205.
15. Petrier C, Lamey MF, Francony A, Benahcene A, David B. Sonochemical degradation of phenol in dilute aqueous solutions. J Phys Chem 1994; 98: 10514–10520.
16. Petrier C, Jeunet A, Luche JL, Reverdy G. Unexpected frequency effects on the rate of oxidative processes induced by ultrasound. J Am Chem Soc 1992; 114: 3148–3150.
17. Kruus P, Entezari MH. Effect of frequency on sonochemical reactions: I. Oxidation of iodine. Ultrason Sonochem 1996; 3:19–24.
18. Hua I, Hochemer RH, Hoffmann MR. Sonochemical degradation of *p*-nitrophenol in a parallel-plate near-field acoustical processor. Environ Sci Technol 1995; 29:2790–2796.
19. Hong Q, Hardcastle JL, McKeoun RAJ, Marken F, Compton RG. The 20 kHz sonochemical degradation of trace cyanide and dye stuffs in aqueous media. New J Chem 1999; 23:845–849.
20. Hua I, Pfalzer-Thompson U. Ultrasonic irradiation of carbofuran: decomposition kinetics and reactor characterization. Water Res 2001; 36:1445–1452.
21. Gutierrez M, Henglein A. Chemical action of pulsed ultrasound: observation of unprecedented intensity effects. J Phys Chem 1990; 94:3625–3628.

22. Hua I, Hoffmann MR. Optimization of ultrasonic irradiation as an advanced oxidation technology. Environ Sci Technol 1997; 31:2237–2243.
23. Destaillats H, Alderson TW II, Hoffmann MR. Applications of ultrasound in NAPL remediation. Sonochemical degradation of TCE in aqueous surfactant solutions. Environ Sci Technol 2001; 35:3019–3024.
24. Suslick KS, Mdleleni MM, Ries JT. Chemistry induced by hydrodynamic cavitation. J Am Chem Soc 1997; 119:9303–9304.
25. Rassokhin DN, Kovalev GV, Bugaenko LT. Temperature effects on the sonolysis of methanol/water mixtures. J Am Chem Soc 1995; 117:344–347.
26. Kondo T, Kirschenbaum LJ, Kim H, Riesz P. Sonolysis of dimethyl sulfoxide–water mixtures: a spin-trap study. J Phys Chem 1993; 97:522–527.
27. Gupta R, Wallace HC. Sonochemical reactions at 640 kHz using an efficient reactor. Oxidation of potassium iodide. Ultrason Sonochem 1997; 4:289–293.
28. Martin PD, Ward LD. Reactor design for sonochemical engineering. Trans Inst Chem Eng 1992; 70:296–303.
29. Destaillats H, Lesko TM, Knowlton M, Wallace H, Hoffmann MR. Scale-up of sonochemical reactors for water treatment. Ind Eng Chem Res 2001; 40: 3855–3860.
30. Taylor E, Cook BB, Tarr MA. Dissolved organic matter inhibition of sonochemical degradation of aqueous polycyclic aromatic hydrocarbons. Ultrason Sonochem 1999; 6:175–183.
31. Hua I, Hoffmann MR. Kinetics and mechanism of the sonolytic degradation of CCl_4: intermediates and byproducts. Environ Sci Technol 1996; 30:864–871.
32. Destaillats H, Hung H-M, Hoffmann MR. Degradation of alkylphenol ethoxylate surfactants in water with ultrasonic irradiation. Environ Sci Technol 2000; 34:311–317.
33. Hueter TF, Bolt RH. Sonics: Techniques for the Use of Sound and Ultrasound in Engineering and Science. New York: John Wiley and Sons, Inc., 1955.
34. Kinsler LE, Frey AR. Fundamentals of Acoustics. 2nd ed. New York: John Wiley and Sons, Inc., 1962.
35. Galiev SU, Romashchenko VA. A method of solving nonstationary three-dimensional problems of hydroelasticity with allowance for fluid failure. Int J Impact Eng 1999; 22:469–483.
36. Thomenius KE. Impact of nonlinear propagation on temperature distributions caused by diagnostic ultrasound. Proc IEEE Ultrason Symp Vols. 1 and 2, 1998: 1409–1413.
37. Richman RH, McNaughton WP. A metallurgical approach to improved cavitation–erosion resistance. J Mater Eng Perform 1997; 6:633–641.
38. Kercel SW, Allgood GO, Dress WB, Hylton JO. An anticipatory model of cavitation. Proceedings of the Society of Photo-Optical Instrumentation Engineers (SPIE): Applications and Science of Computational Intelligence II, Serial Number 0277-786X, 1999; 6:224–235.
39. Hirschi R, Dupont P, Avellan F, Favre JN, Guelich JF, Parkinson E. Centrifugal pump performance drop due to leading edge cavitation: numerical pre-

dictions compared with model tests. J Fluids Eng Trans ASME 1998; 120:705–711.

40. Hordnes I, Green SI. Sea trials of the ducted tip propeller. J Fluids Eng Trans ASME 1998; 120:808–817.
41. Latorre R, Mizina M. Design study for outboard propeller with spoiler. Ocean Eng 1999; 26:727–737.
42. Reisman GE, Brennen CE, Wang YC. Observations of shock waves in cloud cavitation. J Fluid Mech 1998; 355:255–283.
43. Gray DE, Ed. American Institute of Physics Handbook. 3rd ed. New York: McGraw-Hill Book Company, 1972: 3–86.
44. Zhu CF, Zheng RK, Su JR, He J. Ultrasonic anomalies in $La_{0.67}Ca_{0.33}MnO_3$ near the Curie temperature. Appl Phys Lett 1999; 74:3504–3506.
45. Marczak W. Speed of ultrasound, density, and adiabatic compressibility for 2-methylpyridine plus heavy water in the temperature range 293 K to 313. J Chem Eng Data 1999; 44:621–625.
46. Nishikawa S, Huang H, Dong SJ. Ultrasonic relaxation kinetics on fast deuteron transfer reaction. J Phys Chem A 1999; 103:3804–3808.
47. Hung HM, Hoffmann MR. Kinetics and mechanism of the sonolytic degradation of chlorinate hydrocarbons: frequency effects. J Phys Chem A 1999; 103:2734–2739.
48. Hung HM, Hoffmann MR. Kinetics and mechanism of the enhanced reductive degradation of CCl_4 by elemental iron in the presence of ultrasound. Environ Sci Technol 1998; 32:3011–3016.
49. Bhatnagar A, Cheung HM. Sonochemical destruction of chlorinated C1 and C2 volatile organic compound in dilute aqueous solution. Environ Sci Technol 1994; 28:1481–1486.
50. Petrier C, Micolle M, Merlin G, Luche J-L, Reverdy G. Characteristics of pentachlorophenate degradation in aqueous solution by means of ultrasound. Environ Sci Technol 1992; 26:1639–1642.
51. Serpone N, Terzian R, Hidaka H, Pelizzetti E. Ultrasonic induced dehalogenation and oxidation of 2-, 3-, and 4-chlorophenol in air-equilibrated aqueous media. Similarities with irradiated semiconductor particles. J Phys Chem 1994; 98:2634–2640.
52. Weavers L, Ling FH, Hoffmann MR. Aromatic compound degradation in water using a combination of sonolysis and ozonolysis. Environ Sci Technol 1998; 32:2727–2733.
53. Kotronarou A, Mills G, Hoffmann MR. Ultrasonic irradiation of *p*-nitrophenol in aqueous solutions. J Phys Chem 1991; 95:3630–3638.
54. Kotronarou A, Mills G, Hoffmann MR. Decomposition of parathion in aqueous solution by ultrasonic irradiation. Environ Sci Technol 1992; 26:1460–1462.
55. Kang J-W, Hoffmann MR. Kinetics and mechanism of the sonolytic destruction of MTBE by ultrasonic irradiation in the presence of ozone. Environ Sci Technol 1998; 32:3194–3199.
56. Ondruschka B, Lifka J, Hofmann J. Aquasonolysis of ether. Effect of frequency and acoustic power ultrasound. Chem Eng Technol 2000; 23:588.

57. Vinodgopal K, Peller J, Makogon O, Kamat PV. Ultrasonic mineralization of a reactive textile azo dye. Remazol Black B Water Res 1998; 32:3646–3650.
58. Joseph JM, Destaillats H, Hung H-M, Hoffmann MR. The sonochemical degradation of azobenzene and related azo dyes: rate enhancements via Fenton's reaction. J Phys Chem A 2000; 104:301–307.
59. Stock NL, Peller J, Vinodgopal K, Kamal PV. Combinative sonolysis and photocatalysis for textile dye degradation. Environ Sci Technol 2000; 34:1747–1750.
60. Weavers LK, Hoffmann MR. Sonolytic decomposition of ozone in aqueous solution: mass transfer effects. Environ Sci Technol 1998; 32:3941.
61. Olson TM, Barbier PF. Oxidation kinetics of natural organic matter by sonolysis and ozone. Water Res 1994; 28:1383.
62. Weavers LK, Malmstadt N, Hoffmann MR. Kinetics and mechanism of pentachlorophenol degradation by sonication, ozonation and sonolytic ozonation. Environ Sci Technol 2000; 34:1280–1285.
63. Destaillats H, Colussi AJ, Joseph JM, Hoffmann MR. Synergistic effects on sonolysis combined with ozonolysis for the oxidation of azobenzene and methyl orange. J Phys Chem A 2000; 104:8930–8935.

6

Electrochemical Methods for Degradation of Organic Pollutants in Aqueous Media

Enric Brillas and Pere-Lluís Cabot
Universitat de Barcelona, Barcelona, Spain

Juan Casado
Carburos Metálicos S.A., Barcelona, Spain

I. INTRODUCTION

In the last 30 years, a large variety of electrochemical techniques for the destruction of toxic and refractory (i.e., nonbiodegradable) organic pollutants for wastewater treatment have been proposed and developed. Only a few conventional methods are related to the direct electrolysis of pollutants at the electrode. This process can occur either by a direct electron transfer reaction to (reduction) or from (oxidation) the undesired organic, or by a chemical reaction of the pollutant with previously electrogenerated species, which remain adsorbed at the electrode surface. Most electrochemical methods are based on indirect (or mediated) electrolysis in which the target pollutant is removed in the solution by active species produced reversibly or irreversibly at the electrode. The two types of procedures are contrasted in Fig. 1. The use of electrochemical techniques offers the following distinctive advantages for wastewater treatment [1]:

1. Environmental compatibility. The main reactant is the electron, which is a clean reagent.
2. Versatility. Electrolytic treatments can deal with solid, liquid, or gaseous pollutants to generate neutral, positively, or negatively

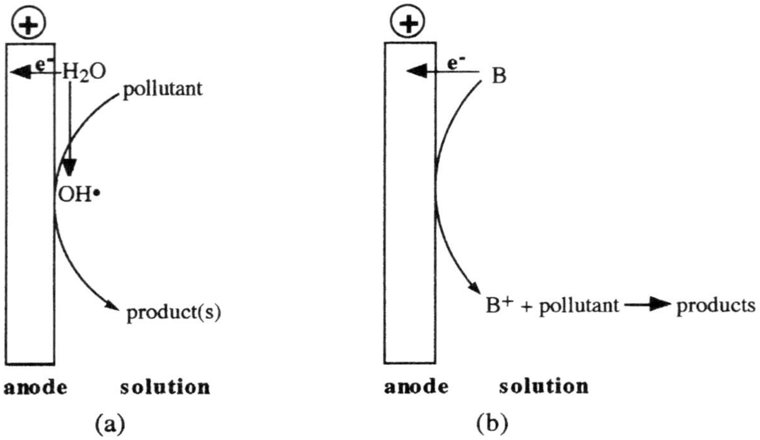

Figure 1 Schemes for different electrochemical treatments of organic pollutants. (a) Direct electrolysis by anodic oxidation in which the pollutant reacts at the electrode surface with adsorbed $OH^{\bullet}$ produced from water oxidation at a high O_2-overpotential anode. (b) Indirect electrolysis where the pollutant reacts in the solution with an irreversibly electrogenerated reagent B^+ produced from the oxidation of inactive B at the anode.

charged inorganic or organic products, also inducing the production of precipitates, gaseous species, pH changes, etc. In addition, a plethora of reactors and electrode materials, shapes, and configurations can be utilized [2]. It is noteworthy that the same reactor can be used frequently for different electrochemical reactions with only minor changes, and that electrolytic processes can be scaled easily from the laboratory to the plant, allowing treatment volumes ranging from milliliters to millions of liters, respectively.

3. Safety. Electrochemical methods are generally safe because of the mild conditions usually employed and the small amount and innocuous nature of the added chemicals.
4. Energy efficiency. Electrochemical processes are amenable to work at low temperatures and pressures, usually below ambient conditions. Electrodes and cells can also be designed to minimize power losses due to poor current distribution and voltage drops. In some instances, the required equipment and operations are simple and, if properly designed, can be made relatively inexpensively.

The electrochemical methods described in this chapter for the destruction of organics in wastewaters are classified in Fig. 2. The direct electrolytic processes include conventional procedures of cathodic reduction and anodic oxidation. The indirect methods deal with the use of redox mediators as reversibly electrogenerated reagents, as well as oxidants as irreversibly electrogenerated reagents at the anode (e.g., O_3, ClO^-, Cl_2, and ClO_2) or the cathode (e.g., H_2O_2). Emerging processes related to electrogenerated Fenton reagent and other electrochemical oxidation processes based on the combined use of iron ions and electrogenerated H_2O_2 are also described. Other indirect electrolytic processes include conventional methods of phase separation, such as electrocoagulation, electroflotation, and electroflocculation. Fundamentals, laboratory experiments, scale-up studies, and environmental/industrial applications for the different electrochemical techniques are discussed in this chapter.

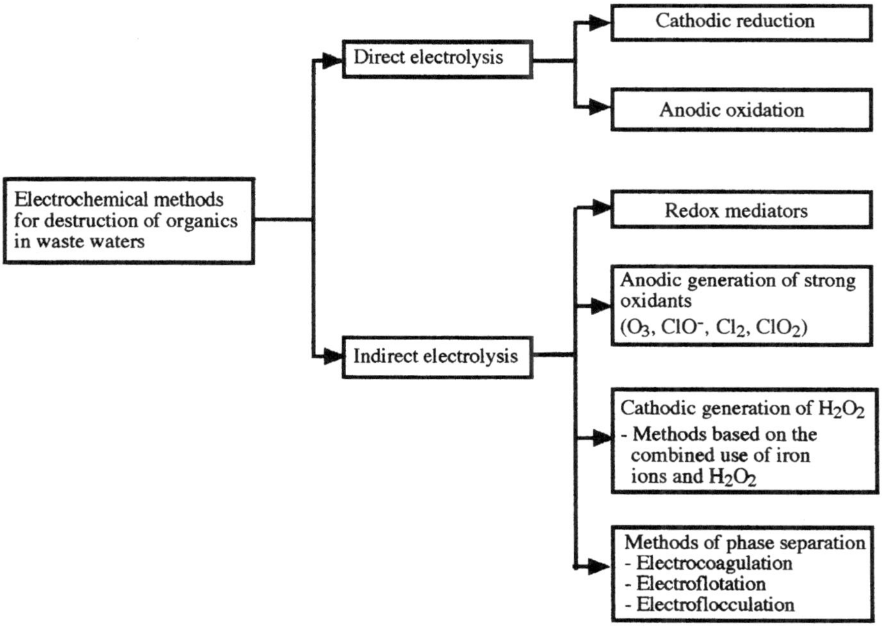

Figure 2 Classification of electrochemical methods for the destruction of organics in aqueous wastes.

II. CATHODIC REDUCTION

The direct electroreduction of organics on suitable cathodes is a method that can be utilized for the dechlorination of pollutants in wastewaters. Many chlorinated organic compounds are produced in the industry and are used as solvents, refrigerants, pesticides, transformer oils, etc. Because most of such compounds are nonbiodegradable, they cause major environmental problems. Often they are found at small concentrations in a wide variety of wastewaters, and are usually decontaminated by concentration techniques such as adsorption on activated carbon or extraction by organic solvents. In fact, the products thus separated have to be further destroyed to avoid increasing pollution in the environment. Some of the existing methods are expensive (e.g., Na treatment), or can produce very dangerous by-products such as dioxins obtained by incineration. The use of cathodic dehalogenation as an alternative method has the following advantages [1,3–5]:

1. The treatment can be performed at ambient temperature.
2. No additional reagents are required.
3. As the chlorine atoms are selectively removed, the resulting dechlorinated compounds can be degraded by a cheaper method such as a biological treatment.

It has to be taken into account that during the cathodic reduction, H_2 evolution is a common side reaction in aqueous media and, therefore, a cathode with high hydrogen overpotential is usually selected to obtain suitable electrodegradation efficiencies [1,6,7]. Moreover, dissolved oxygen can also be reduced (see Sec. VI). Different materials have been utilized as cathodes including carbon electrodes, Pb, Hg, Pt, Cu, Ni, Ni alloys, Ni composites, Ti, TiO_2, and metal hydrides. Several problems have been detected in the use of carbon electrodes. Graphite develops fractures along its basal planes due to the intercalation of ions or organic molecules that migrate under the electrical field through them. In addition, carbon electrodes can suffer from degradation by radicals formed during the electroreduction of dissolved oxygen. Fortunately, these problems are solved using three-dimensional carbonaceous materials made of partially graphitized amorphous carbon and graphite felts [8,9]. Problems of stability have also been found with Pb during the electroreductive dehalogenation of several chlorinated organic compounds [9,10]. Mercury has several drawbacks, including the limitation in current densities, probable metal leaks to the electrolyte, and difficulties in scaling-up with liquid metals.

This section is devoted to the application of cathodic reduction for treating aliphatic and aromatic pollutants at low concentrations, and also to the dechlorination of chlorofluorocarbons (CFCs) in aqueous media. The aromatic pollutants are typically chlorinated compounds that contaminate wastewaters, whereas CFCs are volatile hydrocarbons with chlorine and fluorine atoms, which were primarily used as refrigerants and gas propellants. CFCs destroy the stratospheric ozone layer and contribute to the greenhouse effect. The Montreal Protocol provided an international agreement to stop CFC production beginning in 1996. However, about 2×10^6 tons of CFCs were still stored in freezing devices in 1999. To avoid probable leaks to the atmosphere, they should be degraded or converted to useful chemicals. Because destructive methods such as incineration, UV photolysis, catalytic decomposition at high temperatures, and solar thermal technology are expensive and/or can produce dangerous by-products, the partial conversion of CFCs to useful and harmless compounds is much more attractive.

A. Aliphatic and Aromatic Compounds

Direct electrolysis has been applied successfully to the degradation of chloroalkanes such as $CHCl_3$ [11] and CCl_4 [12]. As these compounds react scarcely with radicals such as $OH^{\cdot}$ and as their toxicity is directly related to their chlorine content, the reductive dechlorination is especially attractive. The electrochemical reduction enables the removal of substituents, especially halogen atoms, along with the hydrogenation of the molecule. The products are less toxic and more biodegradable and/or sensitive to electrochemical oxidation [4,5]. However, an important problem in electrochemical dehalogenation is the low current efficiency.

These reactions take place at room temperature at potentials of about −1 V vs. saturated calomel electrode (SCE). The general reaction per chlorine atom can be written as follows:

$$R - Cl + 2H^+ + 2e^- \longrightarrow R - H + HCl \tag{1}$$

Sonoyama et al. [11] have studied the electroreductive decomposition of chloroform on 15 kinds of metal electrodes using a Pyrex cell divided by a glass filter with a cathode area of 16–18 cm^2. Experiments performed by electrolyzing 200 mL of deaerated aqueous 0.1 M K_2SO_4 solution with 6.20 mM $CHCl_3$ at 1 mA cm^{-2} up to a total charge of 50 C showed a strong dependence of the decomposition efficiency and the main product formed on the metal tested. The hydrogenation of chloroform on Ag, Zn, Pd, and Cu cathodes proceeded at near 100% efficiency and the main product was CH_4. In contrast, 88% dichloromethane was selectively generated on Pb.

Scherer et al. [12] studied the kinetics of CCl_4 dechlorination on an oxide-free iron rotating disk electrode in borate buffer (pH 8.4) at a potential such that an oxide film would not form. The rate of CCl_4 reduction was dominated by reactions at the metal–solution interface, as is the case in oxide-covered granular iron.

The electrolysis of chlorinated aromatics and aliphatics at concentrations ranging from 40 to 560 ppm on carbon fiber gives effective dechlorination [3,13]. Several trials were performed at 10 A using 1 L of aqueous $NaOH + Na_2SO_4$ with small additions of methanol or acetonitrile to enhance the solubility of the pollutant. The cell utilized is illustrated in Fig. 3. The fiber bundles, containing 50,000 single fibers of 8 μm diameter, are clamped at the entrance side. The Pt mesh anode is separated from the cathode by a cation-permeable Nafion® membrane. The current efficiencies obtained were low, although a value of 75% was found for dichlorvos (Table 1). Despite the low current efficiencies, the process is feasible at reasonable costs and yields a high degree of detoxification. The treatment of wastewaters with 50 ppm pentachlorophenol by electrochemical reduction using C fiber electrodes for 30 min decreased its concentration to below the detection limit of 0.5 ppm [13]. During the treatment, the toxicity decreased by a factor of 20. The final reaction products were phenol and, possibly, monochlorophenols.

Many substituted phenols have been electroreduced at Pt electrodes in ethanol–water mixtures [14] or in acid (0.05 M H_2SO_4 or 2 M $HClO_4$) solutions [15], leading to cyclohexanols with a current efficiency close to 100%. The reaction appears to proceed via a surface process involving the adsorption of phenols and hydrogen atoms formed at the cathode. As cyclohexanols are biocompatible, these substituted phenols can be degraded by

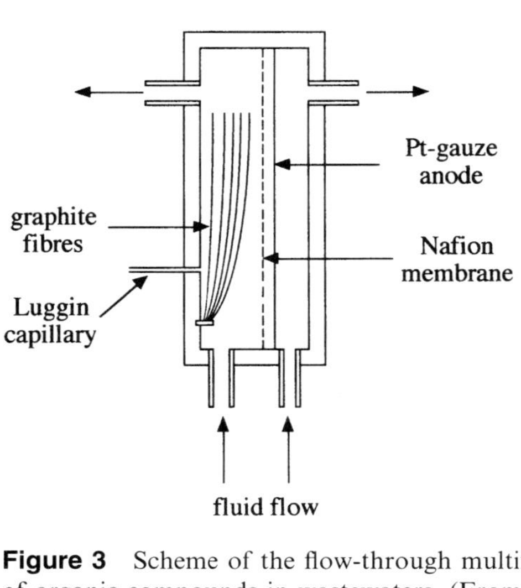

Figure 3 Scheme of the flow-through multifiber cell for the electrodechlorination of organic compounds in wastewaters. (From Ref. 3.)

Table 1 Results Obtained for the Electrochemical Dehalogenation in the Cell Shown in Fig. 3 at 10 A in 1 L of 0.1 M NaOH + 0.1 M Na_2SO_4

Compound	Initial concentration [ppm]	Number of Cl removed	CE [%]	Energy cost [kW hr m^{-3}]
2-NH_2-4-Cl-phenol	100	1	0.7	60
4-Cl-$C_6H_4NO_2$	48	1	0.4	55
DDVP[a]	560	1(2)	75	0.8
C_2Cl_6	50	6	1	77
Pentachlorophenol	50	5	2	36
2,4,5-T[b]	100	3	1	70
C_2Cl_4	74	4	3	70
1,2,4-$Cl_3C_6H_3$	40	3	0.7	70

[a] Dichlorvos or dimethyl-2,2-dichlorovinyl phosphate.
[b] 2,4,5-Trichlorophenoxyacetic acid.
Source: Ref. 3.

electroreduction coupled to biological degradation. However, *p*-nitrophenol and 4-chlorophenol did not yield cyclohexanols. Phenol can be obtained by the electroreduction of 4-chlorophenol on a palladized carbon cloth or palladized graphite cathodes in a divided cell containing acetate buffer [16]. The suggested mechanism involves the adsorption of 4-chlorophenol on the carbon surface near the carbon/Pd interface, followed by its hydrogenation with the hydrogen atoms adsorbed on the Pd surface. A complete dechlorination of 25 mL of 153 ppm 4-chlorophenol in 0.05 M sodium acetate acetic acid buffer at −0.7 V vs. SCE on 2-cm^2 palladized carbon cloth cathode required 15 hr, during which time the current decreased from 2.2 to 0.8 mA. Poorer results were obtained when Pt was used instead of Pd.

Chlorinated hydrocarbons have also been reduced on Cu cathodes in aqueous solutions [7]. In this case, a fixed-bed, flow-through reactor filled with Cu balls, 0.2–0.6 mm in diameter, supported on a Pt gauze was employed. Hexachlorocyclohexane was dechlorinated rapidly and completely. Tetrachloroethylene, trichloroethane, and chlorobenzene were less reactive. However, unsatisfactory results were obtained with a polychlorinated biphenyl (PCB).

The electroreduction of trichloroethylene (0.4 g L^{-1}) on Cu in 0.05 M NaOH was found to be more efficient than on Ag or Cd cathodes [4], with the current efficiency increasing when the applied current density decreased. At a current density of 4 mA cm^{-2}, the current efficiencies for the dehalogenation of monochloroacetic acid, dichloroacetic acid, chloroform, and trichloroethylene were 2%, 10%, 87%, and 29%, respectively. 5-Chlorosalicylic acid could not be dechlorinated on Cu. Nagaoka et al. [17]

reported the quantitative electroreduction of trichloroethylene in water–acetonitrile at composite electrodes consisting of metal particles (stainless steel, Ni, and Cr) and oxidatively treated glassy carbon particles. Trichloroethylene was first reduced to chloroacetylene on the glassy carbon particles, and the latter was further reduced to acetylene on the metal.

o-Chlorobenzoic acid, which is resistant to anodic oxidation, can be cathodically reduced on a Pb cathode to *o*-chlorobenzyl alcohol [18]. This alcohol was further oxidized to *o*-chlorobenzaldehyde at a PbO_2 anode. The extent of its degradation was about 90%. In the presence of $MnSO_4$, both the alcohol and the aldehyde suffer oxidative degradation at the anode during the electrolysis to yield mainly aliphatic acids.

Funabashi et al. [19] electroreduced iodine-containing organic compounds such as iodotyrosine from medical waste solutions to separate iodide, which was adsorbed by Ag-coated Al_2O_3 for effective separation.

Another procedure for the electrochemical dechlorination of pollutants in aqueous media consists of the use of a corrodible metal or a bimetallic system without the application of external current. The reaction proceeds as in corrosion—with the anodic regions being dissolved and with reduction taking place at cathodic regions. The rate of reduction is lower than in the case of the cathodic reduction with imposed DC voltage because the potential of the local cathodes is less negative. However, the dechlorination rate can be increased with the metal surface area exposed to the wastewater. A full-scale column reactor has been described by Sweeny [20,21], and this device has been tested for the treatment of industrial wastewaters using various combinations of catalyzed Zn, Al, or Fe mixed with sand. The detoxification of hexachlorocyclopentadiene, trihalomethanes, chloroethylenes, chlorobenzene, chlordane, atrazine, and nitrophenols was reported. Bachmann et al. [7] employed a suspension of steel grit, covered partially with Cu by cementation. In this case, Fe was oxidized to Fe(II) whereas the chlorinated compound was reduced on Cu releasing Cl^-. Matheson and Tratnyek [22] sequentially dehalogenated carbon tetrachloride via chloroform to methylene chloride on fine-grained iron metal. Trichloroethylene was also dechlorinated by iron, although more slowly than carbon tetrachloride. Grittini et al. [23] have shown the complete dechlorination of PCBs to biphenyl in an aqueous methanol solution in a few minutes by contacting the solution with a Pd/Fe system. In this case, the reduction was assumed to be due to hydrogen adsorbed by Pd during Fe corrosion [16].

B. Chlorofluorocarbons

Electrochemical reduction processes of CFCs leading to partially or completely dehalogenated compounds for synthetic purposes have been described in the literature. Many examples in which the CFCs are converted to hydrochlorofluorocarbons (HCFCs), hydrofluorocarbons (HFCs), and/or fluorocarbons (FCs) have been reported. HCFCs are not as destructive to stratospheric ozone; nevertheless, their production will be gradually reduced to zero in 2020. HFCs and FCs are harmless to stratospheric ozone and there is currently no limitation for their production.

Edison [24] disclosed the conversion of CFCs to HCFCs, HFCs, and FCs using a divided cell with a Hg pool cathode in ethanol (60 vol.%)–water containing potassium acetate. In one example, the conversion of 1,1,2-trichloro-1,2,2-trifluoroethane (CFC 113) to chlorotrifluoroethene (CTFE), an industrial monomer, at 20 mA after passing a total charge of 18,700 C was 64 mol%. The cathodic reaction is:

$$Cl_2FC - CF_2Cl + 2e^- \longrightarrow ClFC{=}CF_2 + 2Cl^- \tag{2}$$

Cabot et al. [25,26] reported CFC 113 electroreduction in Pb and Cd cathodes, combined with a hydrogen diffusion anode in MeOH (50–80 vol.%)–water mixtures with 0.75 M NH_4Cl and 50 ppm $PdCl_2$, with the MeOH content allowing significant CFC solubility. The current efficiency at 80 and 200 mA cm^{-2} was 98%, and difluoroethene and trifluoroethene were the main products in the gas phase. H_2 can be selectively oxidized at the gas diffusion electrode (GDE), and so there is no need for separators, reducing the energy cost. The process has also been extended to CFC 11, and derivatives including fluoromethane have been obtained [27,28].

Inaba et al. [29] have introduced a different cell to work with gaseous compounds (Fig. 4). A metal-plated solid polymer electrolyte (SPE) composite electrode faces the gas to be reduced. On the other side, the SPE is in contact with 0.1 M NaOH in which a Pt wire and an Ag/AgCl reference electrode are immersed. This system permits the electroreduction of insoluble reactants in water without employing organic solvents. For example, 2-chloro-1,1,1,2-tetrafluoroethane (HCFC 124) is transformed into 1,1,1,2-tetrafluoroethane (HFC 134a). The cathodic reaction can be written as follows:

$$CF_3 - CHFCl + H_2O + 2e^- \longrightarrow CF_3 - CFH_2 + Cl^- + OH^- \tag{3}$$

This reaction is considered to be catalyzed by active hydrogen atoms formed on Pd. The product is recovered as a gas mixture, and Cl^- and OH^- ions move to the electrolyte as the counterions of the anion exchange membrane.

Delli et al. [30] have studied the electroreduction of dichlorodifluoromethane (CFC 12) in aqueous solutions on Pd, Au, Cu, and Ag, chemi-

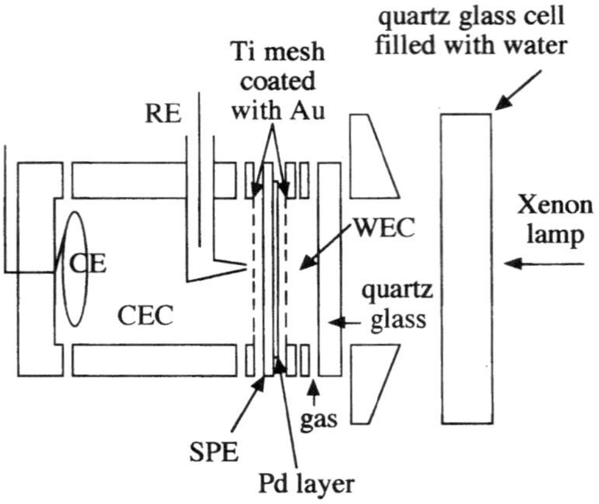

Figure 4 Schematic diagram of the electrolytic cell with a solid polymer electrolyte composite electrode. SPE = Neosepta AM-1; CE = Pt wire; RE = Ag/AgCl; WEC = cathode compartment; CEC = anode compartment. (From Ref. 29.)

cally deposited on Nafion® 117 membranes using a cell similar to that of Fig. 4. CFC 12 circulates over the metal on one side of the membrane, while on the other side, there is a 2-M NaOH solution with the Pt anode and the Ag/AgCl reference electrode. CH_2F_2 (HFC 32, a new refrigerant) and CH_4 were obtained on Ag at −1.4 V vs. Ag/AgCl, with current efficiencies of 60% and 30%, respectively. CH_4 was the main product generated on Au, Pd, and Cu at −1.0 V, with current efficiencies of 14%, 15%, and 47%, respectively.

Wetproofed porous electrodes, applied previously in fuel cells, have also been tested as cathodes for electrosynthesis from gaseous and liquid starting materials with limited solubility in water. The reagent is supplied through the hydrophobic electrode, which is in contact with an aqueous electrolyte. They present some attractive advantages over conventional electrodes because they:

1. Accelerate electrode reactions
2. Lower diffusion limitations significantly
3. Simplify product isolation.

As an example, Table 2 compares the results obtained using a smooth Cd cathode and a hydrophobicized Cd electrode prepared from powdered Cd, carbon, acetylene black A-437E, and polytetrafluoroethylene (PTFE) [31]. The current efficiency and current density increased for smooth Cd

Table 2 Results Obtained for CFC 113 Electrolyses Conducted on a Smooth Cd Electrode and a Hydrophobicized (15 wt.% PTFE) Cd Electrode

Electrode	Electrolyte	Potential[a] [V]	CE for CTFE [%]	j [A m^{-2}]
Smooth	1 M $LiClO_4$	−1.7	9.8	19.5
		−2.1	2.2	271
	1 M $LiClO_4$ + 50% EtOH	−1.7	36.7	306
		−2.1	18.9	450
Hydrophobicized	1 M $LiClO_4$	−1.7	83.6	1200
		−2.1	81.5	1750

[a] Potential vs. saturated Ag/AgCl reference electrode.
Source: Ref. 31.

when ethanol was added to the electrolyte due to the increase in CFC 113 solubility. Higher current densities and efficiencies, however, were found for the hydrophobicized Cd electrode.

Kornienko et al. [32] have utilized wetproofed electrodes of acetylene black containing 40 wt.% PTFE to reduce CFC 113 to CTFE in 3 M LiCl at 35°C. The presence of tetraalkylammonium cations considerably facilitates the reduction process. It is assumed that CFC 113, like other halogenated compounds, forms a positively charged complex with the tetraalkylammonium cation, which is much more easily reduced than the CFC 113 itself. The largest effect was found for tetra-*n*-butylammonium ion (TBA^+), giving a 96% yield of CTFE. The increase in CTFE current yield is explained by the shift of potential to less cathodic values and to the displacement of water molecules by organic cations in the layer next to the electrode because hydrogen evolution is slower.

Sonoyama and Sakata [33] have electroreduced CFC 12 on 12 kinds of metal-supported GDEs in a stainless steel autoclave with 1 M NaOH at 7 atm. Fig. 5 shows the cell employed, which is made of polyvinyl chloride (PVC), due to its high resistance to corrosion by HF. By applying 64 mA cm^{-2}, Cu-, In-, and Pb-supported GDEs gave almost 100% efficiency without producing H_2. Zn-, Ag-, Cu-, and In-supported GDEs produced mainly CH_4. The Pb-supported GDE induced only dechlorination, and 93% HFC 32 is selectively obtained with 74% faradaic efficiency. The evolution of faradaic efficiency of products with current density for Cu- and Pb-supported GDEs is depicted in Fig. 6a and b, respectively. The same authors [34] have also studied the cathodic reduction of CFC 13 by using a 1:1 water–MeOH mixture with 1 M NaOH, and by testing 13 kinds

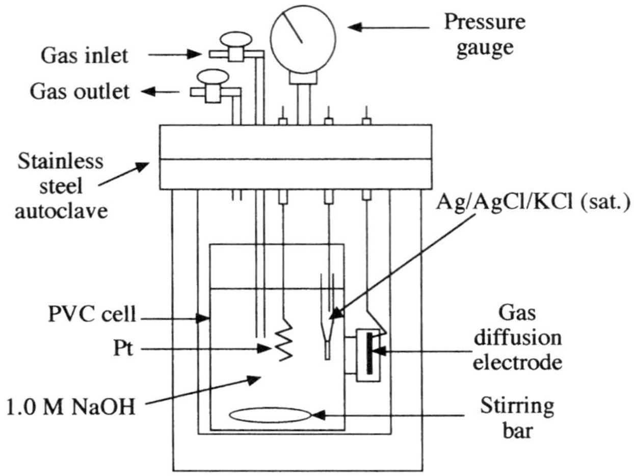

Figure 5 Stainless steel autoclave with the electrolysis cell for the electrochemical reduction of CFC 12 at 7 atm. (From Ref. 33.)

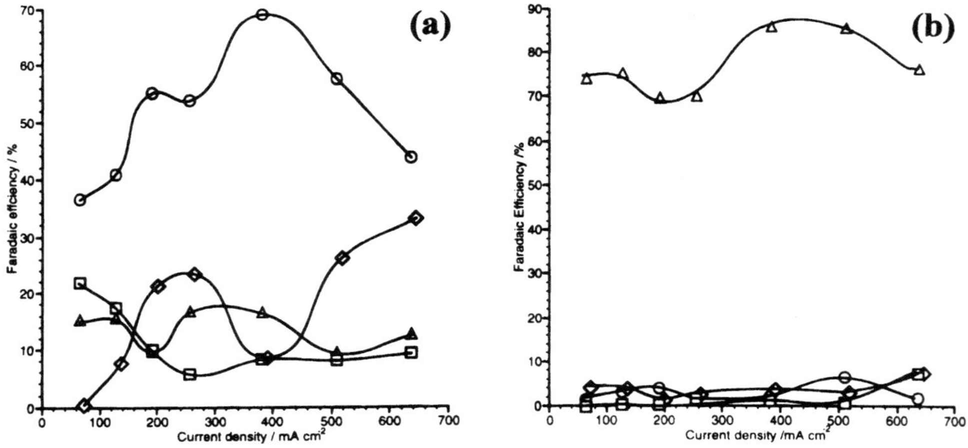

Figure 6 Dependence of faradaic efficiency on current density for products detected in the reduction of CFC 12 using (a) Cu-supported GDE and (b) Pb-supported GDE. (○) Methane, (Δ) difluoromethane, (□) chlorodifluoromethane, and (◇) H_2. (From Ref. 33.)

of metal-supported porous carbon GDEs—the best metals being Cu and Ag. CFC 13 is dechlorinated and defluorinated on the Cu-supported GDE, where CH_4 and CHF_3 (HFC 23) were primarily formed. Dechlorination proceeds selectively on the Ag-supported GDE, giving HFC 23 as the main product. The faradaic efficiencies depend on the current density, pressure of CFC 13, electrolyte composition, and potential applied to the GDE. The optimization of these factors allowed the achievement of 78% faradaic efficiency in the hydrogenation.

III. ANODIC OXIDATION

Direct anodic oxidation is a popular method for the synthesis of many organic and inorganic compounds. The technique is also capable of removing organic pollutants from water streams or reservoirs. Unfortunately, the direct oxidation reactions of organics on inert anodes are very slow as a consequence of kinetic limitations. The pollutant degradation is increased in practice by using a wide variety of electrocatalytic anodes (Pt, IrO_2, RuO_2, undoped and doped PbO_2, doped SnO_2, etc.), usually coating a Ti substrate. When low cell voltages are applied to avoid O_2 evolution, the activity of such anodes decreases with time because of the adsorption of poisoning species on their surfaces. These species can be oxidized only at higher anodic potentials in the region of water discharge with simultaneous O_2 evolution, allowing the regeneration of the anode surface during oxidation of organics.

A. Fundamentals

There are two main approaches for pollution abatement in wastewaters by anodic oxidation [35,36]:

1. The electrochemical conversion method, in which refractory organics are transformed selectively into biodegradable compounds, usually carboxylic acids. These products can be further removed using biological treatment [37].
2. The electrochemical combustion (or electrochemical incineration) method, in which the organics are completely mineralized (i.e., oxidized to CO_2 and inorganic ions).

In both cases, relatively high cell voltages are utilized to produce the simultaneous anodic oxidation of pollutants and water. Experimental results reveal that several anodes (e.g., Pt, IrO_2, and RuO_2) favor the electrochemical conversion with low current efficiency, whereas others (e.g., doped SnO_2 and

PbO_2) allow electrochemical combustion with higher current efficiency. To explain this different behavior, Comninellis [35] has proposed a simple mechanism considering that the anode surface is formed by a metallic oxide MO_x. The oxidation process is initiated by the discharge of H_2O in acid solution (or OH^- in alkaline medium) at the anode to yield adsorbed hydroxyl radical (reaction (4)), which can further undergo an oxygen transfer to the lattice of the metallic oxide, giving the so-called higher metallic oxide MO_{x+1} (reaction (5)).

$$MO_x + H_2 \rightarrow MO_x(OH^*) + H^+ + e^- \quad (4)$$

$$MO_x(OH^*) \rightarrow MO_{x+1} + H^+ + e^- \quad (5)$$

This description predicts the existence of physisorbed (adsorbed $OH^\bullet$) and chemisorbed (MO_{x+1}) "active oxygen" at the anode surface. In the absence of pollutants, both states produce O_2 according to the following reactions:

$$MO_x(OH^\bullet) \rightarrow \frac{1}{2}O_2 + MO_x + H^+ + e^- \quad (6)$$

$$MO_{x+1} \rightarrow \frac{1}{2}O_2 + MO_x \quad (7)$$

When an oxidizable organic species R is present in the solution, the physisorbed "active oxygen" causes predominantly its complete mineralization (reaction (8)), and the chemisorbed "active oxygen" participates in the formation of partially oxidized products RO (reaction (9)):

$$R + zMO_x(OH^\bullet) \rightarrow CO_2 + H_2O + zMO_x + zH^+ + ze^- \quad (8)$$

$$R + MO_{x+1} \rightarrow RO + MO_x \quad (9)$$

Thus, the electrochemical conversion is favored by anodes having a concentration of $MO_x(OH^\bullet)$ near zero. This condition is achieved if the rate of transition of oxygen into the metallic oxide lattice by reaction (5) is much faster than that of hydroxyl radical formation by reaction (4). In contrast, electrochemical combustion takes place in anodes with high surface concentration of hydroxyl radicals because the rate of reaction (5) becomes insignificant. The current efficiency for both methods then depends on the relative rate of reaction (8) or reaction (9) to that of the corresponding oxygen evolution reaction (reaction (6) or reaction (7)).

Electrochemical combustion involves the hydroxylation (reaction (10)) or dehydrogenation (reaction (11)) of organics with hydroxyl radicals. In the last case, O_2 can react with the resulting organic radical $R^\bullet$ to give a very reactive hydroperoxyl radical $ROO^\bullet$ (reaction (12)), which is able to abstract a hydrogen atom from another pollutant R′H (reaction (13)). The resulting organic hydroperoxides ROOH are relatively unstable and decompose, leading to a molecular breakdown with the generation of subsequent intermediates. These scission reactions continue until the final generation of carbon dioxide and inorganic ions:

$$\mathrm{R + MO_x(OH^\bullet) \rightarrow ROH^* + MO_x} \tag{10}$$

$$\mathrm{RH + MO_x(OH^\bullet) \rightarrow R^\bullet + MO_x + H_2O} \tag{11}$$

$$\mathrm{R^\bullet + O_2 \rightarrow ROO^\bullet} \tag{12}$$

$$\mathrm{ROO^\bullet + R'H \rightarrow ROOH + R^{\bullet\prime}} \tag{13}$$

On the other hand, several experimental parameters have been defined to quantify the destruction of an organic species in aqueous medium by anodic oxidation (38–44). Because the main side reaction is O_2 evolution due to water decomposition, the instantaneous current efficiency (ICE) at a given time t for its oxidation can be determined from the O_2 flow rate during electrolysis in the absence (V_0) and the presence ($V_{t,\mathrm{org}}$) of the selected pollutant as follows:

$$\mathrm{ICE} = \frac{V_0 - V_{t,org}}{V_0} \tag{14}$$

This equation assumes that if all the current during electrolysis is used for the oxidation of the initial pollutant and its intermediates, then $V_{t,\mathrm{org}} = 0$ and ICE = 1. When the electrolysis products are soluble in the electrolyte, the ICE can also be calculated from the change in the chemical oxygen demand (COD) [in g O_2 dm^{-3}] using the relation:

$$\mathrm{ICE} = \frac{\left[(\mathrm{COD})_t - (\mathrm{COD})_{t+\Delta t}\right]}{8I\,\Delta t} FV \tag{15}$$

where $(\mathrm{COD})_t$ and $(\mathrm{COD})_{t+\Delta t}$ are the values of the COD at times t and $t + \Delta t$, respectively; I is the applied constant current [in A]; F is the Faraday constant [96,487 C mol^{-1}]; and V is the volume of the electrolyte [in dm^{-3}]. The parameter ICE decreases with time during electrolysis to finally reach a value of about zero in a total time τ. From the ICE–t plot, an average current efficiency, which is called the electrochemical oxidizability index (EOI), is obtained:

$$\mathrm{EOI} = \frac{\int_0^t \mathrm{ICE}\,\mathrm{d}t}{\tau} \tag{16}$$

By calculating the fraction of current that oxidizes EOD species, as opposed to the fraction that oxidizes water, the electrochemical oxygen demand (EOD) [in g O_2 g organic^{-1}] (g_{org}) is determined as follows:

$$\mathrm{EOD} = \frac{8(\mathrm{EOI})I\tau}{Fg_{org}} \tag{17}$$

The degree of oxidation (χ) of the organic pollutant can then be defined from the relation:

$$\chi = \frac{\mathrm{EOD}}{(\mathrm{COD})_0} \times 100 \tag{18}$$

where $(\mathrm{COD})_0$ is the initial COD value [in g O_2 $g_{org}{}^{-1}$]. It can be seen that the anodic oxidation to CO_2 is more complete as χ approaches 100%.

The parameter EOI gives a quantitative estimate of the ease of anodic oxidation of organic pollutants; that is, the larger is the EOI, the easier it is to oxidize the species. Table 3 summarizes EOI values for selected aromatic compounds [40]. Whereas electron-withdrawing groups (–COOH, $-NO_2$, and $-SO_3H$) produce low EOI values indicative of a low electron density available for oxidation, electron-donating groups ($-NH_2$) yield high EOI values due to the increased electron density available. When both types of groups are present, as in *p*-aminotoluenesulfonic acid, the electron-donating group dominates and the benzene derivative has high EOI values. Table 3 also shows that EOD values determined for the degradation of several compounds at pH 12 are in good agreement with those theoretically found if maleic acid is considered as the final product of electrolysis.

It must be borne in mind that anodic oxidation need not go all the way to CO_2 for the process to render the pollutant harmless. Intermediates such as oxalic acid and maleic acid [37–41] are biodegradable, and therefore are acceptable as final products. The reduction by 20% of the initial COD is called a primary degradation because it involves merely a modification of the initial pollutant. In contrast, if $\chi \geq 70\%$, the solution is called intrinsically biodegradable [37]. When the organic pollutant is toxic and/or refractory, an initial electrochemical treatment can be designed to modify it and make it amenable for further biological treatment, thus giving a total process of lower cost. A coupled electrochemical–biological system for water treatment is schematized in Fig. 7.

Different electrochemical reactors have been used for the anodic oxidation of organic pollutants. They are usually simple batch cylindrical, tank, or flow reactors that are designed as undivided cells or with separators between the anolyte (the solution contained in the anode compartment) and the catholyte (the solution filling the cathode compartment). Divided cells

Table 3 EOI and EOD [in g O_2 g Organic^{-1}] of Representative Aromatic Substrates Determined at pH = 12 Using a Pt Anode

Aromatic	EOI	EOD experimental	EOD theoretical[a]
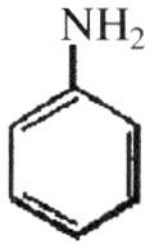	0.56	1.4	1.38
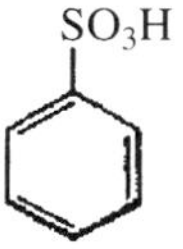	<0.05	—	—
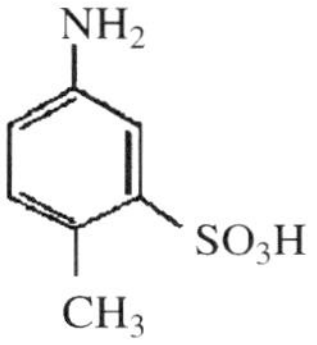	0.58	1.1	1.20
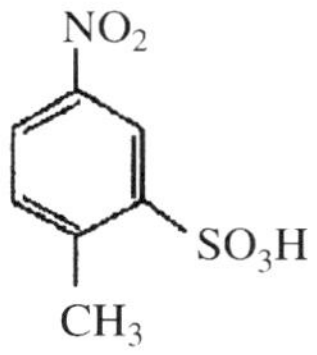	0.10	—	—
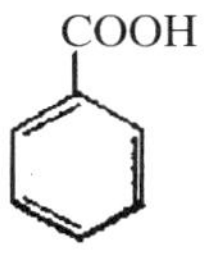	<0.05	—	—
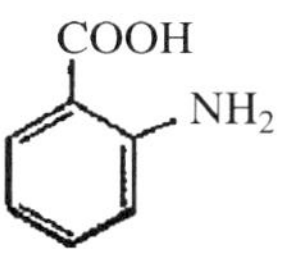	0.55	1.1	0.98
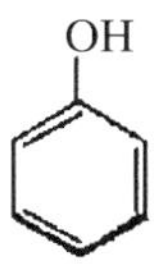	0.20	1.4	1.36

[a] Calculated supposing that maleic acid is the final product of electrolysis.
Source: Ref. 40.

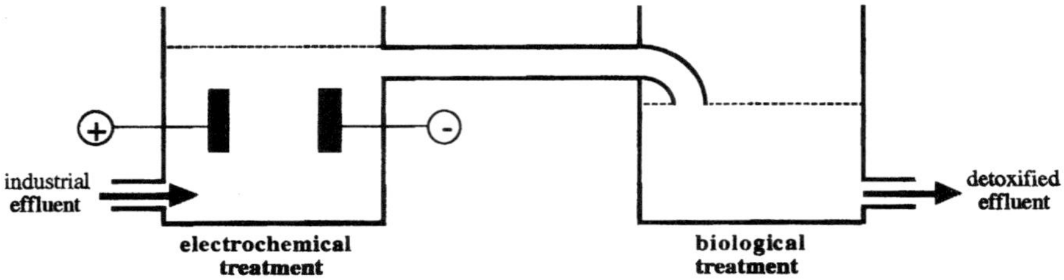

Figure 7 Schematic representation of coupled electrochemical–biological system for wastewater treatment. (From Ref. 37.)

can be useful if the cathodic reaction decreases the overall efficiency. For example, during the oxidation of phenol, *p*-benzoquinone is produced as an intermediate that can be reduced at the cathode to hydroquinone, which can be reoxidized again at the anode, thereby decreasing the efficiency of the process and requiring a separator between the anode and the cathode. Other reactors include packed bed cells and plate-and-frame arrangements with multiple electrodes mounted on a filterpress. When the cell stack contains more than two electrodes, the electrical connection can be monopolar or bipolar [2]. The monopolar connection involves an external electrical contact to each electrode and the application of the cell voltage between each anode and cathode. These electrodes alternate in the cell and both faces of each electrode are active, with the same polarity. The monopolar cell requires a low-voltage, high-current supply. In contrast, the bipolar connection needs only two external electrical contacts to the two end electrodes, and the voltage applied between them causes the polarization of intermediate electrodes (see Fig. 10). The opposite faces of each electrode will then have different polarities. The bipolar cell, in addition to simplicity of electrical connection, has the advantage of producing the equivalent amount of products as monopolar cells using many times lower currents at higher voltages, sometimes leading to a more economic use of power. However, current leaks between adjacent cells are a typical disadvantage.

The selected electrodes have to take into account the composition and the nature of the water to be treated, as well as the stability of the electrode material, its cost, and its environmental compatibility. Because organic pollutants require high potentials for their anodic oxidation, often higher than that for water oxidation, the electrode material has to be chosen carefully to prevent its corrosion under such conditions. Generally, oxidized noble metal surfaces (e.g., Pt, Ir, and Ru) covering Ti substrates are suitable for the degradation of organic substances, although their cost restricts their widespread use. Cheaper substitutes such as oxidized nickel, silver (in alkaline electrolytes), and lead can be used in aqueous media. Doped SnO_2 and PbO_2 anodes with high O_2 evolution potential are also adequate. Commercially available high surface anodes include graphite, reticulated vitreous carbon, stainless steel, nickel, and Ebonex® (a Ti-based ceramic), although most of them are useful in only a limited range of potential and pH. Recently, a synthetic boron-doped diamond thin-film anode suitable for anodic oxidation of carboxylic acids has been described [44].

We shall now proceed to discuss laboratory-scale experiments related to the destruction of different types of organic pollutants with significant environmental applications.

B. Aromatics

Aniline and its derivatives are highly toxic because they can react easily in the blood with hemoglobin, thereby preventing oxygen uptake. These aromatic amines are commonly produced as by-products or wastes in the dye, petroleum, pulp and paper, coal, perfume, and rubber industries. Kirk et al. [45] studied the anodic oxidation of aniline in dilute sulfuric acid at pH 2 for wastewater treatment. Fig. 8 shows the flow system utilized to recirculate the anolyte (bottom) and a cross-section of the fixed packed bed electrochemical reactor (top) containing a PbO_2 anode, a stainless steel cathode, and a Nafion 427 cationic membrane as a separator between anolyte and catholyte. The oxidation of aniline yielded *p*-benzoquinone ($C_6H_4O_2$) and maleic acid ($C_4H_4O_4$) as intermediates, from the following reactions:

$$C_6H_5NH_2 + 2H_2O \rightarrow C_6H_4O_2 + 3H^+ + 4e^- + NH_4^+ \quad (19)$$

$$C_6H_4O_2 + 6H_2O \rightarrow C_4H_4O_4 + 2CO_2 + 12H^+ + 12e^- \quad (20)$$

$$C_4H_4O_4 + 4H_2O \rightarrow 4CO_2 + 12H^+ + 12e^- \quad (21)$$

The treatment of 400 mL of 5.5 mM aniline led to 80% decay within 30 min. After 5 hr, 97% aniline was degraded and 72% was completely mineralized to CO_2. The percentage of oxidized aniline increased with increasing current and pH. Further experiments showed best current efficiencies ($\approx$40%) at 30 min and at pH 11.

Phenols containing one or more hydroxyl groups are produced as wastes in a variety of industries, including plastics, oil refining, pharmaceuticals, and dyes. Normally, biological treatment is preferred, although this is not a viable option for high concentrations or when the effluent has a variable composition. Chemical oxidation with Fenton reagent (H_2O_2 in the presence of Fe^{2+}), ozone, or chlorine is used in these cases. Nonetheless, safety and economic and environmental issues of such chemical oxidizers have led to the study of alternatives based on anodic oxidation [38–43,46–50]. A study by Sharifian and Kirk [39] on the degradation of phenol in

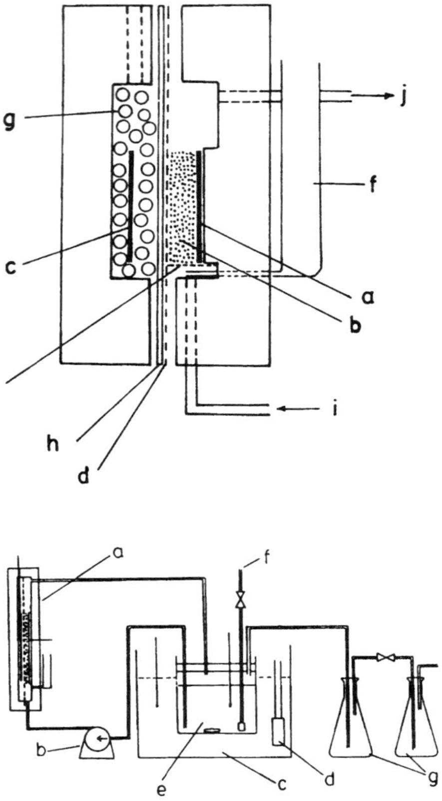

Figure 8 (Top) Electrochemical flow cell for the oxidation of phenol and aniline: (a) Pb anode feeder; (b) packed bed of 1-mm lead pellets; (c) stainless steel cathode plate; (d) Nafion membrane; (e) stainless steel screen; (f) Luggin capillary; (g) glass beads; (h) gasket; (i) reactor inlet; (j) reactor outlet. (Bottom) Schematic of apparatus: (a) electrochemical reactor; (b) peristaltic pump; (c) water bath; (d) heater; (e) anolyte reservoir; (f) gas sparging tube; (g) CO_2 adsorbers. (From Ref. 39.)

dilute sulfuric acid using the packed bed reactor shown in Fig. 8 established the formation of *p*-benzoquinone and maleic acid as major intermediates and CO_2 as the main final product. Note that these species were also obtained for aniline oxidation [45] and a similar sequence to reactions 19–22 can be given to interpret the oxidation of phenol. A higher conversion to CO_2 was achieved by increasing current from 1 to 3 A, but the maximum current efficiency dropped from around 20% to around 12%. The process became faster when sulfuric acid concentration increased from 0.1 to 2 M (promoting the breakdown of the *p*-benzoquinone ring) and when using higher temperatures and dissolved O_2 to accelerate the oxidation rate of intermediates. The production of CO_2, however, is somewhat inhibited as phenol concentration is increased from 3.5 to 56 mM due to the formation of by-products, leading to a polymeric film at the PbO_2 anode. Such polymeric films that hinder the electron transfer at the electrode interface can be prevented when phenol is oxidized more efficiently by incorporation of Bi(V) sites in PbO_2 film anodes [49].

The anodic oxidation of phenol at a Pt anode has been studied by Comninellis and Pulgarin [41]. A yellow-brown, electrically conducting polymeric film is formed at the anode surface at pH > 9, current density < 30 mA cm^{-2}, temperature $\geq 50°C$, and phenol concentration of 50 mM. Under these conditions, the phenolate anion ($C_6H_5O^-$) is oxidized, leading to the formation of a polyoxyphenylene film. The EOI is independent of the current, so that the process is not limited by mass transfer. This suggests that the oxidation occurs by an electrophilic attack of $OH^\bullet$ on the aromatic nucleus, which also explains an increase in EOI from 0.078 to 0.143 and in EOD from 0.99 to 1.41 when the pH increases from near 2 to around 13 at 70°C, because the phenolate ion is more reactive than phenol toward such an attack. Hydroquinone, catechol, and *p*-benzoquinone were initially formed in large amounts and further oxidized into aliphatic acids such as maleic, fumaric, and oxalic acids, which remained stable in the solution. These products are similar to those obtained during the oxidation of phenol with Fentons reagent, yielding about 30% mineralization. Because anodic oxidation allows more mineralization (up to about 60%), the authors proposed the two parallel processes depicted in Fig. 9 involving a chemical reaction of adsorbed $OH^\bullet$ with organics (r_1 path), or a direct reaction to CO_2 of adsorbed organics at the anode (r_2 path).

The use of Ti/SnO_2 anodes doped with Sb for the destruction of phenol has been considered by several authors [43,46,47]. Comninellis and Pulgarin [43] found that at pH 13, this anode had a very high overpotential for O_2 evolution, increasing the rate of phenol oxidation in relation to that of Pt perhaps due to the change in chemical structure of its surface (e.g., by hydration of the—SnO bond during anodic polarization). At the SnO_2 anode, only very small amounts of hydroquinone, catechol, and *p*-benzoquinone were found as aromatic intermediates, whereas fumaric, maleic, and oxalic acids were mineralized rapidly. More than 90% of phenol was oxidized to CO_2. This behavior was explained by a reaction sequence in which the pollutant was preferentially adsorbed at the hydrated electrode surface and further oxidized by adsorbed hydroxyl radicals. The superiority of doped SnO_2 over Pt and PbO_2 with respect to the anodic oxidation of phenols and other aromatics and aliphatics has also been reported by Stucki

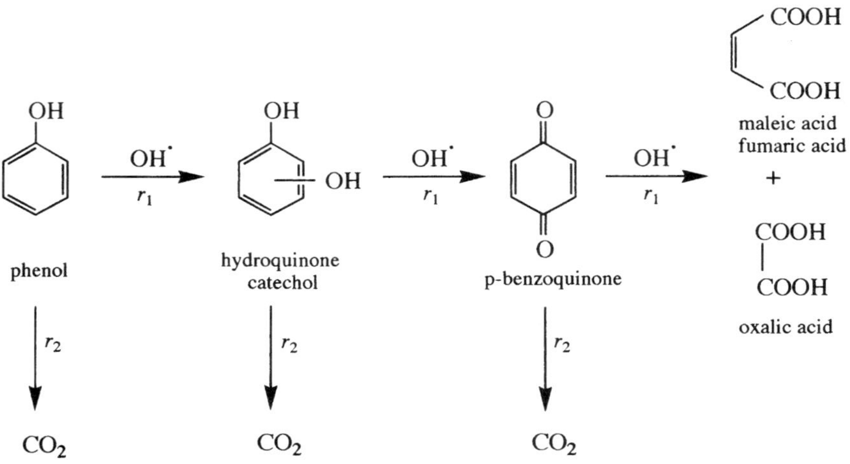

Figure 9 Reaction sequence of the electrochemical combustion of phenol. (r_1) Chemical reaction of adsorbed hydroxyl radicals with the organic molecule. (r_2) Electrochemical cold combustion to CO_2 of adsorbed organic molecules. (From Ref. 41.)

et al. [47], who designed a simple plate-and-frame bipolar reactor with undivided cells for wastewater treatment using doped SnO_2-coated Ti anodes, as illustrated in Fig. 10. The electrodes were platinized on the cathode side to catalyze H_2 evolution and to prevent corrosion in acidic media. Because typical EOI values for the oxidation of organics using SnO_2 anodes were 0.3–0.4, the power consumption of the reactor was 40–50 kW hr for the removal of 1 kg of COD for a cell voltage of 4 V. This treatment was recommended for COD concentrations between 500 and 15000 ppm. A disadvantage could be the low stability of doped SnO_2 anodes.

Several chlorophenols have also been degraded by anodic oxidation. Johnson et al. [51] considered the degradation of 30 mL of a 4-chlorophenol solution at pH 3–6, using a 5.3-cm^2 surface area Pt wire coated with a quaternary oxide film containing Ti, Ru, Sn, and Sb oxide, tightly coiled around a stainless steel cathode with a Nafion membrane as separator. By applying 0.95 A, solution TOC dropped from an initial value of 59 ppm to a value of 1 ppm in 24 hr. Twenty-six intermediates were identified, such as *p*-benzoquinone, 4-chloro-1,2-dihydroxybenzene, maleic acid, succinic acid, malonic acid, and the chloride, chlorate, and perchlorate anions. On the other hand, Gattrell and MacDougall [52] used a flow by-cell with 20.4 cm^3 carbon felts as the anode and cathode (see Fig. 11) to oxidize pentachlorophenol in acetate buffer. By applying 0.35 V vs. Hg/Hg_2SO_4 to the anode, 200 mL of 100 ppm pentachlorophenol was completely degraded

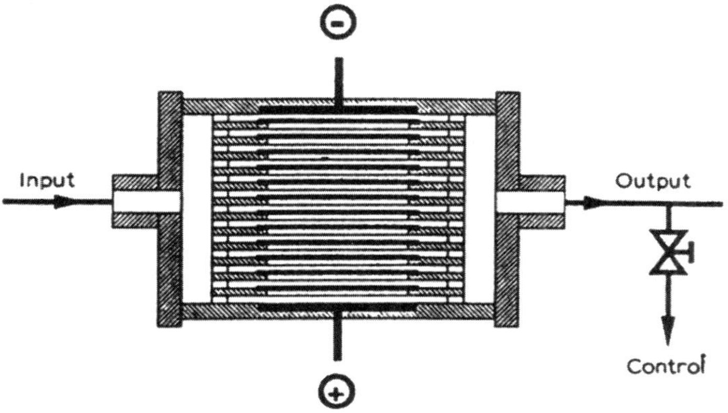

Figure 10 Pilot-scale plate-and-frame bipolar reactor for anodic wastewater treatment. (From Ref. 47.)

in 20 min, whereas about half of the pollutant was removed from 800 mL of the same solution. When the reaction was carried out on a Pt foil, an insoluble dimer, collected as an anodic deposit, was produced. These results suggested that a useful method to remove chlorinated phenols from wastewaters can be its collection onto electrode surfaces by electrochemically driven condensation reactions.

C. *p*-Benzoquinone

The compound *p*-benzoquinone is one of the most toxic xenobiotics and is an intermediate in the course of the oxidative degradation of a wide variety of benzene derivatives. It belongs to an important family of compounds often present in industrial wastewaters, particularly from photographic processes. Pulgarin et al. [53] have studied the detoxification of a *p*-benzoquinone solution at pH 2.5 using Ti/IrO_2 and Sb-doped Ti/SnO_2 anodes. A scheme of the undivided cell used, with a Pt spiral cathode enclosed in a porcelain pot, is depicted in Fig. 12. The EOI values for the anodes were 0.06 and 0.19, respectively—determined from the electrolysis of a 14.8-mM pollutant solution at 50 mA cm^{-2}. The Ti/IrO_2 anode led to electrochemical conversion with formation of aliphatic acids such as maleic, fumaric, mesoxalic, and oxalic acids, which are only minimally oxidized in this system. Because these acids are easily biodegradable and nontoxic, it is hypothesized that a coupled electrochemical–biological system would be effective for treatment (see Fig. 7). In contrast, a complete mineralization was achieved from the Sb-doped Ti/SnO_2 anode with a required electrical charge near

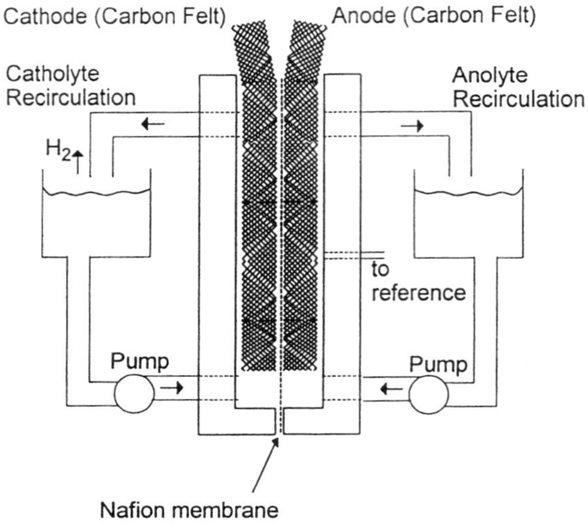

Figure 11 Flow by-cell setup for anodic oxidation of pentachlorophenol. (From Ref. 52.)

40 A hr dm^{-3}. Recently, the electrochemical incineration of *p*-benzoquinone in acetate buffer has been reported by Houk et al. [54]. The cell was similar to that above cited for 4-chlorophenol oxidation (see Sec. III.B), with a Ti or Pt anode coated with a film of the oxides of Ti, Ru, Sn, and Sb. These anodes are stable but somewhat less efficient than an Fe(III)-doped PbO_2 film coated on Ti employed in a previous work [55]. The COD of 50 mL of 100 ppm *p*-benzoquinone decreased from an initial value of 190 to 2 ppm during 64 hr of electrolysis at 1 A. The major intermediate products identified were hydroquinone and aliphatic acids including maleic, succinic, malonic, and acetic acids. The suggested reaction sequence is given in Fig. 13, where succinic acid is obtained from a cathodic reduction of maleic acid, which is formed from the breakdown of the dihydroxylated derivative generated by an attack of adsorbed hydroxyl radicals onto *p*-benzoquinone. Further mineralization of succinic acid occurs via its consecutive oxidation to malonic and acetic acids.

D. Human Wastes

Several investigations [56,57] have been devoted to the electrochemical treatment of human wastes in an attempt to make possible its electrochemical combustion. Tennakoon et al. [57] degraded artificial feces/urine mixtures at 90°C in a "U" tube cell, further scaling up the process to a parallel plate cell

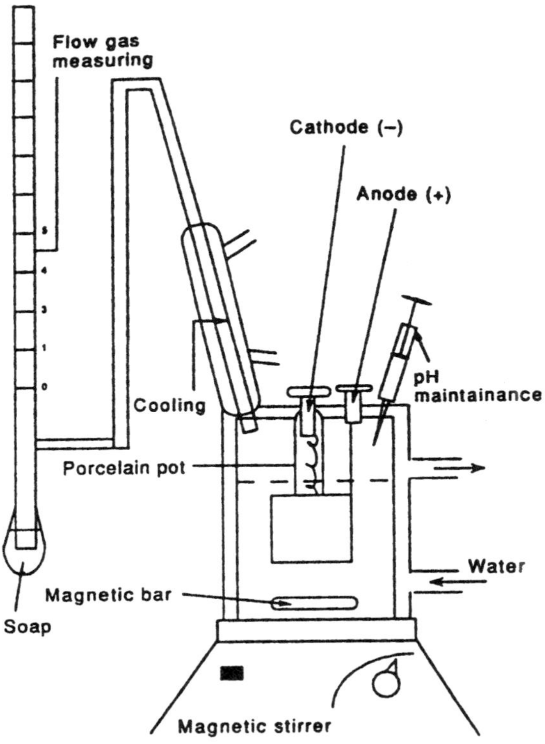

Figure 12 Divided electrolytic cell for the detoxification of *p*-benzoquinone solutions in wastewater treatment. (From Ref. 53.)

with a high solution flow rate and a packed bed cell with a high surface area/volume ratio, as illustrated in Fig. 14. The anodes used were Pt for the "U" cell; graphite, PbO_2, or Ti/RuO_2 for the parallel plate cell; and Ebonex particles coated with SnO_2 doped with Sb_2O_3 for the packed bed cell. The synthetic fecal mixture was composed of cellulose, oleic acid, casein, KCl, NaCl, Ca_2Cl, and the microorganisms *Torpulina* and *Escherichia Coli* being mixed with different proportions of urine. Gases such as CO_2, H_2, O_2, and N_2 were collected as final products in all cases. For the packed bed electrochemical reactor system, particles of 0.5–1.0 mm diameter of coated Ebonex, a solution flow rate of 0.9–1.4 cm sec^{-1} through the packed bed, a bed height of 5–8 cm, and a current density of 5 mA cm^{-2} at 12 V comprise an optimum set of operational parameters. Under these conditions, an energy requirement of 11.4 kW hr was estimated to deal with the waste of one person in 24 hr. This technology could be appropriate for water recycling in space missions.

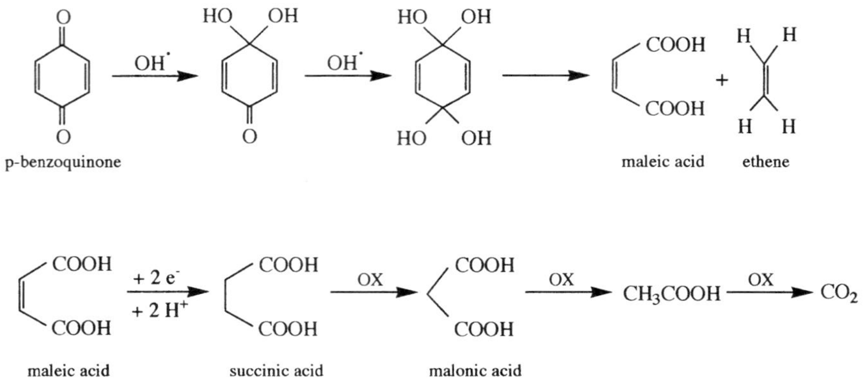

Figure 13 Reaction pathway for the electrochemical incineration of *p*-benzoquinone at a Pt anode covered with a quaternary metal oxide film. (From Ref. 54.)

E. Other Organic Pollutants

Aqueous solutions containing pollutants such as 1,2-dichloroethane [58], dyestuffs [59], benzene [49,60], cyclohexane [60], ethanol [60], methanol [60], carboxylate anions in nuclear wastes [61], and glucose [62] have been treated successfully by anodic oxidation. For the dehalogenation of 1,2-dichloroethane on Pt [58], the reaction path involves oxidation to oxalic acid via the corresponding alcohol and aldehyde, giving CO_2, Cl_2, and $HClO_4$ as final products. In the other cases, however, the mechanistic aspects have not been elucidated. A recent study [44] has reported on the complete mineralization of acetic, formic, and oxalic acids using a one-compartment electrolytic flow cell with a thin boron-doped diamond film on conductive *p*-Si substrate (Si/diamond) disc as the anode and a zirconium disc as the cathode. The anodic oxidation of these carboxylic acids takes place in the potential region of water and/or the supporting electrolyte decomposition, with high current efficiency. As can be seen in Fig. 15, for $0 < \chi < 90\%$, the concentration of acetic acid decreased linearly with the specific electrical charge Q, forming mainly CO_2, whereas its ICE was also reduced linearly from 0.97 to 0.85. In contrast, for $\chi > 90\%$, a rapid decay in ICE was found to be due to the diffusion control of the oxidation process.

IV. REDOX MEDIATORS

For this method of treatment, the electrolysis is performed in the presence of a redox reagent that can be electrochemically reversibly oxidized or reduced to

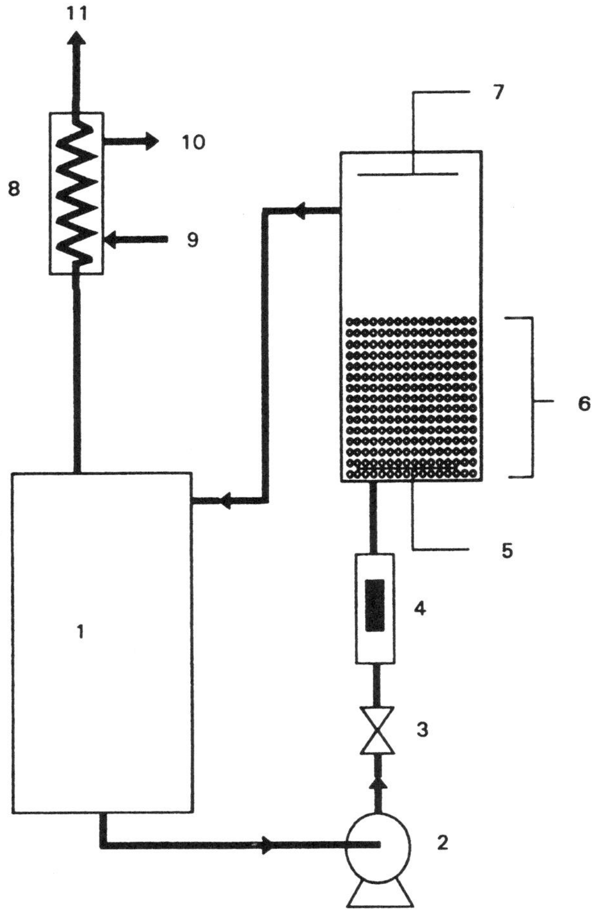

Figure 14 Schematic diagram of the packed bed cell and the flow circuit utilized for the treatment of human wastes: (1) reservoir; (2) pump; (3) valve; (4) flow meter; (5) anode current collector; (6) packed bed anode; (7) cathode; (8) water condenser; (9) water inlet; (10) water outlet; and (11) outlet for gases. (From Ref. 57.)

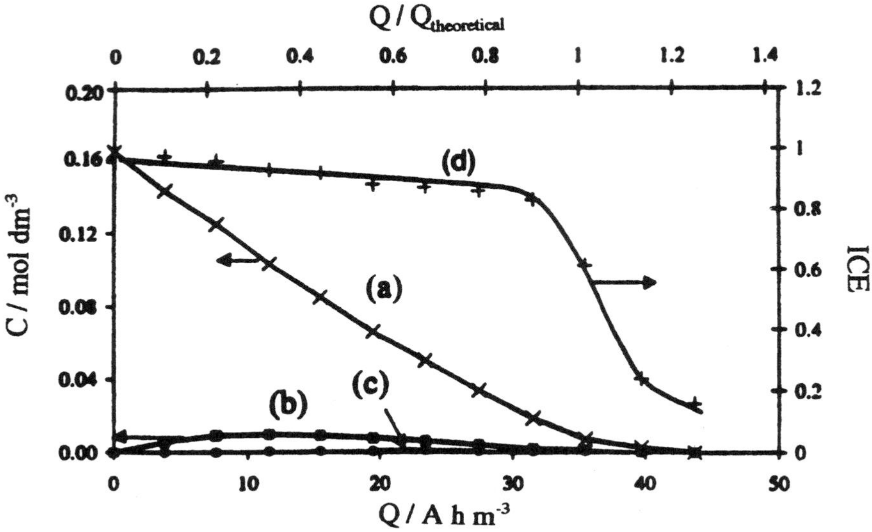

Figure 15 Evolution of (a) acetic acid, (b) formic acid, and (c) oxalic acid concentrations, and (d) ICE during the anodic oxidation of 500 mL of 0.16 mol dm^{-3} acetic acid in 1 mol dm^{-3} H_2SO_4 at 30°C on a 50-cm^2 Si/diamond anode. The applied current is 30 mA cm^2. (From Ref. 44.)

the other form of the redox couple. This species then reacts with the pollutant to produce less harmful products, and the initial reagent is recovered. For example, for pollutant degradation by the oxidized form, the electrolysis is performed in the presence of the reduced form R of the mediator couple, which is anodically oxidized:

$$R \rightarrow O + e^- \tag{22}$$

and the oxidized form O reacts with the pollutant in the solution to recover the redox reagent R:

$$O + \text{pollutant} \rightarrow R + \text{products(s)} \tag{23}$$

Degradation can also be performed by electrolysis in the presence of the oxidized form. Thus, the reduced species formed at the cathode reacts with the organic pollutants, whereas the oxidized form is regenerated:

$$O + e^- \rightarrow R \tag{24}$$

$$R + \text{pollutant} \rightarrow O + \text{product(s)} \tag{25}$$

Note that pollutant degradation will cease when the current is switched off because the species able to react with the pollutants is no longer produced. A direct oxidation of organics is usually diffusion-limited in aqueous media, and high current efficiencies can only be obtained at very low current densities [63]. With a mediator, however, higher current densities can be used. The requirements to apply this method with suitable efficiency are as follows [1]:

1. The potential at which the redox couple is produced must not overlap with the potential for H_2 or O_2 evolution.
2. The rate for reactions 22–25 have to be high relative to possible parallel reactions such as chemical water oxidation or reduction through O or R.
3. The adsorption of pollutants on the electrode should be minimized because the rate of R or O becomes sluggish due to the electrode fouling.

A. Reduction Mediators

Savall et al. [64,65] have studied the electroreduction of CFC 113 to CTFE using Zn, Cd, Ni, and Cu cathodes and MeOH (50–90 vol.%)–water mixtures with NH_4Cl. At low CFC 113 concentration, the process on the Zn cathode is mass transfer-limited. The electrochemical reaction is the same as reaction (2), but it is assumed to be mediated by NH_4^+ via an NH_4^0 adsorbed intermediate:

$$2NH_4^+ + 2e^- \rightarrow 2NH_4^0 \tag{26}$$

$$2NH_4^0 + Cl_2FC - CF_2Cl \rightarrow ClFC{=}CF_2 + 2NH_4^+ + 2Cl^- \tag{27}$$

In the presence of NH_4Cl, the direct reaction of CFC 113 with Zn also takes place:

$$Cl_2FC - CF_2Cl + Zn \rightarrow ClFC{=}CF_2 + ZnCl_2 \tag{28}$$

Zn metal can be regenerated by an electroreduction of Zn^{2+} at the cathode. At 25°C, current densities greater than 300 A m^{-2} were found, and the indirect electrochemical reduction of 2 M CFC 113 in 90 vol.% methanol–water mixture was dominated by zinc mediation.

Chlorinated aliphatic compounds were dechlorinated even in water by electrochemical reduction on a Zn-modified carbon cloth cathode consisting of partly amorphous and partly graphitized carbon material with 10 wt.% Zn [9]. This electrode has good adsorption properties, conductivity, and stability in different solvents, allowing the combination of both adsorption and dehalogenation processes. The process is carried out in a divided cell with an ion exchange membrane and a Pt anode. The indirect electrochemical dechlorination of hexachlorocyclohexane (lindane) on this cathode shows much better efficiency in comparison to its reduction on a carbon cloth. Lindane is essentially reduced by Zn, which is regenerated by the electrochemical reduction of the Zn^{2+} produced. The final Zn^{2+} concentration in the solution after the electrolysis (2.9 ppm) is much lower than that found after the electrodeless stoichiometric reductive dechlorination (40 ppm).

Zhang and Rusling [66] employed a stable, conductive, bicontinuous microemulsion of surfactant/oil/water as a medium for catalytic dechlorination of PCBs at about 1 mA cm^{-2} on Pb cathodes. The major products were biphenyl and its reduced alkylbenzene derivatives, which are much less toxic than PCBs. Zinc phthalocyanine provided better catalysis than nickel phthalocyanine tetrasulfonate. The current efficiency was about 20% for 4,4′-DCB and about 40% for the most heavily chlorinated PCB mixture. A nearly complete dechlorination of 100 mg of Aroclor 1260 with 60% Cl was achieved in 18 hr. Electrochemical dehalogenation was thus shown to be feasible in water-based surfactant media, providing a lower-cost, safer alternative to toxic organic solvents.

Anthraquinoid compounds have also been employed as mediators for reduction of dispersed organic compounds, particularly for dyestuffs used in dyeing of cellulose fibers in 0.1 M NaOH [67]. The reduction efficiency is characterized by comparing the maximum cathodic current of the anthraquinoid solution containing the dyestuff with the cathodic peak current without reducible vat dye. The limiting current density depends on the diffusion transport of the anthraquinoid compound, whereas the addition of dispersed dyestuff has a minor influence.

B. Oxidation Mediators

As indicated by the high value of its standard electrode potential ($E^0 =$ 1.98 V) Ag^{2+} is a powerful oxidizing agent. When Ag^+ is anodically oxidized in nitric acid [63,68], the dark-brown complex $AgNO_3^+$ is formed:

$$Ag^+ + NO_3^- \rightarrow AgNO_3^+ + e^- \tag{29}$$

which reacts with the oxidizable pollutant and even with water, as follows:

$$AgNO_3^+ + \text{pollutant} + H^+ \rightarrow Ag^+ + HNO_3 + nCO_2 \tag{30}$$

$$4AgNO_3^+ + 2H_2O \rightarrow 4Ag^+ + HNO_3 + O_2 \tag{31}$$

The stoichiometry in reaction (30) depends on the pollutant. In the absence of organic waste, the anolyte rapidly becomes dark brown, but it is not possible to convert all Ag^+ ions into $AgNO_3{}^+$ because reaction (31) follows second-order kinetics and rapidly equals the rate of reaction (29). In the presence of organic waste and under steady state conditions, there is no brown color visible, which is expected if the $AgNO_3{}^+$ concentration is very low.

Nitric acid is reduced to nitrous acid at the cathode during electrolysis, although a further reduction to NO_x occurs when the nitrous acid concentration increases. Nitrous acid can be oxidized to nitric acid by O_2 and, therefore, it can be recycled into the system. Other reactions involved in the process are the formation of highly active radicals, such as $OH^\bullet$ and $O^\bullet$, allowing a wide scope for destruction of organics [60,63]:

$$\mathrm{Ag(II)} + \mathrm{H_2O} \rightarrow \mathrm{Ag(I)} + \mathrm{H^+} + \mathrm{OH^\bullet} \qquad (32)$$

$$\mathrm{2Ag(II)} + \mathrm{H_2O} \rightarrow \mathrm{2Ag(I)} + \mathrm{O^\bullet} + \mathrm{2H^+} \qquad (33)$$

The process was developed initially for the nuclear industry and a variety of combustible wastes including rubber, some plastics, polyurethane, various ion exchange resins, hydraulic and lubricating oils, and waste process solvents [63]. The method has also been applied to ethylene glycol [68], benzene [68], kerosene [63], organic acids [69], isopropanol [70], acetone [71] and organophosphorous, organosulphur, and chlorinated aliphatic and aromatic compounds, including PCBs [63]. For example, Choi et al. [69] have obtained 94–96% destruction efficiencies for ethylendiaminetetraacetic and oxalic acids, and 87–90% decays for citric acid and nitrilotriacetic acid, explaining the difference as 100% by the transport of some organics to the catholyte across the cation-selective membrane. Steele [63] has described a continuously run pilot-scale system, with a constant feed of organic pollutants to the anolyte and a constant regeneration of the acid degraded at the cathode (Fig. 16). Several cells were employed with Pt-coated Ti anodes to produce CO_2 and CO. In the case of benzene, one has:

$$\mathrm{C_6H_6} + \mathrm{12H_2O} \rightarrow \mathrm{6CO_2} + \mathrm{30H^+} + 30e^- \qquad (34)$$

$$\mathrm{C_6H_6} + \mathrm{6H_2O} \rightarrow \mathrm{6CO} + \mathrm{18H^+} + 18e^- \qquad (35)$$

CO_2 and CO were vented via an appropriate treatment system to avoid emissions of fumes. The catholyte loop allows a regeneration of nitric acid in a packed reflux column where air is injected. Ag^+ ion is more concentrated in the catholyte than in the anolyte because of its migration through the sulphonated fluoropolymer cation exchange membrane. Ag is not deposited on the cathode under the experimental conditions and no hydrogen evolution is observed. The destruction capacity for a given configuration of the plant depends on the chemical nature of the substrate. For example, a plant for treating 200 kg hr^{-1} of dodecane is capable of destroying 3350 kg hr^{-1} of CCl_4.

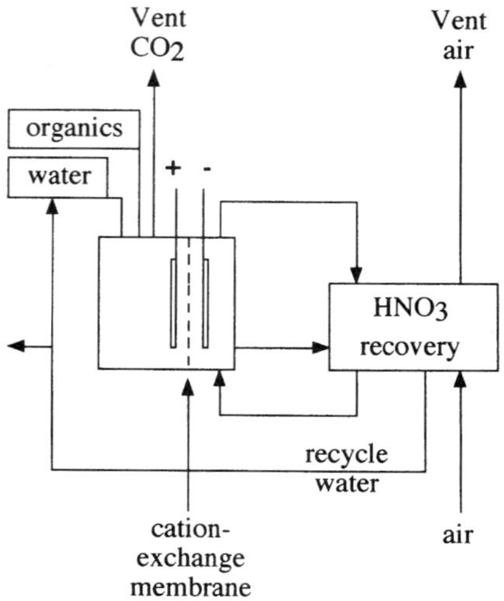

Figure 16 Scheme of the pilot-scale rig for the mediated destruction of organic wastes using the Ag(II)–HNO_3 system. (From Refs. 60 and 63.)

The use of this method presents some problems [1,72]:

1. Silver ions should be totally regenerated or recycled because they are hazardous wastes by themselves.
2. Ag^+ precipitates with Cl^- produced during the dechlorination processes.
3. Silver is expensive.
4. There are corrosion problems of the anode and other cell components in HNO_3/HCl.
5. HNO_2 and NO_x are formed at the cathode.
6. A separator is needed to prevent mixing of the very highly oxidizing anolyte with the reduced species formed at the cathode. Problems inherent to separators are diffusion of mediators, fouling and rupture, leakage of separator seals, and an increase in the electrical cost of the process.

Like silver, Co(III) is also a powerful oxidizing agent with $E^0 = 1.82$ V. Co(II) in HNO_3 has been employed to degrade different organic compounds [70,71,73,74] by using separators to prevent Co electrodeposition. In acidic aqueous media, the oxidation of Co(II) to Co(III) has less than 100% current efficiency because it occurs at a more positive potential than water. Cobalt has the advantage over silver in that cobalt chloride complexes are generally soluble and the precipitation of chloride salts is thus avoided. In addition, the rate for the reaction of Co(III) with water is slow at room temperatures, and the economical and environmental costs of cobalt as a mediator are smaller than for other metals, particularly silver.

Zawodzinski et al. [70] have compared the mediated and the direct electrolyses of isopropanol in 6 M HNO_3 using a Pt anode to yield CO_2 and acetic acid. The reaction proceeds more quickly to completion in the order: Co(III) > unmediated > Ag(II).

Dziewinski et al. [71] have reported current efficiencies greater than 90% at 80°C for isopropanol, acetone, and acetic acid, whereas methanol, chloroform, and carbon tetrachloride were converted into carbon dioxide at room temperature with ca. 100% current efficiency. The predicted order of ease of oxidation is: alcohols < aldehydes, ketones < carboxylic acids. Thus, the complete conversion to carbon dioxide, in particular for large organic molecules, is expected only at elevated temperatures. To avoid the use of a separator, Farmer et al. [72,75] employed $CoSO_4$ in sulfuric acid. A scheme of the system is shown in Fig. 17. Both electrodes, the anode and the cathode, are usually of a Pt-containing surface layer, although other materials such as Au, steel, graphite, Ti, Nb, Ir, or tin oxide can also be used. H_2 is evolved at the cathode because the deposition of the mediator takes place

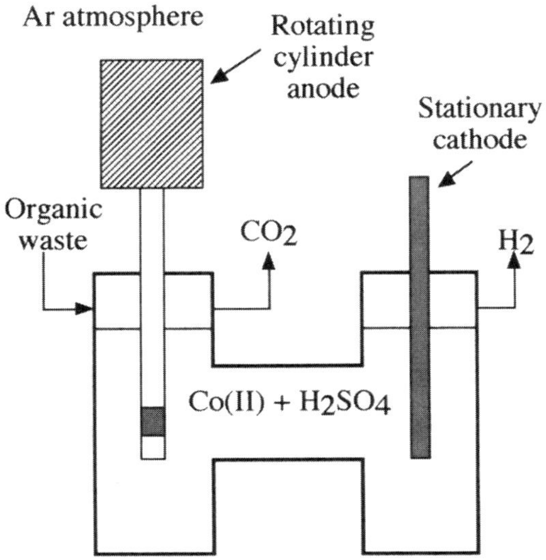

Figure 17 Electrolysis cell, closed in a reaction chamber, for the Co(III)-mediated destruction of organics in sulfuric acid. (From Ref. 75.)

at more cathodic potentials. Co(III) produced from anodic oxidation of Co(II) reacts with the solvent and dissolved organics in a thin layer of electrolyte at the anode surface to produce CO_2, so that no membrane is required to prevent Co(III) migration to the cathode. 1,3-Dichloro-2-propanol and ethylene glycol have been degraded with conversions close to 90% and faradaic efficiencies of 43% and 76%, respectively. Ethylene glycol is probably mineralized by a sequence of reactions involving the production of two molecules of formaldehyde per molecule of ethylene glycol, followed by their oxidation to formic acid.

Steward et al. [73] preferred the Co(III)/HNO_3 system. These authors utilized the cell of Fig. 18 with a separator (which permits concentration of the waste in the anolyte reservoir) and corrosion-resistant electrodes such as Pt. The suggested concentrations are 0.5 M for Co(II) and 4–12 M for HNO_3. This process appears to be able to destroy the vast majority of organic materials. Double bond, alcohol, and carboxylic acid groups greatly facilitate the oxidation process. However, aliphatic hydrocarbons exhibit slow oxidation. Only the CF bond, such as that contained in PTFE, polyvinylidenefluoride, and fluoroelastomers (Viton), is not oxidized. Thus, these polymers are excellent materials for the construction of mediated

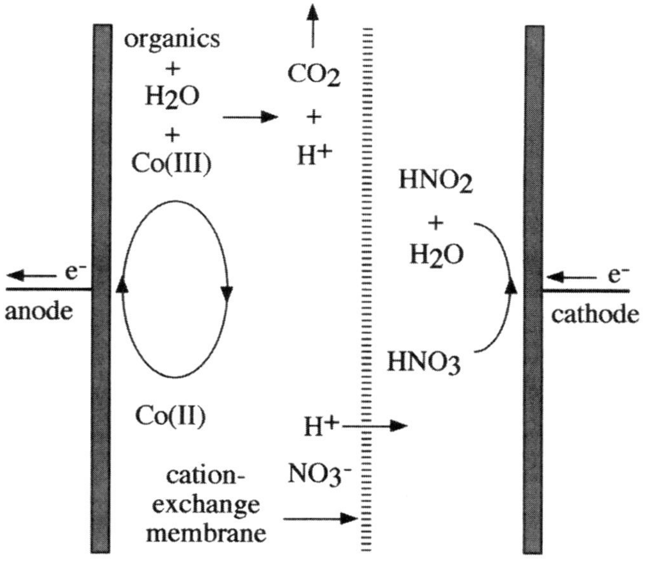

Figure 18 Mediated electrooxidation of organics using the Co(III)/HNO_3 system in a cell with a cation exchange membrane. (From Refs. 71 and 73.)

electrochemical oxidation devices. The process can also be conducted in a neutral pH anolyte at up 2 M Co(II) [74].

The Fe(II)/Fe(III) system presents a much lower standard electrode potential ($E^0 = 0.77$ V) than those considered above, allowing a significant reduction of the electric cost, which represents about 90% of the total cost [76]. However, the process is slow because of the low current densities, and temperatures around 100°C are needed for an efficient operation [1]. Koval et al. [77] have considered reversible complexation and electrolytic regeneration to remove nitrogen and sulfur-containing organic pollutants such as thionaphthene, isoquinoline, and carbon disulfide from a diluted waste to obtain a concentrated hydrocarbon waste. The cathodic compartment initially contained an Fe(II) complex of the sodium salt of tetra(4-sulfonatophenyl)porphyrin [Fe(II)L], which was able to yield a very stable complex, Fe(II)LW, with the nitrogen and sulfur-containing organic waste. Fe(II)LW is then oxidized to Fe(III)LW in the anode compartment. Because the resulting complex has a very small formation constant, it is easily decomposed to Fe(III)L and the organic waste, which is concentrated and partitioned at the anode compartment. To complete the cycle, Fe(III)L is pumped to the cathode compartment where it is reduced to Fe(II)L again.

The Fe(II)/Fe(III) system has been utilized to oxidize carbonaceous materials such as coal slurry in acidic media at the anode compartment [78], giving partially oxidized material or carbon oxides. In contrast, cellulosic materials, fats, urea, manure, sewage sludge, and meat packing plant wastes have been treated in this form with high oxidation efficiencies [76]. Hydrogen is evolved at the cathode and the degradation rate increased by an order of magnitude by means of ions such as Pt, Pd, V, Co, and Ru, added as cocatalysts to the solution. A prototype unit with a 1200-L reaction vessel and two electrochemical cell stacks were operated at current densities 1–10 mA cm^{-2} and cell voltages below 1 V for 200 hr, but improvements in membrane and electrode materials are needed.

Chou et al. [79] have studied the anodic oxidation of anthracene to anthraquinone in a slurry electrolyte containing both a Mn^{3+}/Mn^{2+} redox mediator and dodecylbenzenesulfonate (DBS) as surfactant. No reaction of anthraquinone is found in the absence of DBS. The main factors affecting the current efficiency of the process were sulfuric acid concentration, current density, particle size of anthracene, temperature, and concentrations of DBS and redox mediator.

Davidson et al. [80] have used a ruthenium electrocatalyst to mineralize highly chlorinated and aromatic species such as chlorobenzene, pentachlorophenol, and tetrachloroethylene, with minimum generation of secondary waste and efficient recovery of the ruthenium mediator. This system is similar to that of Ag(II), but it is not affected by the presence of the halogen, particularly chloride. The overall process was performed in a divided cell, very similar to the type used for chlorine production, consisting of two linked cycles involving the $Cl^-/HOCl$ and RuO_2/RuO_4 pairs. Chloride ions were converted into HClO by anodic oxidation and HClO reacted with the organic wastes fed to the anolyte, forming mainly CO_2, CO, and inorganic anions, recovering RuO_2.

Nauflett and Farncomb [81] have described the easy destruction of Otto Fuel II, a three-component liquid monopropellant used for torpedo propulsion, by Ce(IV) in a mineral acid electrolyte. Chung and Park [82] have treated aniline solutions with Ce(IV) and Co(III) as mediators, the final product being CO_2. Parameters affecting current efficiency included oxidation potentials of mediators, their concentrations, and temperature.

V. ANODIC GENERATION OF STRONG OXIDANTS

Oxidizing agents such as ozone, hypochlorite, chlorine, and chlorine dioxide can be produced anodically and used subsequently to destroy pollutants in water streams. These agents are also commonly employed in the disinfecting of drinking water. Reactions of such strong oxidants with organics and their environmental/industrial applications are discussed in other chapters of this book. This section is devoted to describing the reactions involved in the anodic generation of oxidants for the degradation of organic pollutants by indirect electrolysis.

A. Electrogeneration Reactions for Ex Situ Applications

The anodic synthesis of ozone from anodic oxidation of water in sulfuric acid media has been known since 1840. The reaction and its standard potential are:

$$3H_2O \rightarrow O_3 + 6H^+ + 6e^- \quad E^0 = 1.51\text{V} \tag{36}$$

Oxygen evolution occurs at a much lower potential and is therefore favored thermodynamically:

$$2H_2O \rightarrow O_2 + 4H^+ + 4e^- \quad E^0 = 1.23\text{V} \tag{37}$$

The oxidation of evolved oxygen also contributes to ozone production:

$$O_2 + H_2O \rightarrow O_3 + 2H^+ + 2e^- \quad E^0 = 2.07\text{V} \tag{38}$$

O_2 evolution must be inhibited to obtain ozone at significant current efficiencies. Some of the requirements for practical ozone electrogeneration include the use of high oxygen overpotential anodes that are stable in the highly acidic environments produced by the anodic decomposition of water, and electrolytes whose anions and cations do not engage in competitive oxidation or reduction.

On the other hand, it is well known that chlorine gas is generated by the anodic oxidation of the chloride ion present in aqueous sodium chloride using a divided cell. The anodic reaction is:

$$2Cl^- \rightarrow Cl_2 + 2e^- \qquad (39)$$

whereas the cathodic process involves the reduction of water to hydrogen and hydroxide ion as follows:

$$2H_2O + 2e^- \rightarrow H_2 + 2OH^- \qquad (40)$$

In basic solutions, the chlorine generated anodically in the electrolyte can be hydrolyzed by the cathodically produced hydroxide ion to form hypochlorite ion:

$$Cl_2 + 2OH^- \rightarrow ClO^- + H_2O + Cl^- \qquad (41)$$

Chlorine and hypochlorite thus obtained are traditional oxidizing agents of organic pollutants. Chlorine dioxide is another known oxidant for water treatment, which can be generated in divided cells containing sodium chlorite or sodium chlorate. This last compound can also be produced anodically from hypochlorite oxidation:

$$6ClO^- + 3H_2O \rightarrow 2ClO_3^- + 4Cl^- + 3/2O_2 + 6H^+ + 6e^- \qquad (42)$$

Commercial electrolytic systems for the generation of all these species on small and large scales have been described previously in some detail [2].

B. In Situ Degradation of Organics with Electrogenerated Hypochlorite

Several authors [83–90] have explored the behavior of small electrolytic cells with alkaline electrolytes containing Cl^- for treating wastewaters with in situ electrogenerated ClO^-. Do and Yeh [83,84] and Do et al. [85] have used a cell with a sintered glass frit separator to degrade 120 mL of contaminated anolytes in the presence of NaCl using anodes of 3 cm^2 area. The lowest voltage for Cl^- oxidation was obtained on a RuO_2 anode, whereas the highest voltage for O_2 evolution was found on an SPR (SnO_2–PdO–RuO_2–TiO_2) anode. O_2 evolution is then inhibited effectively on SPR and a maximum current efficiency of 99.3% can be attained for the production of hypochlorite (reactions (39) and (41)) by electrolyzing 1 M NaCl solutions at pH 10–13 and 25°C. The degradation fraction of a 3000-ppm formaldehyde solution with ClO^- electrogenerated under these conditions was about 90% after about 2 hr of electrolysis at 75 mA cm^{-2}. Comninellis and Nerini [87] compared the influence of Ti/SnO_2 and Ti/IrO_2 anodes on the oxidation of 150 mL of 10 mM phenol at pH 12–13 in the presence of NaCl. They found that Cl^- catalyzes the degradation of phenol only at Ti/IrO_2 anodes due to the participation of electrogenerated ClO^-. An analysis of intermediates shows the initial formation of 2-chlorophenol, 4-chlorophenol, 2,4-dichlorophenol, and 2,4,6-trichlorophenol, which are, however, further transformed into volatile chloroderivative organics such as chloroform. This product was also reported by Rigordy et al. [88] during the treatment of 150 mL of an industrial nonbiodegradable effluent with 170 ppm of TOC from a flavor manufacturing facility. The presence of 0.4 g L^{-1} NaCl in the wastewater at pH 8 caused an 85% decrease in the initial TOC after 30 hr of electrolysis at 300 mA on a Pt anode of 2.1 cm^2 area. The formation of chloroform was explained by the ClO^- attack on organic compounds such as alkenes, methylketones, and derivatives, present in the wastewater. In this regard, the addition of Cl^- for promoting the electrochemical destruction of pollutants could have a negative environmental effect because of the possible appearance of chlorinated organics (disinfection by-products) that are usually more toxic and less biodegradable than the original pollutants. Chlorinated products, however, are not detected in the electrochemical incineration of 200 mL of 10 g L^{-1} glucose in 0.1 M Na_2SO_4 + 0.01 M NaOH with NaCl up to 5 g L^{-1} using Pt, PbO_2, and SnO_2–Pt composite anodes [89]. Another procedure has been reported in a Spanish patent [90]. It consists of a cyclic process involving an chemical–electrochemical interconversion between ClO^- and Cl^-, which yields high COD abatements up to 80–90% of wastewaters originating from different chemical industries. In the first step, the sample is chemically oxidized with 10% NaClO until this reagent is reduced completely to Cl^-. The resulting solution is further electrolyzed as anolyte in a divided filterpress cell composed of an anode catalyzed for Cl_2 release, a stainless steel cathode, and a Nafion membrane as separator. The electrochemical treatment ends after passing the theoretical current required for the complete oxidation of Cl^- by reaction (39), and chlorine gas obtained is collected in a scrubber as 10% NaClO solution, which can be used again for the chemical treatment of more wastewater.

VI. CATHODIC GENERATION OF HYDROGEN PEROXIDE

Hydrogen peroxide is a "green" chemical that only leaves water or O_2 as by-products. Thus, it is increasingly used to bleach pulp and paper, bleach textiles, clean electronic circuits, and delignify agricultural wastes. Hydrogen peroxide is also used as a disinfectant in medical and industrial applications, and as an oxidant in synthesis and wastewater treatment [91]. Its demand is expected to grow by 10% per annum in the next few years [92], as the use of chlorine-based oxidants is declining. However, H_2O_2 is hazardous due to its thermodynamic tendency to decompose to water and O_2. Peroxide activity decreases through spontaneous decomposition, which is catalyzed by many metal compounds and favored by UV light and high temperature. Therefore, its transport and storage are undesirable, and on-site production should be preferred whenever possible.

Industrially, H_2O_2 was first produced by the anodic oxidation of sulfuric acid or bisulfates to persulfate, followed by hydrolysis and distillation [93]. The original process dates back to 1853 and was dominant for decades in spite of its high power consumption [1]. In 1882, Traube obtained H_2O_2 by a cathodic reduction of dissolved O_2 [2,94]. This is noteworthy because a strong oxidant with $E^0 = 1.78$ V at pH 0 was not expected to be formed at the cathode. The electrogeneration process can be written as follows:

$$O_2 + 2H_2O + 2e^- \rightarrow H_2O_2 + 2OH^- \quad E^0 = -0.146\text{V} \tag{43}$$

The current efficiency of Traube's method was pushed to 92% [94] with the introduction of pressurized electrolysis. In 1939, Rubio [93] developed another process for producing H_2O_2 by the reduction of O_2 in 50% KOH using an active carbon cathode and a Ni anode, with no industrial success. Today, most H_2O_2 are produced chemically by the anthraquinone process, which is unsuitable for small-scale production [91]. Another electrochemical process involving O_2 reduction on carbon-based cathodes developed by Dow Chemical [2,95] has found a marketplace for on-site production of alkaline H_2O_2 for pulp bleaching.

Several works on the electrogeneration of alkaline H_2O_2 for different purposes have been described [96,97], and good reviews on these reports are available [92,98–100]. An elegant method involves the paired synthesis of O_3 and H_2O_2 with a proton exchange membrane as the electrolyte [98,100,101]. However, low current efficiencies and high voltages seem to preclude its industrial application [1]. Other studies report the spontaneous production of H_2O_2 from H_2 and O_2 in a membrane electrode assembly fuel cell [102], or an alkaline fuel cell [103]. Carbon-based cathodes are commonly used because other materials in which H_2O_2 can be generated with high yield are either toxic (Hg) or expensive (Au). Three-dimensional electrodes [92,95] or GDEs [97,99,103,104] have to be utilized to accelerate the process rate, which is limited by mass transport due to the low O_2 solubility in water (~ 1 mM).

As we will see, H_2O_2 is usually needed in acidic media for effluent treatment. This could be achieved by using anion exchange membranes, which transfer electrogenerated hydrogen peroxide from alkaline to acidic or neutral media [99], but such membranes are not stable enough and the process is not competitive.

It has been demonstrated in the laboratory that both GDEs [105,106] and three-dimensional electrodes [107] may also be used to reduce oxygen to H_2O_2 in acid solutions at rates that are appropriate to the needs of synthesis and effluent treatment. As shown below, the application of these cathodes might allow the corresponding processes to be feasible on an industrial scale.

A. Oxidation of Organics by In Situ Electrogenerated H_2O_2

H_2O_2 can be electrogenerated in the cell where oxidation occurs with the added advantage of easy control of dosage by simple means, such as current or voltage presetting.

Moeglich [108] has disclosed a complex reactor for the removal of impurities from water, including packed beds of different conductivities with different pairs of electrodes of graphite. The particle beds were coated with catalyst oxides from Mn, Cr, Bi, Pb, Ni, or Fe. The solution was subjected to alternating current and H_2O_2 was claimed to be produced from water. After treating a solution containing 200 ppm phenol, 13 ppm CN^-, and 2400 ppm COD for 1 hr, phenol, CN^-, and 99.8% COD disappeared.

Porta and Kulhanek [109] patented a process for the destruction of pathogens in wastewaters, which can also be applied to chemicals. Water is passed through a porous fixed bed carbon cathode in the presence of O_2 and the generated H_2O_2 is able to kill the cells in the effluent with the application of low current densities from 0.1 to 13 mA cm^{-2}.

Do and Chen [110] have studied the oxidation of 250–1000 ppm formaldehyde by H_2O_2 electrogenerated on a graphite plate in a divided cell. Factors affecting the process were pH, current density, temperature, and O_2 sparging rate. A maximum current efficiency of 93.5% was found at pH 13, 0.5 mA cm^{-2}, 25°C, and O_2 rate >5 mL sec^{-1}, whereas a degradation of >99% was achieved at 45°C and an O_2 rate of 2 mL sec^{-1}. The kinetics of this process has also been reported [111]. A first-order process was found for O_2 reduction with respect to dissolved oxygen, whereas oxidation was second order for CH_2O and first order for H_2O_2, with an activation energy of 36 kJ mol^{-1}. Model calculations proposed fitted empirical data. These authors further studied this reaction on modified graphite and also using an undivided cell [112]. The current density reached 1.1 mA cm^{-2} upon cathode anodization and a slight fall in degradation fraction took place in the undivided cell.

Recently, Do and Yeh [113] have considered the degradation of formaldehyde by paired oxidation with electrogenerated H_2O_2 and hypochlorite in a cell with two compartments separated by a glass frit with an anodized graphite cathode and an SPR anode. Hypochlorite is generated by the anodic oxidation of chloride ion (see Sec. V). Under these conditions, the degradation of formaldehyde in the catholyte is affected by the volume/cathode area ratio and stirring rate, indicating that the dissolution of O_2 is limiting at sparging rates < 13 mL sec^{-1}. The overall current efficiency reached 62% at 0.75 mA cm^{-2} with a degradation fraction of 93%. A theoretical model was proposed.

Amadelli et al. [114] have studied comparatively the electrochemical combustion of 4-chlorophenol and 4-nitrophenol at PbO_2 anodes by direct electrolysis at constant current, with ex situ use of O_3 or with the combined use of anodically generated O_3 and H_2O_2 produced at a graphite cathode. The last method was more efficient, but a significant role of ozone in the electrooxidation of phenol at the anode was ruled out because organics seem to react competitively with adsorbed oxygen intermediates ($OH^{\bullet}$ and $O^{\bullet}$), leading to an inhibition of O_3 formation.

B. Electrogenerated Fenton Reagent (EFR)

Fenton reagent is a mixture of H_2O_2 and Fe^{2+} utilized commonly for the degradation of organics (see the chapter on Fenton chemistry). In the presence of Fe(II), H_2O_2 decomposes to produce hydroxyl radicals, one of the most powerful and nonselective oxidants available in aqueous solution.

Initially, EFR was used for synthetic purposes. Probably, the first paper on the oxidation of organics by EFR dates back to 1971, when Tomat and Vecchi [115] studied the synthesis of phenol from suspensions of benzene in water at 8.5°C. H_2O_2 and Fe(II) were produced by the reduction of dissolved O_2 and Fe(III), respectively, on a Hg cathode at -0.35 V vs. SCE. The current efficiency for phenol reached 60% (assuming 3F per mole) and traces of higher oxidized products appeared, but no mineralization was found (see Table 4). A maximum yield was found by the addition of 0.5 mM Fe(III). The efficiency decreased with increasing H_2SO_4 concentration and passing charge. Tomat and Rigo [116–118] have applied this method to produce (methyl)benzaldehyde and smaller quantities of (methyl)benzyl alcohol from toluene or polymethylbenzene in water, using either Fe(II), V(IV), V(III), or Cu(I) ions to catalyze Fenton-type reactions. An increase in efficiency was detected in the presence of Cl^- [118]. Although current efficiencies close to 100% were found in some cases, the volume yields and currents were low due to the low O_2 solubility in water. A modification of this approach has been made by Matsue et al. [119] who used graphite cathodes in diluted aqueous solutions to synthesize alkylbenzene derivatives such as benzaldehyde and cresol isomers from toluene, or acetophenone and

Table 4 Relevant Results Obtained for the Oxidation of Selected Substrates Using the Cathodic Generation of Hydrogen Peroxide (in Chronological Order)

Substrate	O_2 feeding	Cell[a]	Cathode	Anode	Fe ion concentration [mM]	pH	j_c[b] [mA cm^{-2}]	Time [hr]	% Degradation	Reference
Benzene-saturated	Bubbled	D	Hg pool	—[c]	0.1–10	<1	—[c]	—[c]	—[c]	115
Alkylbenzenes	Bubbled	D	Graphite	Pt	5–50	0.7–3	ca. 0.01	—[c]	—[c]	119
Phenol, 1–12 mM	Bubbled	D	Graphite 78 cm^2	Pt	0.2–50	1–4	<1	≈20	71	128
Chlorobenzene, 1 mM	In situ	U	RVC	Pt	5	3.8	0.7	—[c]	—[c]	137
Phenol, 4 mM	Evolved		153 cm^2						—[c]	
Formaldehyde,	Bubbled	D	RVC	Pt	0	14	—[c]	3	—[c]	138
0.2–3.8 mM			5×5×1.2 cm^3		0.5	3	—[c]	3	80	
Aniline, 1.1 mM	Diffused	U	ODC[d]	PbO_2	0	11–13	10–242	11	>95	142
4-Chloroaniline, 0.8 mM			3.1 cm^2						>95	
Aniline, 1.1 mM	Diffused	U	ODC[d]	Pt	0–1	3	32	6	14–94[e]	105

Aniline, 1.1 mM			3.1 cm^2	PbO_2					25–96[e]	
Phenol, 4.7–27 mM	Bubbled	D	Graphite	SPR	0–0.3	3	0.75	2.8	15[f]	133
Aniline 1.11–11 mM	Diffused	U	ODC[d]	Fe	High	3–3.5	32–145	0.5	95	146
			3.1 cm^2					2	91	
Chlorophenoxyacid herbicides	Bubbled	D	Hg pool	Pt	2	2	4	2.5	—[c]	134
Phenol, 0.33–0.9 mM	Bubbled	D	RVC	Pt	0.5–2.7	2	12[g]	≈ 8	67	139
Oxalic acid			5×5×1.2 cm^3				14[g]		37	
Phenol, aniline	Diffused	D	ODC[d]	Pt	1	1–3	20–130	0.2–6	>90[f]	106
Acetic acid, etc.			2.3 cm^2							

[a] Divided (D) or undivided (U) cell.
[b] Cathodic current density.
[c] Not specified.
[d] Carbon–PTFE oxygen diffusion cathode.
[e] Irradiated with UVA light.
[f] COD removal.
[g] In milliamperes per cubic centimeter.

ethylphenol isomers from ethylbenzene. By applying cathode potentials from −0.5 to −0.6 V vs. SCE, very low current densities were obtained (see Table 4). Even so, the electrolysis time was limited to 1 hr to avoid further oxidation of reaction products, so that no mineralization occurs. A maximum current efficiency of 63% was found with 5 mM Fe(II). The yields of phenolic products were higher at pH 3 and were decreased by lowering the concentration of Fe(II) from 50 to 5 mM. The proposed reaction pathway involves either hydrogen abstraction or $OH^{\bullet}$ addition as initial steps, along with fast O_2 addition as radical chain propagation. The generation of OH^{*} is considered as the rate-determining step for the oxidation of aromatics by Fenton reagent.

Fleszar and Sobkowiak [120] obtained less than 5% of catechol and hydroquinone from the hydroxylation of benzene and phenol on Hg, Pb, Cu, and Ag cathodes, even in the absence of Fe(II). It appears that the generation of H_2O_2 occurs even in electrodes where it may undergo catalytic decomposition if there are small quantities of adsorbable substances in solution.

A fuel cell with Nafion as electrolyte has been used by Otsuka et al. [121] for the spontaneous production of Fenton-type reagents in aqueous HCl or H_2SO_4 with Fe or Cu ions at Pd or Au cathodes. A diagram of this cell is presented in Fig. 19. The $OH^{\bullet}$ radicals produced in the aqueous phase were utilized to transform benzene into phenol. On the other hand, Tzedakis et al. [122] obtained about 70% current efficiency for phenol by a continuous extraction of phenate ions, using a Hg pool cathode in the presence of

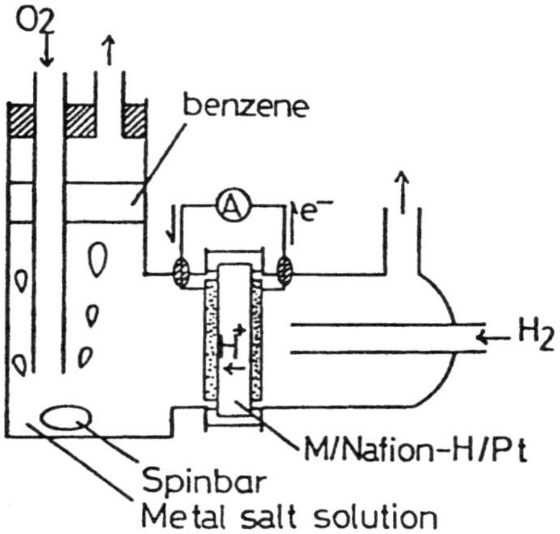

Figure 19 Diagram of the fuel cell-type reactor for the oxidation of benzene to phenol. (From Ref. 121.)

0.1M Fe(III). Related papers devoted to synthetic or mechanistic studies by partial oxidation of aromatics with H_2O_2 [123–127] will not be detailed here. Most of them found low yields and had difficulty stopping the reaction at the desired product due to the nonselectivity of $OH^{\bullet}$ and the increase in hydroxylation rate with the number of OH groups present in the ring [125].

As far as we know, Sudoh et al. [128] were the first to use, in 1986, the EFR for wastewater treatment (see Table 4). In their two-compartment cell with a graphite plate as the cathode and a Pt plate as the anode, the current efficiency for H_2O_2 generation reached 85% at −0.6 V vs. Ag/AgCl. Phenol was degraded to oxalic acid and CO_2 most efficiently at pH 3. Hydroquinone, catechol, *p*-benzoquinone, and muconic and maleic acids were detected as intermediates. The current efficiency (>60% in the range 260–2600 ppm of initial COD) was greater than that of the anodic oxidation process. The current increased with temperature and with O_2 bubbling rate, but phenol degradation was less affected by these factors. The optimum Fe(II) concentration was 2 mM for 2.5 mM phenol.

Provided that there is enough Fe(II) in solution, the half-life of H_2O_2 is less than 0.1 sec [119], so that the thickness of the reaction layer may extend by about 1 mm from the cathode and there is no need for separators to avoid H_2O_2 oxidation at the anode. Under these conditions, one can simultaneously produce $OH^{\bullet}$ radicals at the anode and the cathode reaction layer of an undivided cell, with the total degradation rate being the sum of the rates at each electrode [105]. Sudoh et al. [129] used an undivided bipolar cell with a vertical stack of perforated graphite electrodes to destroy phenol [130]. They compared the evolution of its derivatives in the presence of either N_2 or O_2, confirming the superiority of O_2. More than 95% reduction of COD was reported in the last case. One advantage of the undivided cell is that no acid addition was needed to keep the solution pH near 2 during runs. Energy consumptions were higher for the bipolar cell than for a single cell due to bypass currents, but decreased at high O_2 flow to a minimum of 10^8 J kg^{-1} COD.

Chu and Huang [131] have reported the removal of >90% of 1 mM solutions of chlorophenols, dichlorophenols, or trichlorophenols in a few hours. The expected stoichiometric amount of released Cl^- was not obtained, probably because some chlorinated intermediates were produced. In the absence of O_2, phenol degradation did not proceed, and Fe(II) was regenerated at the graphite cathode at −0.6 V vs. SCE. These authors have also compared EFR in a divided cell with other advanced oxidation processes to remove 1 mM 1,4-dioxane and 2-methyl-1,3-dioxolane [132].

Do and Yeh [133] followed the study on the degradation of phenol by paired oxidation with electrogenerated H_2O_2 and hypochlorite using the same cell and electrodes as in Ref. 113. Surprisingly, the cathodic reaction was slightly affected by pH and Fe(II) concentration. This last effect was attributed to the inactivation of $OH^{\bullet}$ on the large cathodic surface of 330 cm^2 used to overcome the low current density (see Table 4). Maximum phenol degradation fractions for 4.7 mM solutions were 58.8% and 67.2% at the cathodic and anodic chambers, respectively. A 15% decay in COD with a 57% current efficiency was obtained for the cathodic chamber after 2.8 hr at 35°C as the optimum temperature.

Oturan et al. [134] expanded their synthetic work to the degradation of 0.5–1 mM chlorophenoxyacid herbicides by EFR on a Hg pool. Aromatic ring cleavage was found upon exhaustive electrolysis (see Table 4). A number of polyhydroxyphenols and quinones were identified as intermediates. These authors have also considered the degradation of *p*-nitrophenol [135] and the herbicides 2,4-D or 2,4-dichlorophenoxyacetic acid [136] by the same technique using a carbon-felt cathode.

C. Reticulated Vitreous Carbon (RVC)

A new concept within this field was described in 1993 by Hsiao and Nobe [137], who studied the hydroxylation of chlorobenzene and phenol (see Table 4) working in recycle mode in the undivided flow reactor shown in Fig. 20. The oxygen evolved at the anode and Fe(III) were reduced downstream to H_2O_2 and Fe(II), respectively, at the RVC cathode, so that no O_2 was sparged. Unfortunately, the process rate in this system was not high enough for practical purposes, perhaps because of the limitation of current efficiency for H_2O_2 50% because $4e^-$ is needed to produce O_2, whereas only $2e^-$ is used to reduce it to H_2O_2. Phenol and 4-chlorophenol were detected as chlorobenzene derivatives. Phenol yielded catechol, resorcinol, and hydroquinone.

Following the work on formaldehyde reported above, Ponce de Leon and Pletcher [138] considered its oxidation via H_2O_2 electrogenerated at an RVC in a divided cell with different electrolytes. The removal of formaldehyde from 19 to <0.3 ppm was achieved at −0.6 V vs. SCE in 1 M NaOH. Usually, the final product was formic acid, but the addition of 0.5 mM $FeCl_2$ at pH 3 catalyzed the electrochemical combustion of a solution of the aldehyde from 6.8 to 1.3 ppm of TOC (see Table 4), although the current efficiency was as low as 6%. A Nafion membrane between the anodic and cathodic compartments was used to avoid the contribution of anodic oxidation, but required high voltages of around 30V. A further paper from Alvarez-Gallegos and Pletcher [107] describes the production of H_2O_2 with 36–69% current efficiency in a divided flow cell from 10 mM acid solutions. The low efficiencies were explained by $2e^-$ and $4e^-$ reductions of O_2 occurring in parallel. More recently, these authors have confirmed the

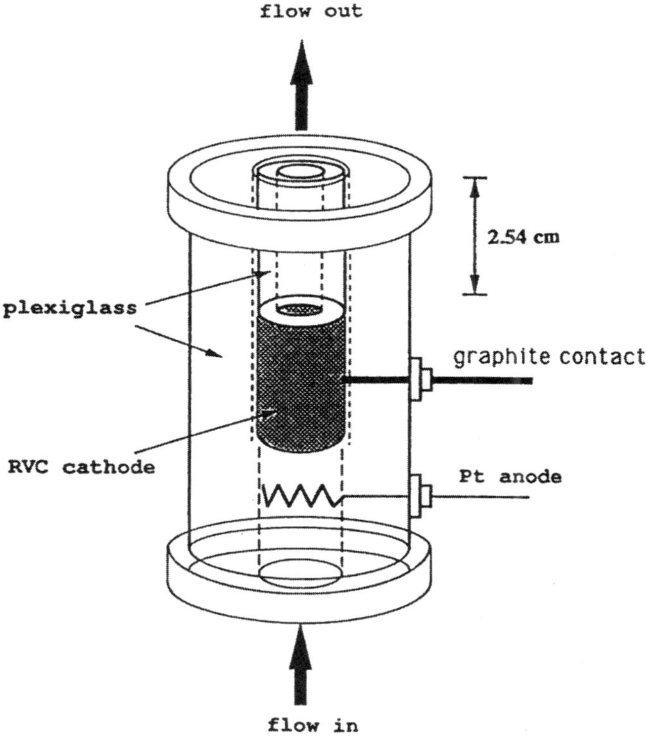

Figure 20 Schematic diagram of the flow reactor working in recycled mode for the hydroxylation of chlorobenzene and phenol. (From Ref. 137.)

general viability of EFR to remove diluted organics as phenol, cresol, catechol, quinone, hydroquinone, oxalic acid, and amaranth dye [139]. COD was reduced from 50–500 ppm to below 10 ppm, normally with current efficiencies $>50\%$ (see Table 4). Incomplete mineralization was attributed to a change in the concentration or speciation of Fe(II). This explanation can be ruled out because the continuous addition of Fe(II) does not change the general behavior of this oxidative system [140]. A more likely explanation is the generation of stable intermediates.

D. Oxygen Diffusion Cathode (ODC)

In 1994, Casado et al. [141,142] filed a patent and published a paper disclosing the conditions for wastewater treatment with in situ electrogenerated H_2O_2 and EFR using an oxygen diffusion cathode. This cathode is a microporous film made by hot-pressing an intimate mixture of carbon black and PTFE onto a metal mesh or carbon paper. A typical flow reactor for implementing such a process is depicted in Fig. 21. The cell contains an anode and an ODC, which separates an O_2 feed chamber from the wastewater circulating in a single compartment [143]. In another configuration, each anode can be surrounded by two ODCs.

The above process was applied initially [142,144] to destroy 100 ppm aniline and 4-chloroaniline in alkaline solutions of pH between 11 and 13 by anodic oxidation in the presence of H_2O_2 electrogenerated at an ODC (54 mM H_2O_2 at 600 mA). Both substrates presented pseudo first-order decays with half-lives less than 30 min at 600 mA. After 11 hr of electrolysis at 300 mA, a TOC decay >95% was found (see Table 4). Nitrobenzene and *p*-chloronitrobenzene were detected, respectively, as intermediates, which degraded via maleic acid. Cl^- was quantitatively released from 4-chloroaniline, and NH_3 was a final product of both substrates. General reaction pathways involving oxidation of organics by $OH^\bullet$ and $HO_2^\bullet$ were proposed.

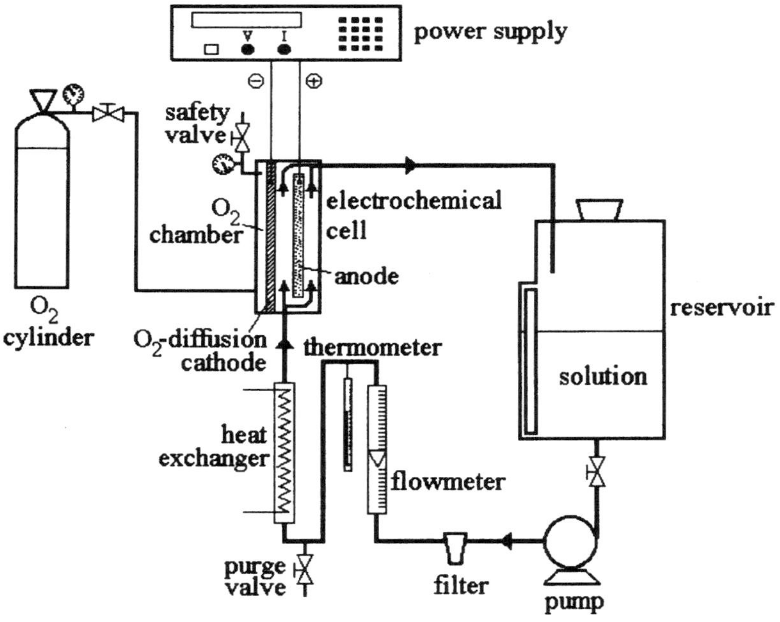

Figure 21 Scheme of the pilot flow reactor for the treatment of 10–75 L of wastewaters by the Electro-Fenton and Peroxi-coagulation processes. (From Ref. 143.)

A further paper [105] was devoted to the catalytic effect of Fe(II) on the anodic oxidation of aniline in a diluted electrolyte with 0.05 M Na_2SO_4 under Electro-Fenton® conditions. This treatment was more efficient than the above alkaline process (see Table 4). The regeneration of Fe(II) by peroxide species, organic radicals, and UVA light was also described. Although PbO_2 and Pt were comparable as anodes, the former was dissolved slightly. The TOC removal using Pt was 76% after 6 hr at 100 mA, but increased to 94% when the solution was UVA-irradiated (Photo-electro-Fenton process). This improvement arises from the increase in the amount of $OH^{\bullet}$ by photo-Fenton reaction (see the chapter on Fenton chemistry) and decomposition of photolabile intermediates [e.g., photo-decarboxylation of Fe(III) complexes with organic acids]. Recently, it has been shown that sunlight after Electro-Fenton is also able to improve the process in the same way [145].

The main advantages of the Electro-Fenton process are:

1. It is faster and goes deeper than classical anodic oxidation.
2. It is safer than anodic oxidation because no H_2 is formed as by-product, and safer than Fenton reagent because no H_2O_2 has to be transported or stored.
3. It is cheaper. Less electricity is used because O_2 reduction is thermodynamically easier than that of H_2O.
4. Neither membranes nor separators are needed, further saving energy costs.
5. Paired electrooxidation is feasible, allowing apparent current efficiencies of 200% and higher.
6. O_2 evolved at the anode favors Fenton's chemistry and could be reused as cathode feeding.

These authors also introduced the Peroxi-coagulation process [146]. The main difference with Electro-Fenton is that Fe(II) is electrogenerated from an Fe-based anode. The solution pH then increases during the process and the excess Fe(III) precipitates as hydrated oxide. In addition to oxidation by $OH^{\bullet}$, some of the intermediates coagulate with the precipitate, which does not occur in electrocoagulation. The Peroxi-coagulation process of a 1000-ppm aniline solution reached 91% TOC removal in 2 hr, but most of the organics remained in the precipitate. A degradation pathway involving *p*-benzoquinone and nitrobenzene as intermediates, with loss of NH_4^+ and NO_3^-, was proposed. Casado and Brillas [147] have also filed a patent disclosing an improvement for these processes by using a cell with two different anodes, one inert and the other one a sacrificial Fe electrode. More than 90% TOC from 1000 ppm aniline solutions and 67% TOC from a real effluent with 3660 ppm TOC could thus be removed. The improved process works faster, allows automatic control of pH and Fe(II) concentration without specific equipment, and consumes less anodic Fe than Peroxi-coagulation.

All these advanced methods have been compared with anodic oxidation for a 178-ppm 4-chlorophenol solution [140]. The mineralization rate was much faster for the new processes and the substrate was completely transformed into CO_2 by the Photoelectro-Fenton method, as can be deduced from data given in Fig. 22. In all cases, the initial intermediate was 4-chloro-1,2-dihydroxybenzene, which was further oxidized to yield maleic and fumaric acids with loss of Cl^-. The subsequent degradation of these acids gave oxalic acid, which complexes with Fe(III). Such complexes are attacked slowly by $OH^\bullet$, but decomposed quickly by UVA light.

A comparison of these new processes with TiO_2 photocatalysis for the degradation of aniline [148] showed that the former are faster. Although all these methods are considered to proceed via $OH^\bullet$ radicals, some different intermediates were detected in the electrochemical and photocatalytic experiments. *p*-Benzoquinone and NH_4^+ appeared in all solutions tested. A recent study [149] reports 87% and 99% TOC removals after 4 hr of

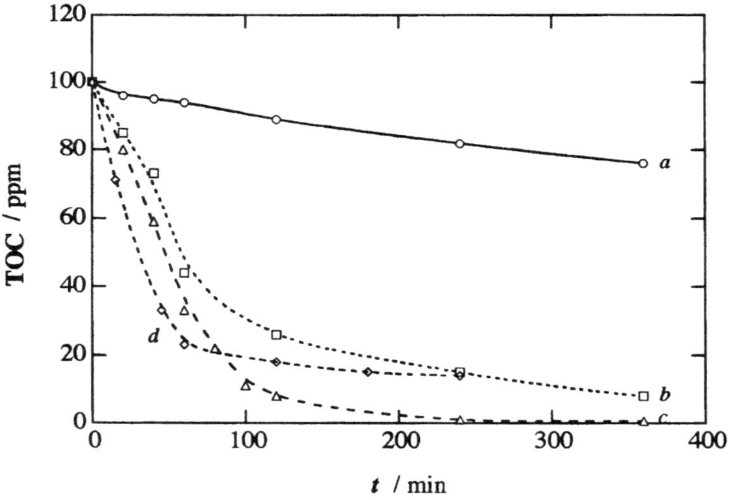

Figure 22 TOC removal vs. electrolysis time for 100 mL of a 178-ppm 4-chlorophenol solution in 0.05 M Na_2SO_4 at pH 3.5 and at 25°C, using a 3.1-cm^2 ODC fed with an O_2 flow of 20 mL min^{-1}. (a) Anodic oxidation in the presence of electrogenerated H_2O_2 with a 10-cm^2 Pt anode. (b) Electro-Fenton process with the Pt anode and 1.0 mM Fe^{2+} in the solution. (c) Photoelectro-Fenton process with the Pt anode, 1.0 mM Fe^{2+}, and UVA irradiation. (d) Peroxi-coagulation process with a 10-cm^2 Fe anode. The applied current is 100 mA. (From Ref. 140.)

electrolysis of a 230-ppm 2,4-D solution at 450 mA by Electro-Fenton and Photoelectro-Fenton processes, respectively. 2,4-D disappeared in 20 min by both methods. Anodic oxidation, even in the presence of H_2O_2, is a much slower process. The proposed reaction pathway involving all intermediates found is shown in Fig. 23. The same chloroderivative products, along with 2,4-dichlororesorcinol, have been determined for 2,4-D degradation by EFR with a carbon felt cathode [136].

Brillas et al. [150] showed that Peroxi-coagulation can be spontaneous in a shorted Fe/O_2 battery, and so reported the destruction of benzoic acid, 4-chlorophenol, nitrobenzene, 2-nitrophenol, phenol, and aniline in this way. Whereas TOC removals within 1 hr ranged from 77% to 92%, parent compounds disappeared in about 20 min. In this system, 95% TOC removal was achieved after 1 hr of treatment of 129 ppm aniline at pH 4 and 35°C [151]. Similarly, 81% degradation was reached after 2 hr for a 1000-ppm aniline solution. In this case, the formation of polymers that coagulate with $Fe(OH)_3$, where most of the initial C and N were found, is suggested.

Casado et al. [142] have also reported results on the degradation of 1000 ppm aniline solutions with a small pilot plant (see Fig. 21) able to treat

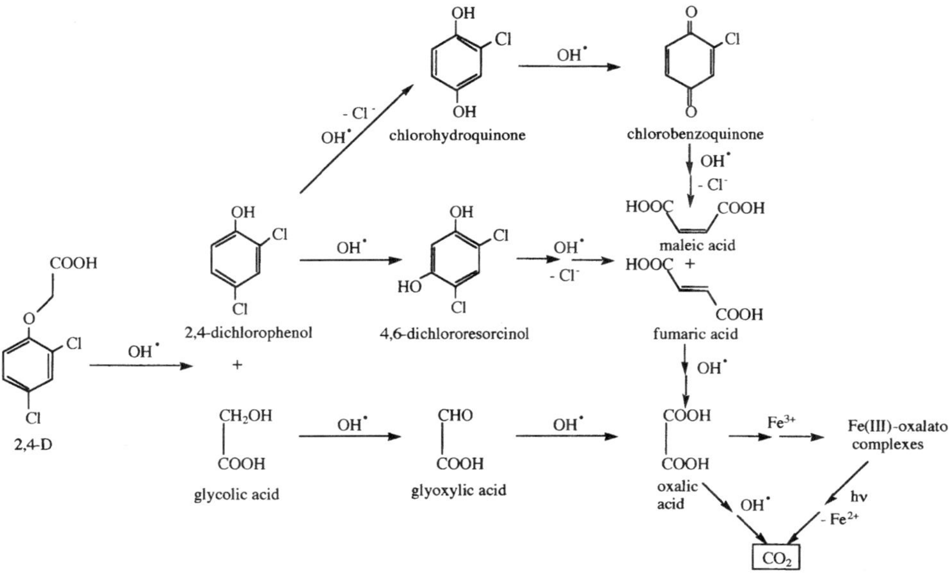

Figure 23 Proposed reaction pathway for the mineralization of 2,4-D at pH 3 by anodic oxidation, anodic oxidation in the presence of electrogenerated H_2O_2, and Electro-Fenton and Photoelectro-Fenton processes. (From Ref. 149.)

from 10 to 75 L of wastewaters using a flow cell of 200 cm^2 electrode area in recycle mode. In the absence of organics and Fe(II), H_2O_2 generation was roughly proportional to the applied current. The Electro-Fenton process gave a violet polymer and soluble TOC was reduced by 61% in 2 hr at 20 A. Peroxi-coagulation allowed the removal of >95% TOC. The energy cost for both processes ranged between 6 and 45 vs. 502 kW hr m^{-3} for anodic oxidation, and decreased with decreasing run times and currents. Lower costs were reached by increasing the solution conductivity. Real wastewaters from pulp and paper and fine chemicals industries have been purified successfully in this plant [144]. A demonstration plant on Electro-Fenton process with a filterpress-type cell of 2400 cm^2 cathode area has been constructed recently and started up by Carburos Metálicos SA. Currents up to 500 A have been applied successfully to the degradation of benzoic acid and *p*-benzoquinone solutions [144].

Results on the Electro-Fenton process have been confirmed and extended by Harrington and Pletcher [106], who degraded 0.1–5 mM solutions of phenol, aniline, formaldehyde, acetic acid, and three azo dyes with 50 mM Na_2SO_4 at pH 2 in a small divided cell. More than 90% of initial COD was destroyed, with current efficiencies >50%, independent of the applied potential (see Table 4). O_2 was found better than air as ODC feeding. The process is quite general, faster than using an RVC of lower energy consumptions (from 1.25 to 14 kW hr m^{-3}). The cost estimations overlap with those quoted above, but are lower because of the more diluted solutions tested and the assumption of a cell voltage of 5 V. The ohmic drop within the pores of the ODC was high in the electrolyte used. No cathode damage was found after repeated reuse.

In summary, the use of ODC or three-dimensional electrodes seems to be the key for the application of processes involving H_2O_2 electrogeneration to wastewater treatment at industrial scale.

E. Related Processes

A different process involving the in situ cathodic regeneration of Fe(II) from Fe(III) and ex situ addition of H_2O_2 to the wastewater has also been confusingly called "Electro-Fenton" [152,153]. Roe and Lemley [154] and Pratap and Lemley [155] have described a similar Fenton electrochemical treatment that doses H_2O_2 and generates Fe(II) from a sacrificial Fe anode in near-stoichiometric amounts. The process was successful in the removal of very diluted pesticides at pH >5, with 55% maximum mineralization [154], but no analysis of the organic content of Fe(III) precipitates was attempted. Near-UV light improved the degradation of malathion only at low iron concentrations. The system did not work when H_2O_2 was added after Fe(II) generation. Essentially the same process is marketed by Andco Environmental Processes for the removal of pollutants such as pentachlorophenol and As at specific U.S. Environmental Protection Agency (EPA) Sites [1]. Other related Fenton process, called FSR® process [156], involves the redissolution of $Fe(OH)_3$ sludge from ex cell Fenton reaction and cathodic Fe(II) regeneration. The treatment of an industrial wastewater of 4000 ppm COD in a pilot plant resulted in about 80% COD reduction. Following previous work on methane and ethylene, Savinova et al. [157] have applied another "Electro-Fenton" system to oxidize ethane to acetaldehyde and ethanol in the gas phase. Catholyte and anolyte were separated by a Nafion membrane and ethane was fed along with O_2 to the gas compartment. Maximum current density was 17 mA cm^{-2}, with current efficiencies from 1% to 3%.

VII. METHODS OF PHASE SEPARATION

Water effluents containing colloidal particles of oil and other organic compounds are produced in the oil industry, printing and paint workshops, food processing industry, fiber manufacture, etc., and need to be treated before discharge to the environment. The traditional treatment is to wait for the phase separation in holding tanks [2]. As this requires time and is expensive because large tanks are needed, it has been common to add coagulating agents such as Fe^{3+} or Al^{3+} for pollutant precipitation (this supposes an additional cost and a reliable dosing procedure is required to avoid environmental hazard), or to use compressed air for pollutant flotation, where a well-designed air sparger is needed. There are electrochemical methods that also produce the same effects, such as electrocoagulation, electroflocculation, and electroflotation. Electrocoagulation refers to the electrochemical generation of destabilization agents that bring about charge neutralization for pollutant separation. In the electroflocculation treatment, agents that promote particle bridging or coalescence are produced electrochemically. Electroflotation is a method in which the pollutants (fats, oils, etc.) are attached to the bubbles of the gas evolved on the electrodes (O_2, H_2) and are transported to the top of the solution where they can be removed. In addition to the mechanical capture of the pollutants, the neutralization of the charge of the colloidal particles (that produce their coalescence) by small charges carried by the gas bubbles evolving on the electrodes seems to occur.

In the electroflotation method, the bubble size is critical to the efficiency of the phase separation and depends on the electrode metal. The bubble sizes formed on cathodes with small overpotentials for hydrogen evolution (such as Pd, W, and Ni) are larger than those formed on high overpotential cathodes such as Pb, Sn, and Cu. Khosla et al. [158] have proposed the use of the pulsed electrogeneration technique to obtain optimum-sized bubbles that are independent of solution conditions.

A major disadvantage of an undivided cell is the fire and explosion hazard caused by electrical sparking in the case of mixing finely dispersed H_2 and O_2 formed during the electrolysis of water. Weintraub et al. [159] utilized porous Fe anodes that are oxidized while air or oxygen is bubbled through the solution to separate oil–water emulsions. The overall process can be explained by the following sequence of reactions [160,161]:

$$4Fe \rightarrow 4Fe^{2+} + 8e^- \tag{44}$$

$$4Fe^{2+} + 10H_2O + O_2 \rightarrow 4Fe(OH)_3 + 8H^+ \tag{45}$$

$$8H^+ + 8e^- \rightarrow 4H_2 \tag{46}$$

In the absence of O_2, Fe is oxidized to Fe^{2+} and $Fe(OH)_2$ is formed. However, $Fe(OH)_3$ precipitates more easily than $Fe(OH)_2$ and, thus, the injection of O_2 facilitates pollutant removal. The injection of gas also helps in the flotation process. Vik et al. [162] have described a process in which H_2 is produced at the cathode and Al is oxidized to Al(III) ions at the anode. The OH^- generation from H_2 evolution produces precipitation of the metal hydroxides and also contributes to the coagulation–flocculation process. In addition, the hydrogen gas bubbles result in the flotation of the sludge formed. The principles of such a process are depicted in Fig. 24.

Based on these fundamentals, many systems and apparatus have been built, being operative on an industrial scale. Different types of reactors have also been designed. The electrodes may be parallel plates [162, 163] or sacrificial Al pellets as anode [164,165]. The feeding of pressurized air has been implemented in many electrocoagulation–electroflotation systems [159,166–168]. Some plants have a press to remove water from the sludge [169,170] and a processing tank with a closed S-shaped one-way flow path [171].

Many advantages for electrocoagulation have been reported [1,160, 163,162,172–174]:

1. Cathode durability.
2. Effective and rapid organic matter separation with current efficiencies close to 90%.
3. pH control is not necessary, except for extreme values.
4. The amount of required chemicals is small.
5. The amount of sludge produced is small when compared with the conventional addition of $FeCl_3$ followed by the addition of NaOH

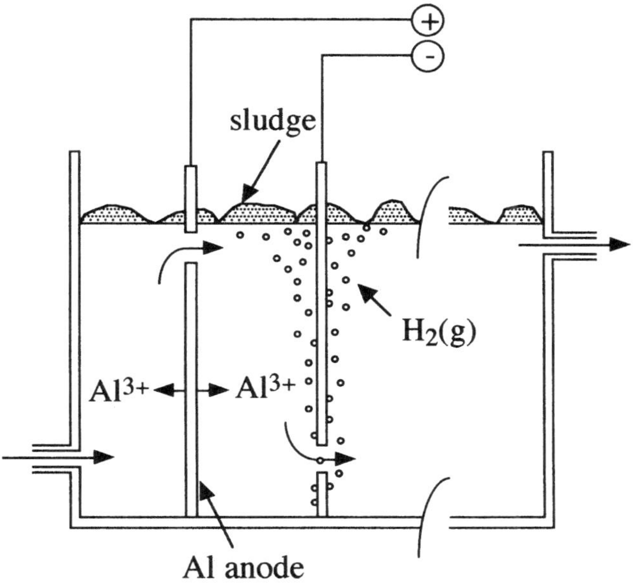

Figure 24 Diagram of the electrocoagulation process using soluble anodes. Several couples of electrodes, connected in parallel, can be used in the tank. (From Ref. 162.)

or lime because the content of dry and hydrophobic solids in the electrocoagulation method is higher. Normally, the decantation times are small and can be reduced using a magnetic field or polymer flocculants such as $AlCl_3$.

6. The operating costs are much lower than in most of the conventional technologies.

In contrast, the major challenges for electrocoagulation are [1,172,175]:

1. H_2 evolution may be a safety hazard and may prevent precipitated matter from settling properly.
2. Anode passivation and sludge deposition on the electrodes can limit the process. Alternating current and promotion of high turbulence prevent this problem, thus increasing the time required for the electrode replacement.
3. The concentrations of iron and aluminum ions in the effluent are increased.
4. Investment costs are relatively high.

A. Oils

Several million of tons of oils from refineries, oil transportation, cutting machines, mills, off-shore platforms, etc., are spilled every year in water reservoirs and the sea. About half of this amount contaminates fresh water and an estimate suggests that humans use almost 4 L of hydrocarbons per person each day in the world [176]. Oils can be present in wastewaters as a supernatant layer, adsorbed on suspended particles, forming emulsions, or even dissolved. Oils produce many changes in water properties: their viscosity and conductivity are altered, and they acquire color and opacity. In addition to a negative esthetic impact and a bad taste, the light necessary for photobiological processes is absorbed.

Oil colloidal particles generally have surface negative charges, due to friction or ionization of the carboxylic groups, which impede the coalescence of colloidal particles. These charges can be neutralized by positive cations proceeding from polyamines, sulfuric acid, $AlCl_3$, or $Fe_2(SO_4)_3$. However, the addition of these chemicals to oil/water emulsions produces large volumes of sludge, and it is not effective for small colloidal particles. For this reason, electrocoagulation is preferred. Ibañez et al. [177] have described a simple laboratory experiment to illustrate the process. The cathodic reaction is H_2 evolution, and electrocoagulation is thus combined with electroflotation. The gas bubbles carry upward flocs or precipitate, and are able to adsorb oil particles. In addition, the solution pH increases, thus facilitating the precipitation or flocculation of metal hydroxides as sludge. The resulting oily layer and the precipitated sludge can be mechanically removed. About 96–99% efficiencies for the oil removal have been reported [159,162,172,176,178]. For example, Biswas and Lazarescu [176] reported 96% oil removal from emulsions using electrocoagulation with direct current and the addition of $Fe_2(SO_4)_3$. Pouet et al. [178] combined the electroflotation–electrocoagulation process with microfiltration to yield irrigation-quality water. In this case, 99% turbidity, 77% COD, and 98% suspended solids were removed.

Il'in and Sedashova [179] have described an electroflotation method and a small-sized plant to remove oil products from effluents based on adhesion of pollutant particles to finely divided bubbles of H_2 and O_2 produced during the electrolysis of water. From an initial petroleum product concentration of 1000 mg L^{-1}, a residual concentration of 1–10 mg L^{-1} was found following electroflotation. This concentration could be further reduced to about 0.01 mg L^{-1} by the addition of inorganic coagulants.

B. Dyes

Large amounts of toxic dyestuffs can be found in wastewaters from the textile industry because dye fixation is sometimes as low as 60%. Most dyes are stable to photodegradation and microbial attack. Effective decolorizing methods are chemical oxidation by Cl_2, hypochlorite, and O_3. As these methods are rather expensive or unsafe, electroflotation–electrocoagulation methods appear to be promising. Ibañez et al. [180] described a laboratory experiment to show the color removal of simulated wastewater by electrocoagulation–electroflotation. Aluminum and iron anodes yielded different results in this combined process [181,182]. With an Al anode, dyes appear to be removed by adsorption on the precipitate, whereas with an Fe anode, some dye derivatives are found, perhaps due to the chemical reduction by Fe^{2+}. The latter process seems to be more effective for reactive dyes, the former for disperse dyes. Fe(III) hydroxide is found to be the best precipitate for color removal [161]. Lee et al. [183] have shown that using sacrificial iron anodes, color removal increases as the number of anodes, voltage, and treatment time increase. Decreasing the distance between electrodes also increased color removal.

Dyes can also be reduced cathodically. Youchun et al. [184] have proposed a coagulation–electroflotation–electrolytic reduction and oxidation method using graphite electrodes, which allows COD removals close to 90%. Takaoka [185] has suggested the addition of NaCl to the wastewater as a source of HClO in the electrocoagulation process. Changhai [186] has used an electroflocculation process for treating, dyeing, and printing wastewaters with polyacrylamide as the coflocculating agent. The COD and color decreased by 88% and 97%, respectively.

C. Other Pollutants

Wastewaters containing other pollutants have also been treated by electrocoagulation. This is the case of phenol with an Fe–C packing, aeration, and H_2O_2 addition [187]; proteins with iron electrodes, followed by neutralization with lime milk [188] and algae and pesticides, apart from organic material and dissolved oils, with a flow-through reactor in which floating solids are collected in channels for removal [189]. Electrocoagulation has also been used successfully for the removal of heavy metals [190] and other dissolved pollutants such as phosphates, nitrites, and sulphates [188].

VIII. CONCLUSIONS

Halogenated organics are usually toxic, and their electroreduction can make them easily biodegradable. This is also a common goal in the electrochemical decontamination of organic pollutants in wastewaters by anodic oxidation. Although modern processes have seriously reduced the operation costs for cathodic reduction and anodic oxidation, both methods are still much too costly to be competitive with biological treatment. Many efforts are now directed to combine electrochemical and biological processes to produce cheaper treatments. So industries having toxic effluents could use electrochemical systems of reasonable cost to produce biodegradable effluents that can be discharged to the public sewage works in a safe way.

Applied research on indirect electrolytic methods has experienced a spectacular evolution in the last decades. More efficient, faster, and/or cheaper processes than classical anodic oxidation are now operative or in pilot trials. Redox mediators such as Ag(I)/Ag(II) and Co(II)/Co(III), and on-site production of strong oxidants like ozone at the anode and EFR at the cathode, are notable examples. These methods have been specially focused on aromatic compounds, probably because they are easily oxidizable, through hydroxyderivatives and quinones, to aliphatic acids. A further photochemical treatment of these products in the presence of Fe(III) allows their mineralization.

In the last years, electrochemical treatments have been notably improved by the use of new effective and stable anodes such as DSA® and SPR. Boron-doped diamond is now a promising anodic material. However, wastewaters usually contain detrimental substances for the processes or the electrodes, promoting fouling, corrosion, inhibition, or passivation. Common ions such as CO_3^{2-}, HCO_3^-, or HSO_4^- are radical scavengers, competing with the pollutants for $OH^{\bullet}$ radicals, the main oxidizing species in environmental electrochemistry. Chloride ion found in most industrial effluents is more problematic because it is a corrosion promoter, a good radical scavenger, and can be anodically discharged, wasting electric energy and producing, in some cases, toxic chlorinated organics. So each effluent should be studied to tailor its specific degradative process, usually including a pretreatment and a biological step. Electrocoagulation, electroflocculation, and electroflotation are mature technologies that can be successfully applied to phase separation and heavy metals precipitation, but they are not efficient in the treatment of soluble organics.

REFERENCES

1. Rajeshwar K, Ibañez J. Environmental Electrochemistry: Fundamentals and Applications in Pollution Abatement. Chaps. 5 and 8. San Diego: Academic Press, Inc., 1997.
2. Pletcher D, Walsh FC. Industrial Electrochemistry. Chaps. 1, 2, 5, and 7. 2nd ed. London: Blackie Academic and Professional, 1993.
3. Schmal D, Van Duin PJ, de Jong AMCP. Electrochemical dehalogenation of

toxic organic compounds in industrial waste water. Dechema Monogr 1991; 124:241–247.
4. Zimmer A. Electrochemical processes for decomposition of organic matter in wastewater. Wiss Z Tech Univ Dresden 1997; 46:80–85.
5. Iniesta J, González-García J, Montiel V, Aldaz A. Electrochemistry and the environment. Electrochemical treatment of organic matter in wastewater. Energía (Madrid) 1998; 24:55–60.
6. Byker HJ. Halogenated aromatic compound removal and destruction process. United States Patent 4,659,443, 1987.
7. Bachmann T, Vermes I, Heitz E. Abbau umweltbelastender halogenierter Kohlenwasserstoffe durch Metalle. Dechema Monogr 1991; 124:221–239.
8. Mazur DJ, Weinberg NL. Methods for electrochemical reduction of halogenated organic compounds. United States Patent 4,702,804, 1987.
9. Kulikov SM, Plekhanov VP, Tsyganok AI, Schlimm C, Heitz E. Electrochemical reductive dechlorination of chloroorganic compounds on carbon cloth and metal-modified carbon cloth cathodes. Electrochim Acta 1996; 41:527–531.
10. Davies G, Maling G. Electrochemical process. European Patent 0,294,108, A2, 1988.
11. Sonoyama N, Hara K, Sakata T. Reductive electrochemical decomposition of chloroform on metal electrodes. Chem Lett 1997; 131–132.
12. Scherer MM, Westall JC, Ziomek-Moroz M, Tratnyek PG. Kinetics of carbon tetrachloride reduction at an oxide-free iron electrode. Environ Sci Technol 1997; 31:2385–2391.
13. Schmal D, Van Erkel J, De Jong AMCP, Van Duin PJ. Electrochemical treatment of organohalogens in process wastewaters. Environ Technol (Proc Euro Conf Dordr Neth)1987; 284–293.
14. Bagnell LJ, Jeffery EA. Electrolytic hydrogenation and hydrogenolysis of phenols at platinum black electrodes. Aust J Chem 1980; 33:2565–2569.
15. Sasaki K, Kunai A, Harada J, Nakabori S. Electrolytic hydrogenation of phenols in aqueous acid solutions. Electrochim Acta 1983; 28:671–674.
16. Cheng F, Fernando Q, Korte N. Electrochemical dechlorination of 4-chlorophenol to phenol. Environ Sci Technol 1997; 31:1074–1078.
17. Nagaoka T, Yamashita J, Takase M, Ogura K. Composite electrodes consisting of metal and oxidized carbon particles for complete degradation of trichloroethylene in acetonitrile. J Electrochem Soc 1994; 141:1522–1526.
18. Watanabe A, Takahashi H, Horimoto Y. Electrochemical treatment of wastewaters containing *o*-chlorobenzoic acid. Anodic oxidation after the cathodic conversion of *o*-chlorobenzoic acid into a readily oxidizable form. Denki Kagaku 1985; 53:207–208.
19. Funabashi K, Ozawa Y, Nishi T, Izumida T, Matsuda M, Miura H, Komori I, Sawa T, Yamazaki Y. Reduction of iodine-containing organic compounds in radioactive wastewaters. Japan Patent JP 6223739 A2, 1987.
20. Sweeny KH. The reductive treatment of industrial wastewaters: I. Process description. AIChE Symp Ser 1981; 77:67–71.

21. Sweeny KH. The reductive treatment of industrial wastewaters: II. Process applications. AIChE Symp Ser 1981; 77:72–78.
22. Matheson LJ, Tratnyek PG. Reductive dehalogenation of chlorinated methanes by iron metal. Environ Sci Technol 1994; 28:2045–2053.
23. Grittini C, Malcomson M, Fernando Q, Korte N. Rapid dechlorination of polychlorinated biphenyls on the surface of a Pd/Fe bimetallic system. Environ Sci Technol 1995; 29:2898–2900.
24. Edison M. Procedimento per la preparazione di fluoro-e/o clorofluoro(idro)-carburi per via elettrolitica. Italian Patent 852,487, 1969.
25. Cabot PL, Centelles M, Segarra L, Casado J. Electrosynthesis of trifluoroethene and difluoroethene from 1,1,2-trichloro-1,2,2-trifluoroethane. J Electroanal Chem 1997; 435:255–258.
26. Cabot PL, Centelles M, Segarra L, Casado J. Palladium-assisted electrodehalogenation of 1,1,2-trichloro-1,2,2-trifluoroethane on lead cathodes combined with hydrogen diffusion anodes. J Electrochem Soc 1997; 144:3749–3757.
27. Casado J, Cabot PL. Process for the reduction of chlorofluorocarbons and production of derivatives thereof in an electrolytic cell, cell for carrying out said reduction and process for removing the byproducts formed within the cell. PCT Int Appl WO 9724162 A1, 1997.
28. Cabot PL, Centelles M, Segarra L, Casado J. Electrodehalogenation of trichlorofluoromethane on lead cathodes using hydrogen diffusion anodes. J Electrochem Soc 2000; 147:3734–3738.
29. Inaba M, Sawai K, Ogumi Z, Takehara Z. Electroreduction of a chlorofluoroethane on a solid polymer electrolyte composite electrode. Chem Lett 1995; 471–472.
30. Delli E, Kouloumtzoglou S, Kyriacou G, Lambrou Ch. Electrochemical reduction of CCl_2F_2 on Nafion® solid polymer electrolyte composite electrodes. Chem Commun 1998; 1693–1694.
31. Kornienko VL. Electrochemical synthesis of organic substances using hydrophobized electrodes. Russ J Electrochem 1996; 32:77–87.
32. Kornienko VL, Kolyagin GA, Kornienko GV, Saltykov YuV. Electrosynthesis of organic substances at wetproofed porous electrodes. Sov Electrochem 1992; 28:507–516.
33. Sonoyama N, Sakata T. Electrochemical decomposition of CFC 12 using gas diffusion electrodes. Environ Sci Technol 1998; 32:375–378.
34. Sonoyama N, Sakata T. Electrochemical hydrogenation of CFC 13 using metal-supported gas diffusion electrodes. Environ Sci Technol 1998; 32:4005–4009.
35. Comninellis C. Electrocatalysis in the electrochemical conversion/combustion of organic pollutants for waste water treatment. Proceedings of the Symposium on Water Purification by Photocatalytic, Photoelectrochemical and Electrochemical Processes. Vol. 94-19. Pennington, NJ: The Electrochemical Society, 1994:75–86.
36. Comninellis C, De Battisti A. Electrocatalysis in anodic oxidation of organics with simultaneous oxygen evolution. J Chim Phys 1996; 93:673–679.

37. Seignez C, Pulgarin C, Peringer P, Comninellis C, Plattner E. Degradation des pollutants organiques industriels. Traitements electrochimique, biologique, et leur couplage. Swiss Chem 1992; 14:25–30.
38. Smith de Sucre V, Watkinson AP. Anodic oxidation of phenol for waste water treatment. Can J Chem Eng 1981; 59:52–59.
39. Sharifian H, Kirk DW. Electrochemical oxidation of phenol. J Electrochem Soc 1986; 133:921–924.
40. Comninellis C, Plattner E. Electrochemical waste water treatment. Chimia 1988; 42:250–252.
41. Comninellis C, Pulgarin C. Anodic oxidation of phenol for waste water treatment. J Appl Electrochem 1991; 21:703–708.
42. Murphy OJ, Hitchens GD, Kaba L, Verostko CE. Direct electrochemical oxidation of organics for wastewater treatment. Water Res 1992; 26:443–451.
43. Comninellis C, Pulgarin C. Electrochemical oxidation of phenol for wastewater treatment using SnO_2 anodes. J Appl Electrochem 1993; 23:108–112.
44. Gandini D, Mahé E, Michaud PA, Haenni W, Perret A, Comninellis C. Oxidation of carboxylic acids at boron-doped diamond electrodes for wastewater treatment. J Appl Electrochem 2000; 30:1345–1350.
45. Kirk DW, Sharifian H, Foulkes FR. Anodic oxidation of aniline for waste water treatment. J Appl Electrochem 1985; 15:285–292.
46. Kötz R, Stucki S, Carcer B. Electrochemical waste water treatment using high overvoltage anodes: Part I. Physical and electrochemical properties of SnO_2 anodes. J Appl Electrochem 1991; 21:14–20.
47. Stucki S, Kötz R, Carcer B, Suter W. Electrochemical waste water treatment using high overvoltage anodes: Part II. Anode performance and applications. J Appl Electrochem 1991; 21:99–104.
48. Gattrell M, Kirk DW. A study of the oxidation of phenol at platinum and preoxidized platinum surfaces. J Electrochem Soc 1993; 140:1534–1540.
49. Kawagoe KF, Johnson DC. Electrocatalysis of anodic oxygen-transfer reactions. Oxidation of phenol and benzene at bismuth-doped lead dioxide electrodes in acidic solutions. J Electrochem Soc 1994; 141:3404–3409.
50. Boscoletto AB, Gottardi F, Milan L, Panocchia P, Tartari V, Tavan M, Amadelli R, De Battisti A, Barbieri A, Pattacchini D, Battaglin G. Electrochemical treatment of bisphenol-A containing wastewaters. J Appl Electrochem 1994; 24:1852–1858.
51. Johnson SK, Houk LL, Feng J, Houk RS, Johnson DC. Electrochemical incineration of 4-chlorophenol and the identification of products and intermediates by mass spectrometry. Environ Sci Technol 1999; 33:2638–2644.
52. Gattrell M, MacDougall B. The electrochemistry of chlorophenols and its implications for waste water treatment. Abstract of the 10th International Forum on Electrolysis in the Chemical Industry. Clear Water Beach, FL: The Electrosynthesis Co., 1996
53. Pulgarin C, Adler N, Peringer P, Comninellis C. Electrochemical detoxification of a 1,4-benzoquinone solution in wastewater treatment. Water Res 1994; 28:887–893.

54. Houk LL, Johnson SK, Feng J, Houk RS, Johnson DC. Electrochemical incineration of benzoquinone in aqueous media using a quaternary metal oxide electrode in the absence of a soluble supporting electrolyte. J Appl Electrochem 1998; 28:1167–1177.
55. Feng J, Houk LL, Johnson DC, Lowery SN, Carey JJ. Electrocatalysis of anodic oxygen-transfer reactions: the electrochemical incineration of benzoquinone. J Electrochem Soc 1995; 142:3626–3632.
56. Kaba L, Hitchens GD, Bockris JO'M. Electrochemical incineration of wastes. J Electrochem Soc 1990; 137:1341–1345.
57. Tennakoon CLK, Bhardwaj RC, Bockris JO'M. Electrochemical treatment of human wastes in a packed bed reactor. J Appl Electrochem 1996; 26:18–29.
58. Beck F, Schulz H, Wermeckes B. Anodic dehalogenation of 1,2-dichloroethane in aqueous electrolytes. Chem Eng Technol 1990; 13:371–375.
59. Ogutveren UB, Kopara S. Electrochemical treatment of water containing dyestuffs: anodic oxidation of Congo Red and Xiron Blue 2R-HD. Int J Environ Stud 1992; 42:41–52.
60. Almon AC, Buchanan BR. Electrochemical oxidation of organic waste. In: Genders JD, Weinberg N, eds. Electrochemistry for a Cleaner Environment. Chap. 15. East Amherst, NY: The Electrosynthesis Co., 1992
61. Hobbs DT. Electrochemical treatment of liquid nuclear wastes. Abstract of the 8th International Forum on Electrolysis in the Chemical Industry. Lake Buena Vista, FL: The Electrosynthesis Co., 1994
62. Bonfatti F, Ferro S, Lavezzo F, Malacarne M, Lodi G, De Battisti A. Electrochemical incineration of glucose as a model organic substrate: I. Role of the electrode material. J Electrochem Soc 1999; 146:2175–2179.
63. Steele DF. Electrochemistry and waste disposal. Chem Ber 1991; 27:915–918.
64. Savall A, Dalbéra S, Abdelhedi R, Bouguerra ML. Réduction électrochimique du trichloro-1,1,2-trifluoroéthane: étude de la corrosion chimique de cathodes de zinc. J Appl Electrochem 1990; 20:1045–1052.
65. Savall A, Abdelhedi R, Dalbéra S, Bouguerra ML. Réduction électrochimique du trichloro-1,1,2-trifluoroéthane: effect catalytique de l'ion ammonium. Electrochim Acta 1990; 35:1727–1737.
66. Zhang S, Rusling JF. Dechlorination of polychlorinated biphenyls by electrochemical catalysis in a bicontinuous microemulsion. Environ Sci Technol 1993; 27:1375–1380.
67. Bechtold T, Burstcher E, Turcanu E. Anthraquinones as mediators for the indirect cathodic reduction of dispersed organic dyestuff. J Electroanal Chem 1999; 465:80–87.
68. Farmer JC, Wang FT, Hawley-Fedder RA, Lewis PR, Summers LJ, Foiles L. Electrochemical treatment of mixed and hazardous wastes: oxidation of ethylene glycol and benzene by silver. J Electrochem Soc 1992; 139:654–662.
69. Choi WK, Kim YM, Lee KW, Oh WZ, Nam H. Destruction of organic decontamination liquid wastes using electro-regenerative Ag(II). Vol. 1. Proceedings of the International Conference on Decommissioning and Decontamination and on Nuclear and Hazardous Waste Management, Denver, 1998; 201–206.

70. Zawodzinski C, Smith W, Martinez K. Kinetic studies of the electrochemical generation of Ag(II) ion and catalytic oxidation of selected organics. Proc Electrochem Soc (Honolulu, HI) 1993; 93(18):170–176.
71. Dziewinski J, Marczak S, Smith W. Electrochemical destruction of mixed wastes. Chemtec 1996; 30–33.
72. Farmer JC, Wang FT, Hickman RG, Lewis PR. Mediated electrochemical oxidation of organic wastes without electrode separators. United States Patent 5,516,972, 1996.
73. Steward GA, Chiba Z, Balazs GB, Nelson N, Lewis PR. Mediated electrochemical oxidation of organic wastes using trivalent cobalt mediator in a nitric acid based system. United States Patent 5,911,868, 1999.
74. Balazs GB, Lewis PR. Mediated electrochemical oxidation of organic wastes using a Co(III) mediator in a neutral electrolyte. United States Patent 5,919, 350, 1999.
75. Farmer J, Wang FT, Lewis PR, Summers LJ. Destruction of chlorinated organics by Co(III)-mediated electrochemical oxidation. J Electrochem Soc 1992; 139:3025–3029.
76. Dhooge PM. Catalyzed, mediated electrochemical oxidation of organic compounds with iron(III). Proc Electrochem Soc (San Francisco) 1994; 94(19): 152–163.
77. Koval CA, Drew SM, Spontarelli T, Noble RD. Concentration and removal of nitrogen and sulfur containing compounds from organic liquid phases using electrochemically reversed chemical complexation. Sep Sci Technol 1988; 23:1389–1399.
78. Clarke RL. Iron(III), a catalyst for the anodic destruction of carbonaceous waste. In: Genders D, Weinberg N, eds. Electrochemistry for a Cleaner Environment. New York: The Electrosynthesis Co., 1992:271–285.
79. Chou TC, Do JS, Cheng CH. Using the mediators/phase transfer catalyst on the anodic oxidation of organic compounds. Proceedings of the International Symposium on Organic Reactions, Tokyo 1992; 283–301.
80. Davidson BL, Quinn Y, Steele DF. Ruthenium-mediated electrochemical destruction of organic wastes. Platin Met Rev 1998; 42:90–98.
81. Nauflett GW, Farncomb RE. Development of an electrochemical waste treatment facility. NASA Conf Publ 1997; 3349:429–436.
82. Chung YH, Park SM. Electrochemical treatment of organic wastes by mediated electrochemical oxidation. Proc Electrochem Soc (Honolulu, HI) 1999; 99(39):67–74.
83. Do JS, Yeh WC. In situ degradation of formaldehyde and benzaldehyde with electrogenerated hypochlorite ion: effect of anodes. J Chin Inst Chem Eng 1994; 25:221–229.
84. Do JS, Yeh WC. In situ degradation of formaldehyde with electrogenerated hypochlorite ion. J Appl Electrochem 1995; 25:483–489.
85. Do JS, Yeh WC, Chao IY. Kinetics of the oxidative degradation of formaldehyde with electrogenerated hypochlorite ion. Ind Eng Chem Res 1997; 36:349–356.
86. Yang CH. Hypochlorite production on Ru–Sn binary oxide electrode and its

application in treatment of dye wastewater. Can J Chem Eng 1999; 77:1161–1168.
87. Comninellis C, Nerini A. Anodic oxidation of phenol in the presence of NaCl for wastewater treatment. J Appl Electrochem 1995; 25:23–28.
88. Rigordy P, Pulgarin C, Kiwi J, Péringer P. Electrochemical versus photochemical pretreatment of industrial wastewaters. Water Sci Technol 1997; 35:293–302.
89. Bonfatti F, Ferro S, Lavezzo F, Malacarne M, Lodi G, De Battisti A. Electrochemical incineration of glucose as a model organic substrate: II. Role of the active chlorine mediation. J Electrochem Soc 2000; 147:592–596.
90. Sánchez G, Pérez JR, Codina GProcedimiento para el tratamiento de aguas residuales y residuos. Spanish Patent ES 2,102,950 A1, 1997.
91. Kinoshita K. Electrochemical Oxygen Technology. New York: John Wiley and Sons 1992:369–372.
92. Pletcher D. Indirect oxidations using electrogenerated hydrogen peroxide. Acta Chem Scand 1999; 53:745–750.
93. Rubio LAElectroquímica: Aplicaciones. Madrid: Editorial Tecnos, 1953:158–159.
94. Ullman's Encyclopedia of Industrial Chemistry. Vol. A13. Weinheim: VCH, 1989:460–461.
95. Haggin J. Trickle-bed electrolytic cell for peroxide developed. Chem Eng News 1984; 12:16.
96. Porta A, Fresnel JM, Kuhlhanek A. Process and reactor for producing a treating aqueous solution containing at least hydrogen peroxide ions and hydroxyl ions according with predetermined concentrations. PCT Int Appl WO 79/00347, 1979
97. Brillas E, Maestro A, Moratalla M, Casado J. Electrochemical extraction of oxygen from air via hydroperoxide ion. J Appl Electrochem 1997; 27:83–92.
98. Tatapudi P, Fenton JM. Electrochemical oxidant generation for wastewater treatment. In: Sequeira CAC, ed. Environmental Oriented Electrochemistry. Amsterdam: Elsevier, 1994:103–130.
99. Foller PC, Bombard RT. Processes for the production of mixtures of caustic soda and hydrogen peroxide via the reduction of oxygen. J Appl Electrochem 1995; 25:613–627.
100. Tatapudi P, Fenton JM. Electrolytic processes for pollution treatment and pollution prevention. In: Gerischer H, Tobias CW, eds. Advances in Electrochemical Science and Engineering. Vol. 4. Weinheim: VCH, 1995:363–417.
101. Murphy JO, Hitchens GD. Method and apparatus for using gas and liquid phase cathodic depolarizers. United States Patent 5,770,033, 1998.
102. Webb SP, McIntyre JA. Generation of hydrogen peroxide in a shorted fuel cell. Proceedings of the Symposium on Oxygen Electrochemistry. Vol. 95-26. Pennington, NJ: The Electrochemical Society, 1996:198–208.
103. Alcaide F, Brillas E, Cabot PL, Casado J. Electrogeneration of hydroperoxide ion using an alkaline fuel cell. J Electrochem Soc 1998; 145:3444–3449.

104. Tseung ACC, Jasem SM. An integrated electrochemical–chemical method for the extraction of O_2 from air. J Appl Electrochem 1981; 11:209–215.
105. Brillas E, Mur E, Casado J. Iron (II) catalysis of the mineralization of aniline using a carbon–PTFE O_2 fed cathode. J Electrochem Soc 1996; 143: L49–L53.
106. Harrington T, Pletcher D. The removal of low levels of organics from aqueous solutions using Fe(II) and hydrogen peroxide formed in situ at gas diffusion electrodes. J Electrochem Soc 1999; 146(8):2983–2989.
107. Alvarez-Gallegos A, Pletcher D. The removal of low level organics via hydrogen peroxide formed in a reticulated vitreous cathode cell: Part 1. The electrosynthesis of hydrogen peroxide in aqueous acidic solutions. Electrochim Acta 1998; 44:853–861.
108. Moeglich K. Apparatus for removal of contaminants from water. United States Patent 4,072,596, 1978.
109. Porta A, Kulhanek A. Procede pour la decontamination electrochimique de l'eau. European Patent 141780-A, 1985.
110. Do JS, Chen CP. In situ oxidative degradation of formaldehyde with electrogenerated hydrogen peroxide. J Electrochem Soc 1993; 140(6):1632–1637.
111. Do JS, Chen CP. Kinetics of in situ degradation of formaldehyde with electrogenerated hydrogen peroxide. Ind Eng Chem Res 1994; 33(2):387–394.
112. Do JS, Chen CP. In situ oxidative degradation of formaldehyde with hydrogen peroxide electrogenerated on the modified graphites. J Appl Electrochem 1994; 24:936–942.
113. Do JS, Yeh WC. In situ paired electrooxidative degradation of formaldehyde with electrogenerated hydrogen peroxide and hypochlorite ion. J Appl Electrochem 1998; 28:703–710.
114. Amadelli R, Bonato T, de Battisti A, Babak A, Velichenko A. A comparative study of the electro-oxidation of some phenolic compounds by electrogenerated O_3 and by direct electrolysis at PbO_2 anodes. Proceedings of the Symposium on Energy and Electrochemical Processing for a Cleaner Environment. Pennington, NJ: The Electrochemical Society, 1998:51–60.
115. Tomat R, Vecchi E. Electrocatalytic production of OH radicals and their oxidative addition to benzene. J Appl Electrochem 1971; 1:185–188.
116. Tomat R, Rigo A. Electrochemical oxidation of toluene promoted by OH radicals. J Appl Electrochem 1984; 14:1–8.
117. Tomat R, Rigo A. Electrochemical production of hydroxyl radicals and their reaction with toluene. J Appl Electrochem 1976; 6:257–261.
118. Tomat R, Rigo A. Oxidation of polymethylated benzenes promoted by hydroxyl radicals. J Appl Electrochem 1979; 9:301–305.
119. Matsue T, Fujihira M, Osa T. Oxidation of alkylbenzenes by electrogenerated hydroxyl radical. J Electrochem Soc 1981; 128(12):2565–2569.
120. Fleszar B, Sobkowiak A. Hydroxylation of benzene and phenol during electroreduction of oxygen. Electrochim Acta 1983; 28:1315–1318.
121. Otsuka K, Hosokawa K, Yamanaka I, Wada Y, Morikawa A. One-step oxidation of benzene to phenol applying a fuel cell system. Electrochim Acta 1989; 34:1485–1488.

122. Tzedakis T, Savall A, Clifton MJ. The electrochemical regeneration of Fenton's reagent in the hydroxylation of aromatic substrates: batch and continuous processes. J Appl Electrochem 1989; 19:911–921.
123. Wellmann J, Steckhan E. Indirect electrochemical processes: 2. Electrocatalytic direct oxidation of aromatic compounds by hydrogen peroxide. Chem Ber 1977; 110:3561–3571.
124. Oturan MA, Pinson J, Deprez D, Terlain B. Polyhydroxylation of salicylic acid by electrochemically generated OH radicals. New J Chem 1992; 16:705–710.
125. Oturan MA, Pinson J. Hydroxylation by electrochemically generated OH radicals. Mono- and polyhydroxylation of benzoic acid: products and isomers distribution. J Phys Chem 1995; 99:13948–13954.
126. Oturan MA, Pinson J, Oturan N, Deprez D. Hydroxylation of aromatic drugs by the electro-Fenton method. Formation and identification of the metabolites of riluzole. New J Chem 1999; 23:793–794.
127. Bartlett PN, Pletcher D, Zeng J. Approaches to the integration of electrochemistry and biotechnology: II. The horseradish peroxidase catalyzed oxidation of 2,4,6-trimethylphenol by electrogenerated hydrogen peroxide. J Electrochem Soc 1999; 146(3):1088–1092.
128. Sudoh M, Kodera T, Sakai K, Zhang JQ, Koide K. Oxidative degradation of aqueous phenol effluent with electrogenerated Fenton's reagent. J Chem Eng Jpn 1986; 19(6):513–518.
129. Sudoh M, Kodera T, Hino H, Shimamura H. Effect of anodic and cathodic reactions on oxidative degradation of phenol in an undivided bipolar electrolyzer. J Chem Eng Japan 1988; 21(2):198–203.
130. Sudoh M, Kodera T, Hino H, Shimamura H. Oxidative degradation rate of phenol in an undivided bipolar electrolyzer. J Chem Eng Jpn 1988; 21(5):536–538.
131. Chu CS, Huang CP. Electrochemical oxidation of halogenated compounds in dilute aqueous solution. Proceedings of the International Conference on Advanced Oxidation Technologies for Water and Air Remediation, London, Ontario, Canada 1994; 118–119.
132. Takiyama MK, Chu CS, Huang YC, Huang CP, Huang HS. The removal of priority pollutants from groundwater by advanced oxidation processes. Hazard Ind Waste 1994; 26:178–185.
133. Do JS, Yeh WC. Paired electrooxidative degradation of phenol with in situ electrogenerated hydrogen peroxide and hypochlorite. J Appl Electrochem 1996; 26:673–678.
134. Oturan MA, Aaron JJ, Oturan N, Pinson J. Degradation of chlorophenoxyacid herbicides in aqueous media, using a novel electrochemical method. Pestic Sci 1999; 55:558–562.
135. Oturan MA, Peiroten J, Chartrin P, Acher AJ. Complete destruction of *p*-nitrophenol in aqueous medium by Electro-Fenton method. Environ Sci Technol 2000; 34:3474–3479.
136. Oturan MA. An ecologically effective water treatment technique using electrochemically generated hydroxyl radicals for in situ destruction of organics

pollutants: application to herbicide 2,4-D. J Appl Electrochem 2000; 30:475–482.
137. Hsiao YL, Nobe K. Hydroxylation of chlorobenzene and phenol in a packed bed flow reactor with electrogenerated Fenton's reagent. J Appl Electrochem 1993; 23:943–946.
138. Ponce de Leon C, Pletcher D. Removal of formaldehyde from aqueous solutions via oxygen reduction using a reticulated vitreous carbon cathode. J Appl Electrochem 1995; 25:307–314.
139. Alvarez-Gallegos A, Pletcher D. The removal of low level organics via hydrogen peroxide formed in a reticulated vitreous cathode cell: Part 2. The removal of phenols and related compounds from aqueous effluents. Electrochim Acta 1999; 44:2483–2492.
140. Brillas E, Sauleda R, Casado J. Degradation of 4-chlorophenol by anodic oxidation, Electro-Fenton, Photoelectro-Fenton and Peroxi-coagulation processes. J Electrochem Soc 1998; 145:759–765.
141. Casado J, Bastida RM, Brillas E, Vandermeiren M. Electrolytic purification of contaminated waters by using oxygen diffusion cathodes. Spanish Patent ES 9400299, 1994; PCT Int Appl WO 95/22509, 1995.
142. Casado J, Brillas E, Bastida RM. Electrodegradation kinetics of aniline and 4-chloroaniline in basic aqueous media. Proceedings of the Symposium on Water Purification by Photocatalytic, Photoelectrochemical and Electrochemical Processes. Vol. 94-19. Pennington, NJ: The Electrochemical Society, 1994:87–98.
143. Brillas E, Casado J. Aniline mineralization by Electro-Fenton® and peroxi-coagulation processes using a flow reactor with an oxygen diffusion cathode. Proceedings of the 2nd International Conference on Oxidation Technologies for Water and Wastewater Treatment, Clausthal-Zellerfeld, Germany, 2000.
144. Brillas E, Bastida RM, Llosa E, Casado J. Electrochemical destruction of aniline and 4-chloroaniline for wastewater treatment using a carbon-PTFE O_2 fed cathode. J Electrochem Soc 1995; 142:1733–1741.
145. Casado J. Advanced oxidation of organic pollutants in waters. The Electro-Fenton® and Helielectro-Fenton processes. EUCHEM 2000 Conference: Organic Electrochemistry in the New Century, Tomar, Portugal, 2000.
146. Brillas E, Sauleda R, Casado J. Peroxi-coagulation of aniline in acidic medium using an oxygen diffusion cathode. J Electrochem Soc 1997; 144:2374–2379.
147. Casado J, Brillas E. Mejoras en el objeto de la patente principal no. 9400299, por "proceso y equipo de depuración electrolítica de aguas contaminadas usando catodos de oxígeno." Spanish Patent ES 9700186, 1997.
148. Brillas E, Mur E, Sauleda R, Sanchez L, Peral J, Domenech X, Casado J. Aniline mineralization by AOPs: anodic oxidation, photocatalysis, electro-Fenton and photoelectro-Fenton processes. Appl Catal B Environ 1998; 16:31–42.
149. Brillas E, Calpe JC, Casado J. Mineralization of 2,4-D by advanced electrochemical oxidation processes. Water Res 2000; 34(8):2253–2262.

150. Brillas E, Sauleda R, Casado J. Destruction of aromatic contaminants in an Fe/O_2 battery. Electrochem Solid-State Lett 1998; 1:168–171.
151. Brillas E, Sauleda R, Casado J. Use of an acidic Fe/O_2 cell for wastewater treatment: degradation of aniline. J Electrochem Soc 1999; 146:4539–4543.
152. Chou S, Huang YH, Lee SN, Huang GH, Huang C. Treatment of high strength hexamine-containing wastewater by Electro-Fenton method. Water Res 1999; 33:751–759.
153. Huang YH, Chou S, Perng MG, Huang GH, Cheng SS. Case study on the bioeffluent of petrochemical wastewater by Electro-Fenton method. Water Sci Technol 1999; 39:145–149.
154. Roe BA, Lemley AT. Treatment of two insecticides in an electrochemical Fenton system. J Environ Sci Health Part B Pestic Food Contam Agric Wastes 1997; 32(2):261–281.
155. Pratap K, Lemley AT. Fenton electrochemical treatment of aqueous atrazine and metolachlor. J Agric Food Chem 1998; 46:3285–3291.
156. Françoisse P, Gregor KH. Application of a new Fenton process (FSR® process) without sludge production for the treatment of non biodegradable wastewater. Chemical Oxidation. Technologies for the Nineties. Vol. 6. Lancaster: Technomic Publishing Co., 1997:208–220.
157. Savinova ER, Kuzmin AO, Frusteri F, Parmaliana A, Parmon VN. Partial oxidation of ethane in a three-phase Electro-Fenton system. In: Parmaliana A, ed. Natural Gas Conversion: V. Studies in Surface Science and Catalysis 1998; 119:429–434.
158. Khosla NK, Venkatachalam S, Somasundaran P. Pulsed electrogeneration of bubbles for electroflotation. J Appl Electrochem 1991; 21:986–990.
159. Weintraub MH, Dzieciuch MA, Gealer RLMethod for breaking an oil-in-water emulsion. United States Patent 4,194,972, 1980.
160. Lin SH, Peng CF. Treatment of textile wastewater by electrochemical method. Water Res 1994; 28:277–282.
161. Wilcok A, Brewster M, Tincher W. Using electrochemical technology to treat textile wastewater: three case studies. Am Dyest Rep 1992; 81:15–22.
162. Vik EA, Carlson DA, Eikum AS, Gjessing ET. Electrocoagulation of potable water. Water Res 1984; 18:1355–1360.
163. Schade H, Peter KProcess and device for separating solid and liquid matter from industrial wastewater by electroflotation. European Patent EP 686603 A1, 1995.
164. Osasa K, Kawanami H, Tukamoto E, Sakai K. Performance of continuous electroflotation of an oil-in-water (O/W) emulsion using a packed-bed-type sacrificial electrode as an anode. Kagaku Kogaku Ronbunshu 1996; 22:1450–1456.
165. Osasa K, Miyazaki T, Sakai K. Performance test of a rectangular electroflotation cell with packed-bed-type sacrificial electrodes. Kagaku Kogaku Ronbunshu 1997; 23:594–596.
166. Beaujean H. Electroflotation or electrocoagulation of wastewaters. PCT Int Appl WO 9515295 A1, 1995.

167. Pouet MF, Grasmick A. Urban wastewater treatment by electrocoagulation and flotation. Water Sci Technol 1995; 31:275–283.
168. Ritter H. Purification of dirty water by electroflotation. German Patent DE 4416973 A1, 1995.
169. Morkovsky PE, Kaspar DD, Petru JM. Process and apparatus for electrocoagulative treatment of industrial wastewater. PCT Int Appl WO 9926887 A1, 1999.
170. Minghi O. Method and device for treating industrial wastewater. PCT Int Appl WO 9816478 A1, 1998.
171. Lee MS. Wastewater treatment by means of electroflotation and electrocoagulation. European Patent EP 794157 A1, 1997.
172. Weintraub MH, Gealer RL, Golovoy A, Dzieciuch MA, Durham H. Development of electrolytic treatment of oily wastewater. Environ Prog 1983; 2:32–35.
173. Persin F, Rumeau M. Le traitement electrochimique des eaux et des effluents. Trib Eau 1989; 82:45–56.
174. Jenke DR, Diebold FE. Electroprecipitation treatment of acid mine wastewater. Water Res 1984; 18:855–859.
175. Moeglich K. Sorbent particulate material and manufacture thereof. United States Patent 4,048,028, 1977.
176. Biswas N, Lazarescu G. Removal of oil from emulsions using electrocoagulation. Int J Environ Stud 1991; 38:65–75.
177. Ibañez JG, Takimoto MM, Vasquez RC, Basak S, Myung N, Rajeshwar K. Laboratory experiments on electrochemical remediation of the environment: electrocoagulation of oily wastewater. J Chem Educ 1995; 72:1050–1052.
178. Pouet MF, Persin F, Rumeau M. Intensive treatment by electrocoagulation–flotation–tangential flow microfiltration in areas of high seasonal population. Water Sci Technol 1992; 25:247–253.
179. Il'in VI, Sedashova ON. An electroflotation method and plan for removing oil products from effluents. Chem Pet Eng 1999; 35:480–481.
180. Ibañez JG, Singh MM, Szafran Z. Laboratory experiments on electrochemical remediation of the environment: Part 4. Color removal of simulated wastewater by electrocoagulation–electroflotation. J Chem Educ 1998; 75:1040–1041.
181. Wilcock A, Tebbens J, Fuss F, Wagner J, Brewster M. Spectrophotometric analysis of electrochemically treated, simulated, dispersed dyebath effluent. Text Chem Color 1992; 24:29–37.
182. Do JS, Chen ML. Decolourization of dye-containing solutions by electrocoagulation. J Appl Electrochem 1994; 24:785–790.
183. Lee BS, Son EJ, Choe EK, Kim JW. Electrochemical treatment of dyeing effluent using sacrificial iron electrodes. J Korean Fiber Soc 1999; 36:329–337.
184. Youchun Z, Dunwen L, Yongqi Z, Jianmin L, Meiqiang L. The study of the electrolysis coagulation process using insoluble anodes for treatment of printing and dyeing wastewater. Water Treat 1991; 6:227–236.
185. Takaoka S. Electrode configuration in dyeing wastewater decolorization by electrolysis. Japan Patent JP 08267073 A2, 1996.

186. Changhai L. Study on decolorization of dyeing and printing wastewater by electroflocculation process with guide shell. Huagong Huanbao 1999; 19:264–268.
187. Zhalsanova DB, Batoeva AA, Ryazantsev AA, Khankhasaeva ST. Oxidative destruction of organic contaminants in electrocoagulation of wastewaters. Khim Interesakh Ustoich Razvit 1998; 6:409–415.
188. Pogrebnaya VL, Ovdienko YI, Khanaev PE, Tarasenko AG, Simonova EA, Burtsev VA. Novel technology for removal of protein substances, nitrites, and sulfates from wastewaters. Khim Neft Mashinostr 1995; 6:23–24.
189. Nogueira W. Electrochemical reactor for removal of algae, pesticides, organic material, and dissolved oils in waters and wastewaters. Brazilian Patent BR 9703015 A, 1998.
190. Daniliuc L, Baland M, Tariot P, Haine T. Performance and economic aspects of the wastewater electroflocculation process DEQUAFLOX. Trib Eau 1995; 48:37–39.

7
The Electron Beam Process for the Radiolytic Degradation of Pollutants

Bruce J. Mincher
Idaho National Engineering & Environmental Laboratory, Idaho Falls, Idaho, U.S.A.

William J. Cooper
University of North Carolina–Wilmington, Wilmington, North Carolina, U.S.A.

I. INTRODUCTION

The ultimate disposal of hazardous organic pollutants is emerging as a priority in the search for innovative treatment technologies. Ultimate disposal is the mineralization of pollutant compounds to inorganic constituents such as water and carbon dioxide. Conventional treatment processes have often focused on removal of a pollutant from a particular location, without concern for its ultimate disposition. Examples include landfilling, deep-well injection, or vapor vacuum extraction with collection on carbon filters. Eventually, the hazardous compound "treated" with these techniques must be dealt with again.

A conventional example of ultimate treatment is incineration. Unfortunately, incineration has met with strong public opposition because of air emissions that potentially contain small amounts of toxic by-products. Presently, incinerators are losing operations licenses, rather than new incinerators being licensed.

Several innovative, ultimate disposal technologies are currently being developed for the treatment of water. These advanced oxidation technologies act as sources of free radicals, principally hydroxyl radical (•OH), which oxidatively decompose pollutants. An excellent source of free radicals for water treatment is ionizing radiation. Irradiation of water produces both reducing and oxidizing species, which allow for a versatile approach to the ultimate treatment of a variety of pollutants. Machine-generated electron beams (e-beams) provide reliable and safe radiation sources for treatment of flowing waste streams on a process scale. Process versatility is provided by continuous, rapid treatment potential and a tolerance for feedstocks of varying quality. Additionally, modern e-beams have excellent operational reliability. They are easily automated and many models are portable. Isotope gamma-ray sources have also been used, but are more important as experimental sources for process design and scale-up for e-beam irradiation.

One of the overlooked aspects of the radiolysis process is that the underlying chemistry is relatively well understood. This chapter will examine the chemistry of free radical generation by radiation, those reactions of radicals with pollutants, which result in mineralization, and the kinetics of reaction from a process chemistry point of view. Two currently operational e-beam processes will be presented.

II. BACKGROUND OF THE TECHNIQUE

A. Generating the e-Beam

An electron accelerator is similar to the common television, except that the accelerating potential is higher. Both emit electrons from a cathode, which are then electrostatically accelerated. The accelerating voltage, commonly referred to as the energy of the electrons, is determined by the design of the accelerator. In the case of electron accelerators that will be used for environmental applications, the lowest potential that is practical is 500 keV. It is likely that potentials of 1–1.5 MeV will be more common and, if the design of several new accelerators is perfected, may reach 10 MeV. The energy of the electron determines its depth of penetration in water. The number of electrons is referred to as the beam current, and is controlled by the cathode size and configuration. Common cathodes used today result in beam currents of from 50 to 100 mA. The power of an accelerator is the energy multiplied by the beam current. Electron accelerators that are used for different purposes typically have power ratings of up to 100 kW.

B. Process Efficiency

A common misconception about the e-beam process is that high-energy electrons mean high energy costs. In fact, the e-beam accelerator (using insulated core transformers) is an energy-efficient means of creating the •OH radical [1]. Many other accelerator designs are available, several of which are discussed in Ref. [2]. Efficiency has been defined as the kilowatt-hours of electricity required to reduce the concentration of a contaminant in 1000 gal of water by one order of magnitude (or 90%). This is termed the electrical energy per order (EE/O in kWh/1000 U.S. gal/order).

For example, if it takes 10 kWh of electricity to reduce the concentration of a contaminant from 100 to 10 mg L^{-1} in 1000 gal of wastewater, then the EE/O is 10 kWh/1000 gal/order. It would then take an additional 10 kWh to reduce the compound one additional order, from 10 to 1 mg L^{-1}, which would be an overall removal efficiency of 99%.

Because the logarithmic relationship between the change in pollutant concentration and e-beam radiation dose is often linear, that slope can be described by the EE/O. This allows for a comparison with the energy costs of competing technologies. However, care should be taken when comparing various processes using the EE/O. In some processes, it is necessary either to take into account all of the energy costs associated with each treatment or to examine both EE/O and operational costs. For example, if H_2O_2 is added during the treatment, there is an electrical cost associated with the production of the peroxide and it needs to be taken into consideration in the comparison, as an addition either to the EE/O calculation or in the operational costs.

The EE/O is determined from a feasibility study. It is specific to the pollutant being treated, its initial concentration, and the nature of the water being tested. Water quality, in particular, may have a great effect on process efficiency, because the presence of various scavenger compounds may remove radicals from solution. Typical EE/O values for common pollutants range from 0.5 to 12 kWh/1000 gal/order. Once an EE/O value has been determined, either through feasibility studies or estimated from a table of values, the e-beam dose (in kW) required for any specific application may be calculated:

$$\text{Dose} = (\text{EE/O}) \times (\log \text{C}_\text{o}/\text{C}) \quad (1)$$

where C_o is the initial contaminant concentration and C is the treatment objective concentration. For waste streams with complex mixtures of contaminants, the energy required for treatment is not additive but is determined for the contaminant with the highest EE/O in the water to be treated.

C. Formation of Reactive Species

To understand how pollutants are decomposed by ionizing radiation, it is necessary to review aqueous-based radiation chemistry. The irradiation of water results in the formation of electronically excited states, free radicals and ions along the path (spur) of the incident particle. The reactive species produced by irradiation may then diffuse away from the spur, or undergo geminate recombination to recreate their parent species. Those that escape recombination may react with solutes, including pollutants present in the water. This is the essence of waste treatment by radiolysis.

Whether the radiation source is a machine-generated e-beam, isotopic gamma rays, or machine-generated x-rays (bremsstrahlung), a continuum of x-rays and electrons is produced as the initial event dissipates its energy in the irradiated medium. This chapter will focus on an e-beam source, as this is the most likely source to be used in a waste-treatment application. It should be noted that most of the comments made here are equally applicable to other sources of radiation.

At 10^{-7} sec, after an electron has passed through water at neutral pH, the products shown in Eq. (2) are present [3]:

$$H_2O \text{ -/\/\/\-> } [0.28]{\bullet}OH + [0.27]e^-_{aq} + [0.06]H{\bullet} + [0.07]H_2O_2 \qquad (2)$$
$$+[0.27]H_3O^+ + [0.05]H_2$$

Unlike in photochemical reactions, a high-energy electron can initiate several thousand secondary reactions as it dissipates its energy. The efficiency of conversion of electron energy to a chemical product is defined as the *G* value [shown in brackets in Eq. (2)]. The *G* value is the micromoles of product formed or lost in a system absorbing 1 J of energy. The most reactive products in Eq. (2) are the oxidizing hydroxyl radical (OH·), and the reducing aqueous electron (e^-_{aq}) and hydrogen atom (H·). These oxidizing and reducing species are produced in approximately equal amounts, although it will be shown that they do not have equal affect. The chemistry of primary interest in the high-energy electron irradiation process is that of these three species. Pollutants with high reaction rates with these species are likely to be amenable to treatment by radiolysis.

Water radiolysis is actually more complex than suggested by Eq. (2). A more complete series of reactions representing the radiolytic decomposition of water may be found in Table 1.

The reactions of the principal reactive species are discussed below.

D. Aqueous Electron (e^-_{aq})

The reactions of the aqueous electron (e^-_{aq}) with specific organic and inorganic compounds have been studied extensively [4–6]. The e^-_{aq} is a powerful reducing agent, with a reduction potential of −2.77 V. The reactions of the e^-_{aq} are single-electron transfer, the general form of which is:

$$e^-_{aq} + S^N \rightarrow S^{N-1} \qquad (3)$$

Table 1 Some Reactions Describing Pure Water Radiolysis and Their Second-Order Rate Constants (k) for Reaction[a]

Reaction	k (L mol^{-1} s^{-1})
$\bullet OH + H_2 \rightarrow H\bullet + H_2O$	4.00×10^{7}
$\bullet OH + H_2O_2 \rightarrow \bullet O_2^{-} + H_2O$	2.70×10^{7}
$\bullet OH + HO_2^{-} \rightarrow H_2O + \bullet O_2^{-}$	7.50×10^{9}
$\bullet OH + \bullet O_2^{-} \rightarrow O_2 + OH^{-}$	1.10×10^{10}
$\bullet OH + H_2O_2^{+} \rightarrow O_2 + H_3O^{+}$	1.20×10^{10}
$\bullet OH + HO_2\bullet \rightarrow O_2 + H_2O$	1.00×10^{10}
$\bullet OH + \bullet OH \rightarrow H_2O_2$	5.50×10^{9}
$\bullet OH + OH^{-} \rightarrow H2O + \bullet O^{-}$	1.30×10^{10}
$\bullet OH + \bullet O^{-} \rightarrow HO_2^{-}$	2.00×10^{10}
$\bullet O^{-} + H_2O \rightarrow \bullet OH + OH^{-}$	9.30×10^{7}
$\bullet O^{-} + HO_2^{-} \rightarrow \bullet O_2^{-} + OH^{-}$	4.00×10^{8}
$\bullet O^{-} + H_2 \rightarrow e^{-}_{aq} + H_2O$	1.20×10^{8}
$\bullet O^{-} + H_2O_2 \rightarrow \bullet O_2^{-} + H_2O$	2.70×10^{7}
$\bullet O^{-} + \bullet O_2^{-} \rightarrow 2OH^{-} + O_2$	6.00×10^{8}
$e^{-}_{aq} + H\bullet \rightarrow H_2 + OH^{-}$	2.50×10^{10}
$e^{-}_{aq} + e^{-}_{aq} \rightarrow 2OH^{-} + H_2$	5.50×10^{9}
$e^{-}_{aq} + O_2 \rightarrow \bullet O_2^{-}$	1.90×10^{10}
$e^{-}_{aq} + H_2O_2 \rightarrow \bullet OH + OH^{-}$	1.20×10^{10}
$e^{-}_{aq} + \bullet O_2^{-} \rightarrow O_2^{2-}$	1.30×10^{10}
$e^{-}_{aq} + H^{+} \rightarrow H\bullet$	2.30×10^{10}
$e^{-}_{aq} + HO_2^{-} \rightarrow \bullet OH + 2OH^{-}$	3.50×10^{9}
$e^{-}_{aq} + \bullet OH \rightarrow OH^{-}$	3.00×10^{10}
$e^{-}_{aq} + \bullet O^{-} \rightarrow 2OH^{-}$	2.20×10^{10}
$H\bullet + O_2 \rightarrow HO_2\bullet$	2.10×10^{10}
$H\bullet + \bullet O_2^{-} \rightarrow HO_2^{-}\bullet$	2.00×10^{10}
$H\bullet + H\bullet \rightarrow H_2$	5.00×10^{9}
$H\bullet + \bullet OH \rightarrow H_2O$	7.00×10^{9}
$H\bullet + HO_2\bullet \rightarrow H_2O_2$	1.00×10^{10}
$H\bullet + H_2O_2 \rightarrow H_2O + \bullet OH$	9.00×10^{7}
$H\bullet + OH^{-} \rightarrow e^{-}_{aq} + H_2O$	2.20×10^{7}
$HO_2\bullet + \bullet O_2^{-} \rightarrow O_2 + H_2O_2 + OH^{-}$	9.70×10^{7}
$H^{+} + \bullet O_2^{-} \rightarrow HO_2\bullet$	4.50×10^{10}
$H^{+} + HO_2^{-}\bullet \rightarrow H_2O_2$	2.00×10^{10}
$H^{+}OH^{-} \rightarrow H_2O$	1.43×10^{11}

[a] About 10^{-7} sec after electron injection.

The e^-_{aq} reacts with numerous organic compounds. Of particular interest to the field of waste treatment are the reactions with halogenated compounds. A generalized reaction is shown below:

$$e^-_{aq} + RCl \rightarrow R\bullet + Cl^- \tag{4}$$

Thus, reactions involving the e^-_{aq} often result in the dechlorination of organochlorine compounds. This result may be sufficient for waste-treatment purposes. However, further reaction of the resulting organic radical (R·) may also be desirable to mineralize the compound. The e^-_{aq} also reacts with many other organic compounds and contributes to the removal of these compounds from aqueous solution. Although aqueous electrons are produced with a *G* value nearly equal to that of oxidizing hydroxyl radicals, they are often less available for reaction. Electrons are scavenged by hydronium ion in acidic water, and by oxygen in solutions exposed to air. This lowers their availability for reactions with pollutants. Interference by competitor species, often called scavengers, is discussed in more detail later.

E. Hydrogen Atom (H·)

The reactions of the hydrogen atom (H·) with organic and inorganic compounds have also been summarized [3,7]. Based on the *G* values shown in Eq. (2), the hydrogen atom accounts for approximately 10% of the total free radical concentration in irradiated water. It too is a powerful reducing agent with a potential of −2.3 V. The H· undergoes two general types of reactions with organic compounds: hydrogen addition and hydrogen abstraction.

An example of a typical addition reaction with an organic solute is that of benzene, shown in Eq. (5). The aromaticity of benzene is destroyed, opening the way for ring-opening reactions.

$$H\bullet + C_6H_6 \rightarrow \bullet C_6H_7 \tag{5}$$

The second general reaction involving the H· is hydrogen abstraction, shown here for the reaction with methanol:

$$H\bullet + CH_3OH \rightarrow H_2 + \bullet CH_2OH \tag{6}$$

The product of Eq. (6) is the methoxy radical, which is also a reducing agent. It is able to participate in reactions with some solutes by electron transfer.

Although less abundant and less reducing than the electron, the relatively small reaction rate constant of H· with the common radical scavengers found in natural waters makes it possible that this radical may be important in removing some pollutants.

F. Hydroxyl Radical (·OH)

This is probably the most important radical species in aqueous solution. Oxidative reactions of the hydroxyl radical, (·OH), with inorganic and organic compounds have been well documented [8]. Compilations of bimolecular (second-order) rate constants have been published [3,8]. The OH· can undergo several types of reactions with species in aqueous solution, including addition, hydrogen abstraction, and electron transfer.

Addition reactions occur readily with aromatic and unsaturated aliphatic compounds. The resulting compounds are hydroxylated radicals:

$$\bullet OH + CH_2{=}CH_2 \rightarrow \bullet CH_2{-}CH_2(OH) \quad (7)$$

Hydrogen abstraction occurs with saturated and many unsaturated molecules, e.g., aldehydes and ketones:

$$\bullet OH + CH_3{-}CO{-}CH_3 \rightarrow \bullet CH_2COCH_3 + H_2O \quad (8)$$

Electron transfer reactions are also common, and occur when aqueous solutions are irradiated with high-energy electrons. For example, reactions involving thioanisole have recently been reported [9]:

$$\bullet OH + CH_3{-}S{-}C_6H_5 \rightarrow [CH_3{-}S{-}C_6H_5]^{\bullet} + OH^- \quad (9)$$

G. Hydrogen Peroxide (H_2O_2)

The G value for the formation of H_2O_2 is 0.07 μmol J^{-1}, and, therefore, the formation of significant concentrations of this relatively stable oxidant are likely. The reaction that results in the formation of most H_2O_2 in irradiated water is the radical–radical recombination involving •OH:

$$\bullet OH + \bullet OH \rightarrow H_2O_2 \quad (10)$$

A second source of H_2O_2 in oxygenated aqueous solutions are the reactions of e^-_{aq} and H• with O_2. Both of these reactions result in the formation of reduced oxygen, the superoxide radical ion, or its conjugate acid:

$$e^-_{aq} + O_2 \rightarrow \bullet O_2^- \quad (11)$$

$$\bullet H + O_2 \rightarrow HO_2\bullet \quad (12)$$

The products of Eqs. (11) and (12) are in equilibrium, with a $pK_a = 4.8$. These products lead to the formation of additional H_2O_2:

$$2[\bullet O_2^-] + 2H^+ \rightarrow H_2O_2 + O_2 \quad (13)$$

$$2[HO_2\bullet] \rightarrow H_2O_2 + O_2 \quad (14)$$

The occurrence of the following reaction:

$$e^-_{aq} + H_2O_2 \rightarrow \bullet OH + OH^- \qquad (15)$$

suggests that the addition of more H_2O_2 to the influent stream may lead to an increase in the •OH concentration. This would be significant if the solute of interest were removed primarily by reactions with the •OH. This is discussed in more detail later.

H. Determining Solute Removal Rates

The rate at which targeted solutes are removed from solution depends on the concentration of the solute, the rate of generation of the required reactive species, the second-order rate constant (k) for the reaction between the two, and the presence of scavenger species that compete. The second-order rate

Table 2 Second-Order Rate Constants (L mol^{-1} s^{-1}) of Selected Organic Chemicals and the Free Radicals Formed in Irradiated Water

Compound	e^-_{aq}	H•	•OH
Benzene	9.0×10^6	9.1×10^8	7.8×10^9
Carbon tetrachloride	1.6×10^{10}	3.8×10^7	NR
Chlorobenzene	5.0×10^8	1.4×10^9	5.5×10^9
Chloroform	3.0×10^{10}	1.1×10^7	5×10^6
o-Cresol	NF	NF	1.1×10^{10}
p-Cresol	4.2×10^7	NF	1.2×10^{10}
1,2-Dichlorobenzene	4.7×10^9	NF	NF
1,3-Dichlorobenzene	5.2×10^9	NF	NF
1,4-Dichlorobenzene	5.0×10^9	NF	NF
trans-1,2-Dichloroethylene	7.5×10^9	NF	6.2×10^9
Ethylbenzene	NF	NF	7.5×10^9
Nitrobenzene	3.7×10^{10}	1.0×10^9	3.9×10^9
Phenol	2.0×10^7	1.7×10^9	6.6×10^9
Pyridine	1.0×10^9	7.8×10^8	3.1×10^9
Tetrachloroethylene	1.3×10^{10}	NF	2.8×10^9
Toluene	1.4×10^7	2.6×10^9	3.0×10^9
Trichloroethylene	1.9×10^9	NF	4.0×10^9
Vinyl chloride	2.5×10^8	NF	1.2×10^{10}
m-Xylene	NF	2.6×10^9	7.5×10^9
o-Xylene	NF	2.0×10^9	6.7×10^9
p-Xylene	NF	3.2×10^9	7.0×10^9

NR = no reaction; NF = not found.

constants for the reactions of many species with e^-_{aq}, H•, and •OH have been reported [3–8]. A selection of important ones is shown in Table 2.

I. Determining the Initial Concentration of Reactive Species

Given radiation chemical yields expressed as *G* values, it is possible to calculate the concentrations of oxidative and reductive species in pure water at a known absorbed dose. The SI unit of dose is the gray (Gy), which equals an energy deposition of 1 J kg^{-1}. For example, the concentration of OH· produced in pure, neutral water by absorbing 1 kGy is:

$$[\bullet OH] = 1000\ J\ kg^{-1} \times 0.28\ \mu mol\ J^{-1} = 280\ \mu mol\ kg^{-1} \quad (16)$$

This calculated value is a maximum concentration. Reactions with solutes and other radiolytically produced species will decrease this concentration via scavenging reactions.

J. Rate Constants for Reaction with Solutes

It is possible to calculate the relative importance of the three transient reactive species on the removal of some organic compounds of interest. These calculations are important in attempting to develop a quantitative understanding of removal efficiency in irradiated waters. It should be noted that the calculations use data that were obtained in laboratory experiments strictly applicable to pure water. The extension of these calculations to natural waters involves additional steps to take into account the reaction of the transient reactive species with naturally occurring scavengers such as oxygen, carbonate/bicarbonate, and others.

If we assume that the only processes responsible for the removal of a solute (R) from an irradiated solution are reactions with the three reactive species e^-_{aq}, H•, and •OH, then the overall removal of solute can be described by the following kinetic expression:

$$\frac{d[R]_t}{dt} = k_1[R][\bullet OH] + k_2[R][e^-_{aq}] + k_3[R][\bullet H] \quad (17)$$

where k_1, k_2, k_3, are the respective second-order rate constants (Table 2). The initial concentrations of each of the three reactive species were found from Eq. (16). Using an absorbed dose of 1 J, relative contributions to solute removal of the three species may be compared. The product of the reactive species concentration and the second-order rate constant (with appropriate unit conversions) for reaction with solute R is a pseudo-first-order rate constant (k'), with units of reciprocal time:

$$G\ (\mu mol\ J^{-1}) \times 1J\ kg^{-1} \times k\ (L\ mol^{-1}\ s^{-1}) = k'\ (s^{-1}) \quad (18)$$

This pseudo-first-order rate constant may be used to compare relative removal rates for the solute R by the reactive species of interest. The total removal rate is given by:

$$\mathrm{Removal}_{[\mathrm{R}]} = (k'_1 + k'_2 + k'_3)[\mathrm{R}] \tag{19}$$

The rate of removal due to an individual reactive species is found when the individual pseudo-first-order rate constant is divided by the sum of the three and converted to percent (by multiplying by 100).

$$\%\mathrm{Removal}_{[\mathrm{R1}]} = 100(k'_1/k'_1 + k'_2 + k'_3)[\mathrm{R}] \tag{20}$$

The pseudo-first order rate constant (k') is sometimes referred to as a "dose constant." Because continuous irradiations are generally performed at constant dose rate, it may also be expressed with respect to absorbed radiation dose (kGy^{-1}) rather than time (s^{-1}). It then represents the concentration of solute removed per unit dose.

The use of the pseudo-first-order dose constant assumes that the removal of solutes is exponential, which is common in waste-treatment applications [10]. For example, the concentration of •OH calculated from Eq. (16) for absorbed doses between 1 and 10 kGy is 0.28 to 2.8 mM. Under these conditions the loss of the solutes (at typical solute concentrations in the micromolar range) is pseudo-first-order, with respect to absorbed dose, and can be described by the following:

$$-\frac{\mathrm{d}[\mathrm{R}]}{\mathrm{d}D} = k[\mathrm{R}][\bullet\mathrm{OH}] = k_0[\mathrm{R}] \tag{21}$$

where [OH·] is hydroxyl radical concentration at dose D, for which the excess concentration remains essentially constant, and k_0, is an empirically determined pseudo-first-order dose constant. The empirical k_0 can be obtained from the slope of the plot of ln[R] vs. dose (D). In an ideal system, the empirical dose constant (k_0) would be the same as the pseudo-first-order dose constant (k'). In real systems the value of k_0 is affected by scavengers, and usually must be determined empirically for a given system.

The half-dose, the dose required for $[\mathrm{R}]_0$ to reach $[\mathrm{R}]_0/2$, can be determined by the following:

$$D_{1/2} = (0.693)/(k_0) \tag{22}$$

Similarly, the dose required to achieve any desired concentration change may be calculated using the first-order rate law:

$$C = C_0 e^{-k_0 D} \tag{23}$$

Another commonly used figure of merit for these pseudo-first order systems is the D_{10}, the dose required to reduce the solute concentration to 10% of its initial value. It is found by rearranging Eq. (23):

$$C/C_0 = 0.1 = e^{-k_0 D_{10}} \tag{24}$$

Eq. (24) is then solved for D to find D_{10}.

K. *G* Values for Solute Removal

The G value (μmol J^{-1}) for solute removal is also empirically determined. Because G values are rates, rather than rate constants, the absorbed dose for which G is reported must be specified. Commonly, G initial (G_0), at zero absorbed dose is chosen. For treatment purposes, G at dose D is defined by the disappearance of the solute in aqueous solutions, and is determined experimentally using the following equation:

$$G = \frac{[C_0 - C](N_A)}{D} \tag{25}$$

The change in organic solute concentration is in micromoles per liter, D is the dose in gray, and $N_A = 6.02 \times 10^{23}$, Avogadro's number.

Alternately, for a solute decreasing in concentration in exponential fashion, G at any dose may be found from the empirical dose constant and Eq. (26):

$$G = k_0 C_0 \tag{26}$$

III. DEGRADATION OF POLLUTANTS

A. Chloroform and Related Compounds

Many undesirable organochlorine compounds are produced by the chlorination of drinking water. Among these are chloroacetic acids and trihalomethanes. Studies have reported the radiolytic decomposition of $CHCl_3$ and related compounds [11–14]. An example is shown in Fig. 1, where the concentration decrease for various bromochloromethanes is plotted vs. absorbed dose. A proposed mechanism for the decomposition of $CHCl_3$ and formation of by-products, involving the three important radical products of water radiolysis is shown:

$$e^-_{aq} + CHCl_3 \rightarrow Cl^- + {\cdot}CHCl_2 \tag{27}$$

$$H{\cdot} + CHCl_3 \rightarrow H_2 + {\cdot}CCl_3 \tag{28}$$

$$\rightarrow HCl + {\cdot}CHCl_2 \tag{29}$$

$$\cdot OH + CHCl_3 \rightarrow H_2O + \cdot CCl_3 \quad (30)$$
$$\cdot CHCl_2 + H_2O \rightarrow \cdot CHO + 2HCl \quad (31)$$
$$\cdot CCl_3 + 2H_2O \rightarrow \cdot COOH + 3HCl \quad (32)$$
$$\cdot COOH + \cdot COOH \rightarrow HOOC - COOH \quad (33)$$
$$\rightarrow HCOOH + CO_2 \quad (34)$$
$$\cdot CHO + HCCl_3 + H_2O \rightarrow \cdot CHCl_2 + HCOOH + HCl \quad (35)$$
$$H\cdot + \cdot CHCl_3 \rightarrow CH_2Cl_2 \quad (36)$$
$$\cdot CHCl_2 + \cdot CHCl_2 \rightarrow CHCl_2CHCl_2 \quad (37)$$
$$\cdot CCl_3 + \cdot CHCl_2 \rightarrow CCl_3CHCl_2 \quad (38)$$
$$\cdot CCl_3 + \cdot CCl_3 \rightarrow CCl_3CCl_3 \quad (39)$$
$$\cdot H + \cdot CHO \rightarrow HCHO \quad (40)$$

The above mechanism describes chloroform decomposition initiated by all three important radicals [Eqs. (27)–(30)]. Whether the actual break-

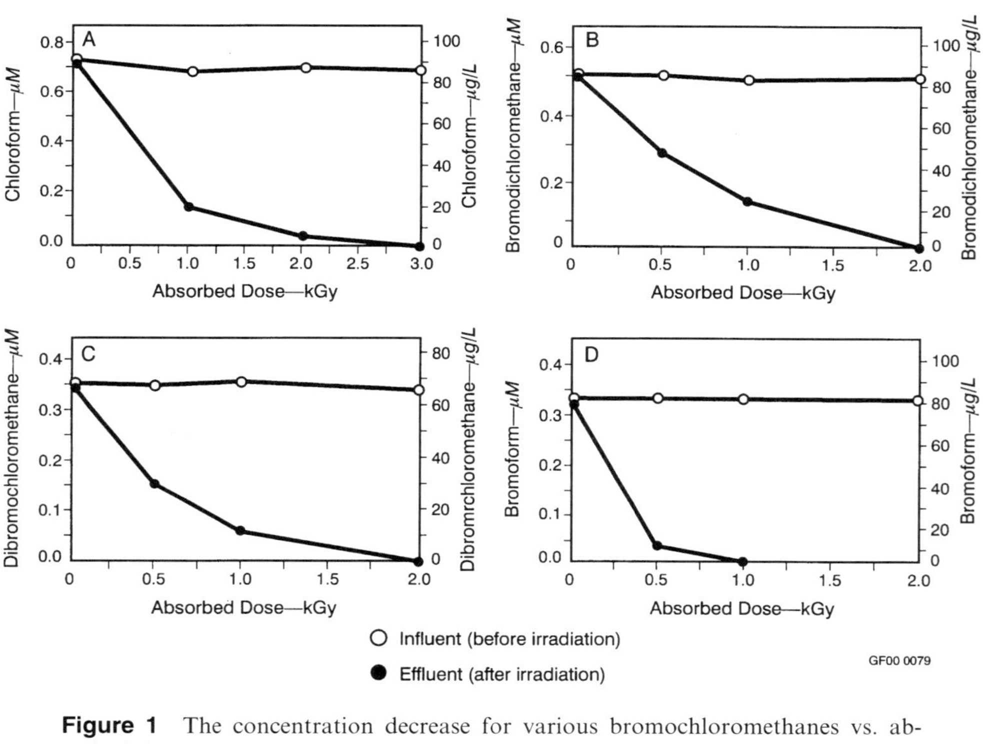

Figure 1 The concentration decrease for various bromochloromethanes vs. absorbed dose.

down process of chloroform in natural waters involves all these reactions may never be known. The importance of Eqs. (27)–(40) is that they provide a point of departure for determining possible reaction products. For example, the production of oxalic acid, formic acid, and formaldehyde are predicted by Eqs. (33), (34), and (40), respectively, as chloroform is mineralized by free radical attack. If the mechanism shown is correct, these products should be detectable in the postirradiation solution.

When chloroform was γ-ray irradiated in water at pH 6.5 by Getoff [14], formaldehyde was generated with $G = 0.025$ μmol J^{-1}. The formaldehyde concentration increased to a maximum at intermediate absorbed dose, and then decreased with increasing irradiation. Presumably, with continued irradiation it was mineralized to CO_2 and H_2O. Only traces of unidentified carboxylic acids were observed as organic products. Similar results were obtained for methylene chloride [14].

When drinking water containing trace bromine is chlorinated, small amounts of mixed bromochloromethanes result. When the bromo- and chloromethanes shown in Fig. 1 were irradiated [13], inorganic halides and formaldehyde were produced stoichiometrically. Brominated compounds were most efficiently destroyed.

Whereas dehalogenation was initiated due to the combination of the reductive and oxidative processes in Eqs. (27)–(30), calculations [Eq. (20)] indicate that > 99% of the chloroform decomposition reactions are initiated by e^-_{aq}. The pseudo-first-order rate constants [k' in Eq. (18)] for the reaction of chloroform with e^-_{aq}, H•, and •OH under the same conditions are shown in Table 3.

Because aqueous chloroform degradation is clearly mostly the result of electron capture, it might be expected that deoxygenation would increase the rate. However, deoxygenated solutions show reduced destruction efficiency [14]. Therefore, a mechanism has been suggested wherein $\cdot CHCl_2$ [product of Eq. (27)] is attacked by oxygen:

$$\cdot CHCl_2 + O_2 \rightarrow \cdot O_2CHCl_2 \quad (41)$$

The product of Eq. (41) is a peroxyl radical. Peroxyl radicals are important intermediates in the oxidative decomposition of many organic compounds.

Table 3 Comparison of Rates of Reaction of e^-_{aq}, H•, and •OH with Chloroform in Pure Water at an Absorbed Dose of 1000 Gy

Species	Concentration (M)	k (L mol^{-1} s^{-1})	k′ (s^{-1})	%
e^-_{aq}	2.7×10^{-4}	3.0×10^{10}	8.1×10^{6}	99.98
H	6.0×10^{-5}	2.4×10^{6}	1.4×10^{2}	0.002
OH	2.8×10^{-4}	5.0×10^{6}	1.4×10^{3}	0.020

The peroxyl radical hydrolyzes in Eqs. (42)–(44). The products are inorganic chloride and formic acid.

$$2\bullet[OOCHCl_2] \rightarrow 2[CHCl_2O\bullet] + O_2 \tag{42}$$

$$CHCl_2O\bullet \rightarrow CHClO + \bullet Cl \tag{43}$$

$$CHClO + H_2O \rightarrow HCO_2H + HCl \tag{44}$$

Equations such as (36)–(40) suggest that radical–radical recombination could result in undesirable organohalogen products. However, such reactions are likely only when the initial halomethane concentrations are very high. Under normal circumstances radical–radical recombination would be unlikely, simply because the radicals are at low concentration. Further, once produced they would decompose by radiolysis under continued irradiation. Halogenated products, other than stoichiometric chloride and bromide ion, have not been detected [13]. Even at intermediate doses, dehalogenation to inorganic halogen anions was stoichiometric.

B. Trichloroethylene (C_2HCl_3) and Perchloroethylene (C_2Cl_4)

Many groundwaters are contaminated with the cleaning solvents trichloroethylene (TCE) and perchloroethylene (PCE). They are two of the most common organochlorine compounds found in Superfund sites. Radiation-induced decomposition of TCE in aqueous solutions has been the subject of several recent studies [15–20]. In most of the referenced studies, the complete destruction of TCE was observed. Dechlorination by a combination of oxidative and reductive radiolysis was stoichiometric. Gehringer et al. [15] and Proksch et al. [18] have characterized the kinetics and mechanism of •OH radical attack on TCE and PCE in γ-ray-irradiated aqueous solution. Trichloroethylene was readily decomposed in exponential fashion, with a reported G value of 0.54 μmol J^{-1}. A 10 ppm (76 μM) solution was decontaminated with an absorbed dose of less than 600 Gy. For each •OH captured, one CO_2 molecule, one formic acid molecule and three Cl^- ions were generated. These products were created by a series of reactions initiated by •OH addition to the unsaturated TCE carbon, which is shown in Eq. (45):

$$C_2HCl_3 + \bullet OH \rightarrow \bullet C_2HCl_3(OH) \tag{45}$$

$$\bullet C_2HCl_3(OH) \rightarrow \bullet CCl_2CHO + H^+ + Cl^- \tag{46}$$

$$\bullet CCl_2CHO + O_2 \rightarrow \bullet OOCCl_2CHO \tag{47}$$

$$2[\bullet OOCCl_2CHO] \rightarrow O_2 + 2[\bullet OCCl_2CHO] \tag{48}$$

$$\bullet OCCl_2CHO \rightarrow COCl_2 + \bullet CHO \tag{49}$$

$$COCl_2 + H_2O \rightarrow CO_2 + 2H^+ + 2Cl^- \quad (50)$$
$$\bullet CHO + H_2O \rightarrow \bullet CH(OH)_2 \quad (51)$$
$$\bullet CH(OH)_2 + O_2 \rightarrow \bullet OOCH(OH)_2 \quad (52)$$
$$2[\bullet OOCH(OH)_2] \rightarrow 2HCOOH + H_2O_2 + \bullet O_2 \quad (53)$$
$$HCOO^- + \bullet OH \rightarrow H_2O + \bullet COO^- \quad (54)$$
$$\bullet COO^- + O_2 \rightarrow \bullet O_2 + CO_2 \quad (55)$$

Peroxyl radical formation [Eq. (47)] is an important step in this mechanism also. Mineralization to chloride ions is shown in Eqs. (46) and (50), and carbon dioxide is shown in Eq. (55). Near mineralization to formic acid is shown in Eq. (53). The rest of the species shown are unlikely to have lifetimes long enough to be observed as permanent products.

In spite of the very high second-order rate constant (1.2×10^{10} L mol^{-1} s^{-1}) for TCE's capture of solvated electrons, Gehringer et al. estimated that no more than 10% of the TCE was decomposed by reductive dechlorination [17]. Reduction of TCE was limited by oxygen scavenging of electrons, which lowers the electron concentration below that calculated from Eq. (16). A proposed mechanism for reductively initiated TCE decomposition is shown in Eqs. (56)–(63):

$$CCl_2{=}CHCl + e^-_{aq} \rightarrow Cl^- + \bullet CCl{=}CHCl \quad (56)$$
$$\bullet CCl{=}CHCl + O_2 \rightarrow \bullet OOCCl{=}CHCl \quad (57)$$
$$2[\bullet OOCCl{=}CHCl] \rightarrow O_2 + 2[\bullet OCCl{=}CHCl] \quad (58)$$
$$\bullet OCCl{=}CHCl \rightarrow COCl - \bullet CHCl \quad (59)$$
$$COCl - \bullet CHCl + O_2 \rightarrow COCl - CHClOO\bullet \quad (60)$$
$$2COCl - CHClOO\bullet \rightarrow O_2 + 2COCl - CHClO\bullet \quad (61)$$
$$COCl - CHClO\bullet \rightarrow Cl\bullet + COCl - CHO \quad (62)$$
$$COCl - CHO + H_2O \rightarrow CHO - COOH + H^- + Cl^- \quad (63)$$

Reductive dechlorination also produces inorganic chloride ion, but glyoxylic acid [Eq. (63)] rather than formic acid. As with the halomethanes, peroxyl radical formation is an important intermediate. Oxygen is clearly important to achieving dechlorination in aqueous solution.

Chloroacetic acids are undesirable by-products that could be postulated in this system. Chloroacetic acids are herbicides, and once formed they are not readily hydrolyzed at normal temperatures and pH, and thus would be persistent pollutants. The production of mono-, and trichloroacetic acid were not observed by Gehringer et al. [17]. Dichloroacetic acid was generated as an intermediate, which disappeared upon continued irradiation, as shown in Fig. 2. The finding that dichloroacetic acid is also

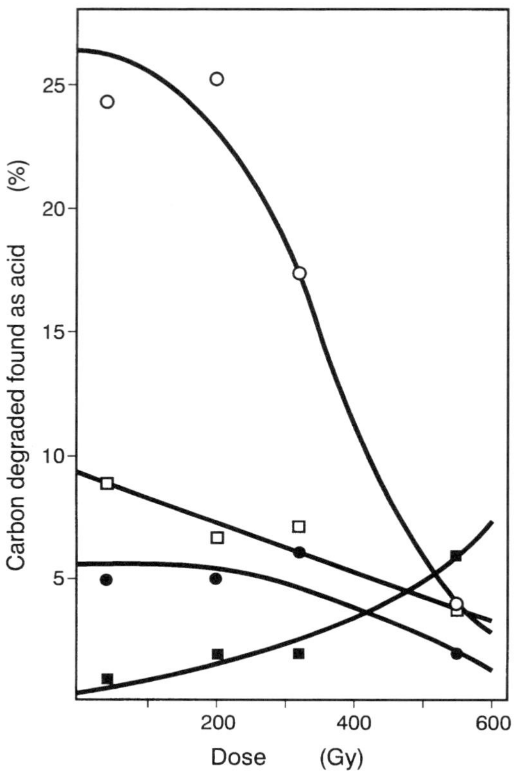

Figure 2 The by-products of trichloroethylene irradiation: ○ formic acid, □ dichloroacetic acid, ● glyoxalic acid, and ■ oxalic acid.

decomposed by radiolysis is important, because chloroacetic acids are also undesirable by-products of conventional water treatment by chlorination.

Aqueous TCE has also been irradiated with γ-rays by Getoff [14], who measured the same by-products. The major, stable reaction products for the irradiation of TCE in aqueous solution are nonhalogenated carboxylic acids. Additional studies are necessary to determine the relative concentration of these known reaction by-products in natural waters.

Considerable research has also been reported on the irradiation of aqueous solutions of PCE [15,16,18–21]. Perchloroethylene, at 10 ppm (60 μM) behaved similarly to TCE, with a lower *G* value of 0.44 $\mu mol\ J^{-1}$ [18]. The lower *G* is expected because of PCE's lower bimolecular rate constant for •OH capture and the slightly lower molar concentration of TCE. The decomposition was first order, trending toward zero order at the highest PCE concentrations, possibly indicating that the radical concentration then became rate limiting. As with TCE, it appears that complete destruction occurs as evidenced by the chloride ion mass balance. A mechanism for the destruction of PCE with •OH, involving peroxyl intermediates has been proposed:

$$CCl_2{=}CCl_2 + {\bullet}OH \rightarrow HOCCl_2{-}C{\bullet}Cl_2 \quad (64)$$

$$HOCCl_2{-}C{\bullet}Cl_2 \rightarrow COCl{-}{\bullet}CCl_2 + H^+ + Cl^- \quad (65)$$

$$COCl{-}C{\bullet}Cl_2 + O_2 \rightarrow COCl{-}CCl_2OO{\bullet} \quad (66)$$

$$2[COCl{-}CCl_2OO{\bullet}] \rightarrow O_2 + 2[COCl{-}CCl_2O{\bullet}] \quad (67)$$

$$COCl{-}CCl_2O{\bullet} \rightarrow {\bullet}COCl + CCl_2O \quad (68)$$

$$CCl_2O + H_2O \rightarrow CO_2 + 2Cl^- \quad (69)$$

$${\bullet}COCl + H_2O \rightarrow {\bullet}COO^- + 2H^+ + Cl^- \quad (70)$$

$${\bullet}COO^- + O_2 \rightarrow {\bullet}O_2^- + CO_2 \quad (71)$$

$$2[COCl{-}C{\bullet}Cl_2] \rightarrow COCl{-}C_2Cl_4{-}COCl \quad (72)$$

$$\rightarrow HOOC{-}C_2Cl_4{-}COOH + 2Cl^- \quad (73)$$

$$COCl{-}C{\bullet}Cl_2 + {\bullet}COO^- \rightarrow COCl - CCl_2{-}COO^- \quad (74)$$

$$\rightarrow HOOC - CCl_2 - COOH + Cl^- \quad (75)$$

$$2[{\bullet}COO^-] \rightarrow HOOC{-}COOH \quad (76)$$

The production of oxalic acid as a stable product is shown in Eq. (76). As with TCE, the principal reaction products at high absorbed doses would be the more oxidized organic aldehydes and acids.

C. Benzene and Substituted Benzenes

Aromatic compounds are especially stable and are, therefore, important persistent pollutants. They include the polyaromatic hydrocarbons (PAHs), and may be halogenated, such as the polychlorinated biphenyls (PCBs) and many pesticides. Also included are the substituted benzenes, such as phenol. A large body of literature has examined aromatic radiation chemistry [22–38]. The discussion that follows examines benzene and substituted benzenes as a model for the radiolysis of more complicated aromatic compounds.

Aromatic rings are susceptible to •OH attack. The oxidation is initiated by addition to the ring, which generates the hydroxycyclohexadienyl radical, shown in Eq. (77). For halogenated benzenes, OH• attack at an unsubstituted carbon is preferred. The hydroxycyclohexadienyl radical may disproportionate to produce phenol, shown in Eq. (78).

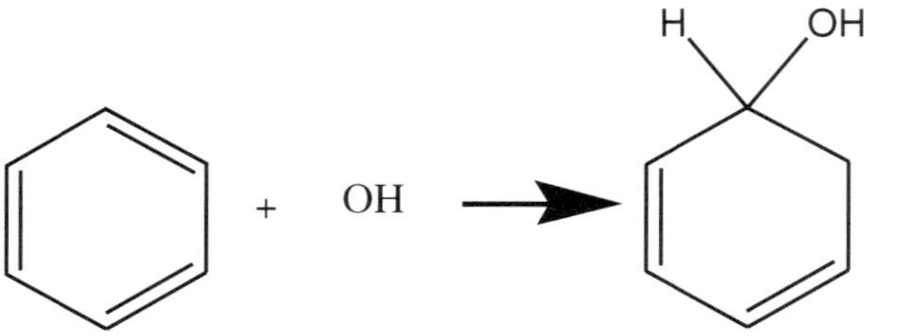

(77)

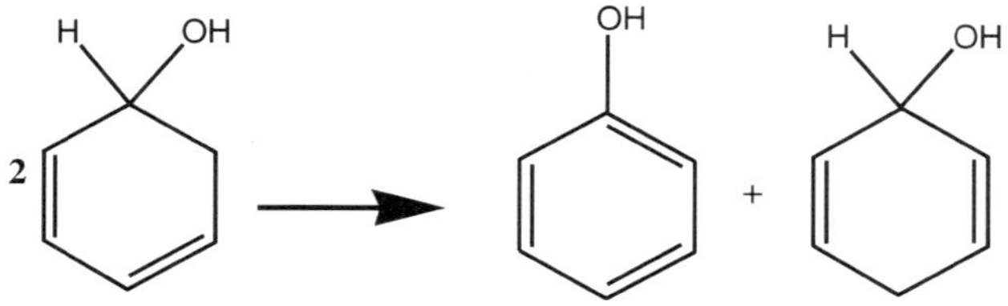

(78)

However, in the presence of oxygen, addition may produce a peroxyl radical from the cyclohexadienyl radical [Eq. (79)], ultimately resulting in decomposition of the aromatic ring, via Eqs. (80) and (81).

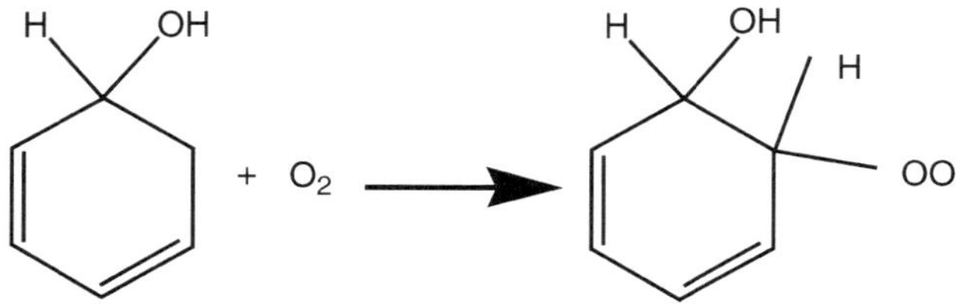

(79)

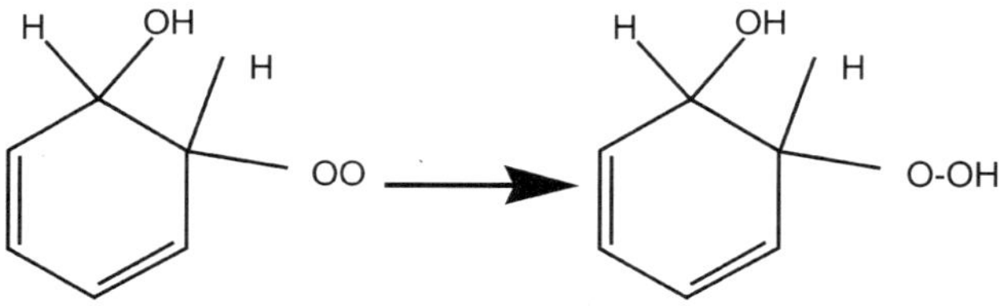

(80)

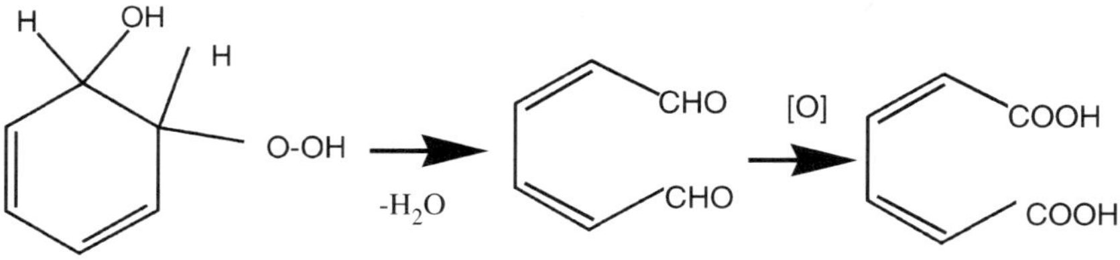

(81)

With decomposition of the aromatic ring, continued oxidation leads to mineralization. However, irradiation of benzene with insufficient O_2 may initially produce a more toxic solution due to phenol formation.

Phenol is also attacked by OH· radicals, and continues to degrade. An example of aqueous phenol radiolysis is given by Getoff [22], who used ^{60}Co γ-rays to produce polyhydroxybenzenes. This again illustrates the importance of a complete understanding of reaction mechanisms. Application of a radiation dose sufficient to destroy a target compound may not be sufficient for total treatment of associated by-products. Increased oxygen concentrations increased phenol radiolysis efficiency in Getoff's work, presumably due to ring-opening reactions via peroxyl intermediates. Continued oxidation in the presence of O_2 predicts mineralization of the target to carbon dioxide. However, various carboxylic acids were the empirically determined products.

PAHs also react with •OH. Removal of PAHs from the atmosphere by photolytic production of •OH may be an important natural remediation mechanism. Because these compounds have limited water solubility, most studies have investigated gas phase reactions. Naphthalene was shown to be subject to a complex series of hydroxylations and peroxyl-induced ring-opening reactions leading to the production of organic acids [37]. Although PAHs have low water solubility, they are often important water pollutants, attached to particles or colloids suspended in solution, or in aqueous sediments. PCBs have been shown to be susceptible to •OH attack, resulting in dechlorination [38].

Many aromatic compounds, including benzene, also have high rates of reaction with the solvated electron. However, the reaction does not typically lead to decomposition of the target compound. Instead, a radical anion is produced from the parent species, as shown in Eq. (82):

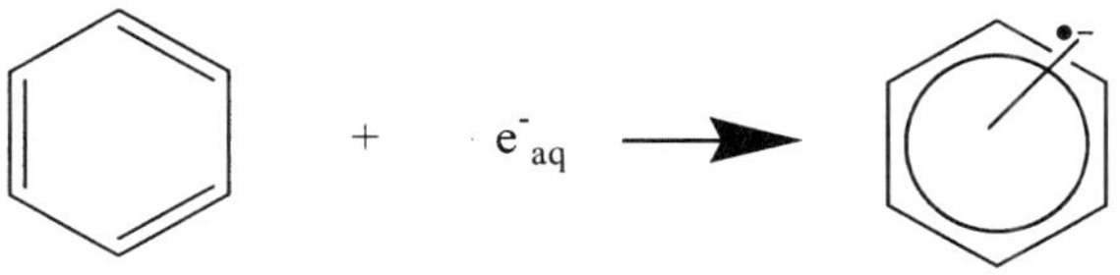

(82)

The radical anion can participate in electron transfer reactions, in which the electron is transferred to a compound with suitable electron affinity. The parent aromatic compound is regenerated. Such a reaction is shown in Eq. (83), where chlorobenzene is dechlorinated by electron transfer from the benzene radical anion:

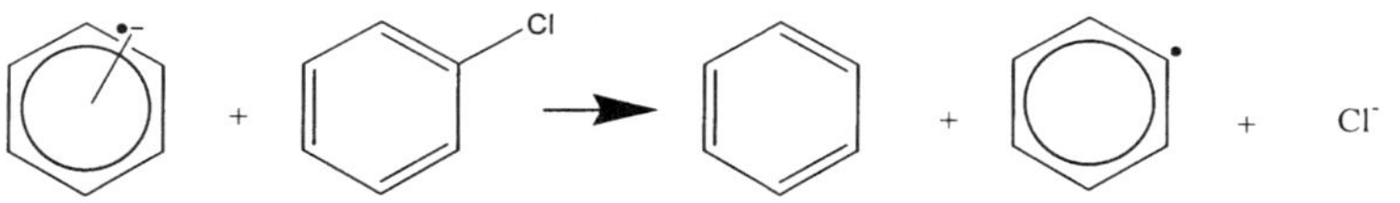

(83)

Chlorinated aromatics such as chlorobenzene may also be dechlorinated by solvated electron capture as shown in Eq. (84). Aromatic halogenated organics are readily reductively degraded by electron capture. Typically, sequential dechlorination occurs until the aromatic skeleton remains in unhalogenated form. The chlorine-containing product is HCl.

(84)

D. Chelating Agents

The previous sections have discussed the radiolytic decomposition of pollutants in water and wastewater. Other applications are possible. Chelating agents were often used in various processes designed to separate radionuclides at nuclear laboratories. The presence of these compounds in the resulting nuclear wastes complicates treatment. The radiolytic degradation of these compounds is an area of current investigation [39].

IV. COMPETITION FROM SCAVENGERS

An important consideration in extending laboratory data to natural waters is the effect of radical scavengers on the removal of the solute of interest. Many naturally occurring species react with their own second-order rate constants, with the reactive species produced in irradiated water. For site remediation, these are generally the natural constituents of the water, while for industrial treatment they may be other organic chemicals not targeted for treatment. The following are the common constituents of natural waters that may affect the efficiency of radiolytic water treatment.

A. pH

The *G* values shown in Eq. (2) for the production of •OH, •H, and e^-_{aq} are those measured in neutral water. At low pH, the electron is scavenged by hydrogen ion, to produce the hydrogen atom. At high pH, •OH dissociates to produce hydrogen ion and oxygen anion [40]. Solution pH also affects •OH concentration through its influence on alkalinity, as discussed below.

B. Carbonate/Bicarbonate Alkalinity

A common •OH scavenger in natural waters is alkalinity. Alkalinity is a measure of the total carbonate concentration. This is complicated by the equilibrium that exists in natural waters produced by the dissolution of atmospheric CO_2 to produce carbonic acid, and the dissociation of carbonic acid.

$$CO_2 + H_2O \rightarrow H_2CO_3 \quad (85)$$

$$H_2CO_3 + H_2O \rightarrow H_3O^+ + HCO_3^- \quad (86)$$

$$HCO_3^- + H_2O \rightarrow H_3O^+ + CO_3^{-2} \quad (87)$$

As can be seen from Eq. (87), which has an equilibrium constant of 4.69×10^{-11}, the carbonate/bicarbonate ion ratio is quite different in various waters depending on pH. For example, in neutral water the ratio of carbonate to bicarbonate is found from Eq. (88):

$$4.69 \times 10^{-11} = 1 \times 10^{-7} \times [CO_3^{-2}]/[HCO_3^-] \quad (88)$$

The result is 0.00047. Bicarbonate is clearly dominant. However, at pH 10, the ratio is 0.47, meaning the two ions are about equally abundant.

The relative effects of these ions on radical scavenging can be found from the equations:

$$\bullet OH + HCO_3^- \rightarrow H_2O + \bullet CO_3^- \quad (89)$$

$$\bullet OH + CO_3^{-2} \rightarrow OH^- + \bullet CO_3^- \quad (90)$$

and the appropriate second-order reaction rate constants. Carbonate ion is the more important hydroxyl radical scavenger. Thus, at unchanged alkalinity, •OH scavenging is more severe in higher pH waters.

Whether the carbonate radical ion, $\bullet CO_3^-$, product of Eqs. (89) and (90), reacts with many solutes is unknown. Therefore, the scavenging of •OH by alkalinity must be considered a net loss of reactive species. It is possible that some solutes are removed by reaction with the carbonate radical anion, but further studies are necessary to determine its effects.

C. Oxygen

Both e^-_{aq} and H· rapidly reduce O_2 to form $O_2\bullet^-$ with second-order rate constants of 1.9×10^{10} and 2.1×10^{10} M^{-1} s^{-1}, shown in Eqs. (11) and (12), respectively [3]. A dose of 10 kGy produces 60 μM H· and 270 μM of e^-_{aq}, in pure water. A dissolved oxygen concentration of 3.7 mg L^{-1} (120 μM) would scavenge approximately a third of the reducing species. In general, processes relying on reductive reactions for treatment are more efficient in the absence of oxygen. Batch irradiations in closed systems rapidly deplete dissolved oxygen, following which reductive attack of target species occurs.

The superoxide radical anion product of oxygen reduction is a reducing agent that may react with some solutes. However, it is relatively inert compared to solvated electrons and the reducing power of an irradiated solution is significantly decreased in the presence of oxygen.

D. Nitrate Ion

The presence of nitrate ion (NO_3^-) in water may effect solute removal efficiency by acting as an e^-_{aq} scavenger. The product, after a series of reactions [Eqs. (91)–(93)], is nitrite, NO_2^- [3].

$$NO_3^- + e^-_{aq} \rightarrow NO_3^{2-} \tag{91}$$

$$NO_3^{-2} + H_2O \rightarrow NO_2 + 2OH^- \tag{92}$$

$$2NO_2 + H_2O \rightarrow NO_2^- + NO_3^- + 2H^+ \tag{93}$$

Thus, decomposition of solutes by reductive means may be suppressed in high nitrate waters. However, the effective concentration of the •OH is increased (by minimizing recombination of the e^-_{aq} and •OH) and oxidative reactions may be enhanced.

The NO_2^- ion is known to react with the •OH and may then result in the nitration of organic solutes. For example, Cooper [41] showed that nitrobenzene was produced in irradiated aqueous benzene solutions containing high nitrate concentrations.

Nitrite ion is an undesirable by-product of radiolysis in waters containing high nitrate concentrations, and regulatory limits exist regarding its acceptable concentration. The presence of nitrate does not preclude radiolytic water treatment. One approach that compensates for nitrite generation in high nitrate waters is ozone addition, discussed later.

E. Dissolved Organic Carbon

Another common component in natural waters is the poorly characterized fraction referred to as dissolved organic carbon (DOC). As a mixture of compounds, there is no data on the reactions of either e^-_{aq} or H•. The reaction of DOC with •OH has been evaluated by Westerhoff et al. [42], and earlier by S. Peyton (Illinois Department of Energy and Natural Resources, Champaign, IL, personal communication). Because of the varied nature of DOC in various waters, it is likely that each source may have slightly different reactivity towards the reactive species. However, it has been shown that DOC does adversely affect the removal efficiency of organic solutes because of the scavenging of •OH.

V. KINETIC MODELING

The underlying assumption for developing a kinetic model is that it will provide a tool to assist engineers in determining important parameters of chemical degradation and the fate of the chemicals in the process [43]. In addition, it serves as a guide to minimize the number of experiments required to obtain the necessary empirical information.

Models have employed a computer code called MAKSIMA-CHEMIST provided by Atomic Energy of Canada [44,45]. The input to the kinetic model includes a list of all known reacting species, their initial empirically obtained concentrations, and the appropriate second-order rate constants. In addition to target solutes, scavengers that exist in the natural water and reaction by-products that may act as scavengers need to be included. The utility of the model depends upon the ability to account for all the existing reactions with proper rate constants and on the accuracy of the measured solute concentrations.

Attempts to model single-solute systems have met with reasonable success. The destruction of CCl_4 has been modeled using the following equations and rate constants:

$$CCl_4 + e_{aq}^- \rightarrow \bullet CCl_3 + Cl^- \quad (1.3 \times 10^{10}\ L\ mol^{-1}\ s^{-1}) \tag{94}$$

$$CCl_4 + H\bullet \rightarrow \bullet CCl_3 + Cl^- + H^+ \quad (3.2 \times 10^{7}\ L\ mol^{-1}\ s^{-1}) \tag{95}$$

$$\bullet CCl_3 + O_2 \rightarrow CCl_3OO\bullet \quad (3.3 \times 10^{9}\ L\ mol^{-1}\ s^{-1}) \tag{96}$$

$$\bullet CCl_3 + \bullet CCl_3 \rightarrow Cl_3CCCl_3 \quad (3.7 \times 10^{8}\ L\ mol^{-1}\ s^{-1}) \tag{97}$$

The kinetic model accurately predicts carbon tetrachloride concentration changes over a range of concentrations and solution pH [46]. In experiments conducted at large scale using the Miami Electron Beam Research Facility (EBRF), methanol was added as an •OH scavenger. Methanol scavenges •OH according to Eq. (98):

$$CH_3OH + \bullet OH \rightarrow \bullet CH_2OH + H_2O \quad (9.7 \times 10^{8}\ L\ mol^{-1}\ s^{-1}) \tag{98}$$

The effects of hydroxyl radical scavenging by methanol were predicted accurately. For more complex compounds and for high concentrations, a detailed knowledge of the reaction mechanism is necessary to obtain good agreement between empirical and modeled results.

VI. OPERATIONAL EXPERIENCE

Two possible radiation sources are available for water treatment. One is isotope gamma-ray sources, and the other is machine-generated e-beams, or the bremsstrahlung produced by colliding the e-beam on a suitable target. Each source has advantages and disadvantages. Isotopes are convenient and uncomplicated sources of radiation and by far, most experimental work has been done using isotope gamma-ray sources, especially ^{60}Co. They are ideally suited to small-scale, batch irradiations. Experiments often involve timed irradiations, scheduled to provide selected exposures based on a known dose-rate. A number of such irradiations may be performed to evaluate kinetic behavior and to determine dose constants or D_{10} values. Accurate dosimetry is essential, and many systems have been employed [47–50].

However, on a process scale isotopes suffer the disadvantages of low dose rates, which limit process flow. A large cobalt source, for example, delivers about 10 kGy hr^{-1}, while rates as high as 1 kGy s^{-1} are common with accelerators. A further problem of isotopes is that the common ^{60}Co source has a 5.27-year half-life. It thus loses 12.5% of its activity annually. Economic analyses have shown that when cobalt replacement costs are accounted for, e-beams are more cost-effective sources for generating the absorbed doses usually necessary for treatment [51].

The combination of factors described above result in a situation where isotope irradiations are usually used to do research, and those results are used for scale-up to process e-beam irradiations.

Occasionally, scale-up may introduce complications not anticipated by experiments. For example, batch irradiations using sealed sample containers favor reducing conditions. In the first few gray of irradiation the natural dissolved oxygen content of the samples is quickly reduced to relatively inert superoxide [Eqs. (11) and (12)], following which the electron population available for reaction with solutes climbs rapidly. By contrast, a process water stream is likely to be irradiated in air. Whether reducing or oxidizing conditions are preferred depends on the nature of the solute to be decomposed. Either option, or both in sequence, may be engineered into an actual system.

A more fundamental difference between isotope and e-beam sources is dose rate. Whereas the high dose rates of radiation provided by e-beams are necessary for cost-effective water treatment, they also introduce complications resulting from the very high radical concentrations produced. High radical concentrations favor radical/radical recombination, resulting in a loss of reactive species. Gehringer [52] has shown a departure from pseudo-first-order kinetics in such situations, due at least in part to dose rate.

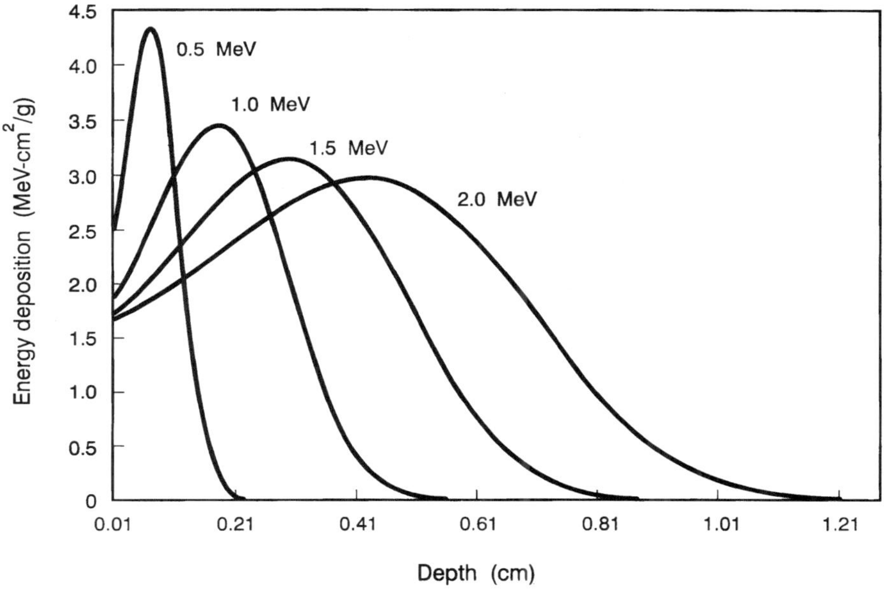

Figure 3 Energy deposition for water irradiated with electrons of various energies.

Further, the increased prevalence of radical/radical reactions has the potential to alter by-product distributions, possibly including the production of undesirable chlorinated products. Proposed treatment regimes should always evaluate by-product generation.

A further difference between isotope gamma rays and electrons is penetrating ability. As charged particles, electrons have short ranges in matter. Fig. 3 shows energy deposition curves for water irradiated with electrons of various energies.

An interesting feature of e-beam systems is the various solutions to this problem that have been engineered. Two of these are discussed below.

A. Miami Electron Beam Research Facility

Originally constructed to irradiate wastewater and sludge for bacterial disinfection, the Electron Beam Research Facility (EBRF) in Miami, FL is the only large-scale e-beam treatment facility in the world [49]. Since 1988, process-scale irradiations have been performed on a large number of solutes. These experiments have included mixtures of solutes in natural waters to simulate anticipated process conditions.

The facility consists of a 1.5-MeV accelerator (Vivirad-High Voltage Corp.) driven by an insulated core transformer (ICT) power supply. The beam current is continuously variable between 0 and 50 mA. The beam may be scanned to provide uniform, continuous irradiation of an area of 122×7.6 cm, and is directed onto a stream of water falling over a weir. At the design flow of 460 L min^{-1}, the thickness of the irradiated falling water stream is 0.38 cm, which results in less than 100% energy deposition of the 1.5-MeV electrons. The absorbed dose is determined by measuring the temperature rise of the water stream [49]. Pre- and postirradiation grab samples are taken from the facility control room. A floor plan for the facility is shown in Fig. 4.

The EBRF can be used in continuous mode to irradiate influent wastewater, or in batch mode while receiving influent from tanker trucks. When operating in continuous mode, experimental contaminants are added to the process stream of wastewater or potable water via injection with positive displacement pumps at a rate chosen to provide a desired concentration. For batch experiments, solutes are dissolved in approximately 20,000-L tank trucks, with mixing by submersible pumps.

Among the many solutes investigated at the EBRF are benzene, toluene, and xylenes [53]. These solutes were irradiated at different initial concentrations in batches of different water types. Decomposition of the

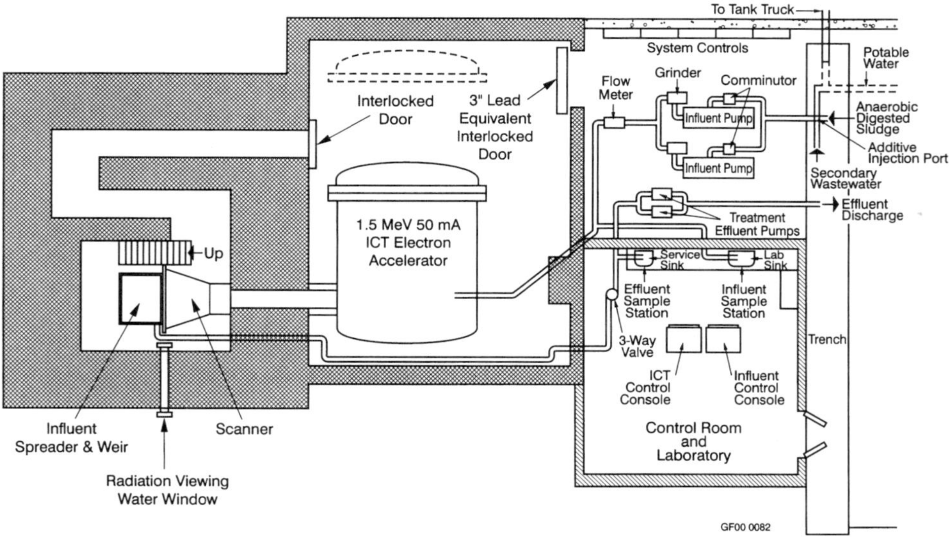

Figure 4 The Miami Electron Beam Research Facility (EBRF).

Table 4 Pseudo-First-Order Dose Constants for the e-Beam Irradiation of Aromatic Compounds in Water (1) and Wastewater (2)

		k_0 (kGy^{-1})	
Solute	Concentration (μM)	1	2
Benzene	1	0.84	0.26
	17	0.31	0.16
Toluene	1.2	0.61	0.38
	17	0.22	0.22
m-Xylene	1	0.64	0.43
	12	0.11	0.13
o-Xylene	1.1	0.62	0.40
	12	0.12	0.12

solutes was found to observe pseudo-first-order kinetics, facilitating the reporting of results as dose constants. Table 4 shows dose constants for benzene, toluene, and xylenes in Miami potable water and in secondary wastewater. Wastewater contained higher alkalinity (but lower pH), higher nitrate, and higher DOC than potable water.

In can be seen from Table 4 that wastewater has similar or lower dose constants for solute degradation than potable water. This was expected due to the higher concentrations of radical scavengers in wastewater. Decomposition of all the studied solutes occurred with higher efficiency at low solute concentrations. At higher solute concentrations, only benzene showed a decreased dose constant in wastewater. This may indicate successful competition for radicals by the higher solute concentration.

To determine the influence of various radicals and radical scavengers on the irradiation of these aromatic solutes, an attempt to model these results may be made using the methods shown in Eqs. (17)–(19). Pseudo-first-order rate constants were calculated and used to determine the percentage of removal of each solute due to each reactive species. The results are shown for pure water in Table 5.

Assuming the usual initial distribution of the three reactive species, it is obvious that •OH radical plays the major role in the decomposition of these solutes. Toluene is also significantly attacked by hydrogen atoms. These data may help to explain the effects shown in Table 4. Benzene decomposition was most severely affected by the changing water quality. This may be due to almost complete reliance on •OH radical attack, and the presence of a sixfold higher DOC concentration competing for •OH in wastewater.

Table 5 Percent Removal of Solutes in Pure Water Due to Individual Reactive Species

Solute	OH (%)	e^{-}_{aq} (%)	H (%)
Benzene	97.5	0.1	2.4
Toluene	83.5	0.4	16.1
m-Xylene	92.9	0.1	7.0
o-Xylene	93.7	0.1	6.2

Toluene and xylenes were removed with similar efficiency at low concentration. At high concentration, toluene was removed at a rate twice that of the xylenes, perhaps due to increased participation by the •H atom. For high concentrations of toluene and xylenes, •OH scavenging may have been compensated for by •H atom attack. This may then have become the dominant mechanism of solute decomposition. Information such as this may be important in predicting by-product distributions when treating water of varying quality.

Surprisingly, in separate experiments, the removal of benzene and toluene from groundwater were relatively unaffected by solution pH [54], despite the fact that the groundwater contained carbonate/bicarbonate alkalinity. It was postulated that the carbonate radical anion [product of Eqs. (89) and (90)] may also degrade these solutes at high solution pHs.

An extensive analysis of the by-products of these solutes was conducted using intermediate doses [53]. This resulted in the production of phenol and the dihydroxy products hydroquinone, catechol, and trace resorcinol. *o*-Cresol was the only hydroxylated product of toluene radiolysis. These were intermediates, however, and continued irradiation produced a variety of more highly oxidized products. Among them were low molecular weight aldehydes, which also decomposed upon continued irradiation. Both solutes generated varying amounts of formaldehyde, acetaldehyde, glyoxal, and methylglyoxal (toluene only) depending on pH. However, all identified products accounted for only about 10% (depending on dose) of the original carbon, suggesting mineralization to CO_2 and water.

Trichloroethylene and PCE have also been irradiated on a process scale at the EBRF [55]. Unlike in the aromatic solute experiments above, increasing pH necessitated increased radiation requirements. Formic acid and smaller amounts of formaldehyde, acetaldehyde, and trace glyoxal were the detectable products. No detectable chloroacetic acids were reported, indicating that if produced they were decomposed by continued irradiation.

Table 6 Typical EE/O Values (and Range) for Contaminant Destruction by the e-Beam Process

Contaminant	EE/O (kWh/1000 U.S. gal/order)
Benzene	0.5–3
Toluene	0.4–3
Xylene(s)	1–3
Phenol	0.4–7
Methylene blue (dye)	0.2–2
Trichloroethylene	0.5–2
Tetrachloroethylene	2–6
Chloroform	2–12
Carbon tetrachloride	0.5–5
Vinyl chloride	0.5–2

Postirradiation chloride analyses indicated complete recovery of solute chlorine as inorganic chloride ion.

Some values of EE/O using the EBRF at a flow rate of approximately 100 gal/min for important pollutants in potable water are shown in Table 6:

B. Austrian Research Center, Seibersdorf Facility

Continuous water irradiation by e-beam is conducted on a bench scale at the Austrian Research Center, Seibersdorf. A 500-keV, 25-mA ICT accelerator (Vivirad-High Voltage Corp.) is used as the electron source [52]. A 3-mm horizontal layer of water is irradiated. Low penetration by the lower energy electrons produced by this smaller accelerator is compensated for by irradiating a turbulent water flow. Dose distribution in the turbulent stream is not uniform, but the overall volume of water treated to an average dose is increased. A schematic of the system is shown in Fig. 5.

Research using gamma-ray irradiation at Seibersdorf has reported pseudo-first-order decomposition of TCE and PCE [17,18]. However, a significant departure from pseudo-first-order kinetics was noted with the e-beam system. This may be due to dose rate effects (1890 Gy s^{-1} maximum vs. 1.5 Gy s^{-1} for ^{60}Co), or due to nonuniform dose distribution, or a combination of both [52]. Fig. 6 shows the departure from first-order behavior seen when the e-beam system is used as the source for the irradiation of drinking water containing TCE.

Because water radiolysis produces both oxidizing and reducing species in equal amounts, the inevitable reactions of these species with each other

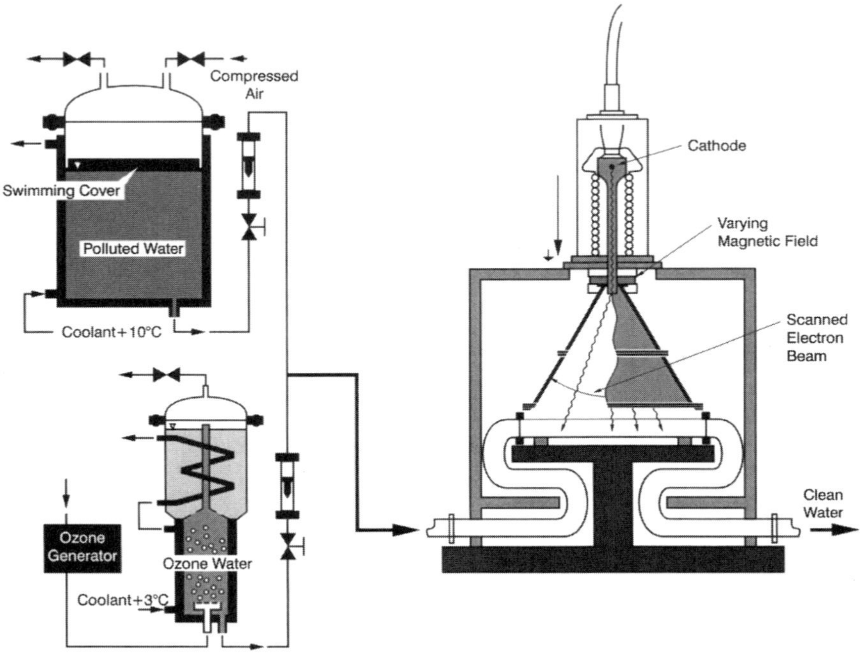

Figure 5 The Seibersdorf Electron Beam Research Facility.

waste e-beam energy, even in scavenger-free waters. Also, because the pollutant degradation mechanisms discussed previously are predominantly oxidative, the Seibersdorf researchers decided that removal of solvated electrons to create a totally oxidizing system would likely increase treatment efficiency. The method chosen to scavenge electrons was ozone addition, shown below:

$$O_3 + e^-_{aq} \rightarrow O_3^- \quad (99)$$

The reaction shown in Eq. (99) is fast, with a bimolecular rate constant of 3.6×10^{10} L mol^{-1} s^{-1}, and the product is the ozonide anion. Ozonation has an additional benefit. Ozonide decomposes [shown in Eqs. (100) and (101)] to hydroxyl radical and oxygen [56], increasing the oxidative power of the solution by a second mechanism:

$$O_3^- + H^+ \rightarrow HO_3 \quad (100)$$

$$HO_3 \rightarrow O_2 + {}^\bullet OH \quad (101)$$

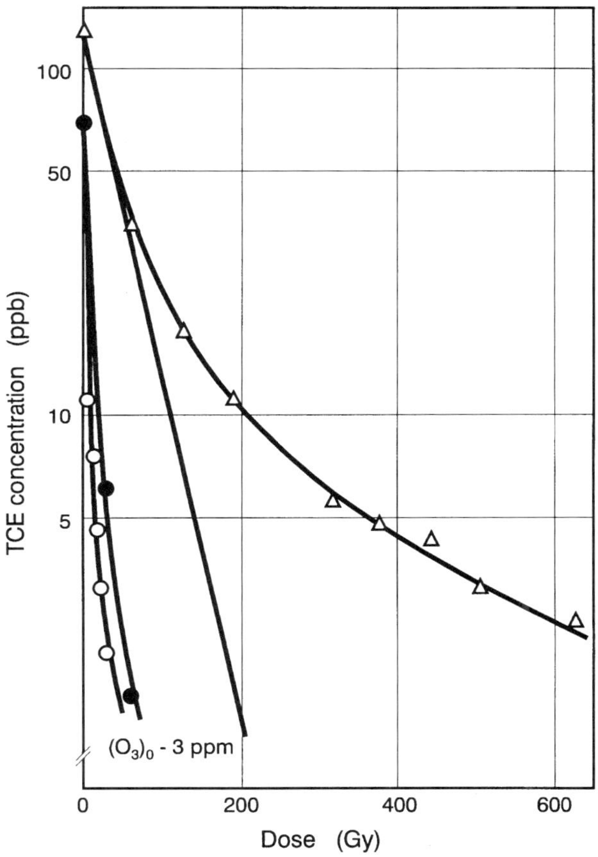

Figure 6 The change in TCE concentration vs. dose using: Δ electron beam alone; – γ-irradiation alone: acid, ● ozone-electron beam; and ○ ozone-γ-irradiation. Note the departure from first-order behavior using the e-beam.

Oxygen also scavenges electrons [Eq. (11)] to produce superoxide, which may also act as an electron-transfer agent initiating decomposition of ozone to hydroxyl radical:

$$O_2^- + O_3 \rightarrow O_3^- + O_2 \tag{102}$$

The addition of ozone before radiolysis thus produces an overwhelmingly oxidizing system. This is probably the most powerful oxidizing system produced by any of the advanced oxidation technologies currently proposed for water treatment [52].

Table 7 The Effect of Ozone Addition on Solute Decomposition in Irradiated Water

	D_{10} (Gy)		
Water type	No O_3	3 ppm O_3	5 ppm O_3
Demineralized	30	—	—
Vienna	165	30	25
Seibersdorf	220	115	65

Gehringer irradiated TCE contaminated waters in the presence and absence of ozone [52]. The use of ozone dramatically increased the rate of 100 ppb TCE decomposition, as shown in Fig. 6 and in Table 7. It was also found that the D_{10} (and thus dose constant) values were independent of initial TCE concentration over a wide range, indicating that the mechanism of decomposition was constant. Thus, the factors for reduction in D_{10} shown in Table 7 are useful predictive process parameters.

Table 7 also illustrates the effect of water quality on radiation requirements. Drinking water at Seibersdorf contains higher amounts of bicarbonate and nitrate than does water at Vienna. The absence of scavengers in the demineralized water resulted in far lower radiation requirements.

The by-products of the combined ozone/e-beam process were also investigated and were found similar to the pure e-beam process. An additional benefit of ozonation is the oxidation of undesirable nitrite to nitrate.

VII. CONCLUSIONS

In conclusion, the electron beam treatment process has several advantages that make it suitable for use as an ultimate treatment process for hazardous organic chemicals. The process is *nonselective* in the *destruction of organic chemicals* because strongly reducing reactive species (e^-_{aq}/H·), and strongly oxidizing reactive species (OH·), are formed at the same time and in approximately the same concentration in solution. This is important because either single compounds or complex mixtures of organic compounds can be destroyed simultaneously to the treatment objectives. Additionally, if the exact nature of the waste changes with time it will not adversely affect the process.

Reactions of organic compounds with the electron-beam-induced reactive species are very rapid, and occur in less than a second. This allows the design of a *flow-through system* with good *process flexibility* at full scale in the event the flows vary with time. For example, the full-scale system would be modular in design, thereby allowing for decreased operational cost if the quality of the waste improves over time (i.e., the concentration of organic compounds decrease). The process can effectively treat *aqueous streams* and slurried *soils*, *sediments*, and *sludges*.

The formation of the three reactive species of water radiolysis is *pH independent* in the range 3–11, and thus changes in pH of the feedstock over time do not affect process efficiency, except in cases where the pollutant itself is pH sensitive.

The process is *temperature independent*. Both ambient temperature and water temperature may vary over time. The electron beam system would be housed in a building and except for the control room, no temperature conditioning is required. Also, variations in water temperature have no effect on the treatment efficiency of the process.

The process produces *no air emissions*. Because this is an aqueous-based technology, no NO_x or SO_x are produced. The only volatilization of contaminants that occurs is that associated with moving water. A design using laminar flow minimizes loss of organics by this mechanism.

Because of the above, the electron beam process can be used efficiently and effectively as *pretreatment for biological processes*. The electron beam process can "break apart" complex organic compounds, making them available for microbiological degradation.

However, there are as yet no commercial facilities using the technology. There are probably two main reasons for this: (1) the process is innovative, and there has been no single problem where the potential costs of using an innovative process outweigh the potential risks to the design engineer and (2) the capital costs of an accelerator are high, and the capital "payback" is long.

ACKNOWLEDGMENTS

WJC acknowledges NSF grant BES 97-29965 for support of this research.

REFERENCES

1. Bolton JR, Valladares JE, Cooper WJ, Waite TD, Kurucz CN, Nickelsen MJ, Kajdi DC. Figures-of-merit for advanced oxidation processes—a comparison of homogenous UV/H_2O_2, heterogeneous TiO_2 and electron beam processes. J AOTs Adv Oxid Technol 1998; 3:174–181.

2. Cooper WJ, Curry RD, O'Shea KE, Eds. Environmental Applications of Ionizing Radiation. New York: John Wiley and Sons, 1998: 722.
3. Buxton GV, Greenstock CL, Helman WP, Ross AB. Critical review of rate constants for reactions of hydrated electrons, hydrogen atoms and hydroxyl radicals (•OH/•O−) in aqueous solution. J Phys Chem Ref Data 1988; 17:513–886.
4. Anbar M, Bambenek M, Ross AB. Selected specific rates of reactions of transients from water in aqueous solution. I. Hydrated electron. Natl Stand Ref Data Ser, Natl Bur Stand (US) 1973; 43: 1998; 43(Suppl): 67 pp.
5. Ross AB. Selected specific rates of reaction of transients from water in aqueous solution. Hydrated electron, supplemental data. Natl Stand Ref Data Ser, Nat Bur Stand (US) 1973; 43(Suppl): 43 pp.
6. Hart EJ, Anbar M. The Hydrated Electron. New York: Wiley-Interscience, 1970: 267.
7. Anbar M, Farhataziz, Ross AB. Selected specific rates of reaction of transients from water in aqueous solution. II. Hydrogen atom. Natl Stand Ref Data Ser, Natl Bur Stand (US) 1975; 51:56.
8. Dorfman LM, Adams GE. Reactivity of the hydroxyl radical in aqueous solution. Natl Stand Ref Data Ser, Natl Bur Stand (US) 1973; 46:59.
9. Tobien T, Cooper WJ, Nickelsen MG, Pernas E, O'Shea KE, Asmus K-D. Odor control in wastewater treatment: the removal of thioanisole from water—a model case study by pulse radiolysis and electron beam treatment. Environ Sci Technol 2000; 34:1286–1291.
10. Mincher BJ, Curry RD. Considerations for choice of a kinetic figure of merit in process radiation chemistry for waste treatment. Appl Radiat Isot 2000; 52:189–193.
11. Rezansoff BJ, McCallum KJ, Woods RJ. Radiolysis of aqueous chloroform solutions. Can J Chem 1970; 48:271–276.
12. Dickson LW, Lopata VJ, Toft-Hall A, Kremers W, Singh A. Radiolytic removal of trihalomethanes from water. Proceedings from the 6th Symposium on Radiation Chemistry, 1986, 173–182.
13. Cooper WJ, Cadavid EM, Nickelsen MG, Lin K, Kurucz CN, Waite TD. Removing THMs from drinking water using high-energy electron-beam irradiation. J Am Water Works Assoc 1993; 85:106–112.
14. Getoff N. Advancements of radiation induced degradation of pollutants in drinking and waste water. Appl Radiat Isot 1989; 40:585–594.
15. Gehringer P, Proksch E, Szinovatz W, Eschweiler H. Der strahlenchemische Abbau von Trichloraethylen- und Perchloraethylenspuren in Trinkwasser. Z Wasser-Abwasser-Forsch 1986; 19:186–203.
16. Gehringer P, Proksch E, Szinovatz W, Eschweiler H. Decomposition of trichlorethylene and tetrachloroethylene in drinking water by a combined radiation/ozone treatment. Water Res 1988; 22:645–646.
17. Gehringer P, Proksch E, Szinovatz W, Eschweiler H. Radiation-induced decomposition of aqueous trichloroethylene solutions. Appl Radiat Isot 1988; 39:1227–1231.

18. Proksch E, Gehringer P, Szinovatz W, Eschweiler H. Radiation-induced decomposition of small amounts of perchloroethylene in drinking water. Appl Radiat Isot 1987; 38:911–919.
19. Proksch E, Gehringer P, Szinovatz W, Eschweiler H. Decomposition of chlorinated ethylenes in drinking water by combined ozone/radiation treatment. International Ozone Symposium, Berlin, 1989.
20. Gehringer P, Proksch E, Szinovatz W. Radiation-induced degradation of trichloroethylene and tetrachloroethylene in drinking water. Int J Appl Radiat Isot 1985; 36:313–320.
21. Koester R, Asmus K-D. Pulse radiolysis studies of halogenated organic compounds in aqueous solution. Proceedings of 3rd Tihany Symposium on Radiation Chemistry, Budapest, 1972, 1095–1106.
22. Getoff N. Radiation induced decomposition of biological resistant pollutants in water. Appl Radiat Isot 1986; 37:1103–1109.
23. Neta P, Dorfman LM. Pulse radiolysis studies. XIII. Rate constants for the reaction of hydroxyl radicals with aromatic compounds in aqueous solutions. Adv Chem Ser 81. Washington, DC: American Chemical Society, 1968:222–230.
24. Vysotskaya NA, Bortun LN, Rekasheva AF. Radiation–chemical conversions of condensed aromatic hydrocarbons in aqueous solutions. Presented at the 5th Symposium on Radiation Chemistry, L.V. Pisarzhevsky Institute of Physical Chemistry, Ukrainian SSR Academy of Sciences, Kiev, USSR, 1982.
25. Phung PV, Burton M. Radiolysis of aqueous solutions of hydrocarbons, benzene, benzene-d_6, cyclohexane[I]. Radiat Res 1957; 7:199–216.
26. Goodman J, Steigman J. Products of the radiolysis of water containing benzene and air. J Am Chem Soc 1958; 62:1020–1022.
27. Stein G, Weiss J. Chemical actions of ionizing radiations on aqueous solutions. Part II. The formation of free radicals. The action of x-rays on benzene and benzoic acid. J Chem Soc 1949; 3245–3254.
28. Daniels M, Scholes G, Weiss J. Chemical action of ionizing radiations in solution. Part XV. Effect of molecular oxygen in the irradiation of aqueous benzene solutions with x-rays. J Chem Soc 1956; 832–834.
29. Michael BD, Hart EJ. The rate constants of hydrated electron, hydrogen atom, and hydroxyl radical reactions with benzene, 1,3-cyclohexadiene, 1,4-cyclohexadiene, and cyclohexene. J Phys Chem 1970; 74:2878–2884.
30. Schested K, Corfitzen H, Christensen HC, Hart EJ. Rates of reaction of O^-, OH, and H with methylated benzenes in aqueous solution. Optical spectra of radicals. J Phys Chem 1975; 79:310–315.
31. Land EJ, Ebert M. Pulse radiolysis studies of aqueous phenol. Water elimination from dihydroxycyclohexadienyl radicals to form phenoxy. Trans Faraday Soc 1967; 63:1181–1190.
32. Hashimoto S, Miyata T, Washino M, Kawakami W. A liquid chromatography study on the radiolysis of phenol in aqueous solution. Environ Sci Technol 1979; 13:71–75.

33. Getoff N, Solar S. Radiation induced decomposition of chlorinated phenols in water. Radiat Phys Chem 1988; 31:121–130.
34. Daniels M, Scholes G, Weiss J. Chemical action of ionizing radiations in solution. Part XV. Effect of molecular oxygen in the irradiation of aqueous benzene solutions with x-rays. J Chem Soc 1956; 832–834.
35. Leoff I, Stein G. Radiation and photochemistry of aqueous solutions of benzene. J Chem Soc 1963; 2623–2633.
36. Mantaka A, Marketos DG, Stein G. Continuous gamma and pulse radiolysis of aqueous benzene solutions: some reactions of the hydroxycyclohexadienyl radical. J Phys Chem 1971; 75:3885–3889.
37. Bunce NJ, Liu L, Zhu J, Lane DA. Reaction of naphthalene and its derivatives with hydroxyl radicals in the gas phase. Environ Sci Technol 1997; 31:2252–2258.
38. Sedlak DL, Andren AW. Aqueous-phase oxidation of polychlorinated biphenyls by hydroxyl radicals. Environ Sci Technol 1991; 25:1419–1426.
39. Toste AP. Gamma-ray-induced destruction of nitrilotriacetic acid in a simulated, mixed nuclear waste: radiolytic and chemical forces. J AOTs Adv Oxid Technol 1998; 3:70–78.
40. Spinks JWT, Woods RJ. Introduction to Radiation Chemistry. 3rd ed. New York: Wiley-Interscience, 1990:262.
41. Cooper WJ, Dougal RA, Nickelsen MG, Waite TD, Kurucz CN, Lin K, Bibler JP. Benzene destruction in aqueous waste. I. Bench-scale gamma irradiation experiments. Radiat Phys Chem 1996; 48:81–87.
42. Westerhoff P, Aiken G, Amy G, DeBroux J. Relationship between the structure of natural organic matter and its reactivity towards molecular ozone and hydroxyl radical. Water Res 1999; 33:2265–2276.
43. Scwharz HA. Determination of rate constants for the radical processes in the radiation chemistry of water. J Phys Chem 1962; 66:255–263.
44. Crittenden JC, Hu S, Hand DW, Green SA. A kinetic model for the H_2O_2/UV process in a completely mixed batch reactor. Water Res 1999; 33:2315–2328.
45. Carver MB, Hanlet DV, Chapin KR. MAKSIMA-CHEMIST, A program for mass action kinetic simulated manipulation and integration using stiff techniques, Chalk River Nuclear Laboratories Report, Atomic Energy Canada Ltd. 6413, 1979:1–28.
46. Mak FT, Zele S, Cooper WJ, Kurucz CN, Waite TD, Nickelsen MG. Kinetic modeling of carbon tetrachloride, chloroform, and methylene chloride removal from aqueous solution using electron beam process. Water Res 1997; 31:219–228.
47. Mincher BJ, Zaidi MK. Calibration of Far West Technology (FWT-60) radiochromic dye dosemeters. Radiat Prot Dosim 1993; 47:1/4.
48. Mincher BJ, Zaidi MK, Arbon RE, McGlaughlin WL, Schwendiman GL. Calibration and performance of Gafchromic DM-100 radiochromic dosemeters. Radiat Prot Dosim 1996; 66:1/4.
49. Kurucz CN, Waite TD, Cooper WJ. The Miami electron beam research facil-

ity: a large scale wastewater treatment application. Radiat Phys Chem 1995; 45:299–308.
50. Spinks JWT, Woods RJ. . Introduction to Radiation Chemistry. 3rd ed. New York: Wiley-Interscience, 1990; 71–126.
51. Cleland MR, Fernald RA, Maloof SR. Electron beam process design for the treatment of wastes and economic feasibility of the process. Radiat Phys Chem 1984; 24:179–190.
52. Gehringer P, Proksch E, Eschweiler H, Szinovatz W. Remediation of groundwater polluted with chlorinated ethylenes by ozone-electron beam irradiation treatment. Appl Radiat Isot 1992; 43:1107–1115.
53. Nickelsen MG, Cooper WJ, Kurucz CN, Waite TD. Removal of benzene and selected alkyl-substituted benzenes from aqueous solution using continuous high-energy electron irradiation. Environ Sci Technol 1991; 26:144–152.
54. Nickelsen MG, Cooper WJ, Lin K, Kurucz CN, Waite TD. High energy electron beam generation of oxidants for the treatment of benzene and toluene in the presence of radical scavengers. Water Res 1994; 28:1227–1237.
55. Cooper WJ, Meacham DE, Nickelsen MG, Lin K, Ford DB, Kurucz CN, Waite TD. The removal of tri- (TCE) and tetrachloroethylene (PCE) from aqueous solution using high energy electrons. J Air Waste Manag Assoc 1993; 43:1358–1366.
56. Glaze WH. Chemical Oxidation. Pontius FW. Water Quality and Treatment. 4th ed. New York: McGraw-Hill, 1990; 763.

8

Solvated Electron Reductions: A Versatile Alternative for Waste Remediation

Gerry D. Getman
Commodore Solution Technologies, Inc., Marengo, Ohio, U.S.A.

Charles U. Pittman, Jr.
Mississippi State University, Mississippi State, Mississippi, U.S.A.

I. INTRODUCTION

Polychlorinated biphenyls (PCBs) and other chlorinated aromatic compounds are distributed in soils, sludges, estuaries, etc., at over 400 sites in the United States alone. Chlorinated aliphatic hydrocarbons (CAHs), widely used for degreasing and cleaning of engines, auto parts, and electronic components, are serious contaminants at 358 major hazardous wastes sites in the United States. CAHs migrate vertically through soils to form dense non-aqueous phase liquids (DNAPLs) on aquifer bottoms. Chlorinated organics are also frequently found in mixed wastes (those containing radioactive contaminants). The Environmental Protection Agency (EPA)'s Emergency Response Notification System recorded almost 3600 accidents involving PCBs between 1988 and 1992. These facts highlight the need to develop methods to decontaminate soils, sludges, and aggregates containing chloroorganic compounds to include both ex situ and in situ methods [1–6]. Portable remediation methods that can be located at the contaminated site are needed to reduce the costs of transporting large volumes of soil to an off-site treatment location.

The remediation of soils and DNAPLs has been a high-priority research area at the EPA, Department of Energy (DOE), and Department of Defense (DOD). To give just one example, the DOE's Hanford site has massive soil and groundwater contamination from a carbon tetrachloride subsurface plume extending for over 70 square miles. More generally, PCBs, CAHs, dioxins, furans, halogenated pesticides, benzenes–toluenes–xylenes (BTX), explosives, chemical warfare agents, and chlorofluorocarbons (CFCs), which are all widely distributed in the environment, must be remediated to meet today's stringent standards. Vast quantities of soil, sludge, job equipment, adsorbents, process liquids, and building materials must be treated to remove these species, which may be present in parts-per-million (ppm) quantities.

Work at the Commodore Solution Technologies, Inc. (Commodore) and the Mississippi State University has now demonstrated a generalized solvated electron technology (SET) to decontaminate (in situ and ex situ) soils or sludges contaminated with PCBs, CAHs, CFCs, explosive wastes, and chemical warfare agents. Furthermore, bulk samples of these chemicals can also be degraded. The early patents of Weinberg et al. [7,8] and the reports by Pittman and Tabaei [9] and Pittman and Mohammed [10] proved that neat PCBs and PCB-contaminated soils containing up to 30% water could be decontaminated in liquid ammonia slurries when treated with Ca/NH_3 or Na/NH_3 at room temperature. PCB destruction efficiencies of >99.9% were achieved in only 30 sec. The products were biphenyl or reduced biphenyls and $CaCl_2$ or NaCl. The Commodore has developed a total systems approach to such remediation, called Solvated Electron Technology (SET™), and has received a nationwide EPA operating permit for the nonthermal destruction of PCBs in soils, oils, surfaces, and solid materials. The SoLV™ process is Commodore's total remediation process incorporating pretreatments and posttreatments applicable to liquids, solids, soils, protective equipment, and job materials. After discussing some basic chemical considerations, this chapter will provide an overview of the technology and specific examples of solvated electron remediations.

II. SOLVATED ELECTRON CHEMISTRY FOR ENVIRONMENTAL REMEDIATION: BACKGROUND AND FUNDAMENTALS

A. General Description

Deep blue solutions of solvated electrons are formed when Li, Na, K, Ca, or other group I and group II metals are dissolved into liquid ammonia (Eq. (1)). These media have long been used to reduce organic compounds. The widely used Birch reduction [11–18], known for 80 years, is employed

$$\mathrm{Na} \xrightarrow{NH_3} \mathrm{Na}^+ + e_s^- \quad (1)$$

routinely on a commercial scale to accomplish a variety of reductions. Among the many functional groups reduced by this process, chloroorganic compounds are the ones reduced at the highest rates (Eq. (2)). However, this chemistry was never applied to environmental (soil, sludge, DNAPLs) cleanups in the past because of the widely accepted belief that the solvated electron would react rapidly with water. Thus, it was thought that water in environmental samples would consume the solvated electrons, leading to prohibitive costs. This perception was put to rest in early studies that dealt with using solvated electrons for reducing contaminants in environmental samples [7–10]. Indeed, the reduction of water is much slower than dechlorination, allowing wet soils to be dechlorinated rapidly without undue consumption of metals (Na) through a reaction with water.

$$\mathrm{RCl} + e^- \longrightarrow \mathrm{R}^\bullet + \mathrm{Cl}^- \quad (2)$$

How can the use of solvated electron solutions to decontaminate soils, sludges, and DNAPLs be feasible in the presence of excess water? The reaction of solvated electrons with water (i.e., $e_{(s)}^- + H_2O \rightarrow 1/2H_2 + {}^-OH$) has a far higher kinetic barrier than electron transfer to chlorinated or nitrated organic molecules. Furthermore, when ammonia is present with water, the half-life of the solvated electron dramatically increases. In pure water, the half-life of the solvated electron is short ($t_{1/2} = \sim 100$ μsec) [19]. However, the transfer of solvated electrons to chlorinated organic compounds is much faster. For a 20% solution of water in ammonia, the half-life of the solvated electron increases to about 100 sec [20,21]. In pure ammonia, $t_{1/2} = \sim 300$ hr [21]. Thus, the desired detoxification reductions of chlorinated organic molecules will occur much faster than side reactions with water when ammonia is used as the solvent. The transfer of an electron to RCl occurs in ~1 μsec vs. the transfer to H_2O to give $1/2H_2$ (in 20% H_2O/80% NH_3) in ~100 sec. One can estimate that chloroorganics are reduced $\sim 10^7$ times faster than water even when the medium contains 20% water.

However, significant barriers to the application of SET might still exist. Typically, oxygen and iron will be present in soils and other hazardous wastes. Both Fe^{3+} and O_2 catalyze the reaction of solvated electrons with NH_3 to produce hydrogen and amide [22,23], as shown in Eq. (3) [13,22,23]. Also, if solvated electrons must diffuse into soil particles, these electrons could be consumed in a variety of reactions in competition with diffusion and mass transfer. However, extensive work has shown that this need not be the case. For example, slurrying the soil in NH_3 first has several benefits. It reduces particle size, swells clay layers, and preextracts contaminants. Thus, mass transport limitations and diffusional barriers into solid particles may be avoided or reduced greatly.

$$\text{M= Ca, Na} \qquad M + NH_3 \xrightarrow{O_2 \text{ or } Fe^{+3}} 1/2\ H_2 + M^+ + {}^-NH_2 \quad (3)$$

B. Detailed Description of Method

Pittman et al. [24] have demonstrated the minimum sodium stoichiometry required to completely dechlorinate a variety of aliphatic, aromatic, and phenolic chlorocarbons in dry NH_3(l) and NH_3(l) containing 5-, 20-, and 50-fold molar excesses of water (relative to chlorine atoms). Even in the presence of a 50-mol excess of water, the incremental amount of Na required was modest (see Table 1). The dechlorination of CAHs and chloroaromatics appeared to be diffusion-controlled in reactions where substoichiometric amounts of Na were used [24]. For example, when CCl_4 was reacted with 2 Eq of Na in NH_3(l), only CH_4 (45%) and CCl_4 (54%) were observed. No monochloromethanes, dichloromethanes, or trichloromethanes were formed, suggesting that the CCl_4 in the vicinity of dissolving Na particles was completely dechlorinated before more CCl_4 could diffuse into the region of the particle (despite rapid stirring). Similarly, treating 3,4-dichlorotoluene with 2 Eq of Na in NH_3 or NH_3/H_2O gave 40% toluene and 60% of the starting material, but no monochloro product was observed [24]. Separate studies have observed that metal consumption efficiencies differ depending on the mode of reaction, stirring rate, and metal particle size [24]. Usually, higher efficiencies are observed when sodium is added to preformed sol-

Table 1 Sodium Consumption Per Chlorine Removed Required to Completely Dechlorinate Model Compounds in Liquid NH_3 at Room Temperature

	Na consumed per Cl removed at complete dechlorination	
Substrate	No H_2O	50 mol H_2O[a]
4-Chlorotoluene	1.5	2.5
1,2-Dichlorobenzene	1.4	2.5
1,2,3,4-Tetrachlorobenzene	1.3	2.2
2,4,6-Trichloroethane	1.5	3.3
1,1,1-Trichloroethane	1.2	1.7
Carbon tetrachloride	1.1	1.6

[a] Moles H_2O per mole of chlorine.

utions of chlorinated compounds. When soils are remediated, higher metal efficiencies are often found if the soil is first slurried in NH_3(l) and then metal is added. Metal efficiencies were similar at 25°C and −55°C, except when substrate solubility limitations occurred at −55°C [24].

Which metal is best suited for remediation work? Both Ca and Na have been examined extensively. Commodore has built recently the SET™ and SoLV processes technology around Na, based on extensive commercial experience. Pittman et al. [24] demonstrated recently that the order of metal efficiencies for the dechlorination of aliphatic and aromatic model compounds at 25°C was Na > K > Ca > Li in dry ammonia and Na > K,Ca,Li in the presence of a 50-mol excess of water. These laboratory studies were carried out in the absence of soil [24]. The presence of water significantly reduced the efficiency of both Ca and Li, whereas Na and K efficiencies were only modestly reduced.

Solvated electrons are extremely powerful reducing agents. In NH_3(l), they cleave C–N, C–O, C–S, N–N, N–O, P–halogen, S–S, C–halogen (C–X), aromatic rings, and other functions [11–18]. Aromatic halide reactions with solvated electrons were described in 1914 by Chablay [25] and in 1963 in a thesis by Hudson [26]. Kennedy [27] demonstrated that all the halogens were completely stripped from 19 different pesticides (such as atrazine, DDT, paraquat, trifluralin) in Na/NH_3. The strength of CX bonds increases in the order of C–I < C–Br < C–Cl < C–F. MacKenzie et al. [28] found that the reactivities of α-halogenated naphthalenes (x = F, Cl, Br) exhibited no obvious differences in the selectivity to Na/NH_3. Pittman et al. [24] showed that 3-fluoro-*o*-xylene required 2 Eq of Na to be completely defluorinated (vs. 1.5 Eq for the chloro derivative) in dry NH_3. Four equivalents of Na were required for a complete defluorination in NH_3 with 50 equivalents of H_2O. Thus, whereas defluorinations were very fast, competitive side reactions required the use of more Na, indicating that defluorination is slower than dechlorination.

Phenols are present as phenoxide ions in Na/NH_3 solutions. Therefore, the transfer of an electron into a π-antibonding orbital of a phenoxide ion should be slower than the transfer to benzene. Both chlorophenols and fluorophenols are dehalogenated more slowly than chlorobenzenes or fluorobenzenes [24]. Fluorophenoxide ions were more difficult to defluorinate. Thus, 4-fluoro-2-chlorophenol could be converted selectively to 4-fluorophenol in Na/NH_3 [24]. In contrast, chlorofluorobenzenes lost both Cl and F at close to diffusion-controlled rates [24], indicating that using a stoichiometric deficiency of Na would not give any selectivity to fluorobenzene.

Chlorinated aromatic hydrocarbons can be reduced directly to the parent aromatic hydrocarbon [13,24,29,30]. The parent aromatic molecule can reduce further to give dihydroaromatics or tetrahydroaromatics [11,12,15], or can react to form dimers and higher molecular weight oligomers [16,31,32] in solvated electron solutions.

Aromatic compounds are dechlorinated by the general mechanism shown in Sch. 1. Electron transfer to a π-antibonding orbital forms an aromatic radical anion, which then ejects Cl^- to give an aromatic radical. This radical picks up a second electron to give a very basic σ-anion, which abstracts a proton either from NH_3 or from a more acidic source like water, when water is present. If water is not present, then an $^-NH_2$ anion can be formed. The presence of $^-NH_2$ can lead to the formation of aminated products via the benzyne mechanism. Aminated products were formed in dry NH_3 but not when water was present [24]. A further reduction via radical anion formation and proton abstraction can give dihydroaromatics or tetrahydroaromatics, or dimerization may occur. In soils, both water and

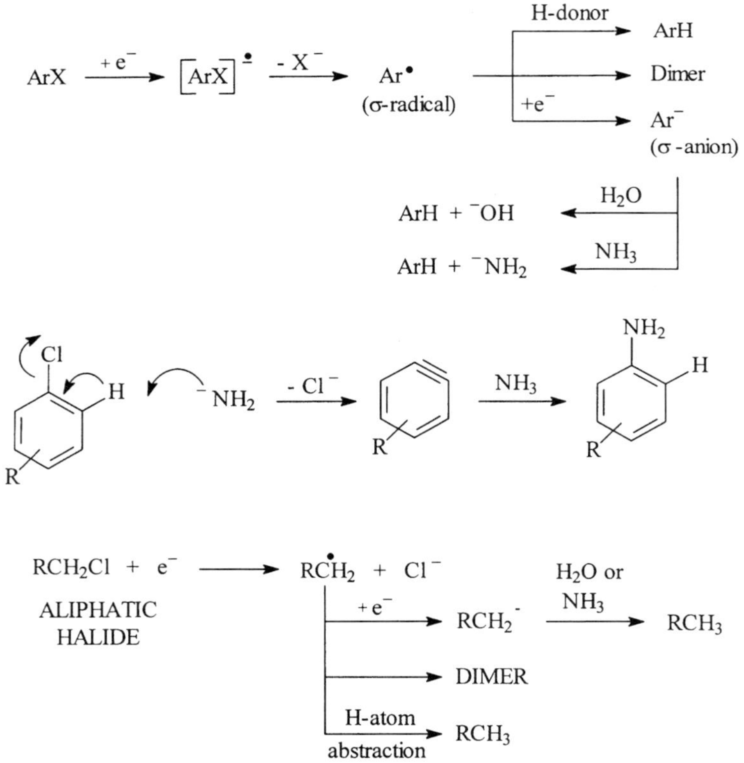

Scheme 1

other functions that are more acidic than NH_3 are present. Thus, halogenated aromatics will not produce aniline derivatives during Na/NH_3 treatments in these systems.

Aliphatic halides are reduced by dissociative electron transfer. A solvated electron is transferred into an antibonding σ-orbital, causing the simultaneous loss of Cl^-. This is a single-step, concerted process as illustrated by the secondary deuterium isotope effect studies of Holm [33]. He and Pittman [29] showed that 1-fluorononane was defluorinated significantly slower than 1-chlorononane in M/NH_3 (M = Na, Ca, Sr, Ba), but this defluorination was accelerated remarkably in the presence of $TiCl_4$. Pittman and He [30,31] have studied the Na/NH_3 remediation of organic soils contaminated with 3000–5000 ppm of such CAHs as CH_3CCl_3. With excess Na, the remediation to a level of 1 ppm is readily possible. The efficiency (moles Na consumed/moles Cl removed) was high at high CH_3CCl_3 concentrations. However, this efficiency drops as the amount of CH_3CCl_3 remaining in the soil decreases. As CH_3CCl_3 decreases to below 20 ppm, competitive reactions require substantial excesses of Na to lower the residual CH_3CCl_3 to 1 ppm or below.

III. TREATMENT OF ENVIRONMENTAL SAMPLES

A. Laboratory-Scale and Commercial-Scale Set™ Experiments

In early works, contaminated soils were slurried in $NH_3(l)$ at ambient temperature and, after premixing, a weighed quantity of solid Na or Ca was dropped directly into the stirring slurry. The metal quickly dissolved. Conductivity and calorimetry showed that the reactions were completed within a few seconds. Reactions of neat PCBs and CAHs are exothermic, but NH_3 reflux can be used to control the exotherm. In soil, sludge, and related decontaminations, the pollutants are diluted in the matrix and are then more highly diluted in the NH_3 slurry. Thus, exotherms are not a problem. Typically, the volume of NH_3 used is twice that of the soil volume.

Table 2 demonstrates the remediation of ~100 g of PCB-contaminated soil samples by Ca/NH_3 (excess Ca). In clay, sandy, or organic soils, the destruction efficiencies were >99.9%. Similar studies were done with sodium. Calcium can be used effectively but becomes far less efficient than sodium as the amount of water in NH_3 increases [24].

Commodore has scaled up these treatments and has developed several process variations depending on the nature of the material being remediated. Modules are tailored to each particular remediation site to achieve the highest cost-effectiveness. Mobile equipment is available at the site in the SoLV process, which eliminates the expense of transporting the hazardous

Table 2 Treatment of PCB-Contaminated Soils with Ca/NH_3 at Room Temperature[a]

Soil matrix	Pretreatment PCB level (ppm)	Posttreatment PCB level (ppm)	Destruction efficiency (%)
Clay	290	0.05	>99.9
Clay	29	<0.06	>99.9
Sand	6200	1.6	>99.9
Organic	660	0.16	>99.9
Organic	83	<0.04	>99.9

[a] Experiments were carried out in a stirred 1.3-L reactor made of steel. Preweighed soil samples were slurried for 10–20 min in liquid NH_3 at ambient temperature. Then a calcium bar was dropped in. The calcium dissolved in a few seconds. The reduction reactions were completed as fast as the calcium dissolved.

substrates. Front-end modules to remove water or extract contaminants can be used. A solvated electron treatment module (SET™) is the centerpiece of each process. Back-end modules are available to recycle NH_3, to adjust pH, and to concentrate or fix the reaction products, depending on client needs. A commercial 1200-L unit is available to treat PCBs, CAHs, etc., in oil, liquid pesticides, or contaminants, which have been extracted from soil, sludges, or other matrices. Most commercial soil decontaminations operate via extraction first, followed by decontamination with Na. A flow diagram of the process is shown in Fig. 1.

A sodium transfer station heats sodium (in shipping drums) to a liquid state and pumps the liquid to the solvator tank. This tank is filled with anhydrous ammonia, which dissolves the sodium. The resulting solvated electron solution is discharged to a reactor vessel, where a volume of approximately 65 gal of the solvated solution is maintained. Contaminated liquid (soil extracts, oil, etc.) is pumped to the reactor vessel, where organics are rapidly destroyed. The solution conductivity is monitored continuously. When the conductivity drops to 200 Mho, the Na/NH_3 feed is stopped. The destruction reaction is essentially diffusion-controlled. Removing ammonia vapor controls the temperature and pressure of the vessel. The feed rate is approximately 1600 lb of soil per day. After the reaction, the solution is transferred to a separator using the natural vapor pressure of the ammonia as the motive force. Ammonia is heated to approximately 125°F and pumped, as a vapor, to a condenser for recycling by a commercial refrigeration subsystem. The treated material is discharged to a storage vessel. After pH adjustment, the product is suitable for on-site or nonhazardous disposal. A more detailed schematic diagram of a multimedia remediation unit is shown in Fig. 2.

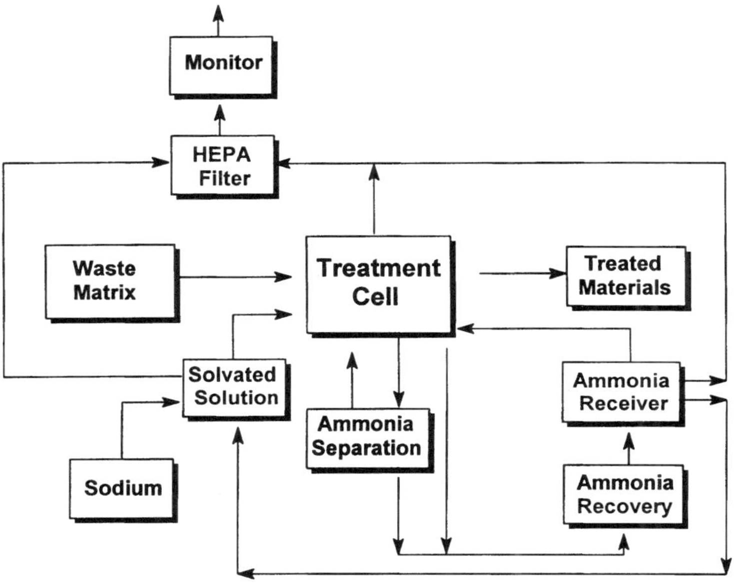

Figure 1 A SET™ process flow diagram.

Solids are treated as-is in a 10 ton/day screw reactor in which the contaminated solid and NH_3(l) are mixed. In most cases, the contaminants are extracted into the ammonia. When sodium dissolves in liquid ammonia, a sodium cation is formed together with a solvated electron. The solutions are deep blue and conducting. Each solvated electron exists over a volume of solution. The solvated electron is best described as a quantum mechanical particle which, due to its tiny mass, is perhaps better thought of in terms of its wave properties. Hence, solvated electrons in NH_3 migrate exceptionally rapidly, penetrating clay layers and other obstructions to reach pollutant molecules that may be otherwise unavailable to less mobile reactants. However, NH_3 also swells soils and sludges and, aided by stirring, effectively extracts pollutants into the solvent medium. Upon the addition of sodium, the extracted pollutant molecules are often reduced at diffusion-controlled rates. Thus, reaction times are very short. Less sodium is consumed by side reactions when pollutants are extracted, leading to higher sodium utility than when the solvated electrons are transported into a solid matrix. Sodium metal is added in both molten or solid form to the NH_3/solids slurry, and the solvated electrons formed proceed to destroy the contaminants. The NH_3 is recycled and the treated solid is returned to the environment. Wet

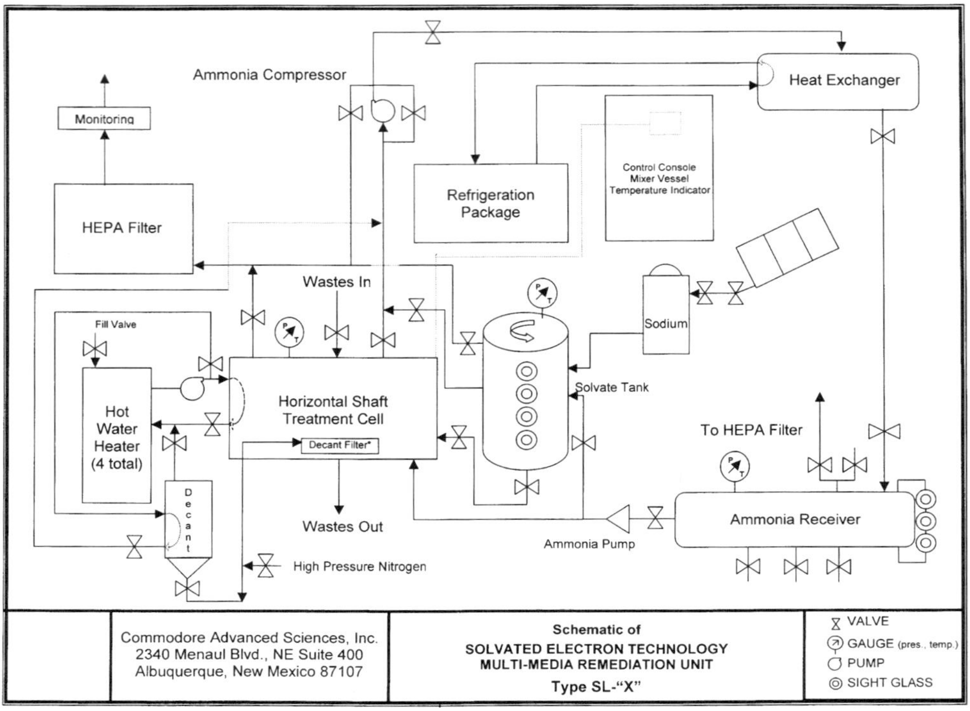

Figure 2 A detailed schematic diagram of a multimedia remediation unit.

sludges often require water removal by drying, or by using a prewash module. The sludge is subsequently remediated. A back-end module removes ammonia. Following pH adjustment, the material can be disposed of in a nonhazardous waste landfill.

B. Scale-Up Remediations of Contaminated Substrates

1. Contaminated Soils and Sludges

A wide variety of soils and sludge have now been treated. Soil characteristics that can impact the SET™ chemistry include the general soil type, which is treated (loam, sand, silt, and clay), the presence of humic material, the pH value, the soil's cation exchange capacity, its particle size, the amount of water present, and the iron content. Processes have been engineered to accommodate this wide range of variables [7,8,34]. Some soils can be treated as-received. Others require preprocessing or postprocessing (e.g., water removal, size reduction, washing, and pH adjustment) to effectively remediate them. The various modules of the SET™ technology are designed to be tailored to each remediation site in the most cost-effective sequence.

The main reactor (treatment cell or module) is the critical component of each process. Pretreatment is sometimes necessary to allow the material to fit in the treatment cell. For example, large rocks may be separated from soils, soils may be dried of excess water, and other substrates may be shredded. The preprocessing of soils usually involves sieving the soil matrix to remove rocks, large stones, and debris. Some soils and sludges may also require drying prior to treatment with solvated electrons.

Metal debris is pretreated with one of two processes. One process is cutting the debris into small pieces, and the second is washing the debris by flooding with ammonia. The ammonia dissolves the organic contaminants. This solution is transferred to the treatment cell after it is used to wash the metal parts. The metal is free of organic contaminant and can be reused or discarded. Porous solids such as concrete, brick, ceramic, and rock are treated by the grinding action, which occurs during mixing in the treatment cell during the slurry process. Alternatively, solids can be size-reduced prior to treatment if the pieces are too large to be placed in the treatment cell. Crushing and sieving before the matrix is placed in the treatment cell can accomplish this. Paper, rags, plastics, and rubber are treated after shredding. Shredding is usually required to prevent the materials from winding around the rotating parts of the treatment cell. A major benefit of the SET™ process is that it can treat a heterogeneous matrix. Examples of heterogeneous debris successfully treated include paint chips, greases, cutting fluids with metal filings, ground corn cobs, charcoal, aluminum capacitor foil, mylar, polyvinyl chloride, Lexan, fiber glass, rubber, and particle board.

Most solvated electron-treated wastes require posttreatment. The first posttreatment involves removing and recovering ammonia from the matrix. This is accomplished by passing hot water or steam through the jacket of the treatment cell and by condensing the ammonia for reuse. Materials such as shredded paper, wood, plastic, rubber, and PPE can be volume-reduced after the SET treatment by using commercially available compacting equipment.

Processes can be modified to deliver targeted remediation levels. Many different soil contaminants have been treated. These include PCBs, polycyclic aromatic hydrocarbons (PAHs), chlorinated solvents, dioxins, furans, pesticides, hexachlorobenzenes, BTXs, volatiles, and semivolatiles. After treatment, the soils pass all toxicity characterization leachate procedure (TCLP) criteria for replacement or nonhazardous waste landfill disposal. Specific examples of soil and sludge treatment will now be given. Table 3 contains data from several PCB remediation projects.

Table 3 Destruction of PCBs in Various Soils with Na/NH_3

Source of soil	Soil type	Pretreatment PCB level (ppm)	Posttreatment PCB level (ppm)
Harrisburg, PA[a]	Sand, clay	777	< 1.0
Los Alamos, NM[b]	Sand, silt, clay	77	< 2.0
New York[c]	Sand, silt	1250	< 2.0
Monroe, LA[d]	Sand, silt, clay	8.8	< 1.0

The range of Na weight percents in NH_3 used was 1.27–3.3% in these examples.
[a]Treatment temperature in liquid NH_3: 32°C.
[b]Treatment temperature in liquid NH_3: 20°C.
[c]Treatment temperature in liquid NH_3: −33°C.
[d]Treatment temperature in liquid NH_3: −33°C.

A New Bedford Harbor Sawyer Street site in Massachusetts has been designated as a superfund site due to PCB contamination of river sediments. Commodore was one of three companies chosen to conduct demonstration studies on-site under contract to Foster Wheeler Environmental Company. The river sediment was first washed with diisopropylamine by the Ionics RCC B.E.S.T™ process [35], producing an oil concentrate. The PCB level in the B.E.S.T concentrate was approximately 32,800 ppm. Dioxins/furans (TEFs) were also present at 47 ppm. This concentrate was reacted with SET™ in the SoLV process to destroy the PCBs and dioxins (Table 4). After treatment, the PCB level was 1.3 ppm, well below regulatory requirements for disposal in nonhazardous waste landfills. Dioxins were also readily remediated. This study also illustrates that the SoLV process can remove metals from substrates. The concentrate received was found to have lead, arsenic, and selenium in high parts-per-billion levels. After treatment with the SoLV process, the levels were below detection limits. The metals were

Table 4 SET™ Treatment of PCB- and Dioxin-Contaminated Sludge from New Bedford Harbor

Contaminant	Pretreatment (ppm)	Posttreatment (ppm)
PCB	32,800	1.3
Dioxin/furan	47	0.012
Mercury	0.93	0.02
Lead	73	0.2
Selenium	2.5	0.2
Arsenic	2.8	0.1

removed from the solid matrix during the transport of liquid ammonia from the reactor vessel. Metals were recovered from the ammonia recycle unit for fixing and disposal.

2. Contaminated Surfaces

Commodore has performed ex situ treatments of PCB-contaminated surfaces with Na/NH_3 (Table 5). In most cases, destruction efficiencies exceeded 99%. Solid materials contaminated with chemical warfare agents have also been treated (Table 6). Mustard (HD), Sarin (GB), and VX (structures presented in Fig. 3) were effectively remediated to nondetectable levels from a variety of metal, wood, cardboard, plastic, glass, concrete, and rubber surfaces. Highly porous wood and concrete require crushing or shredding to increase surface areas prior to treatment. These results suggest that chemical munitions might be opened in liquid NH_3 and the resulting warfare agents might be destroyed effectively by solvated electrons. This would avoid the need to transport unstable and corroding old munitions to a special incinerator facility and would avoid the "not-in-my-backyard" problems associated with incineration.

3. Oils

Contaminated transformer oils and cutting fluids have been remediated readily using Na/NH_3 in a 1200-L equipment (Table 7). Oils containing over 20,000 ppm of PCBs have been detoxified to levels below 0.5 ppm. Typically from 2 to 4 wt.% Na in liquid NH_3 was used. The SET™ process was also used to remediate dioxins in waste oil from the McCormick and Baxter Site in Stockton, CA. As shown in Table 8, dioxins were reduced to parts-per-trillion (ppt) levels.

Table 5 Treatment of PCB-Contaminated Surfaces with Na/NH_3[a]

Surface	PCB level	Destruction efficiencies (%)
Stainless steel	25 mg/cm^2	99.999
Capacitor foil (Al)	60 ppm	99.4
Mylar	60 ppm	99.4
Charcoal	500 ppm	99.98
Ground corn cobs	1270 ppm	99.7

[a] At ~20°C using 2.4–2.8 wt.% Na in NH_3, except for Al surface where 0.9 wt.% Na was used.

Table 6 Na/NH_3 Treatment of Surfaces Contaminated with Chemical Warfare Agents[a]

Coupon material	HD	VX	GB
Wood	ND	ND	ND
Fiberglass	ND	ND	ND
Particle board	ND	ND	ND
Charcoal	ND	ND	ND
Rubber	ND	ND	ND
Brass	ND	ND	ND
Stainless steel	ND	ND	ND
Carbon steel	ND	ND	ND
Aluminum	ND	ND	ND
Copper	ND	ND	ND
PVC	ND	ND	ND
Teflon	ND	ND	ND
Lean	ND	ND	ND
Cardboard	ND	ND	ND

[a] Run at $-33°C$ at 4 wt.% Na in NH_3. The neat warfare agent was added to completely cover the surface, and the coupon was then added to liquid ammonia followed by addition of sodium. After this treatment, the coupon surfaces and the ammonia surfaces were analyzed to try to detect residual HD, VX, or GB.

4. PCBs/Hexachlorobenzene

A common problem at public utilities is soil that becomes contaminated with PCBs near and around transformers. There still are a number of such sites with this problem in the United States. Table 9 gives data from a soil cleanup project from a site in New York State. The soil contained approximately 1200 ppm of PCB (Aroclor 1260) prior to treatment with SET™. After treatment, the PCB level was reduced to 1.4 ppm (Table 9). Aroclor 1260 is particularly hard to bioremediate, so this result is significant. Small quantities of PAHs (pyrene and phenanthrene) were also remediated. The total petroleum hydrocarbons increased, which is what should be expected because the initial

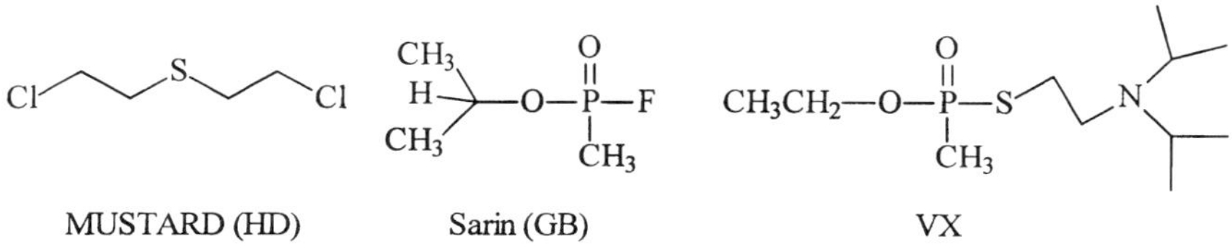

Figure 3 Structures of HD, GB, and VX.

Table 7 Destruction of PCBs in Oils by Na/NH_3[a]

Material (temperature, °C)	Pretreatment (ppm)	Posttreatment (ppm)
Motor oil (16)	23,339	< 1.0
Transformer oil (40)	509,000	20[b]
Mineral oil (40)	5,000	< 0.5
Hexane (40)	100,000	0.5

[a] SET™.
[b] Sodium feed was deficient. It was improved by adding more sodium (SET™ process).

organic product of PCB destruction is biphenyl, which is then further reduced to more highly reduced hydrocarbons. PAHs give higher molecular weight hydrocarbons due to oligomerization/reductions [31,32]. After pH adjustment, the soil could be returned to the site.

Na/NH_3 was very effective in the destruction of hexachlorobenzene in soils. Sandy soil containing 67.6 ppm of hexachlorobenzene was treated with Na/NH_3 in a mobile destruction unit at a site in Las Vegas, NV. Approximately 4 wt.% sodium was used. The treated soil contained < 2 ppm of hexachlorobenzene. GC/MS analysis could not detect any chlorinated species in the treated soil.

5. Pesticides

The destruction of bulk malathion has been carried out in Na/NH_3 at Commodore's Marengo, OH facility in a 1200-L unit in 100-lb quantities. Near-stoichiometric quantities of sodium were able to destroy the malathion. The levels of malathion in the treated material were nondetectable. Most bulk samples of organic pesticides and herbicides containing halogens, phosphorous, or sulfur are amenable to reductive destruction using Na/NH_3. Many waste streams were produced when manufacturing pesticides were remediated using solvated electron reductions.

Pesticides in soils can also be remediated. Table 10 summarizes some results from a project where soils from Hawaii and Virginia, contaminated with DDT, DDD, DDE, and dieldrin, were treated with Na/NH_3. In all

Table 8 Na/NH_3 Treatment of Dioxins in Waste Oil at the McCormick and Baxter Site

Contaminant	Before ppt	After ppt
Dioxins	418,500	2.3
Furans	14,120	1.3

Table 9 Na/NH_3 Treatment of a Transformer Oil Spill at a New York State Utility Site

Contaminant	Pretreatment (ppm)	Posttreatment (ppm)
Aroclor 1260	1200	1.4
Mercury	0.21	0.08
Lead	433	267
Pyrene	1.8	ND[a]
Phenanthrene	1.4	ND[a]

[a] ND = not detectable.

cases, the soils were remediated to nondetectable levels of the respective pesticide. These runs were conducted on-site at Port Hueneme, California Naval Station, using Commodore's mobile demonstration unit. Soils were shipped in from other naval facilities.

6. Chlorofluorocarbons and Halons

The phase out of Class I ozone-depleting compounds under the terms of the Montreal Protocol creates serious disposal concerns for organizations that have large quantities of CFC refrigerants and halons at their facilities. Whereas much of this material is sufficiently pure to be recycled or resold, increasing quantities, which cannot be reused because of their cross-contamination with other compounds, are appearing. As these materials can no longer be buried under land, discharged into water, or released to the atmosphere, responsible parties are left with destruction as their sole means of ultimate disposal. Abel and Mouk [36] and Mouk and Abel [37] patented a process that treats CFC feed streams with solvated electrons, and achieves destruction efficiencies equal to or greater than the United Nations target of

Table 10 Destruction of Pesticides in Soil with Na/NH_3 in a Mobile Unit

	DDT	DDT	DDE	Dieldrin
Barbers Point, HI				
Pretreatment	200	180	69	ND
Posttreatment	< 0.02	< 0.02	< 0.02	ND
Dahlgren, VA				
Pretreatment	9	1.6	ND	15
Posttreatment	< 0.02	< 0.02	ND	< 0.02

99.99% (Table 11). Reaction products were sodium halides and hydrocarbons such a CH_4, CH_3CH_3, and CH_2CH_2. NaCN is a by-product formed during the Na/NH_3 dehalogenation of CFCs. This originates from a pathway in which either $^{-}NH_2$ displaces halogen, or an intermediate carbene such as CF_2 inserts into ammonia. These two paths form amines, which can end up as cyanide by further dehydrohalogenation. Methods have been described for eliminating the formation of cyanides during halofluorocarbon dehalogenations, which are conducted via Na/NH_3 treatments [38,39,42].

7. Polycyclic Aromatic Hydrocarbons

PAHs and soils contaminated with PAHs are readily remediated by solvated electrons in NH_3. Oligomeric reduced products are obtained. These reactions are slower than dehalogenation, as was demonstrated by the rapid formation of benzene, toluene, and naphthalene in Na/NH_3 from their corresponding monochloro derivatives [24,28]. Table 12 summarizes data on the destruction of pure PAHs. Soils contaminated with PAHs have been remediated to below detection levels. Mononuclear aromatics (benzene, toluene, anisole, and nitrobenzene) undergo ring reduction according to the well-known Birch reduction [11–18].

Table 11 Ozone-Depleting Compounds Destroyed by the Na/NH_3 CFC Destruction Process[a]

Compound	Formula	Destruction efficiency (%)
CFC-11	CCl_3F	99.99
CFC-12	CCl_2F_2	99.99
HCFC-22	$CHClF_2$	99.99
HFC-32	CH_2F_2	99.99
CFC-113	CCl_2FCClF_2	99.99
CFC-114	$CClF_2CClF_2$	99.99
CFC-115	$CClF_2CF_3$	99.99
HFC-134a	CH_2FCF_3	99.99
HFC-152a	CH_3CHF_2	99.99
R-500	CFC-12 + HFC-152a	99.99
R-502	CFC-115 + HCFC-22	99.99
Halon 1211	$CBrClF_2$	99.99
Halon 1301	$CBrF_3$	99.99
Halon 2402	$CBrF_2CBrF_2$	99.99

[a] Reductions conducted at 18°C with a 2:1 mole excess of Na (1.2 wt.% in NH_3) on 2400g of pure bulk substrate. Neat samples were treated in each case.

Table 12 Destruction of Neat PAHs in Na/NH_3

PAH	Pretreatment (μg)	Posttreatment (μg)	Destruction efficiency (%)
Acenaphthene	2001	0.012	99.999
Acenaphthylene	2005	ND	99.999
Anthracene	1987	0.37	99.98
Benzo[*a*]anthracene	2005	0.03	99.99
Benzo[*a*]pyrene	2007	0.56	99.97
Benzo[*b*]fluoranthene	2017	0.15	99.99
Benzo[*g,h,i*]perylene	2004	0.22	99.99
Benzo[*k*]fluoroanthene	2024	0.14	99.99
Chrysene	2019	0.20	99.99
Dibenzo[*a,h*]anthracene	2013	0.01	99.999
Fluoranthene	2023	0.06	99.99
Fluorene	2013	0.14	99.99
Indeno[1,2,3-*cd*]pyrene	1998	0.04	99.99
Naphthalene	2011	0.01	99.999
Phenanthrene	2009	0.22	99.99
Pyrene	2012	0.39	99.98

8. Explosives

Most explosives and propellants are nitro or nitrate compounds, respectively. These functional groups are reduced readily by solvated electrons in NH_3. Aromatic nitro compounds may be eventually reduced to the corresponding amino derivatives. Azoxyaryl, azoaryl, and hydrazoaryl dimers are also obtained (Eq. (4)). These dimers are then further reduced to the corresponding monomeric amino compounds. Experience has now been gained at Commodore in detoxifying a number of explosives and propellants including TNT, RDX, nitrocellulose, nitroglycerine, tetryl, PETN, Comp B, and M-28 with Na/NH_3. Explosives and propellants have also been destroyed: (1) neat, (2) when mixed with the chemical warfare agents, (3) from actual armaments, and (4) in soils. No explosive analyte was found after Na/NH_3 treatment upon analysis using EPA method 8330 (revision O, September 1994). The reaction products were found to be oligomeric, with all nitro groups reduced. Similar results were reported in Na/ethylenediamine, Ca/NH_3, and Ca/ethylenediamine, where TNT and dinitrobenzene were destroyed [40]. Na/NH_3 treatments were used to remediate soils from Los Alamos, NM, which had been contaminated with RDX, HMX, and 1,2-dinitrobenzene (1,2-DNB). After treatment with Na/NH_3, no detectable level of explosive was found. Table 13 contains data from this study.

Table 13 Destruction of Explosives in Soil From Los Alamos, NM, Using Na/NH_3[a]

	HMX	RDX	1,2 DNB
Not treated (mg/kg)	1600	3580	9.6
Treated (mg/kg)	0.03	0.03	0.03
Detection limit	0.03	0.03	0.03
Destruction efficiency (%)	99.9999	99.99999	99.99

[a] Na (2.8 wt.%) in NH_3 (1 L) used per 50 g of soil at 39°C.

$$Ar{-}NO_2 \xrightarrow{Na/NH_3} Ar{-}NH_2 + ArNHNHAr + ArN{=}NAr + Ar\overset{O^-}{\overset{|}{N^+}}{=}NAr \qquad (4)$$

9. Chemical Warfare Agents

One of the most exciting applications for solvated electron chemistry is the destruction of chemical warfare agents. A combination of international treaty agreements and U.S. legislation has placed a major responsibility on the U.S. Army to destroy all stockpiled chemical agents over the next decade. Solvated electron chemistry is a highly efficient, cost-effective solution to the global problem of neutralization and disposal of the highly toxic military chemical warfare agents currently stockpiled. Over 300 tests have now been conducted by Commodore on all stockpiled agents (GA, GB, GD, GF, Lewisite, VX, HD, HT, T, HN-1, HN-3, HL, picric acid, CG, and CK). Destruction efficiencies of greater than 99.9999% have been attained consistently. As part of the Army ACWA Program, reaction products have been characterized extensively. These products were tested for acute toxicity and found to be Class 1 or Class 0 level. VX is destroyed by electron transfer to VX, followed by very rapid (transport-limited) P–S (70%) and C–S (30%) cleavage and evolution of ethane. The defluorination of GB is mass transport-limited and proceeds to completion, yielding NaF. The partial removal of the isopropyl group occurs via C–O cleavage, giving propane. A subsequent oxidation by $Na_2S_2O_8$ gives isopropyl alcohol, H_3PO_4, Na_2SO_4, and methanol. These routes are shown in Sch. 2.

10. Mixed Wastes

One particularly vexing problem for waste management professionals is that of mixed wastes (radioactive plus RCRA and/or TSCA waste) disposal. Caught between conflicting regulatory jurisdictions and remediation options, which frequently prove to be mutually exclusive, mixed waste streams and matrices contaminated with two or more types of contaminant represent a

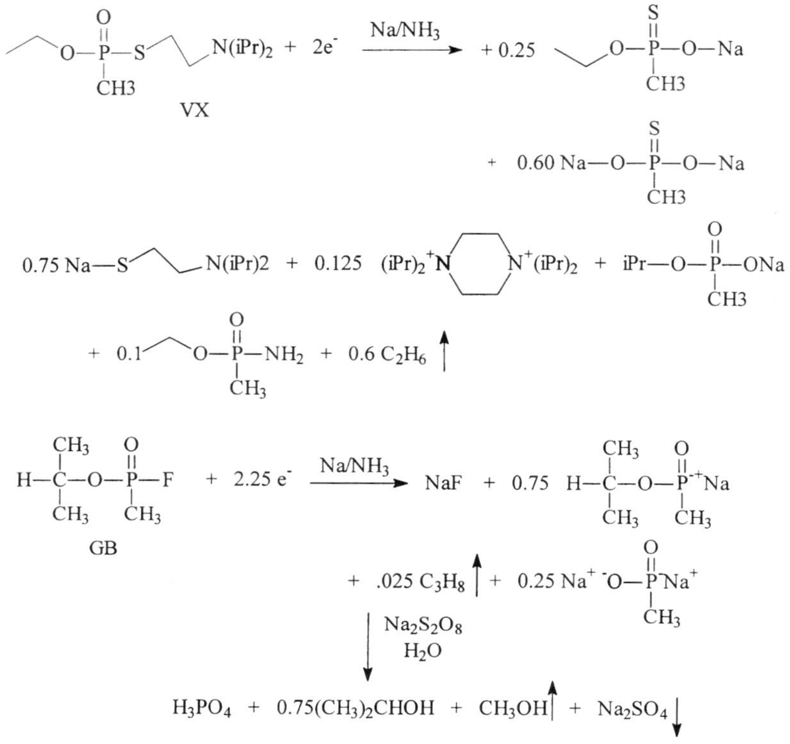

Scheme 2 Teledyne–Commodore mechanism for destruction of VX and GB.

greater level of difficulty and expense to remediate. Soils contaminated with PCBs and with heavy or radioactive metals such as mercury, uranium, or cadmium illustrate this problem. Another example, frequently found at nuclear facilities, is soil contaminated with RCRA-listed organic compounds and low-level radioactivity. Because Na/NH_3 soil decontamination processes are employed at, or below, room temperature, they have special application to mixed waste streams that could not be treated appropriately with thermal processes. Na/NH_3 technology can successfully destroy halogenated organic compounds in soils containing low-level radioactive components, heavy metals, etc., without oxidizing or volatilizing the metallic components [39,41].

Commercial quantities of absorbents contaminated with PCBs and radioactive materials were successfully treated with Na/NH_3 at a DOE site in Weldon Spring, MO (Table 14). The radioactive species remained in the soil for subsequent disposal. Various wash processes can remove radioactive components from the soils to reduce the volume of hazardous material.

11. Metals Removal

Several postprocesses can be used to remove metals that still contaminate soils or other matrices after Na/NH_3 treatment [43,44]. Washing a treated soil with NH_3/H_2O, at high pH using in situ-generated NaOH, was demonstrated in a mobile destruction unit used on-site at Lockheed/Martin Advanced Environmental Systems in Las Vegas, NM. Soil contaminated with 12,000 ppm of Aroclor 1260 and with cobalt (Co), cesium (Cs), and other metals at levels from 85 to 1800 ppm were treated first with Na/NH_3 to destroy the PCBs. PCB destruction efficiencies were in excess of 99.9%. After the first posttreatment $NH_3/H_2O/NaOH$ extraction, 16–93% of each metal species was removed (Table 15). After the second extraction, the removal efficiencies increased from a low of 81% to a high of 97%. Thus, the majority of metals were removed in only two batch extractions.

12. Miscellaneous

Process streams are amenable to detoxification by Na/NH_3 at the "end of the pipe" before the material is categorized as a waste. Products of the reduction can often by recycled back into the process. Examples include chlorinated organics, pesticides, and refrigerants. The decontamination of different types of personnel protective equipment such as gloves, boots, cotton, and coveralls has been performed by Commodore. Many cases require pregrinding or preshredding. After treatment, the residual material can be landfilled. Contaminated gravel and stone can be remediated with Na/NH_3, usually after crushing, to increase surface area and to speed up mass transfer.

The in situ application of solvated electron reductions to remediate and detoxify environmental surfaces, solids, and vadose zones presents two

Table 14 Treatment of PCBs in Low-Level Radioactive Waste from Weldon Spring, MO

Material	Pretreatment (ppm)	Destruction efficiency (%)
Shredded corn cobs	1270	99.8
Unshredded corn cobs	944	97.4
Transformer capacitor parts	6	97.8

Table 15 Metals Removal from Soil by $NH_3/H_2O/NaOH$ Extractions After PCB Remediation with Na/NH_3[a]

	As	Ba	Cd	Cr	Pb	Se	Hg	Co	Cs
Pre (ppm)	182	1831	166	204	193	100	86.7	188	406
Post (ppm)	< 12.5	1370	66.5	172	99.5	< 12.5	26	64.5	150
Percent extracted	93.1	25.2	59.5	15.7	48.4	87.5	70	65.7	63
2nd extracation (ppm)	< 12.5	286	14.9	38.6	30.3	< 12.5	2.9	23	60.5
Percent extracted	NA[b]	79.1	77.6	77.6	69.5	NA	88.7	64.3	59.7
Total percent extraction	93.1	84.3	91	81	84	87.5	96.6	87.8	85

[a] Extractions performed after initial soil slurry in NH_3 (40°C) was cooled to 17°C and treated with Na (3.9 wt.%) in NH_3 to destroy PCBs.
[b] NA = no additional extraction.

major challenges. First, NH_3 vapor containment techniques are needed. Second, the penetration of porous substrates must be controlled. These are primarily engineering, not chemical, challenges. Engineering designs of various in situ processes are underway but not completed.

Solvated electrons, present in or on tiny ice crystals found in polar stratospheric clouds, have recently been proposed to cause Cl^- to dissociate from CFCs via dissociative electron transfer [45]. This process may be contributing to the holes observed in the stratospheric ozone layer. Ionizing radiation can cause electron detachment from atoms in the stratosphere. Upon being solvated in water or ice surfaces, they are rapidly transferred to CCl_2F_2, for example. This causes an immediate cleavage of the vibrationally excited negative Cl^- and $^{\bullet}CClF_2$. Laboratory experiments indicated that NH_3 surfaces gave 10^4 greater cleavage yields than ice surfaces under controlled low-pressure conditions. This agrees with the longer lifetime of the solvated electron in ammonia relative to water.

IV. COMPETING TECHNOLOGIES

There are several technologies that compete with the solvated electron process. These include landfill, incineration, thermal desorption, plasma arc, sodium dispersion, alkali treatment, bioremediation, and washing. The least expensive of all these technologies is landfill. If landfill is a viable alternative, none of the alternate technologies is economically competitive. Next lowest in cost are the thermal desorption and washing technologies. Whereas these processes are economical, they are only effective for volatile organics. Nonvolatile organics such as PCBs cannot be remediated to the required standards using thermal desorption and washing. Incineration, alkali treatment, sodium dispersion, and solvated electron technologies are all cost-competitive. The most expensive process is plasma arc due to its high energy requirements.

SET™, incineration and plasma arc are very effective at achieving regulatory cleanup standards. However, SET™ is a nonthermal process, whereas the other processes are thermal. Thermal processes have the potential to produce hazardous by-products such as dioxins. SET™ does not form dioxins because it is a reduction process. Furthermore, the SET™ process is a closed system without an exhaust stack. When the SET™ process is used in soils, residual ammonia remains in the soil. This is typically on the order of 1–3%. Ammonia is a fertilizer which farmers pump (anhydrous) into their fields, so this residue is usually not harmful.

Bioremediation processes constitute a class by itself, which can often be cost-effective. The primary advantage of bioremediation is that it can be used *in* situ. The negatives of the process are that it is time-consuming and is very sensitive to the feedstock available, temperature, and moisture. Bioremediation works well for materials such as explosives, but is ineffective for materials like PCBs, particularly more highly chlorinated PCBs. Bioremediation, when performed in situ, may work well in one region of the strata but completely fail a short distance away.

It is difficult to provide cost comparison data for the SET™ process vs. alternative technologies because of the wide variability of the waste that must be remediated (soil, sludge, neat, oils, building materials, DNAPLs, etc.) and the nature of the contaminants. Each situation must be examined individually. What can be said is that SET™ is cost-competitive to incineration. Several good cost and effectiveness comparison references for these technologies are available from the Federal Government web sites (www.frtr.gov/costper.html and www.lanl.gov/projects/etcap/intro.html).

V. PRACTICAL ADVANTAGES OF SOLVATED ELECTRON REDUCTIONS

The preceding examples demonstrate that solvated electron reductions are effective for destroying hazardous organic materials. Na/NH_3 treatments, as commercialized in the SET™ and SoLV processes, can remediate halogenated organic compounds (PCBs, CAHs, pesticides, herbicides, PCP, solvents, and chlorinated olefins), organic nitro compounds, nitrates, *N*-nitro and other explosive compounds, PAHs, halons, CFCs, and chemical warfare agents. It can destroy them individually, or in virtually any combination of mixtures. Wet soils, sludges, sediments, contaminated rock, concrete, protective clothing, etc., may all be remediated. The rates of reduction are very fast at room temperature. NH_3 is an ideal liquid for slurrying soils because even soils with high clay content break down into swollen particles, which do not readily agglomerate. Consequently, mass transfer and diffusion barriers are decreased. In addition, handling low-boiling NH_3 is a well-known technology dating back to the era when NH_3 was a standard refrigerant working fluid.

Na/NH_3 reduction processes are nonthermal. Most reactions can be conducted at $-33\,^{\circ}C$ [24], but in practical applications, 10–35 °C temperatures are used most often. These low temperatures protect against volatile emissions. The pollutant destruction process is carried out in a totally closed system. Even the ammonia that is vented when reactors are opened is captured by a scrubber and returned to the reactor during the pH adjustment. During this process, the ammonia is retained for reuse. Minor amounts of hydrogen generated from catalytic sodium degeneration are vented through the scrubber system. Surprisingly little hydrogen is generated in a wide range of reaction applications. Volatile hydrocarbons that may be formed are condensed and made available for fuel use.

One distinguishing feature of the SoLV process is that no portion of the original contaminant molecule is discharged to the atmosphere or to water. The process is reductive in nature and therefore not capable of forming dioxins or furans and similar wastes, which can be found in oxidizing technologies. This is especially beneficial because communities are increasingly watchful of waste facilities as concerns mount over particulate materials that are released to the atmosphere and surrounding water. The end products from Na/NH_3 processes are principally metal salts such as sodium chloride and hydrocarbons. The product streams are not classified as RCRA hazardous and they pass all of the hazard criteria identified in 40 CFR 261.21 through 40 CFR 261.24.

The only raw materials needed for the process are ammonia, sodium, and a neutralizing acid such as sulfuric acid. All of these reactants are commodity chemicals. When considered in light of other processes available, the hardware required to implement the SoLV™ process is simple and compact. All process equipment are off the shelf and engineered to be mobile. Destruction can take place at the site without the cost associated with transporting hazardous cargoes.

ACKNOWLEDGMENTS

We gratefully acknowledge the support of the work at the Mississippi State University by the Water Resources Research Institute of Mississippi,

Department of the Interior, US Geological Survey (grant nos. HQ96G-R02679-1 and HQGR0088) and by the U.S. EPA (grant nos. GAD# R826180 and R-82942101-0).

REFERENCES

1. Subsurface Contaminants Focus Area: Technology Summary, Rainbow Series, U.S. Department of Energy, Office of Science and Technology, August 1996 (Report No. DOE/EM-0296, NTIS Order No. DOE/EM-0296; available through EM Helpline: Tel.: +1-800-736-3282).
2. Dense Non-Aqueous Phase Liquids: A Workshop Summary, Dallas, TX, April 16–18, 1991. Ada, OK: Environmental Protection Agency, February 1992 (EPA/600/R-92/030, NTIS Order No. PB92-178938).
3. FY1995 Technology Development Needs Summary, U.S. Department of Energy, Office of Environmental Management, March 1994, p. 204 (Report No. DOE/EM-0147P, NTIS Order No. DE94012580).
4. LSFA: Landfills Stabilization Focus Area. National Technology Needs Assessment. Working Draft, U.S. Department of Energy, January 31, 1996.
5. PFA: Plumes Focus Area. National Technology Needs Assessment. Review Draft, U.S. Department of Energy, May 17, 1996.
6. Plumes and Landfills Stabilization Focus Areas. Progress Report, U.S. Department of Energy, Winter 1996.
7. Weinberg N, Mazer DJ, Abel AE. Process for decontaminating polluted substances. US Patent 4,853,040, August 1, 1989.
8. Weinberg N, Mazer DJ, Abel AE. Process for decontaminating polluted substances. US Patent 5,110,364, May 5, 1992.
9. Pittman CU Jr, Tabaei SMH. In: Tedder DW, ed. Preprint Extended Abstracts of the Special ACS Symposium: Emerging Technologies in Hazardous Waste Management V, Vol. 11, Atlanta, GA, September 27–29, 1993:557–560.
10. Pittman CU Jr, Mohammed MK. Solvated electron reductions: destruction of PCBs, CFCs, CAHs and nitro compounds by Ca/NH_3 and related systems. Preprint Extended Abstracts of the IC&E Special Symposium, Emerging Technologies in Hazardous Waste Management VIII, Birmingham, AL, September 9–16, 1996:720–723.
11. Havey RG. Metal ammonia reduction of aromatic molecules. Synthesis 1970; 4:161–172.
12. Smith H. Chemistry in Non-Aqueous Solution. New York: Interscience Publishers, 1963:8–24, 151–201.
13. Smith M. Dissolving metal reductions. Augustine RL eds. Reduction: Techniques and Application in Organic Synthesis. New York: Marcel Dekker, 1968:99–170.
14. Schindewolf U. Formation and properties of solvated electrons. Angew Chem (Int Ed) 1968; 7(3):190–193.

15. House H. Dissolving metal reductions. In: Modern Synthetic Reaction. New York: Benjamin Publishers, 1965:50–77.
16. Birch AJ. The reduction of organic compounds by metal–ammonia solution. Quart Rev (London) 1950; 4:69.
17. Birch AJ. Reduction by metal–amine solutions, applications in synthesis and determination of structure. Quart Rev (London) 1958; 12:17.
18. Watt GW. Reactions of organic and organometallic compounds with solutions of metals in liquid ammonia. Chem Rev 1950; 46:317.
19. Gould RF. Advances in Chemistry. Washington, DC: American Chemical Society, 1965:50.
20. Crooks RM, Bard AJ. Electrochemistry in near-critical and supercritical fluids, nitrogen heterocycles nitrobenzene, and solvated electrons in ammonia at temperatures to 150°C. J Phys Chem 1987; 91(5):1274.
21. Schindewolf U. Metal–Ammonia Solutions. Lagowski JJ, Sienko MJ eds. London: Butterworth, 1970:199–218.
22. Eastham JE, Larkin DR. Effect of variables on reductions by alcohol and potassium in ammonia. J Am Chem Soc 1959; 81:3652–3655.
23. Rabideau PW, Wetzeland DM, Young DM. Metal–ammonia ring reduction of aromatic carboxylic acid esters. J Org Chem 1984; 49:1544–1549.
24. Sun GR, He JB, Pittman CU. Destruction of halogenated hydrocarbons with solvated electrons in the presence of water. Chemosphere 2000; 41:907–916.
25. Chablay E. Utilization of metal ammoniums in organic chemistry. Ann Chim 1914; 9:469–519.
26. Hudson FM. A Study of the Reduction of Aromatic Halogenated Compounds by Alkali Metals in Liquid Ammonia. Ph.D. Dissertation, The University of Tennessee, University Microfilms 64-7860, 1963.
27. Kennedy MV, Stojanivic B, Shuman FL. Chemical and thermal aspects of pesticide disposal. Environ Qual 1972; 1:63–65.
28. Mackenzie K, Kopinke KF, Remmie M. Reductive destruction of halogenated hydrocarbons in liquids and solids with solvated electrons. Chemosphere 1996; 33:1495.
29. He J, Pittman CU. Dechlorination of PCBs, CAHs, herbicides and pesticides, neat and in soils at 25°C using Na/NH_3. J Hazard Mater 2002; 92:51–62.
30. Pittman CU, He J. Proceedings of the 31st Mississippi Water Resources Conference April 10–11, 2001:75–84.
31. He J. Solvated Electron Reductions for Environmental Remediation. Ph.D. Dissertation, Mississippi State University, Mississippi State, MS, 2000.
32. Vlieger JJ, Kieboom PG, van Bekkum H. Metal–ammonia reduction of naphthalene at –33°C: formation of oligomeric compounds. J Org Chem 1986; 51:1389.
33. Holm T. Kinetic isotope effects in the reduction of methyl iodide. J Am Chem Soc 1999; 121:515.
34. Abel AE. Pressurized filtration. US Patent 5,582,744, 1996.
35. Jones GR. The B.E.S.T.® solvent extraction process treatment of soil, sediment, and sludges. Environ Prog 1992; 2(3):223–227.

36. Abel AE, Mouk RW. Methods for the destruction of ozone depleting substances. US Patent 5,559,278, 1996.
37. Mouk RW, Abel AE. Non-metallized and substoichiometric metallized reactions with ammonia and other weak bases in the dehalogenation of refrigerants. US Patent 5,414,200, 1995.
38. Mouk RW, Abel AE. Reactions with ammonia and other weak bases in the dehalogenation of refrigerants. US Patent 5,414,200, 1995.
39. Abel MW, Abel AE, Heyduk, Mouk RW. Methods for the elimination of cyanides in dehalogenation of halofluorocarbons. US Patent 5,602,295, 1997.
40. Mohammad M. Solvated electron reductions of aromatic nitro compounds and 2-chloroethylsulfide in ethylenediamine at 116°C. MS Thesis, Mississippi State University, Mississippi State, MS, 1996.
41. Abel AE. Methods of decontaminating nuclear waste-containing soil. US Patent 5,495,062, 1996.
42. Heyduk AF, Abel AE, Mouk RW. Methods of decontaminating substrates with in-situ generated cyanides. US Patent 5,678,231, 1997.
43. Abel AE. Methods of decontaminating mercury-containing soils. US Patent 5,516,968, 1996.
44. Heyduk, Abel AE, Mouk RW. Methods of decontaminating soils containing hazardous metals. US Patent, 5,613,238, 1997.
45. Madey TE, Lu QB. Giant enhancement of electron-induced dissociation of chlorofluorocarbons coadsorbed with water or ammonia ices: implications for atmospheric ozone depletion. J Chem Phys 1999; 111:2861.

9
Permeable Reactive Barriers of Iron and Other Zero-Valent Metals

Paul G. Tratnyek
Oregon Health and Science University, Beaverton, Oregon, U.S.A.

Michelle M. Scherer
University of Iowa, Iowa City, Iowa, U.S.A.

Timothy L. Johnson
AMEC Earth & Environmental, Inc., Portland, Oregon, U.S.A.

Leah J. Matheson
MSE Technology Applications, Inc., Butte, Montana, U.S.A.

I. INTRODUCTION

A. Historical Context

The "modern" history of the use of zero-valent metals (ZVMs) in the remediation of contaminated water has been summarized from several perspectives [1–4]. By most accounts, the critical event was the serendipitous discovery that trichloroethene (TCE) is degraded in the presence of the metal casing materials used in some groundwater monitoring wells [5]. This observation led to recognition that granular iron metal might be applicable to the remediation of groundwater that is contaminated with chlorinated solvents. Around the same time, the possibility of engineering permeable treatment zones for in situ treatment of contaminated groundwater had led to a search for suitable reactive media, and granular iron quickly became the most promising reactive medium for application in permeable treatment zones [6]. The confluence of these two developments (granular iron and permeable treatment zones) made the emergence of reactive barriers containing granular iron into one of the landmark developments in the history of groundwater remediation technology.

The rapid development of this technology over the last decade has been accompanied by a conspicuous increase in the quantity of published information on the reaction of iron metal with organic and inorganic solutes in aqueous systems. With so much activity in the present, it is easy to overlook how much relevant work was done earlier. For example, the electrolytic deposition of dissolved metals onto ZVMs has long been known to chemists, and the potential for application of this chemistry to water treatment was recognized at least as far back as the 1960s [7]. Similarly, the use of ZVMs to perform selective reductions for organic synthesis was already well documented by the 1920s (e.g., Refs. 8 and 9), and environmental applications had been described by the 1980s [10,11]. In fact, prior to 1990, there had already been several detailed "process-level" studies on the removal of organics (e.g., Refs. 12–15) and inorganics (e.g., Refs. 7, 16, and 17). This early work was very widely dispersed, however, and a unified understanding of the processes responsible for contaminant removal by ZVMs has only recently begun to take shape.

B. Scope

The scope of this review is centered around permeable reactive barriers (PRBs) of ZVMs. Among the ZVMs used in remediation applications, iron metal (ZVI or Fe^0) is by far the most important. PRBs of ZVI (sometimes designated FePRBs) are the technology known colloquially as "iron walls." However, as illustrated in Fig. 1, not all PRBs are made from ZVMs and not all remediation applications of ZVMs are PRBs.

1. Permeable Reactive Barriers

Technologies for treatment of subsurface contamination can be divided into "ex situ" methods that involve removal of the contaminated material for treatment at the surface and "in situ" methods where the treatment is applied to the subsurface. In situ treatment technologies include a variety of related methods such as continuous trenches, funnel-and-gates, passive reactive wells, geochemically manipulated zones, and biologically reactive zones. Continuous trenches and funnel-and-gates are the most common types of PRBs [18,19]. At least one formal definition of a PRB has been given [3], but for the present purpose we prefer a slightly narrower and simpler definition: "a permeable subsurface zone constructed of reactive material that is oriented to intercept and destroy or immobilize contaminants." The major elements of a PRB are shown in Fig. 2.

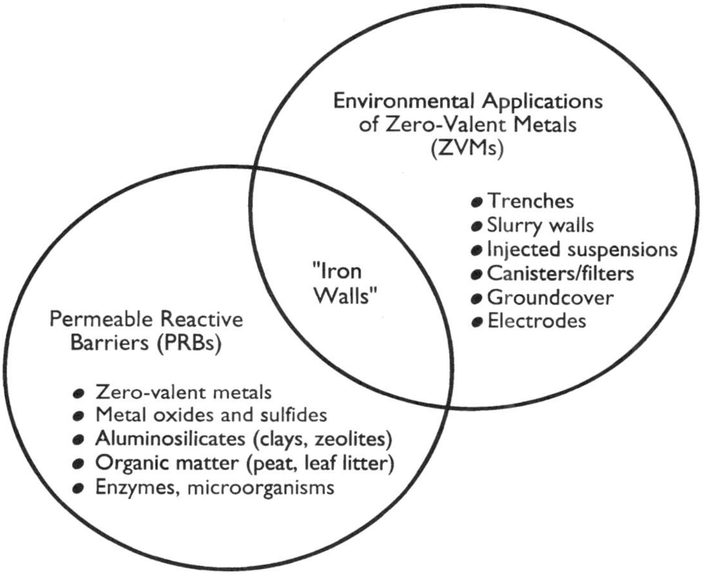

Figure 1 Venn diagram showing the relationship between various types of PRBs and various remediation applications of ZVMs. The intersection of these two categories represents PRBs with ZVI as the reactive medium (i.e., FePRBs or "iron walls").

In contrast to the conventional PRB, a permeable reactive treatment zone (PRTZ) is a geochemically manipulated subsurface zone where aquifer material is altered to promote destruction or immobilization of target chemicals (e.g., flushed with sodium dithionite to create a zone of reduced iron [20–23]). Passive reactive wells (PRWs) are a series of wells or caissons containing a treatment material, through which water flows because of a permeability contrast between the wells and aquifer. A biologically reactive barrier (BRB), sometimes called a "biocurtain," is a subsurface zone where microbiological activity is enhanced or modified to provide treatment of target chemicals.

2. Reactive Media

The core function of a PRB (and many related technologies) is to bring the contaminated material in contact with a reactive material that promotes a process that results in decontamination. The range of reactive materials that can be applied in PRBs is quite diverse, as illustrated by Table 1. The

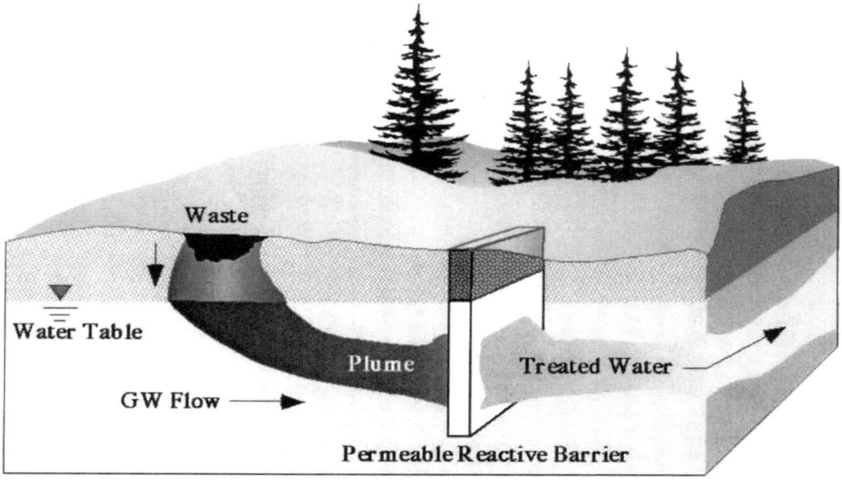

Figure 2 Typical configuration of a PRB, showing the source zone, plume of contamination, treatment zone, and plume of treated groundwater. (Reprinted with permission, Powell and Associates.)

Table 1 Summary of Reactive Media[a]

Type	Composition	Applications	Selected references
Zero-valent metals	Fe, Zn, Sn	TCE, Cr(VI), etc.	Numerous[b]
Bimetallic combinations	Fe/Ni, Fe/Pd	PCBs, chlorophenols, chloromethanes,	[24–26]
Metal oxides	Iron oxides	Cr(VI), U(VI)	[20,21,23,27–30]
Metal sulfides	FeS	Chloromethanes, ethanes, and ethenes	[31,32]
Aluminosilicates	Clays, Zeolites	TCE, Cr(VI)	[33–35]
Calcium phosphates	Apatite, bone char	U(VI), Pb	[36,37]
Carbonaceous materials	Peat, sawdust, leaf litter, ground rubber	Phosphate, BTEX, Acid Mine Drainage	[38–41]

[a] Other tables of this type can be found in Refs. 4, 30, and 42.
[b] Complete list at http://cgr.ese.ogi.edu/ironrefs/.

reactive material can be introduced directly into the subsurface or formed in situ after addition of agents that are not directly involved in reaction with the contaminants. The former is exemplified by ZVMs, whereas the latter is exemplified by the zone of ferrous iron formed by "in situ redox manipulation" [20–23]. In Table 1, we have tried to capture the whole range of reactive media that are currently being used in PRBs, but the remainder of this review will focus on PRBs constructed with ZVMs.

C. Other Sources of General Information on ZVI and PRBs

The rapid increase in interest and knowledge associated with remediation applications of ZVMs and PRBs has led to a number of reviews on these subjects. To date, these include Refs. 2, 3, 18, and 42–53. In general, these reviews do not attempt to provide comprehensive coverage of the primary literature in this field, as it has already become too vast. Fortunately, most of the primary literature is included in several databases that are available on the World Wide Web. These databases can be found at http://cgr.ese.ogi.edu/ironrefs and http://www.rtdf.org.

II. CONTAMINANT-REMOVAL PROCESSES

The processes responsible for contaminant removal by ZVMs and PRBs include both "physical" removal from solution to an immobile phase and "chemical" removal by reaction to form less hazardous products. In the discussion that follows, we will refer to the former as sequestration and the latter as transformation. This distinction has heuristic value, even though sequestration and transformation processes are related for many contaminants.

A. Removal by Sequestration

For the purposes of this review, we have chosen the term *sequestration* to represent contaminant removal by processes that do not involve contaminant degradation. Although the term is most commonly applied to the fate of organic contaminants [54], it can also be applied to metals and other inorganic contaminants. In older literature on removal of contaminant metals, the term *cementation* was commonly used (e.g., Ref. 55), but this term is not used here.

Sequestration by Fe^0 occurs mainly by adsorption, reduction, and coprecipitation, although other processes may be involved such as pore diffusion and polymerization. In most cases, *adsorption* is the initial step and subsequent transformations help ensure that the process is irreversible. In some cases, however, adsorption is the sequestration process of primary importance. This is certainly true with metals that occur as soluble cations, which can be expected to adsorb fairly strongly to iron oxides, but cannot be reduced to insoluble forms by Fe^0: e.g., Mg^{2+}, Mn^{2+}, and Zn^{2+} [56]. It may also be true of toxic heavy metals like Cd, Cu, Hg, Ni, and Pb, which exist predominantly as soluble cations under aerobic conditions, but could be reduced to insoluble species by Fe^0. In some cases, the dominant process is unmistakable, such as in the recovery of Hg^0 using Fe^0 [57–59]. In other cases, however, the relative importance of adsorption vs. reduction is uncertain because most of the available literature either focuses on adsorption without attention to whether the contaminant metal undergoes a change in valence state (e.g., Ref. 60) or assumes sequestration is due to reduction without distinguishing how much is due to adsorption (or coprecipitation) alone (e.g., Refs. 61 and 62).

Of greater recent interest are metals that exist predominantly as soluble, hazardous oxyanions in oxic groundwaters, but that become relatively insoluble species when reduced, making them candidates for remediation by reductive immobilization. These metals include As(V), Cr(VI), Se(VI), Tc(VII), U(VI), and a few others [51,63,64]. In general, a complex and variable mixture of processes is responsible for sequestration of these contaminants by Fe^0. For example, Cr(VI) is at least partially reduced to Cr(III), which is then precipitated as a mixed oxyhydroxide [65–67].

$$Fe^0_{[solid]} + CrO_4^{2-} + 8\,H^+ \longrightarrow Cr^{+3} + Fe^{3+} + 4\,H_2O \quad (1)$$

$$(1-x)\,Fe^{3+} + (x)\,Cr^{3+} + 4\,H_2O \longrightarrow Fe_{(1-x)}Cr_{(x)}OOH_{[solid]} + 3\,H^+ \quad (2)$$

Although further reduction of Cr(III) to Cr^0 is not thermodynamically favorable with Fe^0, reduction of Se(VI) all the way to Se^0 is expected and has been observed [67]. As(V) can also be reduced by Fe^0 to As^0, but sequestration of As(V) seems to involve mainly As(III) under anaerobic conditions [68,69] and adsorbed As(V) under aerobic conditions [70].

Unlike the other metal oxyanions discussed above, the thermodynamic driving force for reduction of U(VI) by Fe^0 is only moderately favorable under conditions of environmental relevance. Because the dominant forms of U(VI) in most groundwaters are carbonate complexes, the following overall (reduction and precipitation) reaction might be expected:

$$Fe^0_{[solid]} + UO_2(CO_3)_2^{2-} + 2\,H^+ \longrightarrow UO_{2[solid]} + 2\,HCO_3^- + Fe^{2+} \quad (3)$$

Reactions of this type could be responsible for the sequestration of U(VI) by Fe^0, as favored by several investigators [63,71,72]. However, adsorption of U(VI) to iron oxides is known to be strong, and evidence that this process is the dominant sequestration mechanism has been provided by other investigators [67,73,74]. Recently, detailed studies of samples from the FePRB at the Y-12 site, Oak Ridge, TN (Fig. 3) have shown that the distribution and speciation of uranium in the Fe^0-bearing zone is complex, and that sampling and characterization of these materials is challenging [75,76].

B. Removal by Transformation

To contrast with the term sequestration, we have chosen *transformation* to represent chemical reactions that convert contaminants to distinct products. The transformation of metals from one valence state to another was included in the previous section because the effect of these transformations is mainly to enhance sequestration. In contrast, there are a few nonmetal inorganic contaminants that are transformed by Fe^0 to soluble but comparatively innocuous products, which are discussed below. Following that,

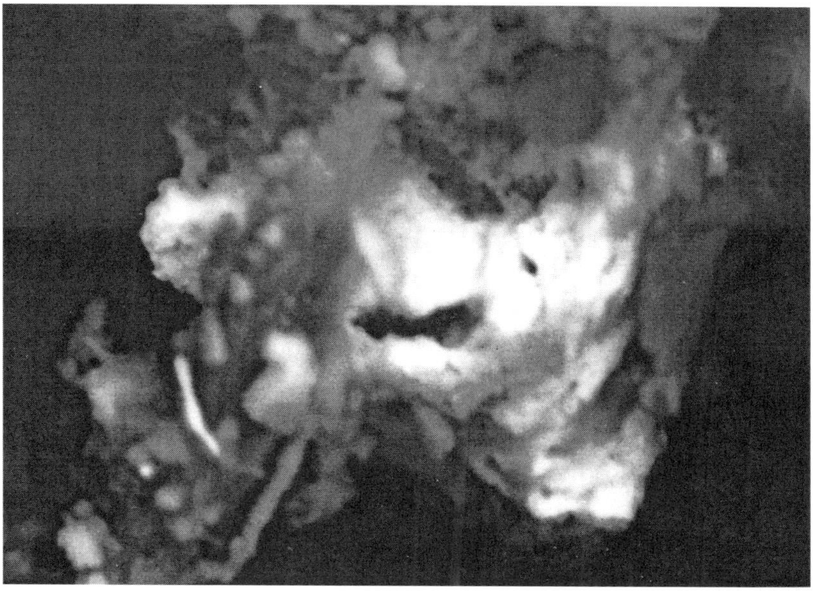

Figure 3 Scanning electron micrograph of an Fe^0 grain taken from an FePRB at the Y-12 site at Oak Ridge, TN. The bright spot is mostly U, showing that these deposits are localized on the Fe^0 surface. These deposits were associated with varying amounts of Fe, S, Si, and Ca. Additional details on the analyses of these samples are in Ref. 76.

we review the reductive transformation of organic contaminants by Fe^0, with emphasis on the two most important pathways: dehalogenation of chlorinated aliphatic or aromatic contaminants and reduction of nitro-aromatic compounds.

1. Inorganic Transformations

The two most notable examples of reductive transformations by Fe^0 that involve nonmetal inorganic compounds are reduction of nitrate [Eq. (4)] and aqueous chlorine [Eq. (5)].

$$4\,Fe^0 + NO_3^- + 10\,H^+ \longrightarrow 4\,Fe^{2+} + NH_4^+ + 3H_2O \tag{4}$$

$$Fe^0 + 2\,HOCl + 2\,H_2O \longrightarrow Fe^{2+} + 2\,H^+ + 2\,Cl^- \tag{5}$$

The reduction of nitrate yields ammonia under most conditions [77–80], but some have suggested that dinitrogen is formed [81]. Possible applications of this process include not only the direct treatment of nitrate-contaminated groundwater, but also the pretreatment of groundwater that is contaminated with both nitrate and radionuclides, in order to allow the development of more strongly reducing biogeochemical conditions (sulfidogenesis or methanogenesis) that are necessary for microbially mediated immobilization of uranium [75].

The reduction of aqueous chlorine (HOCl) to chloride by Fe^0 and other ZVMs [Eq. (5)] has long been known as a major contributor to the decay of residual chlorine disinfectant during distribution in drinking water supply systems that contain metal pipes (e.g., Ref. 82). This reaction can, however, be turned to advantage for the removal of excess residual chlorine, and a variety of proprietary formulations of granular ZVMs are available commercially for this purpose (e.g., KDF Fluid Treatment, Inc. Three Rivers, MI). This application is sometimes called "dechlorination," but should not be confused with the dechlorination of organic contaminants, which is discussed below.

Other nonmetal inorganic compounds that might be usefully transformed by Fe^0 include perchlorate, sulfate, and cyanide. Although the energetics for reduction of these compounds are all favorable, the kinetics appear to be unfavorable in the absence of microbial mediation. In the case of perchlorate, it has been reported that biodegradation can be inhibited by Fe^0 [83]. This means that useful applications of these reactions will have to wait until effective methods of catalyzing these reactions are discovered.

2. Dechlorination

Dehalogenation can occur by several reductive pathways. The simplest results in replacement of a C-bonded halogen atom with a hydrogen, and is known as *hydrogenolysis* or *reductive dehalogenation*. For a general chlorinated aliphatic compound, RCl, hydrogenolysis by Fe^0 corresponds to the overall reaction:

$$Fe^0 + RCl + H^+ \longrightarrow Fe^{2+} + RH + Cl^- \tag{6}$$

This reaction is the dominant dehalogenation pathway in reduction of halogenated methanes [84] and haloacetic acids [85]. In Fig. 4, this reaction is illustrated for perchloroethene (PCE), where complete dechlorination by this pathway requires multiple hydrogenolysis steps. The relative rates of these steps are a critical concern because they determine whether

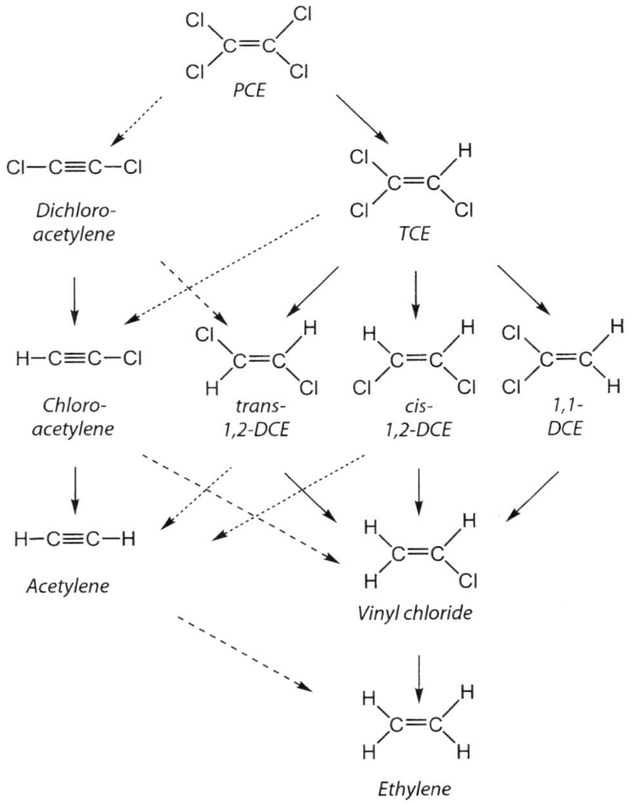

Figure 4 Scheme showing the branching between hydrogenolysis (solid arrows), reductive elimination (fine dashed arrows), and hydrogenation (course dashed arrows) pathways to produce the major products of chlorinated ethene reduction by ZVMs. (Adapted from Ref. 88.)

persistent and hazardous intermediates (such as vinyl chloride, VC) will accumulate [86,87].

In principle, aryl halogens can also be subject to hydrogenolysis, although this reaction is likely to be less facile than hydrogenolysis of most alkyl halogens. In fact, the only confirmed example of hydrogenolysis involving aryl halogens by Fe^0 under environmental conditions is for pentachlorophenol, and the reaction was found similar in rate to literature values for TCE [89]. In contrast, rapid hydrogenolysis of aryl halogens by Fe^0 has been obtained under extreme conditions, such as in supercritical water [90–92] or at high temperature [93]. These variations are not amenable to use in a PRB, but are discussed along with related enhancements in Sec. V.C.

The other major dehalogenation pathway involves elimination of two halogens, leaving behind a pair of electrons that usually goes to form a carbon-carbon double bond. Where the pathway involves halogens on adjacent carbons, it is known as *vicinal dehalogenation* or *reductive β-elimination*. The fine dashed arrows in Fig. 4 illustrate this process for PCE. Note that this pathway can produce alkynes from vicinal dihaloalkenes [88,94,95], as well as producing alkenes from vicinal dihaloalkanes [96,97].

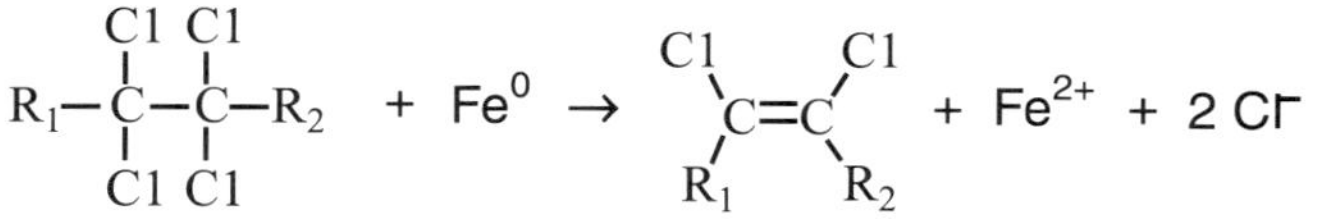
(7)

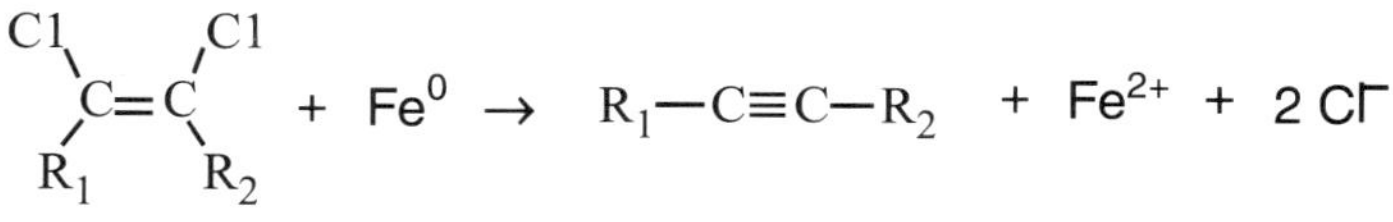
(8)

In addition to the two major reductive pathways for dechlorination, there are two additional reactions that have been observed: *hydrogenation*, which involves addition of hydrogens across a C-C double or triple bond [Eq. (9)] and *dehydrohalogenation*, which involves elimination of H^+ and X^- and creation of a new C-C double bond [Eq. (10)]. Hydrogenation has been invoked to explain the distribution of products observed in several studies involving chlorinated alkenes and Fe^0 [88], and is particularly important where a noble metal like Pd is present to act as a catalyst (see Sec. III.B). Note that we have written H_2(surf) in Eq. (9) to represent all of the various forms of surface-activated hydrogen, and do not mean to imply that the reaction necessarily involves adsorbed diatomic molecular hydrogen. Dehydrohalogenation has not received much attention as a reaction that might contribute to degradation of chlorinated ethenes by Fe^0, even though it can be base catalyzed [98], which might make it favored under the alkaline conditions that can be created by corrosion of Fe^0.

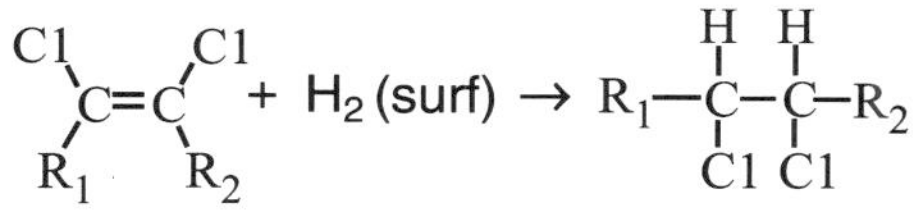 (9)

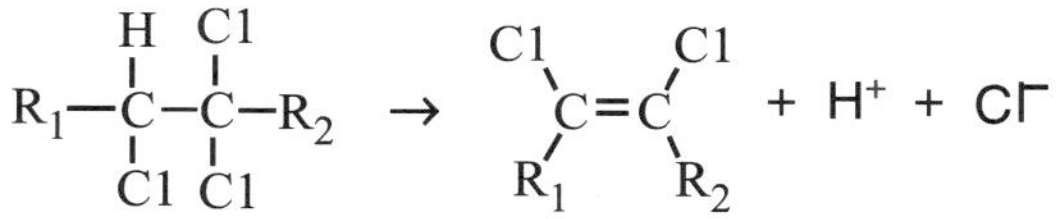 (10)

3. Nitro and Azo Reduction

In general, reduction of aromatic nitro groups occurs in three steps, via nitroso and hydroxylamine intermediates, to the amine. For nitrobenzene and simple substituted nitrobenzenes reacting with Fe^0 in batch model systems, the intermediates have been detected in solution, but the dissolved amine alone is usually sufficient for good mass balance [99–103]. Thus, the net reaction is:

$$3\,Fe^0 + ArNO_2 + 6\,H^+ \longrightarrow 3\,Fe^{2+} + ArNH_2 + 2H_2O \qquad (11)$$

Recently, research on nitro reduction by Fe^0 has been extended to environmental contaminants with multiple nitro groups, such as TNT and RDX [104–107]. As expected, batch experiments show that TNT and RDX are rapidly reduced by Fe^0 to a complex mixture of products (Fig. 5). In contrast, column experiments with TNT have shown a very high capacity to

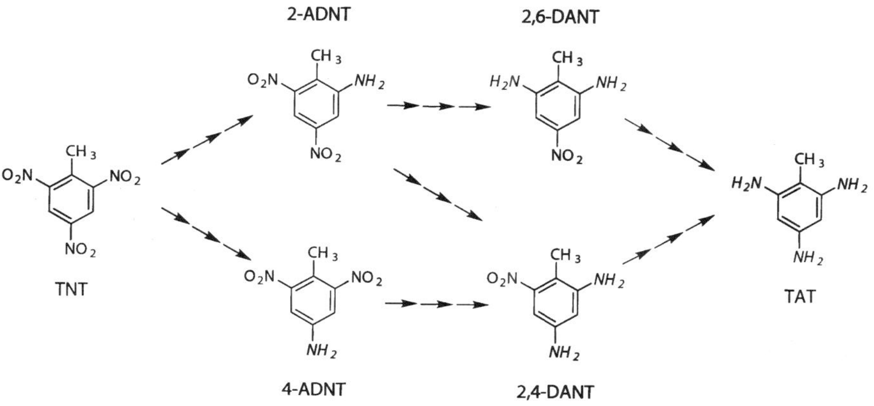

Figure 5 Scheme showing branching among nitro reduction steps for TNT by zero-valent metal. Triple arrows indicate that each step shown presumably proceeds through three steps with nitroso and hydroxylamine intermediates. (Adapted from similar figures for other reducing systems, including Refs. 109,110.)

convert all products to triaminotoluene [108]. This result suggests that earlier work, all of which appears to have been done in batch experiments, may have led to unrepresentative conclusions regarding the formation of soluble reduction products. Despite this simple FePRBs may be sufficient to reach treatment goals for some explosives under real-field conditions.

In principle, the nitroso, hydroxylamine, and/or amine products of nitro reduction might undergo coupling to form azoxy, azo, and/or hydrazo dimers, but no evidence for these products has been found under the conditions that have been studied to date. One reason that these dimers do not accumulate may be that they are rapidly reduced by Fe^0. In fact, Fe^0 reduces azo groups to amines [Eq. (12)] very rapidly [111–113], and this reaction may prove to be useful in the remediation of wastewaters contaminated with azo dyes.

$$2\,Fe^0 + ArN{=}NAr + 4\,H^+ \longrightarrow 2\,Fe^{2+} + 2\,ArNH_2 \tag{12}$$

Like nitro and azo groups, the nitrosamine moiety is subject to reduction by Fe^0 and is present in some important environmental contaminants. One such contaminant is *N*-nitrosodimethylamine (NDMA), which is reduced by Fe^0 via a complex mechanism that gives the following overall reaction [114,115]:

$$3\,Fe^0 + (CH_3)_2NN{=}O + 7\,H^+ \longrightarrow 3Fe^{2+} + (CH_3)_2NH + NH_4^+ + H_2O \tag{13}$$

NDMA is a potent carcinogen that not only occurs in groundwater contaminated with rocket fuels but can be formed from precursors that sometimes occur in groundwater and even drinking water [116]. Another important nitrosamine that is reduced by Fe^0 is the explosive RDX [104–106,117,118]. The products of this reaction are difficult to characterize, but appear to be low molecular weight, polar, N-containing compounds, which are likely to be analogous to the products formed from NDMA [Eq. (13)].

4. Other Organic Transformations

In principle, there are other organic functional groups that might be reduced by Fe^0 under environmental conditions, including aldehyde, ketone, quinone, diamine, nitrile, oxime, imine, sulfoxide, and disulfide moieties [119–121]. Recently, the reduction of quinonoid redox indicators by Fe^0 has been explored in an educational context [122], but we are not aware of any application of FePRBs for remediation of groundwater contaminants that contain these moieties. It is likely, however, that examples will emerge in the future. In addition, it is to be expected that other types of transformations will become accessible as "enhanced" and hybrid technologies involving ZVMs become available. A few of these are discussed in Sec. V.C.

To conclude this section, it is worth noting some of the chemistry that is not expected in association with FePRBs. In general, any compound that is easily oxidized will be a poor candidate for reduction. Such compounds include saturated or aromatic hydrocarbons (including the constituents of gasoline, coal tar, creosote, etc.), ethers and alcohols (including MTBE, glycols, etc.), and phenols (e.g., cresols, various residues from digestion lignin into paper pulp). At the same time, care should be taken not to presume that a contaminant is transformed by reduction just because it is found to be removed by contact with Fe^0. This is illustrated by recent reports that Fe^0 degrades the pesticides carbaryl [123] and benomyl [124], both of which were attributed to reduction. However, these pesticides do not contain any readily reducible functional groups. It is more likely that Fe^0 degrades carbaryl by catalyzing alkaline hydrolysis of the phosphate ester moiety, and benomyl by catalyzing alkaline hydrolysis of the amide moiety.

III. REACTIVE MEDIA AND THEIR PROPERTIES

There are two types of metals that are of interest as reactive media in PRBs: (1) corrodable, base metals, which have equilibrium potentials for dissolution that are below the potential for reduction of water or any strongly oxidizing solutes, and (2) noble, catalytic metals, which are not subject to oxidative dissolution under environmental conditions but which participate in reduction of solutes as catalysts. The corrodable, base metals (Fe, Zn, Sn, etc.) are discussed in Sec. III.A, and the role of noble, catalytic metals (Pd, Ni, etc.) in PRBs is discussed in Sec. III.B.

A. Iron and Other Corrodable Metals

Although the majority of interest in remediation applications of corrodable metals revolves around Fe^0, other possibilities have been investigated, including magnesium, tin, and zinc. The bulk of this work has used Zn^0 as a model system for comparison with Fe^0 (e.g., Refs. 95, 96, 125, and 126), but a few studies have surveyed a range of metals as possible alternatives to Fe^0 in environmental applications other than PRBs (e.g., Refs. 127 and 128).

1. Corrosion Chemistry

The corrosion reaction involving water [Eq. (14)] is slow but presumably ubiquitous, whereas corrosion of Fe^0 by reaction with dissolved oxygen [Eq. (15)] is rapid as long as O_2 is available.

$$Fe^0 + 2H_2O \longrightarrow Fe^{2+} + H_2 + 2OH^- \quad (14)$$

$$2Fe^0 + O_2 + 2H_2O \longrightarrow 2Fe^{2+} + 4OH^- \quad (15)$$

The presence of a reducible contaminant in an Fe^0-H_2O system provides another reaction [in addition to Eqs. (14)–(15)] that can contribute to the overall corrosion rate. This is exemplified in Eq. (6) for the hydrogenolysis of a generic chlorinated hydrocarbon, Eq. (11) for nitro reduction, and Eq. (12) for azo compounds.

For simplicity, we have written and balanced these equations for acidic conditions, but the speciation of iron and some contaminants, as well as the thermodynamic potentials for the associated redox half-reactions, will vary with pH. The most efficient way to represent the effects of pH is in an Eh-pH diagram, such as Fig. 6. This particular diagram shows that reduction of three contaminants (CCl_4, $ArNO_2$, and Cr(VI)) by Fe^0 is thermodynamically favorable over a wide range of pH, even though the speciation of the Fe(II)

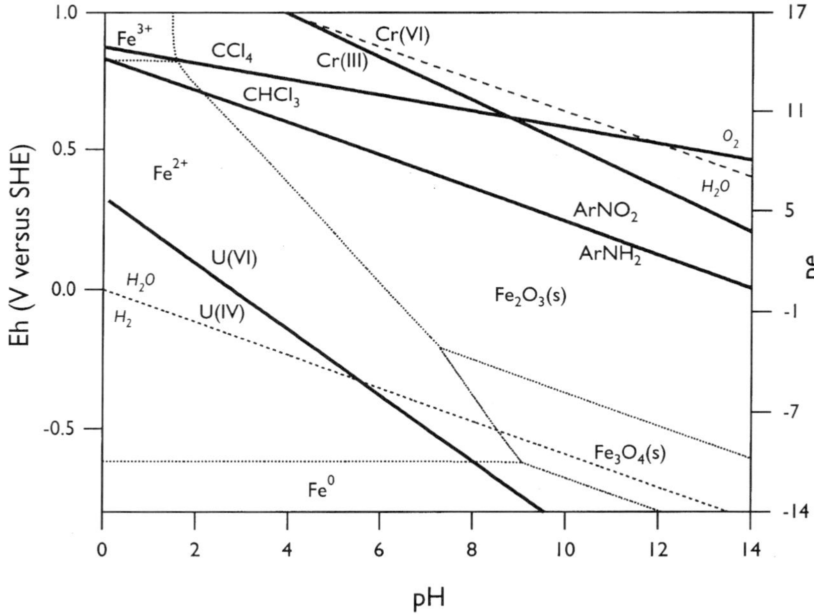

Figure 6 Eh-pH diagram for the Fe^0-H_2O system where total dissolved $Fe = 1 \times 10^{-6}$ M, Fe_3O_4 and Fe_2O_3 are assumed to be the solubility limiting phases, and [ox] = [red] for all redox active species. Other Eh–pH diagrams for Fe^0-H_2O-contaminant systems can be found in Refs. 42, 84, and 129–131.

that is formed changes considerably. In contrast, reduction of U(VI) by Fe^0 switches from favorable to unfavorable where the pH increases above 8.

2. Common Types of Granular Iron

The Fe^0 that has been used for contaminant degradation includes construction-grade material, used primarily in field applications, and reagent-grade material, used primarily in laboratory studies. Construction-grade granular iron is prepared from scrap "gray" or ductile cast iron by grinding and sieving, and annealed under an oxidizing atmosphere. The resulting material usually has a thick outer layer of iron oxide, which sometimes includes considerable amounts of inorganic carbon. Reagent-grade granular iron is usually prepared by electrolytic precipitation, is then ground, sieved, and sometimes annealed under a reducing atmosphere leaving a bright metallic surface.

A great deal of empirical testing has been done to determine which types of iron are most reactive with a particular contaminant, but little of this work has been reported in the peer-reviewed literature. A few studies have summarized readily available properties of a significant range of iron types [126,132], but these efforts fall far short of forming the basis for a systematic understanding of the relative reactivity of granular metals. The role of some physical properties of granular Fe^0 are well established, as discussed in Sec. IV.A.1 and V.B.1, so these properties are summarized in Table 2 for selected construction- and reagent-grade irons.

Table 2 Summary of Iron Properties

Supplier	α_s ($m^2\ g^{-1}$)[a]	ρ_s ($g\ cm^{-3}$)[b]	ρ_b ($g\ cm^{-3}$)[c]
Connelly iron aggregate (ETI CC-1004)	1.8 (2)	7.55	1.9
Peerless cast iron aggregate (ETI 8/50)	0.9±0.7(11)	7.39	2.2
Master builder	1.3±0.7 (14)	7.38	2.7
Fisher electrolytic	0.2±0.2 (9)	9.49	2.6
Fluka filings	0.03±0.06 (5)	8.58	3.8

[a] Average of reported specific surface areas in m^2 per gram of Fe^0 as summarized in Ref. 80. Statistics are based on independently reported values from the literature: uncertainties are one standard deviation and the number of averaged values are given in parenthesis.

[b] Specific density in grams per liter of Fe^0 volume [133]. For comparison, typical literature values are 7.87 $g\ cm^{-3}$ for pure elemental iron, 7.2–7.3 for cast and malleable iron, and 4.9–5.3 for hematite [134].

[c] Bulk density in grams per liter of total volume. Construction-grade Fe^0 can be prepared with bulk densities from 1.4 to 3.5 $g\ cm^{-3}$ (90 to 220 $lb\ ft^{-3}$), but currently available products are about 2.4 $g\ cm^{-3}$ (150 $lb\ ft^{-3}$) (David Carter, Peerless Metal Powders and Abrasives, personal communication).

B. Bimetallic Combinations

In addition to transformation by corrodable metals (such as Fe^0 and Zn^0), bimetallic combinations of a catalytic metal with a corrodable metal (such as Pd/Fe or Ni/Fe) have also been shown to transform a variety of contaminants. In most cases, rates of transformation by bimetallic combinations have been significantly faster than those observed for iron metal alone [26,96,135–139]. Not only have faster transformation rates been observed with bimetallic combinations, but, in some cases, transformation of highly recalcitrant compounds, such as polychlorinated biphenyls (PCBs), chlorinated phenols, and DDT has been achieved [24,140,141]. The mechanism responsible for the enhanced reactivity with bimetallic combinations is still unclear; however, it has been suggested that electrochemical effects, catalytic hydrogenation, or intercalation of H_2 may be responsible. A likely limitation to the full-scale application of bimetallic combinations to groundwater remediation is deactivation of the catalytic surface either by poisoning (e.g., by sulfide) or by formation of thick oxide films [136,142,143].

IV. MICROSCALE PROCESSES

Almost everything that is known about the fundamental processes that are responsible for contaminant removal by ZVMs has been derived from laboratory experiments done with bench-scale model systems. Of this work, the majority has been done in batch reactors consisting of dilute slurries of Fe^0 particles suspended in small bottles. Batch experiments are simple to perform and the results can be easy to analyze, but this method can be limiting, and questions remain about how well it models conditions that are relevant to the field. Recently, a few other small-scale laboratory model systems have been described that offer greater control over key experimental variables, e.g., (rotating) iron disk electrodes [101,144], recirculating batch reactors [100], and small columns operated in "miscible-displacement" mode [145]. Future developments along these lines may greatly improve our understanding of the fundamental chemistry that controls the performance of this technology.

Through the many studies that have now been done in well-controlled model systems, a general conceptual model has emerged of the processes controlling contaminant reduction on Fe^0. Some of the key elements of this model are summarized in Fig. 7, using a generic chlorinated hydrocarbon, RX, as the model contaminant. First, RX must be conveyed to the stagnant boundary layer at the oxide-water interface, then it must diffuse across the boundary layer and form a complex with a reactive site either on or in the oxide film (or directly on the Fe^0 through a defect in the oxide film [146]). Only then can reduction occur (with electrons that ultimately come from the

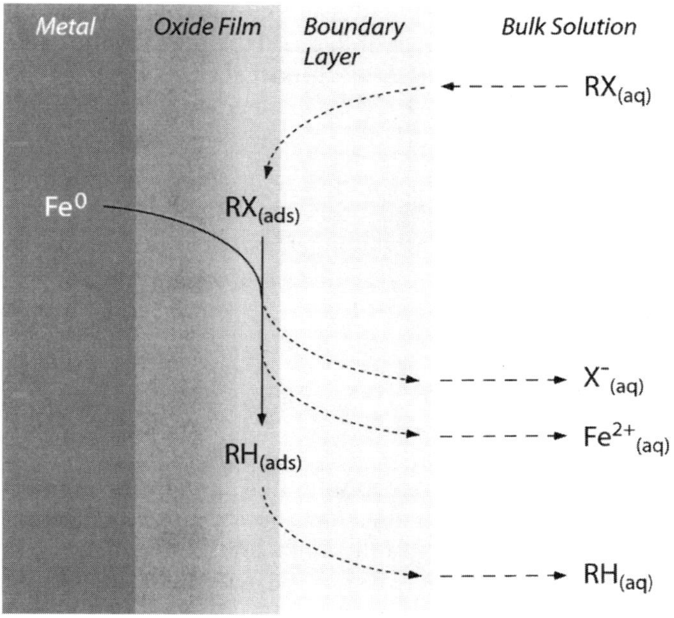

Figure 7 Schematic of the primary steps involved in dehalogenation of RX at Fe^0-oxide-H_2O interface. Coarse dashed arrows represent mass transport between the bulk solution and the particle surface, fine dashed arrows denote diffusion across the stagnant boundary layer and surface complexation, and solid arrows show electron transfer and bond rearrangement on the surface. (Adapted from Ref. 147.)

underlying Fe^0), followed by desorption and diffusion of products away from the surface.

A. Reaction at the Surface

1. Basic Kinetic Model

Most of the primary kinetic data that have been obtained from bench-scale model systems suggest that reaction with Fe^0 is first order in the concentration of solution phase contaminant, C. Thus, we can write the following rate law in differential and integrated forms:

$$-\mathrm{d}C/\mathrm{d}t = k_{\mathrm{obs}}C \tag{16}$$

$$\ln(C_t/C_0) = -k_{\mathrm{obs}}t \tag{17}$$

where k_{obs} is the pseudo-first-order rate constant. Experimental values of k_{obs} are routinely obtained from the slope of the regression line for $\ln(C_t)$

or $\ln(C_t/C_0)$ vs. time. However, k_{obs} will only be constant for a limited range of experimental conditions, as it can be influenced by a host of system properties. Two of the best characterized factors that influence k_{obs} are addressed below. Others, such as pH [84,148–150], are not discussed further here because general models for their effects are not yet available.

Effect of Iron Concentration. Among the factors that influence k_{obs}, the effect of the amount of iron surface area that is accessible to the contaminant has received the most attention. This effect is most often described by a linear relationship:

$$k_{obs} = k_{SA}\rho_a \tag{18}$$

where k_{SA} is the specific reaction rate constant (L hr^{-1} m^{-2}) and ρ_a is the concentration of iron surface area (m^2 L^{-1} of solution). Deviations from this linear relationship have been reported [151–153], but their general significance is not yet well established.

Data for k_{SA} that were available as of November 1995 were summarized in Refs. 86 and 132, but many more data have been reported since then. Selected values of k_{SA} are given in Table 3. In addition, quantitative structure-activity relationships (QSARs) have been reported that may be suitable for estimating values of k_{SA} that have not been measured [87,88, 154–156].

The other term in Eq. (18), ρ_a, can be calculated from:

$$\rho_a = a_s\rho_m \tag{19}$$

where a_s is the specific surface area (m^2 g^{-1}) of a type of Fe^0, usually measured by BET gas adsorption, and ρ_m is the mass concentration of Fe^0.

Table 3 Selected Rate Constants for Reduction by Fe^0

Contaminant	k_{SA}[a] (L hr^{-1} m^{-2})
Carbon tetrachloride (CCl_4)	$1.2(\pm1.5)\times10^{-1}$ (11)[b]; 1.5×10^{-1} [157]
1,1,1-Trichloroethane (TCA)	1.1×10^{-2} [158]; 4.6×10^{-1} [96]
Trichloroethene (TCE)	$3.9(\pm3.6)\times10^{-4}$ (12)[b]; $3.3(\pm5.2)\times10^{-4}$ (4)[c]; 1.1×10^{-3} (2)[d]
Vinyl chloride (VC)	$5.0(+1.5)\times10^{-5}$ (3)[b]; 8.2×10^{-6} (5)[e]
2,4,6-Trinitrotoluene (TNT)	$5.0(\pm0.7)\times10^{-2}$ (5)[f]

[a] For data that are derived from multiple independent experiments, the values in parentheses are the standard deviation of the estimate, followed by the number of experiments.
[b] Represents average of all published data as of November 1998 [86].
[c] Average of values from four types of Fe^0 [69].
[d] From Ref. 152.
[e] From regression of k_{obs} vs. ρ_m [159].
[f] Batch experiments done with Peerless iron [107].

(grams of Fe per liter of solution volume). For batch studies, ρ_m can be calculated from

$$\rho_m = \frac{M_{Fe}}{V_{H_2O}} = \frac{M_{Fe}}{V_{Tot} - M_{Fe}/\rho_s} \tag{20}$$

where M_{Fe} is the mass of Fe^0 (g), V_{H_2O} is the volume of solution (L), V_{Tot} is the total system volume (L), and ρ_s is the density of the iron (g L^{-1} occupied by the Fe^0).

Effect of Temperature. Several studies have shown that the kinetics of contaminant reduction in batch experiments exhibit temperature dependencies that conform to the Arrhenius equation [86,87,132,159,160]. Thus, we can write the following expression relating the rate constants at T_1 and T_2 (in °K):

$$\frac{k_{T_2}}{k_{T_1}} = e^{\left(\frac{-E_a}{R}\right)\left(\frac{1}{T_2} - \frac{1}{T_1}\right)} \tag{21}$$

where E_a is the activation energy (kJ mol^{-1}) and R is the gas constant (8.314 J K^{-1} mol^{-1}). Adapting the approach taken in several previous publications [49,161], we have used Eq. (21) to calculate correction factors (k_{T2}/k_{T1}) as a function of groundwater temperature (T_2), assuming a reference temperature (T_1) of 23°C and appropriate values for E_a. Five values of E_a were selected to generate Fig. 8: 55 kJ mol^{-1} is nearly the value reported for CCl_4 reacting with Fluka Fe^0 [144], 45 kJ mol^{-1} is approximately the value reported for vinyl chloride reacting with Fisher Fe^0 [159], 35 kJ mol^{-1} approximates the average E_a for TCE reacting with both reagent- and construction-grade Fe^0 [160], 25 kJ mol^{-1} represent regimes that are transitional between reaction and mass transfer control [101], and 15 kJ mol^{-1} represents kinetics that are entirely limited by mass transfer [162]. In general terms, Fig. 8 shows that reactions with Fe^0 will occur about half as rapidly in the field as they do at temperatures that are typical of the laboratory. Note that there is less effect of temperature on rates that are influenced by mass transport.

2. Multiprocess Kinetic Models

Competition for Reactive Sites. Recently, it has become widely recognized that k_{obs} can vary with the concentration of the contaminant. In most cases, this effect has been attributed to saturation of reactive sites on the Fe^0 surface. One kinetic model that has been used to describe these data is of the form:

$$-\frac{dC}{dt} = \frac{AC}{B + C} \tag{22}$$

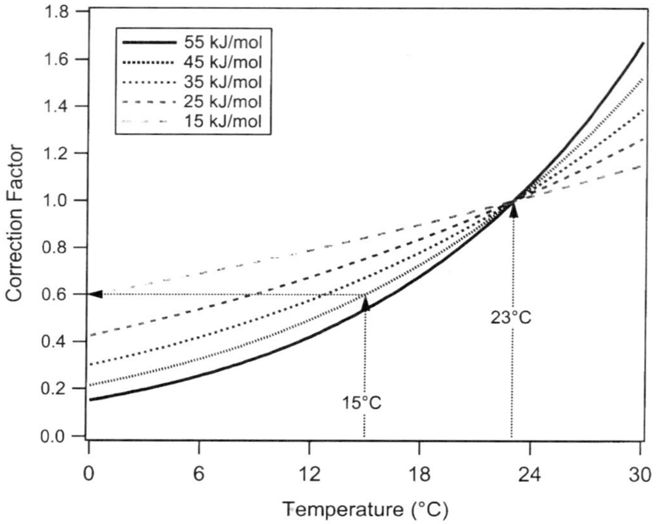

Figure 8 Correction factors for the effect of temperature on the rate of reduction by Fe^0. Arrows indicate the reference temperature around which most laboratory data are obtained (23°C), and a more representative temperature for groundwater (15°C). Assuming $E_a \approx 45$ kJ mol^{-1} for TCE [160], the corresponding correction factor shown is 0.6 (i.e., rates will be slower in the field by 60%).

where A and B are constants [163]. Several studies have equated A and B with V_m and $K_{1/2}$, by analogy to the Michaelis-Menten model for enzyme kinetics [86,111,147]. Other studies have associated A with k_0, the zero-order rate constant observed when surface sites are fully saturated [152], and equated A/B with k_{obs} when $B >> C$ [107,152].

Site saturation kinetics can also be described with a kinetic model of the form:

$$-\frac{dC}{dt} = \frac{DC}{1 + EC} \tag{23}$$

where D and E are constants. Some studies have defined D and E in accord with the Langmuir-Hinshelwood-Hougen-Watson (LHHW) model for surface catalyzed reactions [88,125,164–166], whereas others have defined D and E in terms derived from a surface complexation model [146]. Although the Michaelis-Menten and LHHW models were derived for different systems and conditions, the mathematical forms of these models [represented by Eqs. (22) and (23)] are essentially equivalent. Consequently, it can be shown that $A = D/E$ and $E = 1/B$.

In addition to competition by the contaminant for a limited supply of reactive surface sites (*intraspecies competition*), there is evidence for competition among different species for surface sites (*interspecies competition*). Interspecies competition effects can arise between combinations of contaminants (e.g., Ref. 167) or between contaminants and other adsorbates [102,146,147,159,168–170]. The kinetics that arise from interspecies competition have been modeled by adding appropriate terms to Eq. (22) or Eq. (23). However, in most cases the parameterization of these models has been rather preliminary and their sensitivity to uncertainties in these parameters has not yet been thoroughly investigated.

Competition Among Reactive and Nonreactive Sites. In addition to competition among different adsorbates for reactive sites, there is also competition among different surface sites for adsorbates. In principle, it is easy to imagine that granular Fe^0 under environmental conditions might present surface sites with energies that vary widely for adsorption and reaction. To date, however, most of the available data have been explained using a simple binary model that assumes surface sites are either reactive or nonreactive (i.e., just adsorptive) and that the distribution of reactant between these sites and the solution phase is in quasi-equilibrium. Assuming that contaminant transformation rates are dependent on its aqueous phase concentration and that most adsorption is to nonreactive sites, then a kinetic model for transformation that accounts for sorption can be written

$$-\mathrm{d}C_{\mathrm{T}}/\mathrm{d}t = k_{\mathrm{a}}C_{\mathrm{a}}^{N} \tag{24}$$

where C_{T} and C_{a} are the total system and aqueous phase concentrations, k_{a} is the rate constant for transformation, and N is the reaction order [171]. This model has been applied for PCE and TCE [171,172] and cis- and trans-1,2-dichloroethene [173]. The results are easily interpreted for PCE and TCE, because both gave $N \approx 1$. However, the dichloroethenes both gave $N > 1$, which suggests that Eq. (24) was not entirely adequate to describe the system behavior.

A more complete and mechanistically explicit model has been described that allows for competitive adsorption to reactive and nonreactive sites on Fe^0, as well as partitioning to the headspace in closed experimental systems and branching among parallel and sequential transformation pathways [174,175]. This model represents the distinction between reactive and nonreactive sites by a parameter called the "fractional active site concentration." Simulations and sensitivity analysis performed with this model have been explored extensively, but application of the model to experimental data has been limited to date.

In cases where sorptive equilibrium is reached rapidly and transformation is much slower, the aqueous phase concentration of contaminant may show a rapid initial decrease due to adsorption followed by a slower decline due to transformation. Under these conditions, the kinetic model represented by Eq. (16) is sufficient to describe the kinetics of transformation after the initial data have been excluded. This approach has been taken for TCE [168], vinyl chloride [176], and probably in many other studies where the exclusion of initial rate data was not clearly documented.

Competition Among Parallel Reaction Pathways. The third form of competition that complicates the kinetic description of contaminant degradation by Fe^0 involves branching among parallel pathways (and/or mechanisms) of transformation by the contaminant. Simple manifestations of this effect—such as the transformation of TCE to form chloroacetylene, *trans*-1,2-dichloroethene, *cis*-1,2-dichloroethene, or 1,1-dichloroethene from TCE—can be described with "branching ratios" [132]. However, a more general approach is to divide rate constants for reactant disappearance into separate rate constants for each product formation pathway. Because first-order rate constants are additive (for reactions occurring in parallel), we can write the following for disappearance of TCE by the four pathways noted above:

$$k_{SA} = k_{\text{chloroacetylene}} + k_{trans-1,2-\text{DCE}} + k_{cis-1,2-\text{DCE}} + k_{1,1-\text{DCE}} \tag{25}$$

This approach has been taken for the reaction of chlorinated ethenes with Zn^0 [125,165] and Fe^0 [88,166], resulting in separate rate constants for all the reactions shown in Fig. 3. Care must be taken in using these parameters in predictive modeling, however, as it is not yet known how sensitive the relative values of these rate constants are to pH, thickness and composition of the oxide film, etc. The same caution applies where the approach represented by Eq. (25) is used to describe parallel mechanisms of transformation. For example, it has recently been reported that several experimental factors influence the relative contributions of dissociative electron transfer, hydrogen atom transfer, and reductive elimination to the dechlorination of carbon tetrachloride and TCE by Fe^0 [177].

B. Mass Transport to the Surface

The overall reaction occurring at an Fe^0 surface involves a series of steps including: (1) mass transport to the reactive site, (2) chemical reaction at the surface (e.g., sorption, electron transfer, etc.), (3) desorption, and (4) mass transport to the bulk solution (recall Fig. 7). Any one of these steps can limit the rate of contaminant removal by Fe^0, so the observed rate (k_{SA}) can be represented as a series of resistances due to transport and reaction:

$$\frac{1}{k_{SA}} = \frac{1}{k_{rxn}} + \frac{1}{k_{mt}} \tag{26}$$

where k_{SA} is the overall surface area-normalized rate coefficient, k_{mt} is the mass transport coefficient, and k_{rxn} is the first-order heterogeneous reaction-rate coefficient. Mass transport resistances can be due to either

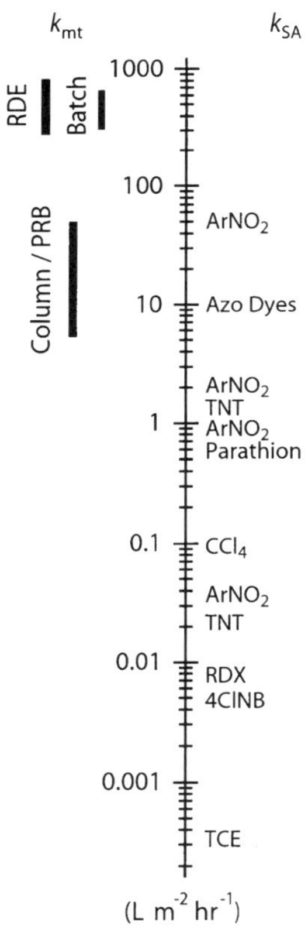

Figure 9 Comparison of previously reported values of k_{SA} for reduction by Fe^0 with external mass transport coefficients estimated for batch, column, and rotating disk electrode reactors. References for the overall rate coefficients are given in Fig. 1 of Ref. 101. Mass transport coefficients were estimated for the batch and column reactors based on empirical correlations discussed in Refs. 125 and 101. Mass transport coefficients for the RDE were calculated using the Levich equation [178].

external transport to the surface or internal transport through the oxide layer (i.e., pore diffusion). Various forms of Eq. (26) have been used to evaluate the role of external mass transport in contaminant reduction by Fe^0 and Zn^0 [101,111,144,165].

Because rates of reduction by Fe^0 vary considerably over the range of treatable contaminants, it is possible that there is a continuum of kinetic regimes from purely reaction controlled, to intermediate, to purely mass transport controlled. Fig. 9 illustrates the overlap of estimated mass transport coefficients (k_{mt}) and measured rate coefficients (k_{SA}). The values of k_{SA} are, in most cases, similar to or slower than the k_{mt} values estimated for batch and column reactors. The slower k_{SA} values suggest that $k_{rxn} < k_{mt}$, and therefore removal of most contaminants by Fe^0 should be reaction limited or only slightly influenced by mass transport effects (i.e., an intermediate kinetic regime).

Direct evidence of mass transport limitations for the more highly reactive contaminants has been observed in RDE experiments and batch and column reactors with Fe^0. The measured k_{rxn} of $ArNO_2$ reduction at a bare Fe^0 electrode was about 10 times faster then the mass transport coefficient estimated in a PRB [101]. In addition, evidence for mass transport effects have been observed in both batch and column Fe^0 experiments where rates of nitro aromatic [99,100] and azo dye [111] removal were dependent on mixing speed or flow rate. The observed dependence of reduction rate on mixing intensity and the similarity between rates of surface reaction and mass transport suggest that mass transport may limit removal rates of these highly reactive contaminants in FePRBs. An interesting implication of these results is that for highly reactive compounds (such as $ArNO_2$ and TNT), hydraulic designs that increase mass transport rates (e.g., funnel-and-gate systems) may be useful for improving contaminant removal rates by FePRBs.

V. MACROSCALE PROCESSES

The microscale processes reviewed in the previous section may be sufficient to describe the behavior of well-mixed model systems, but packed bed systems (including columns, canisters, and PRBs) are also characterized by processes that are manifest on length scales of meters and time scales of hours. Progress toward understanding the macroscale processes associated with FePRBs has been comparatively slow, in part because it has to be built on a thorough understanding of the microscale processes occurring at the metal-water interface, and the latter is still emerging. On the other hand, the ultimate objective of FePRBs is remediation on the aquifer scale, so advancing our understanding of biogeochemical processes at the macro scale is where many of the most significant advances can be expected as the FePRB technology matures in the future.

A. Geochemical Gradients and Zones

Unlike well-mixed batch systems, columns and field conditions result in the development of steep chemical gradients within the iron-bearing zone, at the interfaces between the iron-bearing zone and the surrounding material, and downgradient where the plume of treated water interacts with the native aquifer material. Although some early work recognized that these gradients could be significant (e.g., Refs. 179–181), further characterization of these geochemical gradients has been needed at the field scale before their effects on contaminant fate could be accurately assessed. Recently, considerable progress on this topic has been made through integrated monitoring and modeling studies associated with several field sites, including Moffett Field in Mountain View, CA [182–184], the U.S. Coast Guard Support Center in Elizabeth City, NC [185–188], and the Y-12 uranium processing plant in Oak Ridge, TN [75,131,189,190].

The major geochemical gradients that have been observed associated with FePRBs are summarized in Fig. 10. They involve (1) dissolved oxygen, which is completely removed within a few millimeters of where groundwater enters the iron-bearing zone, (2) dissolved H_2, which rises over the width of the iron-bearing zone to near saturation, (3) pH and dissolved Fe(II), both of which usually rise rapidly inside the wall and then decline gradually in the downgradient region, (4) dissolved CO_2, which precipitates near the upgradient interface as iron carbonates, (5) NO_3^-, which is abiotically reduced to ammonia, and (6) SO_4^{2-}, which is reduced by anaerobic bacteria to sulfide, much of which then precipitates as iron sulfides. Note that lateral diffusion is very slow into the plume of treated groundwater, so reoxygenation by this mechanism is expected to be minimal and the anaerobic plume may eventually extend a considerable distance downgradient from an FePRB.

As a consequence of the gradients in groundwater geochemistry described above, zones of authigenic precipitates develop along the flow path of FePRBs and columns designed to simulate these conditions. A considerable amount of research has been done on the iron oxides and carbonates that accumulate near the upgradient interface because these solids can cement grains and decrease porosity and thereby prevent contaminated groundwater from flowing through the treatment zone [131,189,193–196]. The effect of these precipitates on overall rates of contaminant reduction is not entirely clear, however, because most field

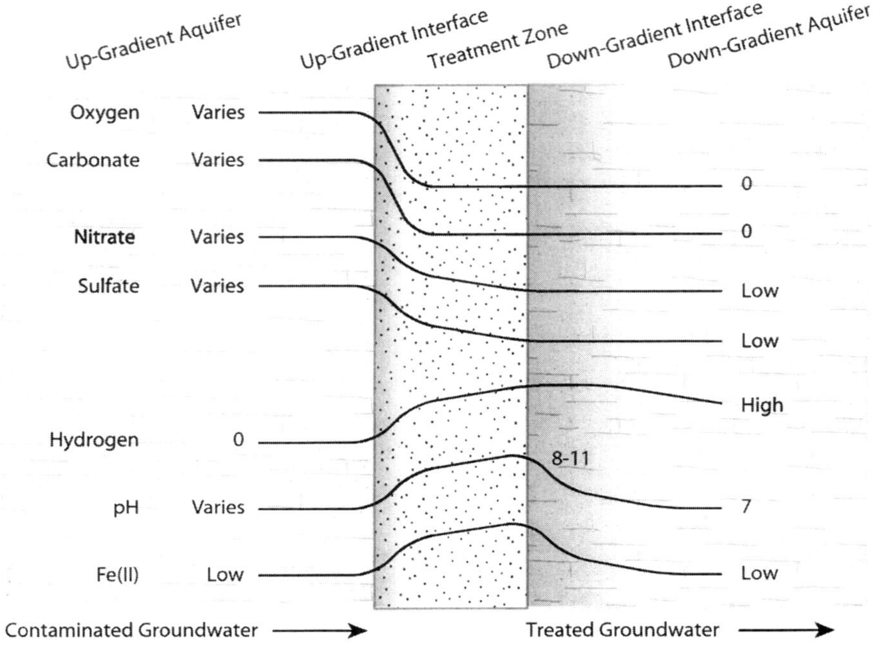

Figure 10 Schematic of a cross section of an FePRB showing the major gradients in groundwater geochemistry, zones of precipitation, and expected regions of microbiological influence. (Adapted from figures in Refs. 75, 184, 191, and 192.)

data suggest that contaminant reduction occurs mainly near the upgradient interface, where precipitation of oxide and carbonates might contribute to passivation of Fe^0 surfaces and, therefore, slower overall rates of corrosion.

The zone of precipitation that develops on the native aquifer material beyond the downgradient interface has received comparatively little attention to date. It is known, however, that most of the dissolved iron that is released by the treatment zone precipitates on the downgradient aquifer material, resulting in the accumulation of Fe(II)-containing oxyhydroxides (and favoring a decrease in pH). These changes minimize undesirable changes in groundwater geochemistry that might be caused by an FePRB. In addition, the accumulation of highly-reactive forms of Fe(II) creates a zone that may result in further contaminant degradation by abiotic and biologically mediated pathways (e.g., Refs. 197–201).

B. Design and Scaling of Conventional PRBs

In the last few years, a number of documents have appeared that are largely devoted to providing guidance for the design of FePRBs [161,202–204]. These guidelines reflect a mixture of scientific, engineering, and regulatory considerations that is certain to evolve as the relevant science and technology matures. The following addresses some of the major issues involved in preliminary calculations for scaling from the batch to the field.

1. Steady-State Design Calculations

From an engineering perspective, a central consideration in the design of an FePRB is how to ensure enough contact between contaminated groundwater and the treatment zone to reach the required treatment goal. This can be expressed as

$$W = V_{\mathrm{lin}}\, t_{\mathrm{c}} \tag{27}$$

where W is the necessary barrier width, V_{lin} is the linear velocity of groundwater, and t_{c}, is the required contact time. One way to get t_{c} is by solving Eq. (13) to give

$$t_{\mathrm{c}} = \frac{-\ln\left(C_{\mathrm{eff}}/C_{\mathrm{inf}}\right)}{k_{\mathrm{SA}}\rho_{\mathrm{A}}} \tag{28}$$

where C_{inf} and C_{eff} are the concentrations of contaminant in the influent and effluent, respectively. For preliminary design calculations, C_{inf} is usually taken to be the maximum concentration in site groundwater and C_{eff} is the treatment goal required by local regulations. As described above, ρ_{a} can be calculated from a_{s} and ρ_{m} using Eq. (19). However, for packed systems, ρ_{m} may be more conveniently estimated from:

$$\rho_{\mathrm{m}} = \frac{\rho_{\mathrm{s}}\rho_{\mathrm{b}}}{(\rho_{\mathrm{s}} - \rho_{\mathrm{b}})} \tag{29}$$

where ρ_{b} is the bulk density of the iron (g of Fe per L of total volume). Some values of ρ_{s} and ρ_{b} are given in Table 2.

2. Reactive Transport Modeling

The reactive transport of contaminants in FePRBs has been modeled using several approaches [179,184,186,205–208]. The simplest approach treats the FePRB as an ideal plug-flow reactor (PFR), which is a steady-state flow reactor in which mixing (i.e., dispersion) and sorption are negligible. Removal rates (and therefore required wall widths, W) can be estimated based on first-order contaminant degradation and residence times calculated from the average linear groundwater velocity [Eq. (27)]. The usefulness of

PFR models are limited, however, because of the slow velocities encountered in groundwater aquifers and the tendency for many contaminants (particularly hydrophobic organic compounds) to sorb. More appropriate but more complex models based on various forms of the advection-dispersion equation (ADE) have been used by several researchers to incorporate more processes, such as dispersion, sorption, mass transfer, sequential degradation, and coupled chemical reactions.

One of the simplest forms of the ADE that has been applied to an FePRB includes both dispersion and sorption [205]. A one-dimensional steady-state ADE was used to estimate W for 1000-fold reduction in contaminant concentrations at a groundwater velocity of 1 ft day^{-1}. Applying the model to chlorinated aliphatic compounds (using rate coefficients summarized in Ref. 86) gave the results shown in Fig. 11. These estimates,

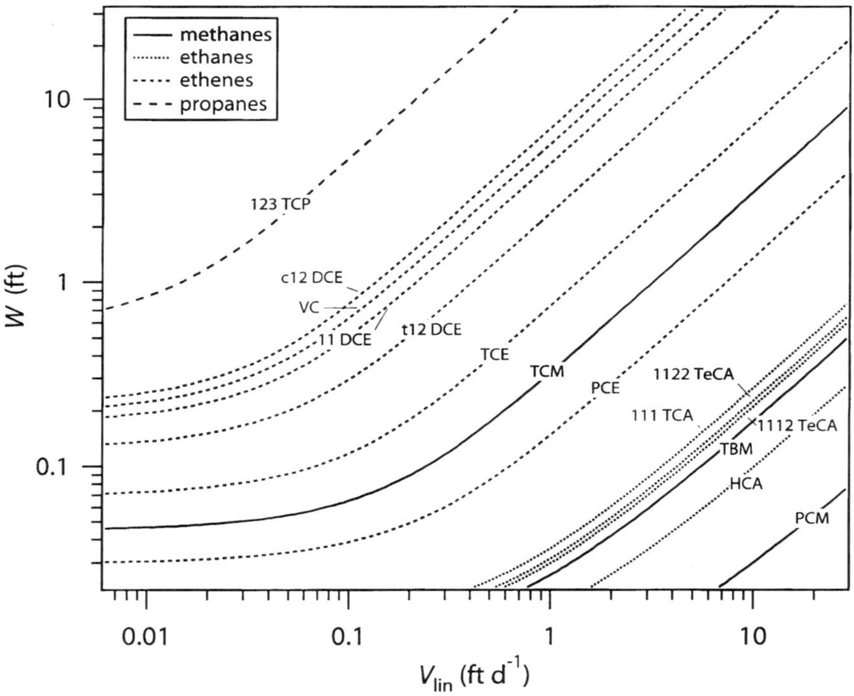

Figure 11 Calculated width of an FePRB (W) required to reduce contaminant concentrations 1000-fold, as a function of linear velocity of groundwater (V_{lin}). (Based on representative values of k_{SA} from Ref. 86. A simplified version of this plot was published previously [132]. This version used with permission of J. Eykholt.)

however, are based solely on the degradation rate of the parent compound and do not consider more complex degradation schemes involving multiple sequential and parallel reactions (e.g., Fig. 3). One approach that can be used to accommodate the complexity of multiple reactions is to design the wall for the least reactive product [132] and apply a safety factor [209].

Another approach has been to model sequential reactions by using multiple advection-dispersion equations [207]. The use of multiple ADEs provides a more realistic model where each reactant can degrade, sorb, and disperse. Simulations using this type of model reveal that breakthrough of degradation products could occur despite complete removal of the parent compound, TCE [207]. Additional simulations were used to explore the effect of slow sorption (i.e., nonequilibrium sorption), and the results suggest that it is reasonable to assume that an FePRB will reach steady-state conditions under typical field conditions.

Nonequilibrium sorption due to mass-transfer limitations (including slow external or internal diffusion) and sorption to two different sorbents have been incorporated into a single ADE to evaluate the conditions under which mass-transfer processes may be important [206]. Simulations with this model, using mass-transfer parameters estimated from empirical correlations, reveal nonequilibrium conditions (i.e., mass-transfer limitations) when groundwater velocities increase (such as those that might occur in a funnel-and-gate system).

The most sophisticated models applied to FePRBs to date combine multiple ADEs (i.e., multicomponent transport) with coupled chemical reactions [184,186,208]. These multicomponent reactive transport models were used to simulate the geochemical evolution in FePRBs for the treatment of TCE [184] and for remediating mixtures of Cr(VI) and chlorinated solvents [186,208]. The models are capable of reproducing the spatial distribution of field-observable parameters such as the concentrations of the chlorinated solvents, pH, Eh, alkalinity, Mg^{2+}, SO_4^{2-}, and NO_3^- [184,186,208], although some discrepancies were observed for Mn^{2+} [186] and Ca^{2+} [184]. These discrepancies were attributed to processes that were not incorporated into the models, such as coprecipitation, sorption, ion exchange [186], and surface complexation [184]. The selection of secondary minerals was found to be the critical factor for accurately simulating the inorganic geochemistry throughout the wall [184]. Concentration profiles of the chlorinated solvents, however, were adequately described by reaction with Fe^0 only, and it was not necessary to model the reaction of the solvents with secondary minerals as separate processes [186,208].

Multicomponent reactive transport models can also be used to estimate the potential for barrier clogging due to secondary mineral formation [186,208]. In addition, processes such as the microbially mediated degradation of sulfate and degassing of hydrogen gas within the barrier, as well as geochemical interactions of treated groundwater with the aquifer matrix downgradient of the barrier were considered [208].

All of the above modeling has assumed uniform flow fields, even though heterogeneity in the hydraulic conductivity of the barrier and surrounding aquifer can lead to preferential flow patterns. Some of the implications of preferential flow have been investigated by varying input parameters in the models described above [209–211].

C. Variants and Integrated Technologies

1. Coupling with Natural Attenuation (Microbiological Effects)

Early studies of Fe^0 systems found little microbial contribution to contaminant degradation [191], and this caused most investigators to focus on abiotic processes during the early development of FePRBs. There are, however, many ways by which microbial activity could influence the performance and longevity of FePRBs, and evidence for some of these effects is being reported with increasing frequency. Most of what was known about this topic as of 1999 was reviewed by Scherer et al. [42], but there have already been significant advances since that time (e.g., Refs. 117, 190, and 212–214).

The main types of effects that microorganisms might have on FePRBs involve (1) biocorrosion, (2) cometabolism, (3) dissolution/precipitation, and (4) biofouling. It is expected that these effects will be localized in and around an FePRB in accord with the strong gradients in chemical conditions that are created by the barrier. For example, the high pH within a treatment zone containing Fe^0 will inhibit most types of microbial activity, whereas the effluent from the treatment zone might stimulate microbial activity downgradient due to elevated levels of dissolved hydrogen and iron. Some of the key spatial relationships are represented in Fig. 10 for a conventional FePRB; however, microbial effects can also be engineered into sequences of treatment zones, as described in the section below.

Early work on biocorrosion led to the notion that H_2-utilizing bacteria accelerate anaerobic corrosion by "cathodic depolarization", i.e., utilizing the H_2 produced by anaerobic corrosion of Fe^0 [Eq. (14)] and thereby favoring further corrosion by this reaction. Several studies have shown that microbial utilization of the H_2 from anaerobic corrosion of Fe^0 can lead to accelerated cometabolism of contaminants. For example, combining Fe^0 with active (H_2-consuming) methanogenic consortia significantly enhances both the rate and extent of transformation of chlorinated methanes, ethanes, and ethenes [215–218]. Experiments conducted with pure cultures of methanogens showed that H_2 can enhance microbial reduction of chlorinated solvents even when H_2 does not serve as growth substrate [219]. Some of the effects of *Methanosarcina thermophila* on degradation of carbon tetrachloride and chloroform may be due to a proteinaceous extracellular factor that is excreted when the bacteria are exposed to Fe^0 [220]. It has also been shown that autotrophic denitrification is synergistic with the abiotic reduction of nitrate by Fe^0 [221].

Surprisingly little attention has been given to the possibility that the precipitation and dissolution of iron oxides downgradient from an FePRB will be strongly influenced by microbial activity. Although the basic geomicrobiology of Fe(II)-oxidizing and Fe(III)-reducing bacteria is well known [222], the environment downgradient from an FePRB provides an exceptional opportunity for synergistic effects between microbial iron metabolism and contaminant transformation reactions. Of particular interest are dissimilatory iron-reducing bacteria (DIRB), which can use a variety of Fe(III)-containing minerals as electron acceptors, resulting in reductive dissolution of iron and cometabolism of a variety of contaminants [223]. Evidence for both of these effects in the presence of Fe^0 has been reported recently in studies with *Shewanella alga* BrY and CCl_4 [212].

Until recently, evidence for *biofouling* of in situ FePRBs has been notably absent. Recently, a study of the in situ FePRB at the Y-12 Plant Site, Oak Ridge, TN [131,190] concluded that microbial biomass was 1–3 orders of magnitude higher in the Fe^0-bearing zone than in the surrounding aquifer based on phospholipid fatty acid analysis (PLFA) and DNA probe results. This result suggests that some conditions may favor microbial growth within FePRBs; although the collective experience with in situ FePRBs to date suggests that macroscopic biofilms or other gross manifestations of biofouling are not observed.

2. Other Variations and Modifications

For the most part, the scope of this chapter has been restricted to core issues related to the "classical" FePRB, consisting of a single in situ treatment zone containing granular, construction-grade, zero-valent iron. In parallel with developments on classical FePRBs, however, a wide range of variations have been described. The growing diversity of variations on the classic FePRB makes them difficult to classify, but the better known variations can be grouped into general categories, as shown in Table 4. Many of these variations are designed to overcome limitations of the classical FePRB by extending the range of contaminants that are treatable by Fe^0 and/or extending the range of situations were Fe^0 can be applied. Other variations are simply efforts to make the classical FePRB more efficient or cost-effective.

Table 4 Variations and Enhancements of the Basic FePRB

Variation	Application/benefit	References
Acid mine drainage	Lower pH, adsorbs metals	[224]
Cyclodextrins	Solubilizes organics, prevents settling	[225,226]
Electroenhanced	Faster reduction	[227–232]
Fenton (Fe^0/H_2O_2)	Oxidative degradation	[233–239]
Vapor phase	PCBs	[93,240,241]
High temperature	PCBs	[93]
Nanoparticles	Enhanced reactivity, avoids settling	[153,242–249]
Photoenhancement	Faster reduction, alters products	[250]
Slurry walls	Barrier to diffusing contaminants	[251]
S(II) or pyrite amended	Chlorinated ethenes, chlorinated phenols	[94,252–254]
Sub/supercritical solvents	PCBs	[90–92,255]
Surfactants	Solubilizes contaminants	[34,168,170,256]
Sequential reactive treatment zones (SRTZs)	Recalcitrant intermediates, decreased risk	[257,258]
Ultrasound	Fast degradation, alters products, restores activity of passivated metals	[157,259–261]
Vadose zone soils	Pesticides	[262]

VI. LIST OF SYMBOLS

ρ_a	surface area concentration (m^2 of Fe surface area L^{-1} of solution volume)
ρ_b	bulk density (g of Fe L^{-1} of total volume)
ρ_m	mass concentration (g of Fe L^{-1} of solution volume)
ρ_s	specific density (g of Fe L^{-1} of Fe volume)
Γ	surface concentration of reactive sites (mol m^{-2})
A, B, D, E	constants defined in Eqs. (22) and (23)
a_s	specific surface area (m^2 Fe surface area g^{-1} of Fe)
C_a	concentration of contaminant in the aqueous phase
C_0	concentration at $t=0$
C_{eff}	concentration in effluent
C_{inf}	concentration in influent
C_t	concentration at $t=t$
C_T	total concentration of contaminant (dissolved and adsorbed)
E_a	activation energy (kJ mol^{-1})
k_0	zero-order rate constant
k_1	first-order rate constant
k_2	second-order reaction rate constant (L mol^{-1} hr^{-1})

k_a	rate constant for transformation
k_{mt}	mass transport coefficient
k_{obs}	observed pseudo-first-order disappearance rate constant (hr^{-1})
$k_{overall}$	overall surface area-normalized rate of reaction
k_{RXN}	first-order heterogeneous reaction rate constant
k_{SA}	observed reaction rate (k_{obs}) normalized to surface area ($L\ m^{-2}\ hr^{-1}$)
k_{T_1}	rate constant at temperature T_1
k_{T_2}	rate constant at temperature T_2
$K_{1/2}$	concentration of P at $V_m/2$ (mol L^{-1})
M_{Fe}	mass of Fe^0 (g)
N	reaction order
$N_{1/2}$	number of half-lives needed to reach a treatment goal
[P]	molar concentration of contaminant P (mol L^{-1})
R	gas constant (8.314 J K^{-1} mol^{-1}).
t_c	contact time necessary to achieve a treatment goal
V_m	maximum reaction rate (mol L^{-1} s^{-1})
V	velocity of groundwater
V_{H_2O}	volume of solution (L)
V_{Tot}	total system volume (L)
W	width of barrier

ACKNOWLEDGMENTS

We thank Ulrich Mayer for his help with the sections on modeling. John Zachara (Pacific Northwest National Laboratory) and the U.S. Department of Energy provided partial support to PGT during writing of this manuscript.

REFERENCES

1. Gillham RW. Resurgence in research concerning organic transformations enhanced by zero-valent metals and potential application in remediation of contaminated groundwater. 209th National Meeting, Anaheim, CA, American Chemical Society 1995; 35(1):691–694.
2. Tratnyek PG. Putting corrosion to use: remediation of contaminated groundwater with zero-valent metals. Chem Ind (London) 1996; 499–503.
3. Powell RM, Powell PD. Iron metal for subsurface remediation. In: Myers RA, ed. Encyclopedia of Environmental Analysis and Remediation. New York: Wiley, 1998; 8:4729–4761.

4. Warner SD, Sorel D. Ten years of permeable reactive barriers: lessons learned and future expectations. 221st National Meeting of the Chemical Society, San Diego, CA, American Chemical Society 2001; 41(1):1104–1112.
5. Reynolds GW, Hoff JT, Gillham RW. Sampling bias caused by materials used to monitor halocarbons in groundwater. Environ Sci Technol 1990; 24:135–142.
6. Gillham RW, Burris DR. Recent developments in permeable in situ treatment walls for remediation of contaminated groundwater. Proceedings of the Subsurface Restoration Conference, Dallas, TX. Houston, TX: Rice University, 1992:66–68.
7. Young GK, Bungay HR, Brown LM, Parsons WA. Chemical reduction of nitrate in water. J Water Pollut Control Fed 1964; 36:395–398.
8. Rhodes FH, Carty JT. The corrosion of certain metals by carbon tetrachloride. Ind Eng Chem 1925; 17:909–911.
9. Lyons RE, Smith LT. Die reduktion von nitroverbindungen mit eisen und löslichen chloriden. Chem Ber 1927; 60:173–182.
10. Sweeny KH. Reductive degradation treatment of industrial and municipal wastewaters. Water Reuse Symposium. Denver, CO: American Water Works Association Research Foundation 1979; 2:1487–1497.
11. Butler LC, Staiff DC, Sovocool GW, Davis JE. Field disposal of methyl parathion using acidified powdered zinc. J Environ Sci Health 1981; B16:49–58.
12. Senzaki T, Yasuo K. Removal of chlorinated organic compounds from wastewater by reduction process. I. Treatment of 1,1,2,2-tetrachloroethane with iron powder. Kogyo Yosui 1988:2–7.
13. Senzaki T, Yasuo K. Removal of chlorinated organic compounds from wastewater by reduction process. II. Treatment of trichloroethylene with iron powder. Kogyo Yosui 1989:19–25.
14. Senzaki T. Removal of chlorinated organic compounds from wastewater by reduction process. III. Treatment of trichloroethylene with iron powder. Kogyo Yosui 1991; 391:29–35.
15. Khudenko BM. Feasibility evaluation of a novel method for destruction of organics. Water Sci Technol 1991; 23:1873–1881.
16. Gould JP. The kinetics of hexavalent chromium reduction by metallic iron. Water Res 1982; 16:871–877.
17. Khudenko BM. Mechanism and kinetics of cementation processes. Water Sci Technol 1985; 17:719–732.
18. Day SR, O'Hannesin SF, Marsden L. Geotechnical techniques for the construction of reactive barriers. J Hazard Mater 1999; 67:285–297.
19. Starr RC, Cherry JA. In situ remediation of contaminated ground water: the funnel-and-gate system. Ground Water 1994; 32:465–476.
20. Amonette JE, Szecsody JE, Schaef HT, Templeton JC, Gorby YA, Fruchter JS. Abiotic reduction of aquifer materials by dithionite: a promising in-situ remediation technology. Proceedings of the 33rd Hanford Symposium on Health & the Environment In-Situ Remediation: Scientific Basis for Current

and Future Technologies, Pasco, WA. Richland, WA: Battelle Pacific Northwest Laboratories, 1994; 2:851–881.

21. Fruchter JS, Cole CR, Williams MD, Vermeul VR, Amonette JE, Szecsody JE, Istok JD, Humphrey MD. Creation of a subsurface permeable treatment zone for aqueous chromate contamination using in situ redox manipulation. Ground Water Monit Remediat 2000; 20:66–77.
22. Istok JD, Amonette JE, Cole CR, Fruchter JS, Humphrey MD, Szecsody JE, Teel SS, Vermeul VR, Williams MD, Yabusaki SB. In situ redox manipulation by dithionite injection: intermediate-scale laboratory experiments. Ground Water 1999; 37:884–889.
23. Szecsody JE, Williams MD, Fruchter JS, Vermeul VR, Evans JC, Sklarew DS. Influence of sediment reduction on TCE degradation. In: Wickramanayake GB, Gavaskar AR, Chen ASC, eds. Chemical Oxidation and Reactive Barriers: Proceedings of the Second International Conference on Remediation of Chlorinated and Recalcitrant Compounds, Monterey, CA, 20–25 May 2000. Columbus, OH: Battelle Press, 2000; C2–6:369–376.
24. Grittini C, Malcomson M, Fernando Q, Korte N. Rapid dechlorination of polychlorinated biphenyls on the surface of a Pd/Fe bimetallic system. Environ Sci Technol 1995; 29:2898–2900.
25. Graham LJ, Jovanovic G. Dechlorination of p-chlorophenol on a Pd/Fe catalyst in a magnetically stabilized fluidized bed; Implications for sludge and liquid remediation. Chem Eng Sci 1999; 54:3085–3093.
26. Wan C, Chen YH, Wei R. Dechlorination of chloromethanes on iron and palladium–iron bimetallic surface in aqueous systems. Environ Toxicol Chem 1999; 18:1091–1096.
27. James BR. The challenge of remediating chromium-containing soil. Environ Sci Technol 1996; 30:248A–251A.
28. Benjamin MM, Sletten RS, Bailey RP, Bennett T. Sorption and filtration of metals using iron–oxide-coated sand. Water Res 1996; 30:2609–2620.
29. Martinez CE, McBride MB. Coprecipitates of Cd, Cu, Pb and Zn in iron oxides: solid phase transformation and metal solubility after aging and thermal treatment. Clays Clay Miner 1998; 46:537.
30. Morrison SJ, Spangler RR. Chemical barriers for controlling groundwater contamination. Environ Prog 1993; 12:175–181.
31. Kriegman-King MR, Reinhard M. Transformation of carbon tetrachloride by pyrite in aqueous solution. Environ Sci Technol 1994; 28:692–700.
32. Butler EC, Hayes KF. Kinetics of the transformation of trichloroethylene and tetrachloroethylene by iron sulfide. Environ Sci Technol 1999; 33:2021–2027.
33. Kriegman-King MR, Reinhard M. Transformation of carbon tetrachloride in the presence of sulfide, biotite, and vermiculite. Environ Sci Technol 1992; 26:2198–2206.
34. Li Z, Jones HK, Bowman RS, Helferich R. Enhanced reduction of chromate and PCE by pelletized surfactant-modified zeolite/zerovalent iron. Environ Sci Technol 1999; 33:4326–4330.

35. Bowman RS, Li Z, Roy SJ, Burt T, Johnson TJ, Johnson RL. Pilot test of a surfactant-modified zeolite permeable barrier for groundwater remediation. In: Smith JA, Burns SE, eds. Physicochemical Remediation of Contaminated Subsurface Environments. New York: Kluwer, 2000:161–185.
36. Chen XB, Wright JV, Conca JL, Peurrung LM. Effects of pH on heavy metal sorption on mineral apatite. Environ Sci Technol 1997; 31:624–631.
37. Hodson ME, Valsami-Jones E, Cotter-Howells JD. Bonemeal additions as a remediation treatment for metal contaminated soil. Environ Sci Technol 2000; 34:3501–3507.
38. James BR, Rabenhorst MC, Frigon GA. Phosphorus sorption by peat and sand amended with iron oxides or steel wool. Water Environ Res 1992; 64:699–705.
39. Kershaw DS, Crouthamel Kulik B, Pamukcu S. Ground rubber: sorption media for ground water containing benzene and o-xylene. J Geotech Geoenviron Eng 1997; 123:324–334.
40. Guerin TF, Horner S, McGovern T, Davey B. An application of permeable reactive barrier technology to petroleum hydrocarbon contaminated groundwater. Water Res 2002; 36:15–24.
41. Benner SG, Blowes DW, Gould WD, Herbert RB Jr, Ptacek CJ. Geochemistry of a permeable reactive barrier for metals and acid mine drainage. Environ Sci Technol 1999; 33:2793–2799.
42. Scherer MM, Richter S, Valentine RL, Alvarez PJJ. Chemistry and microbiology of reactive barriers for in situ groundwater cleanup. Crit Rev Environ Sci Technol 2000; 30:363–411.
43. Gillham RW, Blowes DW, Ptacek CJ, O'Hannesin S. Use of zero-valent metals for in situ remediation of contaminated ground water. Proceedings of the 33rd Hanford Symposium on Health & the Environment In-Situ Remediation: Scientific Basis for Current and Future Technologies, Pasco, WA. Richland, WA: Battelle Pacific Northwest Laboratories 1994; 2:913–930.
44. Shoemaker SH, Greiner JF, Gillham RW. Permeable reactive barriers. In: Bodocsi A, Ryan ME, Rumer RR, eds. Barrier Containment Technologies for Environmental Remediation Applications (Final Report of the International Containment Technology Workshop, Baltimore, MD, 29–31 August 1995). New York: Wiley, 1995:301–353.
45. Gillham RW. In situ treatment of groundwater: metal-enhanced degradation of chlorinated organic contaminants. Recent Advances in Ground-Water Pollution Control and Remediation, NATO Advanced Study Institute, Kemer, Antalya, Turkey. New York: Kluwer Academic, 1996:249–274.
46. Palmer PL. Reactive walls. In: Nyer EK, Kidd DF, Palmer PL, Crossman TL, Fam S, Johns FJ II, Boettcher G, Suthersan SS, eds. In Situ Treatment Technology. Boca Raton, FL: CRC, 1996:271–288.
47. Gillham RW, Burris DR. Recent developments in permeable in situ treatment walls for remediation of contaminated groundwater. In: Ward CH, Cherry JA, Scalf MR, eds. Subsurface Restoration. Chelsea, MI: Ann Arbor, 1997:343–356.

48. Grathwohl P, Dahmke A. Direkte sanierung verschmutzter grundwässer. Spektrum Wiss 1998; (April):89–94.
49. Powell RM, Puls RW, Blowes DW, Vogan JL, Gillham RW, Powell PD, Schultz D, Landis R, Sivavic T. Permeable Reactive Barrier Technologies for Contaminant Remediation, U.S. Environmental Protection Agency, EPA/600/R-98/125, Ada, OK, 1998.
50. Yin Y, Allen HE. In Situ Chemical Treatment, Ground-Water Remediation Technologies Analysis Center (GWRTAC), TE-99-01, Pittsburgh, PA, 1999.
51. Blowes DW, Ptacek CJ, Benner SG, McRae CWT, Bennett TA, Puls RW. Treatment of inorganic contaminants using permeable reactive barriers. J Contam Hydrol 2000; 45:123–137.
52. U.S. Environmental Protection Agency. Field Applications of In Situ Remediation Technologies: Permeable Reactive Barriers, USEPA, Office of Solid Waste and Emergency Response, EPA-542-R-99-002, Washington, D.C., 1999.
53. Bigg T, Judd SJ. Zero-valent iron for water treatment. Environ Technol 2000; 21:661–670.
54. Luthy RG, Aiken GR, Brusseau ML, Cunningham SD, Gschwend PM, Pignatello JJ, Reinhard M, Traina SJ, Weber WJ Jr, Westall JC. Sequestration of hydrophobic organic contaminants by geosorbents. Environ Sci Technol 1997; 31:3341–3347.
55. Khudenko BM, Gould JP. Specifics of cementation processes for metals removal. Water Sci Technol 1991; 24:235–246.
56. Pourbaix M. Atlas of Electrochemical Equilibria in Aqueous Solutions. Oxford: Pergamon, 1966:644.
57. Grau JM, Bisang JM. Removal and recovery of mercury from chloride solutions by contact deposition on iron felt. Chem Biotechnol 1995; 62:153–158.
58. Biester H, Schuhmacher P, Müller G. Effectiveness of mossy tin filters to remove mercury from aqueous solution by Hg(II) reduction and Hg(0) amalgamation. Water Res 2000; 34:2031–2036.
59. Bostick WD, Beck DE, Bowser KT, Bunch DH, Fellows RL, Sellers GF. Treatability study for removal of leachable mercury in crushed fluorescent lamps, Oak Ridge National Laboratory, K/TSO-6, Oak Ridge, TN, 1996.
60. Smith EH. Uptake of heavy metals in batch systems by a recycled iron-bearing material. Water Res 1996; 30:2424–2434.
61. Makhloufi L, Saidani B, Hammache H. Removal of lead ions from acidic aqueous solutions by cementation on iron. Water Res 2000; 34:2517–2524.
62. Ku Y, Chen C-H. Kinetic study of copper deposition on iron by cementation reaction. Sep Sci Technol 1992; 27:1259–1275.
63. Cantrell KJ, Kaplan DI, Wietsma TW. Zero-valent iron for the in situ remediation of selected metals in groundwater. J Hazard Mater 1995; 42:201–212.
64. Morrison SJ, Metzler DR, Dwyer BP. Removal of As, Mn, Mo, Se, U, V and Zn from groundwater by zero-valent iron in a passive treatment cell: reaction progress modeling. J Contam Hydrol 2002; 56(1–2):99–116.

65. Pratt AR, Blowes DW, Ptacek CJ. Products of chromate reduction on proposed subsurface remediation material. Environ Sci Technol 1997; 31: 2492.
66. Blowes DW, Ptacek CJ, Lambor IL. In-situ remediation of Cr(VI)-contaminated groundwater using permeable reactive walls: laboratory studies. Environ Sci Technol 1997; 31:3348–3357.
67. Qiu SR, Lai HF, Roberson MJ, Hunt ML, Amrhein C, Giancarlo LC, Flynn GW, Yarmoff JA. Removal of contaminants from aqueous solution by reaction with iron surfaces. Langmuir 2000; 16:2230–2236.
68. Lackovic JA, Nikolaidis NP, Dobbs GM. Inorganic arsenic removal by zero-valent iron. Environ Eng Sci 2000; 17:29–39.
69. Su C, Puls RW. Arsenate and arsenite removal by zerovalent iron: kinetics, redox transformation, and implications for in situ groundwater remediation. Environ Sci Technol 2001; 35:1487–1492.
70. Farrell J, Wang J, O'Day P, Conklin M. Iron mediated reductive precipitation of arsenic from contaminated groundwater. 220th National Meeting, Washington, DC, American Chemical Society 2000; 40(2):794–796.
71. Gu B, Liang L, Dickey MJ, Yin X, Dai S. Reductive precipitation of uranium(VI) by zero-valent iron. Environ Sci Technol 1998; 32:3366–3373.
72. Morrison SJ, Metzler DR, Carpenter CE. Uranium precipitation in a permeable reactive barrier by progressive irreversible dissolution of zerovalent iron. Environ Sci Technol 2001; 35:385–390.
73. Fiedor JN, Bostick WD, Jarabek RJ, Farrell J. Understanding the mechanism of uranium removal from groundwater by zero-valent iron using X-ray photoelectron spectroscopy. Environ Sci Technol 1998; 32:1466–1473.
74. Charlet L, Liger E, Gerasimo P. Decontamination of TCE- and U-rich waters by granular iron: role of sorbed Fe(II). J Environ Eng 1998; 124:25–30.
75. Gu B, Watson DB, Phillips DH, Liang L. Biogeochemical, mineralogical, and hydrological characteristics of an iron reactive barrier used for treatment of uranium and nitrate. In: Naftz DL, Morrison SJ, Davis JA, Fuller CC, eds. Groundwater Remediation Using Permeable Reactive Barriers. San Diego: Academic Press, 2002:305–342.
76. Matheson LJ, Goldberg WC, Bostick WD, Harris L. Analysis of uranium-contaminated zero-valent iron media sampled from permeable reactive barriers installed at U.S. Department of Energy sites in Oak Ridge, Tennessee and Durango, Colorado. In: Naftz DL, Morrison SJ, Davis JA, Fuller CC, eds. Groundwater Remediation Using Permeable Reactive Barriers. San Diego: Academic Press, 2002. In press.
77. Cheng IF, Muftikian R, Fernando Q, Korte N. Reduction of nitrate to ammonia by zero-valent iron. Chemosphere 1997; 35:2689–2695.
78. Kielemoes J, DeBoever P. Influence of denitrification on the corrosion of iron and stainless steel powder. Environ Sci Technol 2000; 34:663.
79. Rahman A, Agrawal A. Reduction of nitrate and nitrite by iron metal: implications for ground water remediation. 213th National Meeting, San Francisco, CA American Chemical Society 1996; 37(1):157–159.

80. Alowitz MJ, Scherer MM. Kinetics of nitrate, nitrite, and Cr(VI) reduction by iron metal. Environ Sci Technol 2002; 36:299–306.
81. Chew CF, Zhang TC. Abiotic degradation of nitrates using zero-valent iron and electrokinetic processes. Environ Eng Sci 1999; 16:389–401.
82. Özdemir M, Tüfekci M. Removal of chlorine residues in aqueous media by metallic iron. Water Res 1997; 31:343–345.
83. Shrout JD, Parkin GF. Inhibition of anaerobic perchlorate biotransformation by Fe(0). In: Wickramanayake GB, Gavaskar AR, Gibbs JT, Means JL, eds. Case Studies in the Remediation of Chlorinated and Recalcitrant Compounds: Proceedings of the Second International Conference on Remediation of Chlorinated and Recalcitrant Compounds, Monterey, CA, 20–25 May 2000. Columbus, OH: Battelle Press, 2000; C2–7:107–113.
84. Matheson LJ, Tratnyek PG. Reductive dehalogenation of chlorinated methanes by iron metal. Environ Sci Technol 1994; 28:2045–2053.
85. Hozalski RM, Zhang L, Arnold WA. Reduction of haloacetic acids by Fe^0: implications for treatment and fate. Environ Sci Technol 2001; 35:2258–2263.
86. Johnson TL, Scherer MM, Tratnyek PG. Kinetics of halogenated organic compound degradation by iron metal. Environ Sci Technol 1996; 30:2634–2640.
87. Scherer MM, Balko BA, Gallagher DA, Tratnyek PG. Correlation analysis of rate constants for dechlorination by zero-valent iron. Environ Sci Technol 1998; 32:3026–3033.
88. Arnold WA, Roberts AL. Pathways and kinetics of chlorinated ethylene and chlorinated acetylene reaction with Fe(0) particles. Environ Sci Technol 2000; 34:1794–1805.
89. Kim Y-H, Carraway ER. Dechlorination of pentachlorophenol by zero valent iron and modified zero valent irons. Environ Sci Technol 2000; 34:2014–2017.
90. Yak HK, Wenclawiak BW, Cheng IF, Doyle JG, Wai CM. Reductive dechlorination of polychlorinated biphenyls by zerovalent iron in subcritical water. Environ Sci Technol 1999; 33:1307–1310.
91. Wu QX, Majid A, Marshall WD. Reductive dechlorination of polychlorinated biphenyl compounds in supercritical carbon dioxide. Green Chem 2000; 2:127–132.
92. Hinz DC, Wai CM, Wenclawiak BW. Remediation of a nonachloro biphenyl congener with zero-valent iron in subcritical water. J Environ Eng 2000; 2:45–48.
93. Chuang F-W, Larson RA, Scully Wessman M. Zero-valent iron-promoted dechlorination of polychlorinated biphenyls. Environ Sci Technol 1995; 29:2460–2463.
94. Campbell TJ, Burris DR, Roberts AL, Wells JR. Trichloroethylene and tetrachloroethylene reduction in a metallic iron–water-vapor batch system. Environ Toxicol Chem 1997; 16:625–630.
95. Roberts AL, Totten LA, Arnold WA, Burris DR, Campbell TJ. Reductive elimination of chlorinated ethylenes by zero-valent metals. Environ Sci Technol 1996; 30:2654–2659.

96. Fennelly JP, Roberts AL. Reaction of 1,1,1-trichloroethane with zero-valent metals and bimetallic reductants. Environ Sci Technol 1998; 32:1980–1988.
97. Totten LA, Jans U, Roberts AL. Alkyl bormides as mechanistic probes of reductive dehalogenation: reactions of vicinal dibromide stereoisomers with zerovalent metals. Environ Sci Technol 2001; 35:2804–2811.
98. Roberts AL, Jeffers PM, Wolfe NL, Gschwend PM. Structure–reactivity relationships in dehydrohalogenation reactions of polychlorinated and polybrominated alkanes. Crit Rev Environ Sci Technol 1993; 23:1–39.
99. Agrawal A, Tratnyek PG. Reduction of nitro aromatic compounds by zero-valent iron metal. Environ Sci Technol 1996; 30:153–160.
100. Devlin JF, Klausen J, Schwarzenbach RP. Kinetics of nitroaromatic reduction on granular iron in recirculating batch experiments. Environ Sci Technol 1998; 32:1941–1947.
101. Scherer MM, Johnson K, Westall JC, Tratnyek PG. Mass transport effects on the kinetics of nitrobenzene reduction by iron metal. Environ Sci Technol 2001; 35:2804–2811.
102. Klausen J, Ranke J, Schwarzenbach RP. Influence of solution composition and column aging on the reduction of nitroaromatic compounds by zero-valent iron. Chemosphere 2001; 44:511–517.
103. Mantha R, Taylor KE, Biswas N, Bewtra JK. A continuous system for Fe^0 reduction of nitrobenzene in synthetic wastewater. Environ Sci Technol 2001; 35:3231–3236.
104. Singh J, Comfort SD, Shea PJ. Iron-mediated remediation of RDX-contaminated water and soil under controlled Eh/pH. Environ Sci Technol 1999; 33:1488–1494.
105. Singh J, Comfort SD, Shea PJ. Remediating RDX-contaminated water and soil using zero-valent iron. J Environ Qual 1998; 27:1240–1245.
106. Hundal LS, Singh J, Bier EL, Shea PJ, Comfort SD, Power WL. Removal of TNT and RDX from water and soil using iron metal. Environ Pollut 1997; 97:55–64.
107. Tratnyek PG, Miehr R, Bandstra JZ. Kinetics of reduction of TNT by iron metal. Groundwater Quality 2001: Third International Conference on Groundwater Quality, Sheffield, UK. IAHS Press, 2002; 275:427–433.
108. Miehr R, Bandstra JZ, Po R, Tratnyek PG. Remediation of 2, 4,6-trinitrotoluene (TNT) by iron metal: kinetic controls on product distributions in batch and column experiments. 225th National Meeting, New Orleans, LA, American Chemical Society 2003; 43(1).
109. Elovitz MS, Weber EJ. Sediment-mediated reduction of 2,4,6-trinitrotoluene and fate of the resulting aromatic (poly)amines. Environ Sci Technol 1999; 33:2617.
110. Hofstetter TB, Heijman CG, Schwarzenbach RP. Complete reduction of TNT and other (poly)nitroaromatic compounds under iron-reducing subsurface conditions. Environ Sci Technol 1999; 33:1479–1487.
111. Nam S, Tratnyek PG. Reduction of azo dyes with zero-valent iron. Water Res 2000; 34:1837–1845.

112. Cao J, Wei L, Huang Q, Wang L, Han S. Reducing degradation of azo dyes by zero-valent iron in aqueous solution. Chemosphere 1999; 38:565–571.
113. Weber EJ. Iron-mediated reductive transformations: investigation of reaction mechanism. Environ Sci Technol 1996; 30:716–719.
114. Gui L, Gillham RW, Odziemkowski MS. Reduction of N-nitrosodimethylamine with granular iron and nickel-enhanced iron. 1. Pathways and kinetics. Environ Sci Technol 2000; 34:3489–3494.
115. Odziemkowski MS, Gui L, Gillham RW. Reduction of N-nitrosodimethylamine with granular iron and nickel-enhanced iron. 2. Mechanistic studies. Environ Sci Technol 2000; 34:3495–3500.
116. Tomkins BA, Griest WH. Determinations of N-nitrosodimethylamine at part-per-trillion concentrations in contaminated groundwaters and drinking waters featuring carbon-based membrane extraction disks. Anal Chem 1996; 68:2533–2540.
117. Wildman MJ, Alvarez PJJ. RDX degradation using an integrated Fe(0)-microbial treatment approach. Water Sci Technol 2001; 43:25–33.
118. Oh B-T, Just CL, Alvarez PJJ. Hexahydro-1,35-trinitro-1,3,5-triazine mineralization by zerovalent iron and mixed anaerobic cultures. Environ Sci Technol 2001; 35:4341–4346.
119. Rinehart KL Jr. Oxidation and Reduction of Organic Compounds. Englewood Cliffs, NJ: Prentice-Hall, Inc., 1973; 148.
120. Macalady DL, Tratnyek PG, Grundl TJ. Abiotic reduction reactions of anthropogenic organic chemicals in anaerobic systems. J Contam Hydrol 1986; 1:1–28.
121. Tratnyek PG, Macalady DL. Oxidation–reduction reactions in the aquatic environment. In: Mackay D, Boethling RS, eds. Handbook of Property Estimation Methods for Chemicals: Environmental and Health Sciences. Boca Raton, FL: Lewis, 2000:383–415.
122. Tratnyek PG, Reilkoff TE, Lemon A, Scherer MM, Balko BA, Feik LM, Henegar B. Visualizing redox chemistry: probing environmental oxidation–reduction reactions with indicator dyes. Chem Educ 2001; 6:172–179.
123. Ghauch A, Gallet C, Charef A, Rima J, Martin-Bouyer M. Reductive degradation of carbaryl in water by zero-valent iron. Chemosphere 2001; 42:419–424.
124. Ghauch A. Degradation of benomyl, picloram, and dicamba in a conical apparatus by zero-valent iron powder. Chemosphere 2001; 43:1109–1117.
125. Arnold WA, Ball WP, Roberts AL. Polychlorinated ethane reaction with zero-valent zinc: pathways and rate control. J Contam Hydrol 1999; 40:183–200.
126. Su C, Puls RW. Kinetics of trichloroethene reduction by zerovalent iron and tin: pretreatment effect, apparent activation energy, and intermediate products. Environ Sci Technol 1999; 33:163–168.
127. Boronina T, Klabunde KJ, Sergeev G. Destruction of organohalides in water using metal particles: carbon tetrachloride/water reactions with magnesium, tin, and zinc. Environ Sci Technol 1995; 29:1511–1517.
128. Schlimm C, Heitz E. Development of a wastewater treatment process:

reductive dehalogenation of chlorinated hydrocarbons by metals. Environ Prog 1996; 15:38–47.
129. Hardy LI, Gillham RW. Formation of hydrocarbons from the reduction of aqueous CO_2 by zero-valent iron. Environ Sci Technol 1996; 30:57–65.
130. Agrawal A, Tratnyek PG, Stoffyn-Egli P, Liang L. Processes affecting nitro reduction by iron metal: mineralogical consequences of precipitation in aqueous carbonate environments. 209th National Meeting, Anaheim, CA, American Chemical Society 1995; 35(1):720–723.
131. Liang L, Korte N, Gu B, Puls R, Reeter C. Geochemical and microbial reactions affecting the long-term performance of in situ "iron barriers". Adv Environ Res 2000; 4:273–286.
132. Tratnyek PG, Johnson TL, Scherer MM, Eykholt GR. Remediating ground-water with zero-valent metals: kinetic considerations in barrier design. Ground Water Monit Remediat 1997; 17:108–114.
133. Miehr R, Tratnyek PG, Bandstra JZ, Scherer MM, Alowitz M, Bylaska EJ. The diversity of contaminant reduction reactions by zero-valent iron: role of the reductate. In preparation.
134. Weast RC, ed. Handbook of Chemistry and Physics 56th ed. Cleveland, OH: CRC Press, 1975.
135. Liang L, Korte N, Goodlaxson JD, Clausen J, Fernando Q, Muftikian R. Byproduct formation during the reduction of TCE by zero-valence iron and palladized iron. Ground Water Monit Remediat. Winter 1997:122–127.
136. Muftikian R, Nebesny K, Fernando Q, Korte N. X-ray photoelectron spectra of the palladium–iron bimetallic surface used for the rapid dechlorination of chlorinated organic environmental contaminants. Environ Sci Technol 1996; 30:3593–3596.
137. Muftikian R, Fernando Q, Korte N. A method for the rapid dechlorination of low molecular weight chlorinated hydrocarbons in water. Water Res 1995; 29:2434–2439.
138. Appleton EL. A nickel–iron wall against contaminated groundwater. Environ Sci Technol 1996; 30:536A–539A.
139. Mackenzie K, Koehler R, Weiss H, Kopinke F-D. Dechlorination of chlorohydrocarbons in ground-water using novel membrane-supported Pd catalysts. In: Wickramanayake GB, Gavaskar AR, Chen ASC, eds. Chemical Oxidation and Reactive Barriers: Proceedings of the Second International Conference on Remediation of Chlorinated and Recalcitrant Compounds, Monterey, CA, 20–25 May 2000. Columbus, OH: Battelle Press 2000; C2-6:331–338.
140. Neurath SK, Ferguson WJ, Dean SB, Foose D, Agrawal A. Rapid and complete dehalogenation of chlorinated phenols by Fe–Pd bimetallic reductants in bench-scale reactors: implications for soil and ground water remediation. 213th National Meeting, San Francisco, CA, American Chemical Society 1997; 37(1):159–161.
141. Engelmann MD, Doyle JG, Cheng IF. The complete dechlorination of DDT by magnesium/palladium bimetallic particles. Chemosphere 2001; 43:195–198.

142. Sivavec TM, Mackenzie PD, Horney DP. Effect of site groundwater on reactivity of bimetallic media: deactivation of nickel-plated granular iron. 213th National Meeting, San Francisco, CA, American Chemical Society 1997; 37(1):83–85.
143. Gui L, Gillham RW. Preparation and regeneration of nickel–iron for reduction of organic contaminants. 221st National Meeting, San Diego, CA, American Chemical Society 2001; 41(1):1132–1137.
144. Scherer MM, Westall JC, Ziomek-Moroz M, Tratnyek PG. Kinetics of carbon tetrachloride reduction at an oxide-free iron electrode. Environ Sci Technol 1997; 31:2385–2391.
145. Casey FXM, Ong SK, Horton R. Degradation and transformation of trichloroethylene in miscible displacement experiments through zerovalent metals. Environ Sci Technol 2000; 34:5023–5029.
146. Scherer MM, Balko BA, Tratnyek PG. The role of oxides in reduction reactions at the metal–water interface. In: Sparks DL, Grundl TJ, eds. Mineral–Water Interfacial Reactions: Kinetics and Mechanisms. Washington, DC: American Chemical Society, 1998; ACS Symp. Ser. 715:301–322.
147. Johnson TL, Fish W, Gorby YA, Tratnyek PG. Degradation of carbon tetrachloride by iron metal: complexation effects on the oxide surface. J Contam Hydrol 1998; 29:377–396.
148. Chen J-L, Al-Abed SR, Ryan JA, Li Z. Effects of pH on dechlorination of trichloroethylene by zero-valent iron. J Hazard Mater 2001; 83:243–254.
149. Hu H-Y, Goto N, Fujie K. Effect of pH on the reduction of nitrite in water by metallic iron. Water Res 2001; 35:2789–2793.
150. Zawaideh LI, Zhang TC. The effects of pH and addition of an organic buffer (HEPES) on nitrate transformation in Fe^0–water systems. Water Sci Technol 1998; 38:107–115.
151. Gotpagar J, Grulke E, Tsang T, Bhattacharyya D. Reductive dehalogenation of trichloroethylene using zero-valent iron. Environ Prog 1997; 16:137–143.
152. Wüst WF, Köber R, Schlicker O, Dahmke A. Combined zero- and first-order kinetic model of the degradation of TCE and cis-DCE with commercial iron. Environ Sci Technol 1999; 33:4304–4309.
153. Choe S, Chang YY, Hwang KY, Khim J. Kinetics of reductive denitrification by nanoscale zero-valent iron. Chemosphere 2000; 41:1307–1311.
154. Tratnyek PG, Scherer MM. Kinetic controls on the performance of remediation technologies based on zero-valent iron. Proceedings of the 1998 National Environmental Engineering Conference: Water Resources in the Urban Environment, Chicago, IL. American Society of Civil Engineers 1998:110–115.
155. Burrow PD, Aflatooni K, Gallup GA. Dechlorination rate constants on iron and the correlation with electron attachment energies. Environ Sci Technol 2000; 34:3368–3371.
156. Perlinger JA, Venkatapathy R, Harrison JF. Linear free energy relationships for polyhalogenated alkane transformation by electron-transfer mediators in model aqueous systems. J Phys Chem 2000; 104:2752.

157. Hung H-M, Hoffmann MR. Kinetics and mechanism of the enhanced reductive degradation of CCl_4 by elemental iron in the presence of ultrasound. Environ Sci Technol 1998; 32:3011–3016.
158. Gillham RW, O'Hannesin SF. Enhanced degradation of halogenated aliphatics by zero-valent iron. Ground Water 1994; 32:958–967.
159. Deng B, Burris DR, Campbell TJ. Reduction of vinyl chloride in metallic iron–water systems. Environ Sci Technol 1999; 33:2651–2656.
160. Su C, Puls RW. Temperature effect on reductive dechlorination of trichlorethene by zero-valent metals. In: Wickramanayake GB, Hinchee RE, eds. Physical, Chemical, and Thermal Technologies: Proceedings of the First International Conference on Remediation of Chlorinated and Recalcitrant Compounds, Monterey, CA, 18–21 May 1998. Columbus, OH: Battelle Press, 1998; 1(5):317–322.
161. Gavaskar A, Gupta N, Sass B, Janosy R, Hicks J. Design Guidance for Application of Permeable Barriers for Remediate Dissolved Chlorinated Solvents. Columbus, OH: Battelle Press, Final, 2000.
162. Lasaga AC. Rate laws of chemical reactions. In: Lasaga AC, Kirkpatrick RJ, eds. Kinetics of geochemical processes. Washington, DC: Mineralogical Society of America, 1981; 8:1–68.
163. Scherer MM, Tratnyek PG. Dechlorination of carbon tetrachloride by iron metal: effect of reactant concentrations. 209th National Meeting, Anaheim, CA, American Chemical Society 1995; 35(1):805–806.
164. Arnold WA, Roberts AL. Development of a quantitative model for chlorinated ethylene reduction by zero-valent metals. 213th National Meeting, San Francisco, CA, American Chemical Society 1997; 37(1):76–77.
165. Arnold WA, Roberts AL. Pathways of chlorinated ethylene and chlorinated acetylene reaction with Zn(0). Environ Sci Technol 1998; 32:3017–3025.
166. Arnold WA, Roberts AL. Inter- and intraspecies competitive effects in reactions of chlorinated ethylenes with zero-valent iron in column reactors. Environ Eng Sci 2000; 17:291–302.
167. Devlin JF, Morkin M, Repta C. Incorporating surface saturation effects into iron wall design calculations. In: Wickramanayake GB, Gavaskar AR, Chen ASC, eds. Chemical Oxidation and Reactive Barriers: Proceedings of the Second International Conference on Remediation of Chlorinated and Recalcitrant Compounds, Monterey, CA, 20–25 May 2000. Columbus, OH: Battelle Press, 2000; C2–6:393–400.
168. Tratnyek PG, Scherer MM, Deng B, Hu S. Effects of natural organic matter, anthropogenic surfactants, and model quinones on the reduction of contaminants by zero-valent iron. Water Res 2001; 35:4435–4443.
169. Deng B, Hu S, Burris DR. Effect of iron corrosion inhibitors on trichloroethylene reduction. In: Wickramanayake GB, Hinchee RE, eds. Physical, Chemical, and Thermal Technologies: Proceedings of the First International Conference on Remediation of Chlorinated and Recalcitrant Compounds, Monterey, CA, 18–21 May 1998. Columbus, OH: Battelle Press, 1998; 1(5): 341–346.

170. Loraine GA. Effects of alcohols, anionic and nonionic surfactants on the reduction of PCE and TCE by zero-valent iron. Water Res 2001; 35:1453–1460.
171. Burris DR, Campbell TJ, Manoranjan VS. Sorption of trichloroethylene and tetrachloroethylene in a batch reactive metallic iron–water system. Environ Sci Technol 1995; 29:2850–2855.
172. Burris DR, Allen-King RM, Manoranjan VS, Campbell TJ, Loraine GA, Deng B. Chlorinated ethene reduction by cast iron: sorption and mass transfer. J Environ Eng 1998; 124:1012–1019.
173. Allen-King RM, Halket RM, Burris DR. Reductive transformation and sorption of cis- and trans-1,2-dichloroethene in a metallic iron–water system. Environ Toxicol Chem 1997; 16:424–429.
174. Gotpagar JK, Grulke EA, Bhattacharayya D. Reductive dehalogenation of trichloroethylene: kinetic models and experimental verification. J Hazard Mater 1998; 62:243–264.
175. Gotpagar J, Lyuksyutov S, Cohn R, Grulke E, Bhattacharyya D. Reductive dehalogenation of trichloroethylene with zero-valent iron: surface profiling microscopy and rate enhancement studies. Langmuir 1999; 15:8412–8420.
176. Deng B, Campbell TJ, Burris DR. Kinetics of vinyl chloride reduction by metallic iron in zero-headspace systems. 213th National Meeting, San Francisco, CA, American Chemical Society 1997; 37(1):81–83.
177. Li T, Farrell J. Rate-limiting mechanisms for carbon tetrachloride and trichloroethylene reactions at iron surfaces. 221st National Meeting, San Diego, CA, American Chemical Society 2001; 41(1):1154–1159.
178. Levich VG. Physicochemical Hydrodynamics. Englewood Cliffs, NJ: Prentice-Hall, 1962.
179. Fryar AE, Schwartz FW. Modeling the removal of metals from groundwater by a reactive barrier: experimental results. Water Resour Res 1994; 30:3455–3469.
180. Johnson TL, Tratnyek PG. A column study of carbon tetrachloride dehalogenation by iron metal. Proceedings of the 33rd Hanford Symposium on Health & the Environment In-Situ Remediation: Scientific Basis for Current and Future Technologies, Pasco, WA. Richland, WA: Battelle Pacific Northwest Laboratories 1994; 2:931–947.
181. Tratnyek PG, Johnson TL, Schattauer A. Interfacial phenomena affecting contaminant remediation with zero-valent iron metal. Emerging Technologies in Hazardous Waste Management VII. Atlanta, GA: American Chemical Society, 1995:589–592.
182. Gupta N, Sass BM, Gavaskar AR, Sminchak JR, Fox TC, Snyder FA, O'Dwyer D, Reeter C. Hydraulic evaluation of a permeable barrier using tracer tests, velocity measurements, and modeling. In: Wickramanayake GB, Hinchee RE, eds. Designing and Applying Treatment Technologies: Proceedings of the First International Conference on Remediation of Chlorinated and Recalcitrant Compounds, Monterey, CA, 18–21 May 1998. Columbus, OH: Battelle Press, 1998; 1(6):157–162
183. Sass BM, Gavaskar AR, Gupta N, Yoon W-S, Hicks JE, O'Dwyer D, Reeter C. Evaluating the Moffett Field permeable barrier using groundwater

monitoring and geochemical modeling. In: Wickramanayake GB, Hinchee RE, eds. Designing and Applying Treatment Technologies: Proceedings of the First International Conference on Remediation of Chlorinated and Recalcitrant Compounds, Monterey, CA, 18–21 May 1998. Columbus, OH: Battelle Press, 1998; 1(6):169–175.

184. Yabusaki S, Cantrell K, Sass B, Steefel C. Multicomponent reactive transport in an in situ zero-valent iron cell. Environ Sci Technol 2001; 35:1493–1503.
185. Puls RW, Paul CJ, Powell RM. The application of in situ permeable reactive (zero-valent iron) barrier technology for the remediation of chromate-contaminated groundwater: a field test. Appl Geochem 1999; 14:989–1000.
186. Blowes DW, Mayer KU. An in situ permeable reactive barrier for the treatment of hexavalent chromium and trichloroethylene in ground water: Volume 3, Mulitcomponent Reactive Transport Modeling, U.S. Environmental Protection Agency, EPA/600/R-99/095c, Ada, OK, 1999.
187. Blowes DW, Puls RW, Gillham RW, Ptacek CJ, Bennett TA, Bain JG, Hanton-Fong CJ, Paul CJ. An in situ permeable reactive barrier for the treatment of hexavalent chromium and trichloroethylene in ground water: Volume 2, Performance Monitoring, U.S. Environmental Protection Agency, EPA/600/R-99/095b, Ada, OK, 1999.
188. Blowes DW, Gillham RW, Ptacek CJ, Puls RW, Bennett TA, O'Hannesin SF, Hanton-Fong CJ, Bain JG. An in situ permeable reactive barrier for the treatment of hexavalent chromium and trichloroethylene in ground water: Volume 1, Design and Installation, U.S. Environmental Protection Agency, EPA/600/R-99/095a, Ada, OK, 1999.
189. Phillips DH, Gu B, Watson DB, Roh Y, Liang L, Lee SY. Performance evaluation of a zerovalent iron reactive barrier: mineralogical consequences. Environ Sci Technol 2000; 34:4169–4176.
190. Gu B, Watson DB, Wu L, Phillips DH, White DC, Zhou J. Microbial characteristics in a zero-valent iron reactive barrier. Environ Monit Assess 2002; 77(3):293–309.
191. Matheson LJ. Abiotic and Biotic Reductive Dehalogenation of Halogenated Methanes. Ph.D thesis, Oregon Graduate Institute, 1994.
192. Johnson TJ. Surface Mediated Reduction of Chlorinated Solvents by Zero-Valent Iron. Ph.D thesis, Oregon Graduate Institute, 1997.
193. Vogan JL, Butler BJ, Odziemkowski MS, Friday G, Gillham RW. Inorganic and biological evaluation of cores from permeable iron reactive barriers. In: Wickramanayake GB, Hinchee RE, eds. Designing and Applying Treatment Technologies: Proceedings of the First International Conference on Remediation of Chlorinated and Recalcitrant Compounds, Monterey, CA, 18–21 May 1998. Columbus, OH: Battelle Press, 1998; 1(6):163–168.
194. Mackenzie PD, Horney DP, Sivavec TM. Mineral precipitation and porosity losses in granular iron columns. J Hazard Mater 1999; 68:1–17.
195. Roh Y, Lee SY, Elless MP. Characterization of corrosion products in the permeable reactive barriers. Environ Geol 2000; 40:184–194.

196. Gu B, Phelps TJ, Liang L, Dickey MJ, Roh Y, Kinsall BL, Palumbo AV, Jacobs GK. Biogeochemical dynamics in zero-valent iron columns: implications for permeable reactive barriers. Environ Sci Technol 1999; 33:2170–2177.
197. Klausen J, Tröber SP, Haderlein SB, Schwarzenbach RP. Reduction of substituted nitrobenzenes by Fe(II) in aqueous mineral suspensions. Environ Sci Technol 1995; 29:2396–2404.
198. White AF, Peterson ML. Reduction of aqueous transition metal species on the surfaces of Fe(II)-containing oxides. Geochim Cosmochim Acta 1996; 60:3799–3814.
199. Williams AGB, Scherer MM. Kinetics of chromate reduction by carbonate green rust. 220th National Meeting, Washington, DC, American Chemical Society 2000; 42(2):666–668.
200. Schultz CA, Grundl TJ. pH dependence on reduction rate of 4-Cl-nitrobenzene by Fe(II)/montorillonate systems. Environ Sci Technol 2000; 34:3641–3648.
201. Amonette JE, Workman DJ, Kennedy DW, Fruchter JS, Gorby YA. Dechlorination of carbon tetrachloride by Fe(II) associated with goethite. Environ Sci Technol 2000; 34:4606–4613.
202. Suthersan SS. Remediation Engineering: Design Concepts. Boca Raton, FL: Lewis, 1997:362.
203. Gavaskar AR, Gupta N, Sass B, Janosy R, O'Sullivan D. Permeable Barriers for Groundwater Remediation: Design, Construction, & Monitoring; Columbus, OH: Battelle, 1998:188.
204. Interstate Technology and Regulatory Cooperation Work Group (ITRC). Regulatory Guidance for Permeable Reactive Barriers Designed to Remediate Inorganic and Radionuclide Contamination, Interstate Technology and Regulatory Cooperation Work Group (ITRC), Regulatory Guidance, 1998. See http://www.itreweb.org.
205. Eykholt GR, Sivavec TM. Contaminant transport issues for reactive-permeable barriers. In: Acar YB, Daniel DE, eds. Geoenvironment 2000, Vol 2, Characterization, Containment, Remediation, and Performance in Environmental Geotechnics. New York: American Society of Civil Engineers, 1995:1608–1621.
206. Hatfield K, Burris DR, Wolfe NL. Analytical model for heterogeneous reactions in mixed porous media. J Environ Eng 1996; 122:676–684.
207. Khandelwal A, Rabideau AJ. Transport of sequentially decaying reaction products influenced by linear nonequilibrium sorption. Water Resour Res 1999; 35:1939–1945.
208. Mayer KU, Blowes DW, Frind EO. Reactive transport modeling of an in situ reactive barrier for the treatment of hexavalent chromium and trichloroethylene in groundwater. Water Resour Res 2001; 37:3091–3104.
209. Eykholt GR. Uncertainty-based scaling of iron reactive barriers. In: Evans J, Reddi L, eds. In Situ Remediation of the Geoenvironment. New York: American Society of Civil Engineers, 1997:41–55.

210. Benner SG, Blowes DW, Molson JWH. Modeling preferential flow in reactive barriers: implications for performance and design. Ground Water 2001; 39:371–379.
211. Eykholt GR, Elder CR, Benson CH. Effects of aquifer heterogeneity and reaction mechanism uncertainty on a reactive barrier. J Hazard Mater 1999; 68:73–96.
212. Gerlach R, Cunningham AB, Caccavo FJ. Dissimilatory iron-reducing bacteria can influence the reduction of carbon tetrachloride by iron metal. Environ Sci Technol 2000; 34:2461–2464.
213. Schäfer A, Bouwer EJ. Toluene induced cometabolism of cis-1,2-dichloroethylene and vinyl chloride under conditions expected downgradient of a permeable Fe(0) barrier. Water Res 2000; 34:3391–3399.
214. Sfeir HA, Reinhart RD, Chopra M, Clausen C, Geiger C. Biotic attentuation and zero-valent iron permeable barrier technology. In: Wickramanayake GB, Gavaskar AR, Chen ASC, eds. Chemical Oxidation and Reactive Barriers: Proceedings of the Second International Conference on Remediation of Chlorinated and Recalcitrant Compounds, Monterey, CA, 20–25 May 2000. Columbus, OH: Battelle Press, 2000; C2–6:323–329.
215. Weathers LJ, Parkin GF, Alvarez PJ. Utilization of cathodic hydrogen as electron donor for chloroform cometabolism by a mixed, methanogenic culture. Environ Sci Technol 1997; 31:880–885.
216. Gregory KB, Mason MG, Picken HD, Weathers LJ, Parkin GF. Bioaugmentation of Fe(0) for the remediation of chlorinated aliphatic compounds. Environ Eng Sci 2000; 17:169–181.
217. Chiu PC, Lee M, Cha DK, Radosevich M, Rhine ED. Characterizing a culture that dechlorinates TCE with Fe(0). In: Wickramanayake GB, Gavaskar AR, Chen ASC, eds. Chemical Oxidation and Reactive Barriers: Proceedings of the Second International Conference on Remediation of Chlorinated and Recalcitrant Compounds, 20–25 May 2000, Monterey, CA. Columbus, OH: Battelle Press, 2000; C2–6:425–432.
218. Lampron KJ, Chiu PC, Cha DK. Reductive dehalogenation of chlorinated ethenes with elemental iron: the role of microorganisms. Water Res 2001; 35:3077–3084.
219. Novak PJ, Daniels L, Parkin GF. Enhanced dechlorination of carbon tetrachloride and chloroform in the presence of elemental iron and *Methanosarcina barkeri*, *Methanosarcina thermophila*, or *Methanosaeta concillii*. Environ Sci Technol 1998; 32:1438–1443.
220. Novak PJ, Daniels L, Parkin GF. Rapid dechlorination of carbon tetrachloride and chloroform by extracellular agents in cultures of *Methanosarcina thermophila*. Environ Sci Technol 1998; 32:3132–3136.
221. Till BA, Weathers LJ, Alvarez PJJ. Fe(0)-Supported Autotrophic Denitrification. Environ Sci Technol 1998; 32:634–639.
222. Ehrlich HL. Geomicrobiology. 2nd ed. New York: Marcel Dekker, 1990:646.
223. Lovley DR. Dissimilatory Fe(III) and Mn(IV) reduction. Microbiol Rev 1991; 55:259–287.

224. Shokes TE, Möller G. Removal of dissolved heavy metals from acid rock drainage using iron metal. Environ Sci Technol 1999; 33:282–287.
225. Bizzigotti GO, Reynolds DA, Kueper BH. Enhanced solubilization and destruction of tetrachloroethylene by hydroxypropyl-β-cyclodextrin and iron. Environ Sci Technol 1997; 31:472–478.
226. Navon D, Loehr RC, Liljestrand HM, Daniel DE Jr. Impact of biodegradable trenching slurry on iron treatment wall performance. Water Sci Technol 1998; 38:49–53.
227. Lin CH, Tseng SK. Electrochemically reductive dechlorination of pentachlorophenol using a high overpotential zinc cathode. Chemosphere 1999; 39:2375–2389.
228. Roh Y, Lee SY, Elless MP, Moon HS. Electro-enhanced remediation of trichloroethene-contaminated groundwater using zero-valent iron. J Environ Sci Health, Part A 2000; 35:1061–1076.
229. Roh Y, Lee SY, Elless MP, Cho KS. Electro-enhanced remediation of radionuclide-contaminated groundwater using zero-valent iron (A35, pg. 1043, 2000). J Environ Sci Health, Part A 2000; 35:1995.
230. Roh Y, Lee SY, Elless MP, Cho KS. Electro-enhanced remediation of radionuclide-contaminated groundwater using zero-valent iron. J Environ Sci Health, Part A 2000; 35:1043–1059.
231. Scherer MM, Westall JC, Tratnyek PG. An electrochemical interpretation of carbon tetrachloride reduction at an oxide-free iron electrode. 214th National Meeting, Las Vegas, NV, American Chemical Society 1997; 37(2): 247–248.
232. Ho SV, Athmer C, Sheridan PW, Hughes BM, Orth R, McKenzie D, Brodsky PH, Shapiro A, Thornton R, Salvo J, Schultz D, Landis R, Griffith R, Shoemaker S. The Lasagna technology for in situ soil remediation. 1. Small field test. Environ Sci Technol 1999; 33:1086–1091.
233. Deng N, Luo F, Wu F, Xiao M, Wu X. Discoloration of aqueous reactive dye solutions in the UV/Fe^0 system. Water Res 2000; 34:2408–2411.
234. Lücking F, Köser H, Jank M, Ritter A. Iron powder, graphite and activated carbon as catalysts for the oxidation of 4-chlorophenol with hydrogen peroxide in aqueous solution. Water Res 1998; 32:2607–2614.
235. Doong R-A, Chang W-H. Photodegradation of parathion in aqueous titanium dioxide and zero valent solutions in the presence of hydrogen peroxide. J Photochem Photobiol A Chem 1998; 116:221–228.
236. Doong R-A, Chang W-H. Photoassisted iron compound catalytic degradation of organophosphorus pesticides with hydrogen peroxide. Chemosphere 1998; 37:2563–2572.
237. Pulgarin C, Schwitzguebel JP, Péringer P, Pajonk GM, Bandara J, Kiwi J. Abiotic degradation of atrazine on zero-valent iron activated by visible light. 209th National Meeting, Anaheim, CA, American Chemical Society 1995; 35(1):767–770.
238. Balarama Krishna MV, Chandrasekaran K, Karunasagar D, Arunachalam J. A combined treatment approach using Fenton's reagent and zero valent iron

for the removal of arsenic from drinking water. J Hazard Mater 2001; B84:229–240.
239. Arienzo M, Chiarenzelli J, Scrudato R, Pagano J, Falanga L, Connor B. Iron-mediated reactions of polychlorinated biphenyls in electrochemical peroxidation process (ECP). Chemosphere 2001; 44:1339–1346.
240. Uludag-Demirer S, Bowers AR. Adsorption/reduction reactions of trichloroethylene by elemental iron in the gas phase: the role of water. Environ Sci Technol 2000; 34:4407–4412.
241. Uludag-Demirer S, Bowers AR. Gas phase reduction of chlorinated VOCs by zero valent iron. J Environ Sci Health Part A 2001; 36:1535–1547.
242. Wang C-B, Zhang W-X. Synthesizing nanoscale iron particles for rapid and complete dechlorination of TCE and PCBs. Environ Sci Technol 1997; 31:2154–2156.
243. Zhang W-X, Wang C-B. Rapid and complete dechlorination of TCE and PCBs by nanoscale Fe and Pd/Fe particles. 213th National Meeting, San Francisco, CA, American Chemical Society 1997; 37(1):78–79.
244. Lien H-L, Zhang W-X. Complete reduction of chlorinated ethylenes by nanoscale bimetallic particles. 215th National Meeting, Dallas, TX, American Chemical Society 1998; 38(1):29–30.
245. Lien H-L, Zhang W-X. Transformation of chlorinated methanes by nanoscale iron particles. J Environ Eng 1999; 125:1042–1047.
246. Schrick B, Ponder SM, Mallouk TE. Remediation of chlorinated hydrocarbons using supported zero valent nickel–iron nanoparticles. 220th National Meeting, Washington, DC, American Chemical Society 2000; 42(2):639–640.
247. Lien H-L, Zhang W-X. Nanoscale iron particles for complete reduction of chlorinated ethenes. Colloids Surf A Physicochem Eng Asp 2001; 191:97–105.
248. Ponder SM, Darab JG, Mallouk TE. Remediation of Cr(VI) and Pb(II) aqueous solutions using supported, nanoscale zero-valent Iron. Environ Sci Technol 2000; 34:2564–2569.
249. Elliott DW, Zhang W-X. Field assessment of nanoscale bimetallic particles for groundwater treatment. Environ Sci Technol 2002.
250. Balko BA, Tratnyek PG. Photoeffects on the reduction of carbon tetrachloride by zero-valent iron. J Phys Chem B 1998; 102:1459–1465.
251. Rabideau AJ, Shen P, Khandelwal A. Feasibility of amending slurry walls with zero-valent iron. J Geotech Geoenviron Eng 1999; 125:330–333.
252. Butler EC, Hayes KF. Factors influencing rates and products in the transformation of trichloroethylene by iron sulfide and iron metal. Environ Sci Technol 2001; 35:3884–3891.
253. Hassan SM. Reduction of halogenated hydrocarbons in aqueous media: I. involvement of sulfur in iron catalysis. Chemosphere 2000; 40:1357–1363.
254. Lipczynska-Kochany E, Harms S, Milburn R, Sprah G, Nadarajah N. Degradation of carbon tetrachloride in the presence of iron and sulphur containing compounds. Chemosphere 1994; 29:1477–1489.
255. Yak HK, Lang Q, Wai CM. Relative resistance of positional isomers of

polychlorinated biphenyls toward reductive dechlorination by zerovalent iron in subcritical water. Environ Sci Technol 2000; 34:2792–2798.
256. Alessi DS, Li Z. Synergistic effect of cationic surfactants on perchloroethylene degradation by zero-valent iron. Environ Sci Technol 2001; 35:3713–3717.
257. Sequenced Reactive Barriers for Groundwater Remediation. In: Fiorenza S, Oubre CL, Ward CH. eds. Boca Raton, FL: Lewis, 2000.
258. Morkin M, Devlin JF, Barker JF, Butler BJ. In situ sequential treatment of a mixed contaminant plume. J Contam Hydrol 2000; 45:2302–2833.
259. Hung H-M, Ling FH, Hoffmann MR. Kinetics and mechanism of the enhanced reductive degradation of nitrobenzene by elemental iron in the presence of ultrasound. Environ Sci Technol 2000; 34:1758–1763.
260. Clausen CA, Geiger CL, Reinhart DR, Ruiz N, Farrell K, Toy P, Lau Chan N, Cannata M, Burnwinkle S, Quinn J. Ultrasonic regeneration of permeable treatment walls: laboratory/field studies. In: Wickramanayake GB, Gavaskar AR, Chen ASC, eds. Chemical Oxidation and Reactive Barriers: Proceedings of the Second International Conference on Remediation of Chlorinated and Recalcitrant Compounds, Monterey, CA, 20–25 May 2000. Columbus, OH: Battelle Press, 2000; C2–6:385–392.
261. Ruiz NE, Reinhart DR, Clausen CA, Geiger CL, Lau N. Enhanced zero-valent iron degradation of chlorinated solvents using ultrasonic energy. In: Wickramanayake GB, Hinchee RE, eds. Designing and Applying Treatment Technologies: Proceedings of the First International Conference on Remediation of Chlorinated and Recalcitrant Compounds, Monterey, CA, 18–21 May 1998. Columbus, OH: Battelle Press, 1998; 1(6):71–76.
262. Comfort SD, Shea PJ, Machacek TA, Gaber H, Oh B-T. Field-scale remediation of a metolchlor-contaminated spill site using zerovalent iron. J Environ Qual 2001; 30:1636–1643.

10
Enzymatic Treatment of Waters and Wastes

James A. Nicell
McGill University, Montreal, Quebec, Canada

I. INTRODUCTION

The implementation of increasingly stringent standards for the discharge of wastes into the environment has motivated the development of alternative processes for the production of goods and for the treatment and disposal of wastes. Ultimately, these processes are developed to meet one or more of the following objectives: (1) to improve the efficiency of utilization of raw materials, thereby conserving resources and reducing costs; (2) to recycle waste streams within a given facility to minimize the need for effluent disposal; (3) to reduce the quantity and maximize the quality of effluent waste streams that are created during production of goods; and (4) to transform wastes into marketable products. There are multitudes of ways in which the transformation of wastes and pollutants can be carried out. Most of these methods may be classified as being chemical or biological in nature.

Chemical transformations involve the application of reagents and reaction conditions to transform target species. Such processes involve a string of events that are usually well defined and can often be controlled to maximize efficiency. However, chemical processes often require the presence of excess quantities of reagents to accomplish the transformation to the desired extent. In addition, particularly harsh conditions (e.g., high temperature or extremes of pH) are sometimes required to facilitate the chemical transformations. This can present a problem once the desired transformation has taken place because the resulting stream may be a low-quality mixture that cannot be disposed into the environment or reused without subsequent treatment. Finally, many chemical treatment processes are not highly selective in terms of the types of pollutants that are transformed during treatment. Consequently, such processes are usually more economical for the treatment of dilute wastewaters and are often used as a polishing step before waste discharge into the environment [1].

Biological processes have been used with much success for many decades. These processes are designed to take advantage of the biochemical reactions that are carried out in living cells. Such processes make use of the natural metabolism of cells to accomplish the transformation or production of chemical species. The metabolic processes occur as a result of a sequence of reactions conducted inside the cell that are catalyzed by proteins called enzymes. An important advantage of biological systems is that they can be used to carry out processes for which no efficient chemical transformations have been devised. In addition, biological processes can often be conducted without the harsh conditions that are necessary during chemical transformations. However, the use of microorganisms is beset with many rate-limiting factors. For example, costly and time-consuming methods may be necessary to produce microbial cultures that can degrade the targeted pollutant. Furthermore, severe conditions such as chemical shock, extremes of pH and temperature, toxins, predators, and high concentrations of the pollutants, intermediates, and products may irreversibly damage or metabolically inactivate microbial cells. Thus, the sensitivity of microorganisms to changes in their environment can make these processes difficult to control over the long term, and may subject them to frequent upsets. They also require a supply of macro- and micronutrients for the support of microbial growth, and often result in the formation of large quantities of biomass that ultimately must be discarded into the environment. In addition, the biochemical reactions occur at a rate that is limited by the metabolism of the microorganism and, thus, are often slower than chemical processes. Moreover, whereas biological systems are commonly used to remove the bulk organic load in wastewaters, these systems often have difficulty in removing toxic pollutants to consistently low levels [1]. Therefore, conventional biological processes may not be able to improve water quality sufficiently to meet wastewater discharge criteria.

In an attempt to overcome some of the problems associated with chemical and biological systems, recent research has focused on developing environmental applications of enzymes that have been isolated from their parent organisms. This concept of using enzymes in waste-treatment applications is not new. In fact, applications were suggested as long ago as the 1930s [2]. However, recent interest in the development of enzymatic waste-treatment systems has grown for several reasons. Firstly, the rate of introduction of recalcitrant organic pollutants into the environment is on the rise, and it is becoming increasingly difficult to achieve an acceptable degree of removal of these pollutants using conventional chemical and biological processes. Secondly, there is a need for the development of alternative treatment methods that are faster, cheaper, more reliable, and simpler to implement than current processes. Thirdly, there is a growing recognition that enzymes can be used to target specific pollutants for treatment. And, finally, recent biotechnological advances foreshadow the production of cheaper and more readily available enzymes through genetic manipulation of microbial and plant cells and improved efficiency of isolation and purification procedures. Much of this work remains very new and substantial hurdles must be overcome before full-scale industrial application of enzymes can become a reality.

II. BACKGROUND AND FUNDAMENTALS OF ENZYMATIC PROCESSES

A. Introduction to Enzymes

Enzymatic systems fall between the two traditional categories of chemical and biological processes, since they involve chemical reactions based on the action of biological catalysts. Specifically, enzymes are catalysts that regulate the multitude of chemical reactions that occur in living cells whether they are plant, animal or microbial. They carry out such cellular processes as energy conversion, food digestion, and biosynthesis. They have an orderly structure and the catalytic action occurs within a particular region of the enzyme protein. The active site usually consists of several amino acids with a specific conformation. Thus, they exhibit specificity and are characterized by showing an optimum temperature and pH for their actions. Similar to other catalysts, they accelerate the chemical reaction rate by lowering the activation energy for a particular reaction. As shown in Fig. 1, the activation energy for the enzyme-catalyzed reaction is much smaller than that for the nonenzymatic reaction. This results in a much faster reaction rate. For example, the hydrolysis of sucrose by yeast invertase at 37°C is a trillion times faster than that caused by hydrogen ions alone [3].

The reactants of enzyme-catalyzed reactions are termed "substrates" and each enzyme is quite specific in character, acting on a particular substrate or substrates to produce a particular product or products. The names of enzymes usually indicate the substrate involved. For example, hydrogen peroxide oxidoreductase is an enzyme that uses hydrogen peroxide as its substrate to carry out the oxidation of organic substrates. Such formal names are often abbreviated such as in the contraction of hydrogen peroxide oxidoreductase to peroxidase. However, in the interest of avoiding

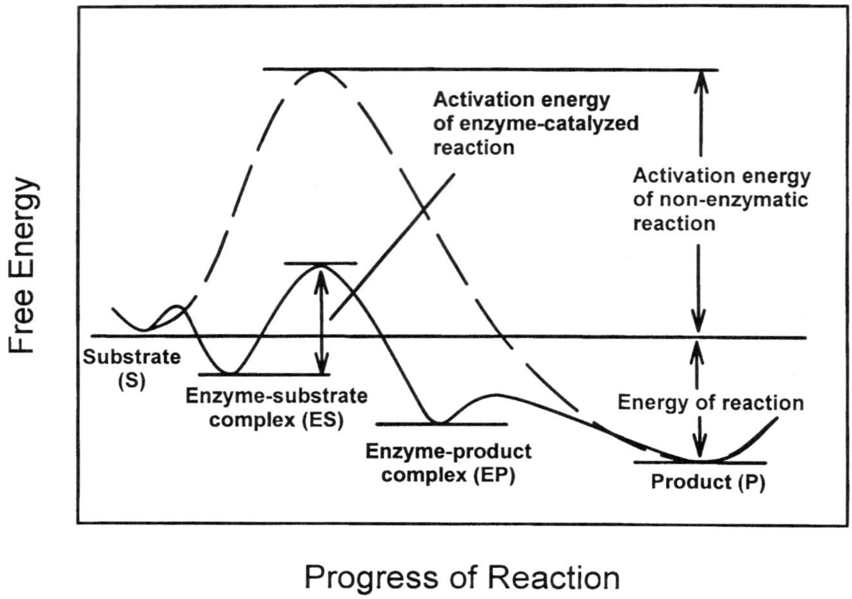

Figure 1 The activation energy of an enzyme-catalyzed reaction compared to a noncatalyzed reaction.

confusion between enzymes with similar names, a numerical classification system is used to specifically identify the nature of the enzyme. This EC classification arose out of the work of the Enzyme Commission of the International Union of Biochemistry, which classifies enzymes into six general groups: (1) oxidoreductases, catalyzing oxidation–reduction reactions; (2) transferases, catalyzing the transfer of functional groups; (3) hydrolases, catalyzing hydrolysis reactions; (4) lyases, catalyzing the addition of groups to double bonds; (5) isomerases, catalyzing intramolecular rearrangements; and (6) ligases, catalyzing the condensation of two molecules coupled with the cleavage of a pyrophosphate bond of ATP or a similar triphosphate. The code consists of a sequence of four punctuated numbers: EC x.x.x.x. The first digit identifies the main class of the enzyme (as numbered above). The second and third digits further describe the kind of reaction being catalyzed. Enzymes catalyzing very similar but nonidentical reactions, e.g., the hydrolysis of different carboxylic esters, will have the same first three digits in their code. The fourth digit distinguishes them by defining the actual substrate, e.g., the actual carboxylic acid ester being hydrolyzed.

However, it should be noted that isoenzymes, i.e., different enzymes catalyzing identical reactions, would have the same four-digit classification. The classification system provides only the basis for a unique identification of an enzyme; the particular isoenzyme and its source still have to be specified. For example, peroxidases isolated from soybeans and horseradishes have the same classification, i.e., EC 1.11.1.7. Currently, there are approximately 3200 enzymes that have been listed and assigned classification numbers.

The kinetics of many enzymatic reactions are often described using a simple model of enzyme catalysis, known as Michaelis–Menten kinetics. This model is based on an assumed string of events in which the enzyme E combines with the substrate S to form a complex ES (see also Fig. 1) according to

$$\mathrm{E} + \mathrm{S} \underset{k_{-1}}{\overset{k_1}{\rightleftarrows}} \mathrm{ES} \tag{1}$$

which then dissociates into product P and free enzyme E according to

$$\mathrm{ES} \xrightarrow{k_2} \mathrm{P} + \mathrm{E} \tag{2}$$

where k_1 and k_{-1} are the rate constants for the forward and reverse components of the first reaction and k_2 is the rate constant for the second reaction.

It can be shown that in a well-mixed closed vessel, under conditions in which the ratio of the concentration of enzyme (E_o) to substrate (S) is very small, the enzyme remains fully active, and the forward and reverse components of the first reaction are in steady state, the rate of change of substrate and product concentrations (P) can be written as [4]:

$$\frac{dS}{dt} = -\frac{dP}{dt} = \frac{-V_{max}S}{K_m + S} \tag{3}$$

where t is time, V_{max} is called the maximum or limiting velocity and is equal to k_2E_o, and K_m is known as the Michaelis constant and is equal to $(k_{-1}+k_2)/k_1$. The constants, V_{max} and K_m, are evaluated by fitting the Michaelis–Menten equation to experimentally determined initial rates. Initial-rate data are used since under initial conditions the influence of accumulating products on the enzyme is negligible, and since the reaction conditions including enzyme (E_o) and substrate (S_o) concentrations are known best at time zero.

This equation is consistent with qualitative observations that hold true for many enzyme systems. That is, the rate of reaction is first order in substrate concentration at low values of concentration but approaches zero order as the substrate concentration increases. In addition, the rate of reaction is proportional to the total amount of enzyme present. This equation is useful in that it can be used, for instance, to compare the relative rates of various enzymes carrying out the same reaction. However, its application in engineering systems is somewhat limited. For example, the Michaelis–Menten equation can be integrated to give:

$$V_{\max}t = S_o - S + K_m \ln\left(\frac{S_o}{S}\right) \tag{4}$$

from which the substrate concentration, S, corresponding to any time, t, can be determined. While inactivation is not usually a significant problem in vivo (in the living organism) where enzyme synthesis compensates for any loss of previously active enzymes, enzyme deactivation in vitro (isolated from a living cell) cannot be overlooked in kinetic studies or reactor design. Thus, if the quantity of enzyme participating in the reaction declines through inactivation or inhibition processes, this equation cannot be used to adequately describe the performance of the reactor. In addition, if it is intended that the reactor system achieve a high degree of conversion of the substrate (which is often the case during the transformation of pollutants), as the reaction progresses the ratio of enzyme to substrate will become relatively large and the terms under which the Michaelis–Menten equation was derived will be violated. Of course, a major impediment to the application of this equation is that the enzyme reactions of interest may not follow the simple sequence of events described above. Therefore, studies of the mechanisms and kinetics of enzyme systems can be crucial for the design of efficient reactor systems. The importance of such studies will be demonstrated later in this chapter.

B. Advantages and Limitations of Enzyme Applications for Pollutant Transformation

When accomplishing the transformation of chemical species, cell-free enzymes (i.e., those that have been isolated from their parent organisms) are often preferred over intact organisms containing the enzyme because they act with greater specificity, their activity can be better standardized, they are easier to handle and store, and enzyme concentration is not dependent on bacterial growth rates [5]. This can lead to some important advantages of enzymatic processes over biological systems. For example, enzymes can be applied to transform targeted contaminants including many of those that may resist biodegradation. This catalytic action can be carried out on, or in the presence of, many substances that are toxic to microbes. In addition, some enzymes can operate over relatively wide temperature, pH, and salinity ranges compared to cultures of microorganisms. They can also be used to treat contaminants at high and low concentrations and are not susceptible to shock loading effects associated with changes in contaminant concentrations that can often irreversibly damage or metabolically inactivate microbial cells. Consequently, there are fewer delays associated with shutdown/startup periods that are normally required to acclimatize biomass to waste streams. Importantly, the catalytic action of enzymes enables the development of smaller systems of lower capital cost due to the high reaction rates associated with enzymatic reactions. In addition, because bacterial growth is not required to accomplish waste transformations, sludge production is reduced because no biomass is generated. Many of the advantages listed above allow for the development of stable enzymatic treatment systems with simpler process control [6].

Enzymes can also offer several advantages over conventional chemical processes. In particular, the significance of enzymes lies in their ability to carry out processes that are impractical or impossible through nonbiological chemistry. Their high degree of specificity allows enzymes to remove target pollutants selectively, which precludes undesirable or unnecessary reactions that would otherwise increase reactant consumption and, correspondingly, increase the cost of treatment. They also make efficient use of chemical reagents and typically are characterized by a high reaction velocity. Their selectivity and high reaction rates make them ideal for the treatment of compounds that are present in trace quantities (i.e., micropollutants) or that cannot be removed by traditional physicochemical processes. Additionally, they operate at low temperature conditions, thereby reducing energy requirements for processes normally conducted at elevated temperatures (e.g., thermal oxidation). They also operate under mild pH conditions, thereby reducing the impact of corrosion on reaction vessels and avoiding the need for waste neutralization (e.g., Fenton's oxidation). Finally, an inherent advantage of enzymes is their compatibility with the environment and their nonhazardous nature. Thus, enzyme residues that may be present following waste treatment are of low pollution potential.

Whereas the above advantages are indeed significant, it should be noted that the majority of chemical and biological processes are not candidates for replacement by enzymatic processes. That is, biological processes and some chemical processes (e.g., chemical or thermal oxidation) have a fundamental advantage over enzymatic systems, i.e., their ability to simultaneously transform a broad range of compounds. For example, many municipal, agricultural, and industrial wastes consist of a mixture of organic compounds usually classified under the broad categories of biological oxygen demand (BOD) or chemical oxygen demand (COD). Once released into receiving water bodies, these collections of organic compounds result in the depletion of dissolved oxygen in the water column as a result of natural microbial processes. In many instances, the majority of these compounds can be efficiently degraded through the combined action of mixed cultures of microorganisms. In contrast, enzymes are biological catalysts whose actions are tailored to exclusively act upon specific chemical species. Thus, enzymatic treatment will not result in the removal of a broad range of compounds from a waste stream, but will only accomplish the transformation of an individual compound or class of compounds. If the product of the reaction is less toxic, more readily degraded than the reactant, or has a commercial value, then an enzyme-based strategy could be effective. This limits the application of enzymatic processes to accomplish the transformation of target species that are either problematic due to their toxicity or that have been identified as the raw materials from which enzymes can produce value-added products. In addition, because enzymatic reactions often require cosubstrates, these cosubstrates must be readily available. Water- or oxygen-requiring enzymes (e.g., hydrolases and oxygenases, respectively) are the most obvious candidates for waste applications, whereas enzymes that require cell-generated cosubstrates (e.g., ATP or NAD(P)H) would be less practical. Finally, it must be emphasized that some enzymes are quite fragile and their catalytic activities are sensitive to changes in ionic strength, the presence of metal ions, solvents, and other inactivating or inhibiting species. Therefore, wastewater characteristics will play a significant role in determining the feasibility of enzymatic treatment.

III. POTENTIAL APPLICATIONS OF ENZYMES

Recent research has focused on the development of enzymatic processes for the treatment of wastewaters, solid wastes, hazardous wastes, and soils. The environmental applications may be classified according to their objectives. For example, some processes are specifically designed to accomplish the transformation of target pollutants in wastewater streams to reduce toxicity. Alternatively, the conversion of waste materials can sometimes be achieved in a manner that produces a product with commercial value. Some applications that have recently been identified will be outlined below.

A. Enzymatic Treatment to Improve Waste Quality

Because of their high specificity to individual species or classes of compounds, enzymatic processes can be designed to specifically target selected compounds that are detrimental to the environment. Compounds that are candidates for this type of treatment are usually those that cannot be treated effectively or reliably using traditional techniques. Alternatively, enzymatic treatment can be used as a pretreatment step to remove one or more compounds that can interfere with subsequent downstream treatment processes. For example, if inhibitory or toxic compounds can be removed selectively, the bulk of the organic material could be treated biologically, thereby minimizing the cost of treatment. Many enzymes are susceptible to inactivation in the presence of other chemicals. Therefore, it is likely that enzymatic treatment will be most effective in those streams that have the highest concentration of the target contaminant and the lowest concentration of other contaminants that may tend to interfere with enzymatic treatment. The following situations are those where the use of enzymes might be most appropriate [1]:

- removal of specific chemicals from a complex industrial waste mixture before on-site or off-site biological treatment;
- removal of specific chemicals from dilute mixtures, for which conventional mixed-culture biological treatment might not be feasible;
- polishing of a treated wastewater or groundwater to meet limitations on specific pollutants or to meet whole effluent toxicity criteria;
- treatment of wastes generated infrequently or in isolated locations, including spill sites and abandoned waste-disposal sites; and
- treatment of low-volume, high-concentration wastewater at the point of generation in a manufacturing facility to allow reuse of the treated process wastewaters, to facilitate recovery of soluble products, or to remove pollutants known to cause problems downstream when mixed with other wastes from the plant.

Some potential applications of enzymes that have been identified for the improvement of waste quality include the transformation of toxic and color-causing aromatic compounds, cyanide, pesticides, surfactants, and heavy metals. In addition, some physical modifications in waste characteristics have been achieved through the mixing of solid wastes with enzymes. These applications are summarized in Table 1, and some promising applications are described in the following.

1. Aromatic Pollutants

Aromatic compounds constitute one of the major classes of pollutants and are heavily regulated in many countries. They are found in the wastes of a variety of industries including coal conversion, petroleum refining, resins and plastics, wood preservation, metal coating, dyes and other chemicals, textiles, mining and dressing, and pulp and paper. Most aromatic compounds are

Table 1 Cell-Free Enzymes Used to Accomplish the Transformation of Pollutants

Enzymes[a]	Proposed application
Acylamidase	Transformation of pesticides including acid analides (e.g., propanil, karsil) and phenoxyacetates (e.g., 2,4-D, MCPA) [53]
Alcalase	Hydrolysis of tannery waste for chrome recovery [110]
Amylases (in the presence of proteases) (EC 3.2.1)	Reduced treatment time for activated sludges from food processing wastes [63,111]
Cellulase glucanglucanohydrolase (EC 3.2.1.4) glucan exo-cellobiohydrolase (EC 3.2.1.91) B-glucosidase (EC 3.2.1.21)	Degradation of sludge from municipal wastewater treatment plants [61]; treatment of hazardous wastes [112]
Chloroperoxidase (EC 1.11.1.10)	Oxidation of phenolic compounds [29,85,102]
Cyanidase[b]	Cyanide decomposition [56,57]
Cyanide hydratase (EC 4.2.1.66)	Cyanide hydrolysis [57,58]
Dehalogenases	Degradation of agrochemicals, solvents, degreasers, flame retardants, and chemical intermediates used in the production of high volume chemicals [113]; detoxification of chlorophenols [61]
Esterase	Transformation of pesticides including phenylcarbamates (e.g., CIPC, IPC) and organophosphates (e.g., malathion, paraoxon, parathion) [53]
Laccase (EC 1.10.3.2)	Oxidation of phenols, dyes, and polycyclic aromatic hydrocarbons [48,49], decolorization of Kraft bleaching effluents, binding of phenols and aromatic amines with humus [47]
Lignin peroxidase	Transformation of phenols, aromatic amines, polyaromatic hydrocarbons, and other aromatic compounds, decolorization of Kraft bleaching effluents, treatment of dioxins, pyrene [86–89,114]
Lipase (EC 3.1.1.3) (with carbohydrase and protease)	Improved sludge dewatering [59]
Lysozyme (EC 3.2.1.17)	Cell lysis resulting in sludge degradation and improved dewatering [61]
Mn peroxidase	Oxidation of monoaromatic phenols and aromatic dyes [2]
Muramidase	Dewatering of pulp and paper sludges [61]
Parathion hydrolase	Hydrolyzation of organophosphate pesticides [53,55]
Pectinesterase (EC 3.1.1.11)	Degradation of pectin [64]
Peroxidase (EC 1.11.1.7)	Transformation of phenols and aromatic amines [7–24], decolorization of Kraft bleaching effluents [21], dewatering of phosphate slimes [60]
Perma-Zyme[b]	Improve the strength and stability of pure clayey soils and soil–fly ash mixtures [62]
Proteases	Degradation of proteins, improving sludge dewatering [61]
Tyrosinase (EC 1.14.18.1)	Transformation of phenols [42,44]
Ureases	Treatment of urinous waste waters [53]

[a] EC classifications are provided wherever they were specified in literature sources.
[b] Commercial enzyme preparation that may consist of more than one enzyme.

toxic and must be removed from wastes before they are discharged into the environment. Several methods of treatment based on the use of peroxidases and polyphenol oxidases have been proposed as potential alternatives to conventional methods for the treatment of such compounds.

Peroxidase enzymes are produced in the cells of many microorganisms and plants. They catalyze a variety of reactions, but all require the presence of oxidants such as hydrogen peroxide to activate them. They can carry out the oxidation of a wide range of toxic aromatic compounds including phenols, biphenols, anilines, benzidines, and related heteroaromatic compounds. Peroxidases that have been used for the laboratory-scale treatment of aqueous aromatic contaminants include horseradish peroxidase, soybean peroxidase, chloroperoxidase, manganese peroxidase, and lignin peroxidase. Of these, horseradish peroxidase (HRP) is undoubtedly one of the most studied enzymes in the relatively new area of enzymatic waste treatment. The mechanism of action of HRP has been studied extensively and is shown in Fig. 2. During the reaction, peroxidase passes through a series of oxidation steps (HRP, HRP_i, HRP_{ii}) in which it is first oxidized by hydrogen peroxide and subsequently oxidizes two molecules of aromatic substrate (AH_2). The free radical products (AH·) polymerize through a nonenzymatic process resulting in the formation of water-insoluble precipitates that can be

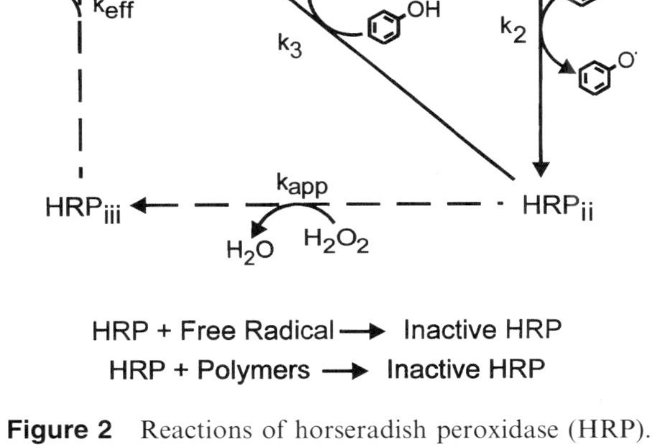

Figure 2 Reactions of horseradish peroxidase (HRP). (Adapted from Ref. 92.)

removed from the aqueous phase by sedimentation or filtration. The overall enzymatic reaction can be summarized as follows:

$$2AH_2 + H_2O_2 \rightarrow 2AH\bullet + 2H_2O \quad (5)$$

Other peroxidases have a similar mechanism, but peroxidases vary significantly in terms of their rate constants and their susceptibility to side reactions that may cause temporary or permanent inactivation.

Studies have been performed to demonstrate the wide range of substrates that can be treated by peroxidase [7,8] and to define the conditions of pH and temperature at which the enzyme remains active [9]. In addition, extensive work has been conducted to develop kinetic models to describe the mechanisms of catalysis and inactivation of the enzyme and to use models to optimize reactor design [10–12]. Furthermore, in the interest of reducing biocatalyst costs, research has been conducted involving the use of protective additives such as polyethylene glycol, sodium borate, chitosan, and gelatin to preserve enzyme activity during treatment [13–15]. Immobilization of peroxidases on solid supports has also been explored [16] because the advantages of this technique can include the preservation of enzyme activity [17], the potential for continuous flow treatment [18], and the possibility of enzyme reuse [19]. Thus, this methodology has been pursued during the development of many different types of enzymatic processes, as will be seen below. Finally, laboratory-scale experiments have demonstrated the ability of peroxidase to treat wastewaters from such processes as coal conversion [20,21], pulp and paper [22], petroleum refining [23], and foundries [24].

Another peroxidase, called lignin peroxidase (LiP) and also known as ligninase, was first reported in 1983 [25,26]. It is produced by the white-rot fungus *Phanerochaete chrysosporium* [2,27]. In contrast to other peroxidases, LiP was shown to be capable of mineralizing a variety of recalcitrant aromatic compounds (i.e., breaking them down into smaller molecules rather than achieving their polymerization) including dibenzo-*p*-dioxin [28], polycyclic aromatic compounds, and phenols [29]. LiP's role in breaking down lignin has also been confirmed [30,31]. Immobilization of LiP on porous ceramic supports did not adversely affect LiP's stability and showed a good potential for degradation of environmentally persistent aromatics [31].

The Kraft process that is widely used in wood pulping leaves behind 5% to 8% by weight of residual modified lignin in the pulp. This residual is responsible for the characteristic brown color of the pulp and is commercially removed by the use of bleaching agents such as chlorine and chlorine oxides [32]. Bleaching operations produce dark-brown effluents that contain toxic and mutagenic chlorinated products that constitute an environmental hazard [33]. Several studies have focused on the use of microorganisms in the treatment of bleaching effluents and there has been an interest in the potential use of enzymes such as peroxidases and laccases (see below) to perform the same task. The use of HRP and LiP in color removal from Kraft mill effluents has been reported [34]. Both enzymes were found to have considerable potential and it was observed that the immobilized form of the enzymes was invariably more efficient than the free form. LiP from *P. chrysosporium* was also found to play a positive role in color removal of bleaching effluents and it was explained that LiP was thought to degrade lignin by catalyzing the oxidation of aromatic units to cation radicals that can spontaneously decompose [35]. Pulp mill effluents have also been decolorized with LiP and laccase from *P. chrysosporium* that had been immobilized on Amberlite IRA-400 resin [36]. The immobilized enzyme was very efficient in removing color and phenolic species from a Kraft effluent with insignificant adsorption of colored species on the resin support. However, during the decolorization of bleaching effluent and black liquor, although substantial decolorization was achieved, a significant fraction of decolorization was found to be due to adsorption on the resin surface. Such fouling of the solid support is likely to be a common problem associated with the treatment of wastes using immobilized enzymes.

Whereas peroxidases have the potential to treat a variety of compounds over wide ranges of pH and temperature, one of the major concerns regarding the use of peroxidase is their prohibitive cost. A potential alternative to peroxidase is polyphenol oxidase, which also catalyzes the oxidation of aromatic compounds but which uses oxygen as an oxidant. This represents an important advantage over peroxidases due to the difficulty in safely transporting and handling hydrogen peroxide, and the ready availability and low cost of oxygen. Two types of polyphenol oxidases have been used for the treatment of aqueous phenols, namely, tyrosinase and laccase.

Because of the availability of tyrosinase (EC 1.14.18.1) in a variety of sources such as fruits, vegetables, and seafood products [37], this enzyme may eventually represent a less expensive alternative to peroxidases for the treatment of phenolic wastes. As shown in Fig. 3, tyrosinase catalyzes two distinct oxidation reactions of phenols. Molecular oxygen binds to the initial state of tyrosinase (E_{deoxy}) bringing it to an oxygenated state (E_{oxy}). Thereafter, monophenols or *o*-diphenols are oxidized by tyrosinase to *o*-quinones in cycle I (E_{oxy}, $E_{oxy\text{-}M}$, and $E_{met\text{-}D}$) or cycle II (E_{oxy}, $E_{oxy\text{-}D}$, E_{met}, and $E_{met\text{-}D}$), respectively. Quinones are mostly unstable and undergo nonenzymatic polymerization to yield substances of reduced solubility [38–41]. In contrast to treatment conducted with peroxidase and hydrogen peroxide during which polymeric products spontaneously precipitate from the aqueous phase, phenol solutions treated with tyrosinase changed from colorless

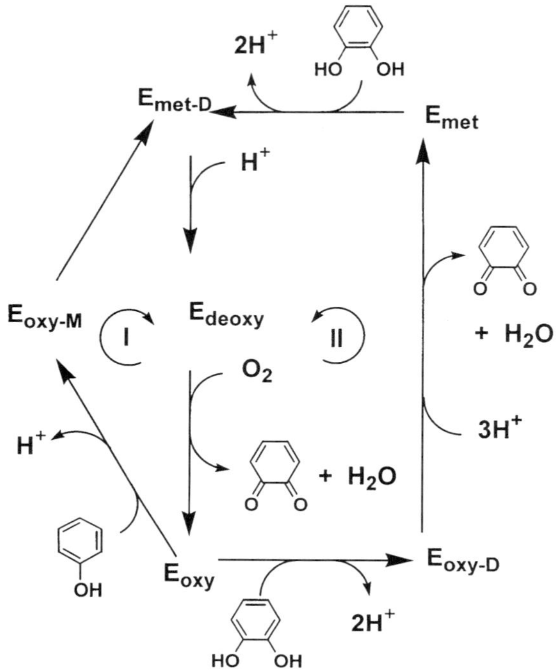

Figure 3 Catalytic cycle for the (I) hydroxylation of monophenols and (II) dehydrogenation of *o*-diphenols to *o*-quinones by tyrosinase. M = monophenol and D = diphenol bound forms. (Adapted from Ref. 42.)

to dark brown with no resulting precipitate [40,42]. However, it was discovered that chitin and chitosan could be used to coagulate and precipitate reaction products [39,43]. Chitin is an abundant and cheap polysaccharide available as shellfish waste, whereas chitosan is a product derived from chitin. In addition to its role as a coagulant, it has been suggested that the amino groups of chitosan may interact with the carbonyl groups of reaction products resulting in the formation of bonds that alter the characteristics and the solubility of chitosan [40]. It was also shown in previous studies that chitosan had the ability to protect peroxidase enzymes during the oxidation of phenols [15]. However, whereas it was confirmed that chitosan induced the precipitation of products and depressed color generation, it had no effect on the quantity of phenol that could be transformed by a limiting quantity of tyrosinase [42]. Therefore, it was concluded that, unlike the case of peroxidase, chitosan fails to protect tyrosinase from inactivation and does not enhance the extent of the tyrosinase-catalyzed oxidation of phenols.

Although chitosan was better at removing the colored reaction products than chitin, both adsorbed the products quickly. One advantage to using such an approach is the conversion of a waste product from the shellfish industry to a useful product. However, if chitosan is to be stripped of adsorbed material and reused, the cost of such an operation might be high in view of the strong adsorption of quinones to chitosan [40]. It was also reported that the use of chitosan in combination with tyrosinase accomplished very good detoxification of solutions of aqueous phenols [44]. Immobilized tyrosinase was also used along with chitosan to treat phenols and resulted in 100% phenol removal within 2 hr [39]. The immobilization of tyrosinase has the advantage of retaining enzymes in the reactor and protecting them from inactivation by reaction with quinones [39,45]. It should be noted, however, that despite the savings associated with using oxygen as an oxidant, the cost of tyrosinase is currently very high.

Laccase (EC 1.10.3.2) is produced by several fungi and seems capable of decreasing the toxicity of phenolic compounds through a polymerization process that is quite similar to that of peroxidases [46]. Moreover, because of its relative nonspecificity, laccase can induce the cross-coupling of pollutant phenols with naturally occurring phenols. In fact, laccase can oxidize phenolic compounds to their corresponding anionic free radicals, which are highly reactive [47]. In a study performed on laccase from the fungus *Rhizoctonia praticola*, the ability of the enzyme to detoxify some of the solutions of phenolic compounds was studied [46]. Detoxification of a particular phenol appeared to be dependent on the ability of the enzyme to transform the compound, as demonstrated by the disappearance of the parent phenol. However, the reaction products were not identified. It was concluded that the ability of laccase to detoxify a solution containing phenols appears to be a function of the particular compound being treated, the source of the enzyme and other environmental factors [46]. The use of laccase for the treatment of a "flushing liquor" from a coke plant (pH 8.6, 400 mg/L phenols) resulted in 94% conversion of phenols [38]. In the same study, laccase was used to treat a wastewater (pH 7.8, 100 mg/L phenols) from a plant that produces triarylphosphates (used as flame retardants). Purified and nonpurified enzyme resulted in 99% total phenol removal within 5 hr, with stirring under air. Ninety-eight percent removal of aniline (50 mg/L) that had been dissolved in the wastewater sample was also accomplished. Laccase of *Trametes versicolor* was also able to oxidize a variety of polycyclic aromatic hydrocarbons consisting of between 3 and 5 aromatic rings [48]. The same enzyme was implicated in the decolorization of synthetic dyes including anthraquinone, azo, and indigo [49]. The decomposition of such dyes is important, not only for the color that is imparted to receiving water bodies, but also because of their decomposition products, which can be carcinogenic [49].

Laccase has also been cited as a possible candidate for the treatment of bleaching plant effluent [50,51]. One study demonstrated that aeration in the presence of laccase led to a reduction of over 90% in the amount of phenolic compounds in debarking wastewater [52]. This included the removal of 75% to 99% of chlorophenols, depending on the group of chlorophenols analyzed. In a study on the removal of chlorophenols and chlorolignins from bleaching effluents by precipitation, it was reported that laccase could polymerize low-molecular-mass phenols, thereby facilitating their removal by reaction and precipitation with polyethyleneimine [50]. In another study, it was noted that intracellular enzymes (not identified) of the fungus *T. versicolor*, known at that time as *Coriolus versicolor*, had some effect on effluent decolorization [33]. Whereas enzymes used in their free form caused a slight decolorization, immobilized enzymes had a more pronounced result.

It has also been suggested that it is possible to enhance the natural process of the binding and incorporation of xenobiotics into humus by adding enzymes such as laccase to contaminated soils [47]. Laccases from *T. versicolor*, *R. praticola* and other fungi have been used to enhance the binding through oxidative coupling of various chlorinated phenols and aromatic amines with phenolic humic constituents. The advantages of such a process would be to immobilize and detoxify hazardous compounds. Binding of pollutants to humic material would decrease the amount of pollutant available to interact with the flora and fauna, reduce the toxicity of the pollutants through the coupling process, and prevent leaching of chemicals through the formation of insoluble precipitates. The xenobiotics seem to be stable once incorporated into the soil. They are released minimally and gradually, which should not constitute any health hazard, and the released compounds can be mineralized or bound again to humus by natural processes.

2. Pesticides

Pesticides, which include herbicides, insecticides, and fungicides, are widely used throughout the world today for crop protection. The potential adverse effects that the pesticide industry can have on the environment arise from the disposal of wastes formed during production and formulation of pesticides, detoxification of pesticide containers and spray tanks, and the pollution of surface and groundwater by pesticide runoff [19,53]. Common treatment methods include incineration, chemical methods, and landfilling. However, these systems have serious limitations including high cost, production of hazardous byproducts, disposal of chemical reagents, and the susceptibility of sensitive, biological treatment systems to sudden changes in contaminant concentrations [18].

It has been proposed that parathion hydrolase, also known as organophosphorous phosphotriesterase, be used for pesticide detoxification as an alternative to more common treatment methods. Parathion hydrolase is produced by a number of bacteria including *Pseudomonas* sp., *Flavobacterium* sp. and a recombinant *Streptomyces* [17,54]. It has been shown to hydrolyze some of the most widely used organophosphate pesticides such as methyl and ethyl parathion, diazinon, fensulfothion, dursban, and coumaphos [18,19]. It should be noted that organophosphate pesticides constitute the major proportion of agricultural pesticides used at present and are implicated in an estimated 800,000 pesticide poisoning cases every year [18]. Hydrolysis accomplishes the detoxification of the pesticide and makes the products amenable to biological treatment.

It has also been proposed that parathion hydrolase be used for the cleanup of small spills of organophosphate pestidice that occur during their handling, pouring and mixing [55]. The enzyme was isolated from an overproducing strain of bacterium, *Pseudomonas diminuta* and was immobilized on a polyurethane sponge. The sponge could be used to soak up a spill and then would be sealed in a plastic bag to carry out the hydrolysis reaction. On completion of the reaction, the sponge could be wrung into an appropriate container and reused or thrown away as a nonhazardous waste.

Several researchers have studied the hydrolysis of the insecticide coumaphos by parathion hydrolase. Coumaphos is used to kill a cattle disease-causing tick by dipping the cattle in large vats of the pesticide. When anaerobic dechlorination of coumaphos occurs (as is the case in the dip vats used), a product called potasan is formed. Because potasan is toxic to cattle, the contents of the vats used for cattle dipping must be renewed frequently, thereby generating a large volume of coumaphos-based waste [54]. The hydrolysis products of coumaphos and potasan are chlorferon and 4-methylumbelliferon, respectively [54]. These hydrolysis products are much more soluble than the initial substrates and can be treated using ultraviolet light in combination with ozonation. Parathion hydrolase has been shown to successfully hydrolyze coumaphos much more rapidly than chemical hydrolysis and it was further shown to selectively hydrolyze potasan while preserving coumaphos, when specific amounts of enzymes were involved [17]. The advantage of this latter process is the prolongation of the useful life of coumaphos and subsequent decrease in coumaphos waste generation. In laboratory experiments, free parathion hydrolase was stable at temperatures up to 45–50°C, and over the pH range 5.5 to 10.0. Organophosphate insecticides were also successfully hydrolyzed using parathion hydrolase immobilized on glass [19].

3. Cyanide Wastes

It is estimated that more than 3 million tons of cyanide are used annually throughout the world in different industrial processes including the production of chemical intermediates, synthetic fibers, rubber, and pharmaceuticals, as well as in ore leaching, coal processing, and metal plating [56]. In addition, many plants, microbes, and insects can release hydrogen cyanide from the enzymatic hydrolysis of certain compounds they produce [57]. Finally, effluents of food and feed production also contain substantial amounts of cyanide whose source is the cyanogenic glycosides of various crop materials such as apricots [56]. Because cyanide is a metabolic inhibitor and can be lethal to humans and other organisms, it is essential that it be removed from effluents before discharge [56, 57].

Cyanidase, a commercial enzyme preparation, can convert cyanide in industrial wastewaters to ammonia and formate in what appears to be a single-step reaction [56], as follows:

$$HCN + 2H_2O \rightarrow HCOO^- + NH_4^+ \quad (6)$$

Cyanidase is based on certain gram-negative bacterial isolates from the genus *Alcaligenes denitrificans* and it is prepared by proprietary methods. It is characterized by a high affinity and a high stability toward cyanide and is able to remove the latter down to very low levels, i.e., < 0.02 mg/L CN^- [56]. Cyanidase activity was neither affected by the common ions normally present in wastewaters (e.g., Fe^{2+}, Zn^{2+}, and Ni^{2+}), nor by organic substrates such as acetate, formamide, acetamide, and acetonitrile. In a study done on the detoxification by cyanidase of a cyanide-containing extract from debittering apricot seeds in the food industry, a diffusional-type flat membrane reactor (FMR) was developed, and its performance was noted to be superior relative to stirred tank reactor and fixed bed reactor configurations [57]. The advantages of an FMR include the protection of the enzyme from interfering particles and large molecules to avoid attrition and shear damage to the immobilization support. Cyanide diffuses through a semipermeable membrane to react with the entrapped enzyme behind the membrane and reaction products diffuse back across the membrane to the solution. It was noted that cyanidase was only capable of degrading free cyanide in apricot-seed extracts. However, cyanide in bound form was successfully degraded by cyanidase in the presence of β-glucosidase in the seed extract.

Cyanide hydratase, also known as formamide hydrolyase, has been reported to hydrolyze cyanide to formamide [57,58]. Cyanide hydratase is produced by a variety of fungi and is inducible upon preexposure of the fungi to low concentrations of cyanide. It was noted that cyanide hydratase was more stable when immobilized and that the enzyme from *Gloeocerco-*

spora sorghi was much more stable than that from *Stemphylium loti* [58]. Researchers concluded that immobilized fungal cyanide hydratase could be useful for the treatment of industrial effluents containing sodium cyanide or potassium cyanide.

4. Solid Wastes

Enzymes have also been used to enhance the dewatering of sludges that are produced as a by-product of wastewater treatment. Dewatering is essential to reduce the mass of water contained in sludges that are to be incinerated or discarded into landfills. In one study, an enzyme product containing carbohydrase, lipase, and protease activities was used to improve the dewaterability of sewage sludge and to increase the quantity of water released during pressing to achieve a reduction in sludge volume [59]. The effect of enzyme addition was to cause the cleavage of the water-binding macromolecules. However, whereas an enzyme dose between 2.5 and 5 mg/L gave good results, an increased dose led to the creation of more hydrophilic end groups that negatively affected the dewatering process. Attempts to dewater phosphate slimes using peroxidases have also been reported [60]. Phosphatic slimes contain considerable amounts of swelling claylike material and their fine particle size, along with their slow settling behavior, makes them particularly difficult to dewater. Peroxidase pretreatment of the slimes induced a higher mechanical binding among slime particles and significantly promoted the growth of algae and mold with the beneficial effect of enhanced aggregation of the particles and very high viscosity and gel strength. Such high viscosity and gel strength would help in postsedimentation handling. Cellulase and the bacterial enzyme lysozyme, or muramidase, were also used for sludge dewatering [61]. Whereas cellulase was used with penicillin and gave rather poor results, lyzozyme was allegedly able to alter floc matrices and to cause a dramatic increase in dewatering rates.

Cellulase has also been used for the treatment of low-level radioactive, mixed, and hazardous/chemical wastes. A system was developed that involved a sequence of technologies including air stripping and biodegradation of volatile organics, enzymatic digestion of cellulosics, and metal ion extraction. The system was effective in treating a radiologically contaminated heterogeneous paint-stripping waste. Cellulase digestion of bulk cellulose reduced the final solid waste volume by approximately 80%. During the enzymatic stage, metals and radioactive species were released into solution and the solid residues were of sufficient quality that they passed regulatory criteria for metals content. As a result of this treatment system, the metals and radioactivity were concentrated into less than 5% of the original waste volume.

The effect of enzyme addition on the strength and stability of pure clayey soil and soil–fly ash mixtures has also been investigated [62]. The results obtained were encouraging. The strength of the fly ash–kaolinite soil mixture (5% fly ash) increased with enzyme concentration. At 10% enzyme concentration, the highest level investigated, the unconfined compressive strength of the mixture had increased by about 53%. The advantages of using enzymes in soil–fly ash stabilization is to help avoid the necessity for the disposal of fly ashes while at the same time increasing the strength of the soil. Fly ash constitutes around 90% of coal-combustion residues and it is estimated that around 65 million tons are produced yearly in the United States alone. In this context, enzymatic stabilization of fly ash compares well to other stabilization processes and may actually be cheaper [62]. The enzyme that was used was a commercial enzyme product known as Perma-Zyme. The authors mentioned that it was a product of fermentation and that it had no corrosive or toxic effects but they did not identify the specific enzyme activities it contained.

B. Enzymatic Treatment for the Production of Value-Added Materials

Waste-treatment processes can be a significant financial burden to those companies or communities who operate them. Therefore, growing emphasis is being placed on the development of waste-valorization strategies based on the recognition that waste products can be used as a source of raw materials for the production of goods. Such a strategy has some very important advantages. First, by replacing conventional sources of raw materials with a waste source, resources are conserved. Secondly, the conversion of such wastes results in a decrease in the quantity of materials that must be discarded into the environment. Thirdly, this strategy can be used to produce new materials of significant monetary value that can be a source of profit or can be used to offset waste-treatment costs. Recent research has shown how enzymes can fit into a waste-valorization strategy by chemically converting target components of various waste matrices into food products, sugars, alcohols, and commodity chemicals. Examples of these types of applications are summarized in Table 2 and are discussed below.

1. Food Processing Wastes

The food processing industry produces extremely large quantities of wastes. It has been suggested that this industry could benefit from enzymatic processes for (1) more efficient utilization of raw materials; (2) processing waste streams to produce marketable by-products; (3) processing of raw

Table 2 Cell-Free Enzymes Used to Convert Wastes to Value-Added Products

Enzymes[a]	Proposed applications
Amylases: α-amylase (EC 3.2.1.1) glucoamylase (EC 3.2.1.3)	Hydrolysis of starches found in rice and grain wash waters to produce glucose [64]
Cellulolytic enzymes: cellulase (EC 3.2.1.4) cellobiohydrolase (EC 3.2.1.91) cellobiase (EC 3.2.1.21) exo-1,4-β-D-glucosidase (EC 3.2.1.74)	Hydrolysis of cellulosic materials from pulp and paper and municipal solid wastes to produce alcohol; degradation of grain, fruit and vegetable wastes to produce glucose syrups [64]
Chitinase (EC 3.2.1.14)	Bioconversion of shellfish waste to *N*-acetyl glucosamine for yeast production [69]
L-Galactono-lactone oxidase (EC 1.1.3.24)	Conversion of galactose from whey hydrolysis to L-ascorbic acid [64]
Pectinases	Improved juice yields and dissolved solids resulting in less waste [115]
Pectin lyase (EC 4.2.2.10)	Degradation of apple pomace to produce butanol and animal feed [64]
Lactase (EC 3.2.1.108)	Whey hydrolysis and dairy waste processing for production of food ingredients [116]; production of glucose, galactose, substrates for alcohol and ascorbic acid production [64]
Proteases	Solubilization of fish, meat, and leather remains, production of protein hydrolysates [64,65]

[a] EC classifications are provided wherever they were specified in literature sources.

materials; and (4) decontamination of food waste by-product streams [63,64]. Notably, whereas the wastes generated by other industries are generally deleterious and must be rendered innocuous by appropriate treatment schemes, food wastes have the advantage of being amenable to conversion into food, feed, or nonfood products with added value.

Proteases are a group of hydrolases that are widely used in the food industry in processing fish and meat waste [65]. There is a need for recovering proteins and for improving conservation and reuse of water to minimize waste volumes and conserve energy [63]. Proteases can solubilize proteins in waste streams, resulting in recoverable liquid concentrates or dry solids of nutritional value for fish or livestock [63]. For example, alkaline protease from *Bacillus subtilis* was used in the processing of waste feathers from poultry slaughterhouses [66]. Feathers constitute 5% of the body weight of poultry and can be considered as a high-protein source for food and feed provided their rigid keratin structure is completely destroyed. Total solubilization of the feathers was achieved after pretreatment with sodium hydroxide, mechanical disintegration, and enzyme hydrolysis. The end product was a heavy, grayish powder with a very high protein content that could be used mainly as feed constituent. In another application, cells from *Bacillus megaterium*, *Pseudomonas marinoglutinosa* and *Acromonas hydrophila* were immobilized in calcium alginate [67]. The extracellular proteases secreted by these cells were used to solubilize fish meat.

Amylases are polysaccharide hydrolases that have been used in the simultaneous saccharification and fermentation of starch [64] and the treatment of starch-containing food wastewaters [63]. It was reported that amylases could be used to produce alcohol from rice-processing wastewater [63]. It was also reported that amylases seemed to enhance activated sludge wastewater treatment by reducing the treatment time. In another interesting application, α-amylase and glucoamylase enzymes were used to support the production of photodegradable and biodegradable plastics [68]. The process involved the conversion of the starchy material contained in cheese whey or potato waste from commercial food processing to truly biodegradable plastics. α-Amylase was first used to break down the long starch molecule into smaller fragments. The latter were then attacked by glucoamylase to produce glucose through saccharification. During this process more than 90% of the starch was converted to glucose. The glucose was then fermented to lactic acid, which was recovered, purified, and used in the production of photo- and biodegradable plastics whose rate of decay can be controlled by judicious mixing of various lactic acid isomers and other compounds. The end product was 95% lactic acid and 5% of an environmentally safe product. Besides reducing starchy food wastes and producing a valuable product, photo-and biodegradable plastics can be used in the preparation of mulch films and compost bags and can also be used as programmable fertilizer and pesticide delivery systems, thus offering a possible solution to pesticide runoff that result from the excessive quantities applied [68].

Pectinesterase from *Clostridium thermosulfurogenes* and pectin lyase from *Clostridium beijerinckii* have been reported to degrade pectin, which is a water-soluble substance that binds adjacent cell walls in plants. It has been shown that apple pomace (a food processing waste) can be degraded using pectinesterase to yield butanol [64]. The enzyme L-galactonolactone oxidase from the yeast *Candida norvegensis* can be used to bioconvert galactose resulting from the hydrolysis of the lactose contained in whey to L-ascorbic acid, which is a valuable commodity chemical [63]. Lactases have also been used in dairy waste processing, mainly to produce value-added products [63,64]. Whey protein concentrate can be separated from permeate solids that contain high levels of lactose whose degradation and transformation is sought through enzymatic action [64]. Billions of kilograms of whey are produced yearly, and therefore any beneficial treatment of whey could have a significant positive impact on the environment [64]. Chitinase from *Serratia marcescens* QMB1466 has also been cited for its ability to induce chitin degradation. In fact, the bioconversion of chitin to yeast single-cell protein has been proposed as an alternative to the disposal of shellfish waste, which has a high chitin content [69]. The process consists of first pretreating shrimp waste by size reduction, deproteination, and demineralization to produce a chitin material that can be easily bioconverted by chitinase to the monomer *N*-acetyl glucosamine. The latter serves as a substrate for the single-cell protein production.

An unspecified enzyme was also used to treat leather wastes [70]. The degradation of leather wastes yielded a water-soluble hydrolysate that could be concentrated and dried to produce fine flour, which was claimed to be used for a variety of commercial purposes.

2. Solid Wastes

For the past decade, there has been an increasing interest in the enzymatic hydrolysis of cellulose, a major component of paper [71]. This interest stems from the advantages that such a process would offer; namely, the conversion of lignocellulosic and cellulosic wastes to a useful energy source through the production of sugars, ethanol, biogas, or other energetic end products [72,73]. However, the industrial production of sugars from cellulose has been plagued with several constraints pertaining to enzyme production and cost and substrate preparation, among others. Still, there have been numerous reports on the possible ways of improving the enzymatic hydrolysis of the cellulose contained in the organic fraction of municipal solid wastes (MSW) to produce fermentable sugars and eventually ethanol or butanol. For example, cellulases from *Trichoderma reesei* CL847, *Trichoderma penicillium* strain CLD20 and the thermophilic fungus CL240 have been used to obtain hydrolysates from the organic fraction of MSW that would be directly used as base for fermentation media [71]. Sugar concentrations as high as 45 g/L were produced in a packed-column reactor and the hydrolysates obtained could reportedly be used in anaerobic fermentations such as acetone–butanol or organic acid production.

A study was conducted to examine the effects of several parameters on the yields of enzymatic hydrolysis for two major MSW components; namely, newspaper and corrugated cardboard [74]. The primary aim was to try to determine the modifications of MSW that would lead to increased conversions to ethanol. A full complement cellulase system from *T. reesei* QM 9414G was used. The enzyme system consisted of endoglucanase, cellobiohydrolase, and cellobiase. Another study involving the fungal enzyme preparation Econase was conducted to investigate the effect of cellulolytic enzymes on MSW degradation [73]. Econase consists mainly of endo-1,4-β-D-glucanase, cellobiohydrolase, and exo-1,4-β-D-glucosidase in addition to a number of other enzymes. The use of Econase seemed to enhance the degradation of MSW as well as cellulose degradation.

There have also been some reports on the use of cellulolytic enzymes in the treatment of sludges from pulping and deinking operations. For example, studies were performed to investigate the possibility of hydrolyzing the highly cellulosic sludges that result from pulp and paper operations to produce an energy source such as ethanol [75]. Because an average of 60 kg of primary sludge are generated per tonne of pulp produced, ethanol production is evidently an attractive way of decreasing the amount of sludges that have to be handled and disposed while producing a marketable product. The enzymes used included a blend of cellobiohydrolase, cellulase, and β-glucosidase. Another study focused on the conversion of low-value cellulosic substrates from fiber recycling and deinking operations to fermentable sugars [76]. The enzymes used were not inhibited by high ink content and it was observed that the presence of surfactants enhanced the rate of enzymatic hydrolysis especially in the early stages of the reaction. In a related application, cellulolytic enzymes were used in combination with surfactants to accomplish the deinking of complex recovered paper mixtures [77]. The deinked mixture could then be used to produce products such as fine printing paper, writing-grade paper, and tissue.

IV. BARRIERS TO FULL-SCALE APPLICATION

Enzymes have undoubtedly stimulated the research community's interest and have been investigated for their potential use in many types of applications. Applications have been demonstrated that aid in the efficient use of raw materials, improve the quality and reduce the quantities of wastes for disposal, and catalyze the transformation of waste compounds into value-added products. Other applications, not discussed in detail here, include the use of enzymes to replace processes that have significant environmental impacts. For example, paper manufacturers have investigated the use of enzymes as a means of reducing energy consumption during deinking operations [78]. Alternatively, in an approach that has already been adopted by industry, enzymes such as xylanase have been used to reduce or replace the use of chlorine in bleaching systems [78]. In a similar manner, enzymes are now used in many household detergents. In addition, the use of enzymatic cleaning solutions has been reported for the cleaning of milk heaters in Germany [79].

It was estimated that this could eliminate the need for the disposal of approximately 1000 tons of nonbiodegradable cleaning chemicals per year. These approaches demonstrate the application of enzymes as a means of pollution prevention, not just pollutant transformation.

However, despite the number of proposed applications of enzymes, few of them have been developed sufficiently for full-scale application. This delay is due to some very important issues that remain to be resolved. Some of the most significant hurdles that must be addressed before the full potential of enzymes may be realized are discussed below.

A. Enzyme Cost and Availability

Whereas many authors have recognized the potential for enzymatic treatment systems, the development of these processes from an engineering perspective is conspicuously lacking. The primary reason for this appears to be the cost of enzymes that have traditionally been very expensive to produce in the quantities that are required at an industrial scale. Enzymes are expensive because of the cost of their isolation, purification, and production.

Very recently, the development of molecular farming techniques has dramatically improved the potential for application of enzymes at the industrial scale. For example, new techniques involving transgenic manipulation of plants have created the potential for using plants as factories for large-scale production of enzymes. Transgenic techniques involve the genetic engineering of plant material aimed at improving the productivity of the crop and increasing the levels of natural or foreign enzymes in the plant cells. In comparison to typical fermentation techniques, it has been claimed that plants offer lower unit production costs, a better safety profile, shorter development time, simpler commercial scale-up, relative genetic stability, convenient storage, and potentially easier purification [80]. For products where large volumes are wanted but a high degree of purification is not required (such as in waste-treatment operations), plants appear to offer a clear advantage. It is expected that the use of transgenic plants could reduce the price of bulk enzyme production by as much as 2 to 4 orders of magnitude below that of current techniques based on animal cell culture and microbial fermentation technologies [80]. Tobacco has recently been identified as a very suitable plant that can be manipulated to produce rich sources of industrial enzymes [81]. Several different enzymes have already been produced in the tobacco and, as such, researchers are launching into an extensive program to develop this alternate use of tobacco. This could have profound implications for the tobacco industry and farmers who have been experiencing a decline in tobacco income. Other plants can also be manipulated to produce enzymes.

It should be recognized that public pressure arising from uncertainties associated with the cultivation of genetically modified plants might forestall the development of technologies that rely on transgenic plants for the production of enzymes. Such is also the case when considering the use of genetically modified organisms (GMO) in biological treatment systems. For example, whereas the use of GMOs has great potential for achieving the transformation of important contaminants, applications of recombinant organisms have been limited due to restrictions on the release of such organisms into the environment [17]. However, organisms can instead be genetically engineered to express enzymes that can be used for the transformation of pollutants. In this case, the enzyme could be first isolated from the organism and then applied directly to the waste material. This may alleviate many of the concerns associated with the release of GMOs into the environment. Thus, enzyme-based treatment strategies may be considered as more environmentally acceptable and may represent the first feasible applications of genetic engineering to solve waste problems.

Although these developments foreshadow the production of plentiful and inexpensive sources of enzymes, the current high cost of enzymes should by no means hamper the efforts to carry out more extensive research to identify the most promising enzymes and determine the optimal conditions for their use. In fact, the results of such research would provide the incentive for commercial development to eventually achieve large-scale production of the enzymes at a much lower cost.

B. Enzyme Efficiency Under Waste-Treatment Conditions

Numerous studies have shown that the enzymatic process can be used for the treatment of synthetic wastewaters consisting of selected contaminants dissolved in aqueous, and often buffered, solutions. However, actual wastewaters vary a great deal in terms of aromatic pollutant concentrations, suspended solids, oil and grease, pH, bulk organic concentration (BOD, COD), and concentrations of specific chemical species that may interfere in the treatment process. Some of these contaminants may inhibit or destroy the catalytic activity of the enzyme. Alternatively, the enzyme might be destroyed by microbes contained in the waste or might become adsorbed to solid particles. In addition, even though there are many advantages to using immobilized enzymes, the surfaces upon which enzymes are immobilized are likely to become fouled with solid materials or biomass over time. This could result in inactivation of the enzyme or may reduce mass transfer rates to unacceptably low levels.

Despite these concerns, very little information has been collected with respect to the performance of enzymes under actual waste treatment conditions. For the most part, proposed applications were studied using synthetic wastes of controlled composition. However, there have been some demonstrations of successful applications of enzymes to treat pollutants in real wastewater matrices. For example, a variety of industrial wastewaters containing phenols and aromatic amines have been successfully treated at the laboratory scale using peroxidases and polyphenol oxidases [20–24,38]. A pesticide waste obtained from cattle dip was successfully treated with parathion hydrolase [17]. In addition, feather wastes from poultry slaughterhouses were converted into a protein concentrate using alkaline proteinase [66]. Cyanidase has been used to treat the waste resulting from apricot-seed extracts [57]. Interestingly, problematic interferences of wastewater components with cyanidase were resolved by separating the enzyme and waste by a membrane that was permeable to the target contaminant. This is very likely to become a common approach for reducing the susceptibility of enzymes to components of the waste matrix.

These studies do indicate that some enzymes are capable of exerting their catalytic action in waste matrices. However, it is very likely that each specific application of enzymes will require tailored studies to determine if enzymatic treatment is a feasible option for a given waste.

C. Fate and Disposal of Reaction Products

The main objective of using enzymes in waste treatment is to act on specific pollutants to transform them into innocuous products. Complete mineralization of contaminants is usually preferred to other transformations but, for example, the production of ammonia and formate from cyanide wastes or the conversion of organophosphate pesticides to easily removable products is also acceptable. However, it must be recognized that in certain cases the products of enzymatic action might be more toxic than the parent compounds, thereby defeating the purpose of enzyme use. Ultimately, it is desirable that the reaction products are less toxic or more biodegradable than the original pollutants, or otherwise be more amenable to subsequent treatment. However, no single enzyme can be expected to catalyze very extensive transformations of pollutants and, thus, reaction products are likely to retain some of the characteristics of the parent compound [1]. Therefore, it is necessary to characterize reaction products to assess their impact on downstream processes or on the environment into which they are released.

Unfortunately, only a few studies have been done to investigate this matter mainly because of the difficulty of identifying the products of enzymatic reactions and the difficult task of assessing their toxicity. For example, it was demonstrated that solutions of phenols and chlorophenols that were treated with tyrosinase followed by the addition of chitosan had substantially lower toxicities [44]. In addition, a 60- to 200-fold reduction was reported in the toxicities of aqueous solutions of pesticides including parathion and paraoxon following their hydrolysis by enzymes [53]. These results for the detoxification of solutions of phenols and pesticides are encouraging but similar studies should be performed for other enzymes that are being considered for use in waste treatment.

Waste-treatment processes commonly result in the production of solid wastes that must be disposed of safely. Enzymatic treatment is no exception. For example, although enzymatic treatment may not produce as large a quantity of solid products as does biological treatment, some solid residues may be formed, e.g., the polymer precipitates formed during the treatment of phenols with peroxidases, spent adsorbents such as talc, chitin, or activated carbon that are used to eliminate the soluble products of enzymatic reactions, or residues of plant materials such as raw soybean hulls when they are used in place of purified enzymes during treatment. Perhaps, the polymers and adsorbents could be incinerated to recover some energy if the emission of dangerous combustion by-products can be controlled or prevented. The residues of plant materials could potentially be composted and used as soil conditioners, provided that pollutants do not leach from them at substantial rates. To date, none of these disposal problems have been addressed adequately.

V. CASE STUDY: TREATMENT OF AROMATIC POLLUTANTS USING PEROXIDASE ENZYMES

Several important criteria should be met before an enzyme can be judged feasible for any particular waste-treatment application [1]. These criteria include the following: (1) the ability of the enzyme to selectively act upon the target pollutant must be confirmed; (2) the enzyme must exhibit reasonable catalytic activity under typical treatment conditions; (3) the enzyme must be relatively stable under the required reaction conditions; (4) reactor systems that are used to implement enzymatic reactions must be relatively simple in order to be accepted by potential users of the technology; (5) reaction products must be less toxic, more biodegradable or more amenable to subsequent treatment than the targeted pollutant; and (6) enzymes must be available commercially. Satisfying these criteria can represent significant barriers to the development of enzymatic systems. Unfortunately, there are no examples of enzymatic treatment systems that have been studied sufficiently to confirm that all of these criteria have been met. However, many of these issues have been addressed during the development of technologies for the treatment of aqueous phenols using peroxidase enzymes, particularly those isolated from horseradishes and soybeans. Therefore, such enzymes will be used in the discussion below as an example of fundamental and ongoing research that is required when developing enzymatic processes for eventual full-scale application.

A. Candidate Wastes and Enzymes

Aromatic compounds, including phenols and aromatic amines, are present in the wastewaters of industries that produce petroleum products, petrochemicals, plastics, textiles, iron and steel, pulp and paper, among others. They are also present in many contaminated soils that resulted from spills, leaks, and industrial activities. Many of these compounds are toxic and some are known or suspected carcinogens. In fact, the United States Environmental Protection Agency has classified the majority of phenols as toxic priority pollutants. The concentrations of aromatic compounds in waste streams vary from dilute (< 1 mg/L) to concentrated ($> 10{,}000$ mg/L). Effective treatment technologies for many of these compounds currently exist. However, industries often encounter difficulties in employing traditional treatment technologies for particular wastes. For instance, synthetic dyes are often highly resistant to biological degradation. This is a major source of difficulty for the textile industry that uses over 700,000 tons/yr of dyes, of which 10% to 15% are lost in the effluent [82]. In addition, excess reagents are often required when using chemical oxidants such as ozone or hydrogen peroxide to oxidize aromatic compounds in wastewaters with a high organic content. In addition, high fuel costs are incurred during the incineration of highly contaminated waters that arise from the production of phenolic resins. In addition, petroleum and petrochemical industries frequently encounter problems with biological treatment systems resulting from fluctuations in contaminant concentrations. Therefore, research has been conducted in recent years to overcome some of these problems through the development of enzymatic treatment systems that specifically target the transformation of aromatic contaminants.

Oxidoreductases comprise a large class of enzymes that catalyze biological oxidation/reduction reactions. Because so many chemical transformation processes involve oxidation/reduction processes, the idea of developing practical applications of oxidoreductase enzymes has been a very attractive, but quite elusive, goal for many years [83]. Applications have been sought for the production of pharmaceuticals, synthesis and modification of polymers, and the development of biosensors for a variety of clinical and analytical applications [83]. In recent years, the use of oxidoreductive enzymes to catalyze the removal of aromatic compounds from wastewaters has been investigated. In particular, peroxidase enzymes have been the subjects of many studies.

Peroxidase, also known as hydrogen peroxide oxidoreductase (EC 1.11.1.7), was the first enzyme that was investigated for its ability to accomplish the treatment of aqueous aromatic compounds [7]. The system of reactions involved in the catalytic cycle of several peroxidases has been studied extensively and is summarized in Fig. 2. The catalytic cycle involves a progression of the enzyme through three oxidation states, i.e., native state (HRP), Compound I (HRP_i) and Compound II (HRP_{ii}), following its reaction with hydrogen peroxide and its subsequent reactions with an aromatic substrate (AH_2). The products of this cycle are water and highly reactive free radicals (AH•). It is these free radicals that spontaneously combine to produce insoluble precipitates.

Although several peroxidase enzymes obtained from plant, animal, and microbial sources have been investigated for their ability to catalyze the removal of aromatic compounds from wastewaters, the majority of studies have focused on using HRP. In particular, it has been shown HRP can transform phenol, chlorophenols, methoxyphenols, methylphenols, aminophenols, resorcinols, and various binuclear phenols [7]. HRP was also used for the treatment of contaminants including anilines, hydroxyquinoline, and arylamine carcinogens such as benzidines and naphthylamines [7,8]. In addition, it has been shown that HRP has the ability to induce the formation of mixed polymers resulting in the removal of some compounds that are either poorly acted upon or not directly acted upon by peroxidase [7]. This phenomenon, termed coprecipitation or copolymerization, has important practical implications for wastewaters that usually contain many different pollutants. This principle was demonstrated when it was observed that polychlorinated biphenyls (PCBs) could be removed from solution through coprecipitation with phenols [20]. However, this particular application of HRP does not appear to have been pursued in any subsequent research.

Although HRP is commercially available, it is not currently available in bulk quantities or at a cost that that is sufficiently low to make full-scale applications of enzymatic treatment feasible. It has been suggested that the use of a less expensive and more ubiquitous source of peroxidase could circumvent the problem of enzyme cost. For example, in 1991, the seed coat of the soybean was identified as a rich source of a single peroxidase isoenzyme [84]. Because the seed coat of the soybean is a waste product of the soybean industry, soybean shells could provide an inexpensive and abundant source of peroxidase. Thus, soybean peroxidase (SBP) has the potential of being a cost-effective alternative to HRP for wastewater treatment.

Several fungal peroxidases have also been investigated. A fungal peroxidase from *Coprinus macrorhizus* was used to treat aqueous phenols in batch reactors. Chloroperoxidase from the fungus *Caldariomyces fumago* has been reported to oxidize several phenolic compounds [85]. In addition, it has been shown to catalyze certain oxygen-transfer reactions such as the oxidation of ethanol to acetaldehyde or the oxidation of chloride ions [29]. This latter reaction might lead to the formation of a different range of products (which may be more toxic) when chloride ions are added to reaction mixtures containing chloroperoxidase [29]. The wood-rotting fungus *P. chrysosporium* possesses some unique and remarkable biodegradative capabilities. Research has shown it can degrade a wide variety of recalcitrant and environmentally persistent pollutants found at many hazardous waste sites. The extraordinary biodegradative abilities of this organism are due in part to its lignin degrading enzyme system. Manganese peroxidase, produced by this fungus, has been observed to catalyze the oxidation of several phenols and aromatic dyes but these reactions depend on the presence of both manganese and certain types of buffers [2]. Unfortunately, the enzyme's requirement for high concentrations of manganese makes its feasibility for wastewater treatment applications doubtful [29]. LiP, another enzyme from the same fungus, has been studied to evaluate its potential application in industrial waste treatment. The enzyme was observed to catalyze the oxidation of a variety of phenolic pollutants, including phenol, *o*-cresol, 2-nitrophenol, 2-chlorophenol, and pentachlorophenol using hydrogen peroxide as a cosubstrate. It has also been show to catalyze the transformation of dioxins, pyrene, and pentachlorophenol [86], the dechlorination of polychlorinated phenols [87], the demethylation of crystal violet dye [88], and the oxidation of thianthrene [89]. Because of its ability to degrade many hazardous organic compounds, *P. chrysosporium* has been identified as a promising candidate for applications in hazardous waste treatment. However, its commercial application appears to have been limited by the lack of suitable bioreactor systems that yield consistent production of high levels of the enzymes responsible for catalyzing the degradation of hazardous pollutants [27].

B. Enzyme Activity and Stability

In order to maximize treatment efficiency, it is necessary to select an enzyme that can be used to treat a variety of contaminants, can maintain its catalytic activity over extended periods of time, and is active under conditions that can be found in many industrial wastewaters. Such wastewaters are highly variable in terms of their composition, pH, and temperature. Therefore, studies must be conducted to compare the effectiveness of various enzymes.

For example, in experiments involving the treatment of aqueous phenols, it has been shown that SBP and HRP can be used to treat a variety of methyl- and chlorophenols over a range of pHs [9,90]. It can be seen in Fig. 4 that the optimal activities of HRP and SBP are near neutral pH conditions and they exhibit excellent activity (>90% of optimal activity) under slightly acidic and slightly alkaline conditions. However, the range of pH under which significant activity (>10% of optimal activity) is demonstrated is also very wide. These studies demonstrate that HRP is slightly less susceptible to pH changes than SBP and it is probably more suitable for the treatment of alkaline wastes.

HRP and SBP are quite stable and active at or near room temperature. However, they can be susceptible to inactivation at elevated temperatures, which may be important in certain waste-treatment applications. The stability of enzymes is often first order with time and can be expressed as follows:

$$A(t) = A_o 10^{-t/D} \tag{7}$$

where $A(t)$ is the activity at time t, A_o is the initial activity, and D is the inactivation decimal reduction value. The decimal reduction value is essentially a measure of the time it takes for enzyme activity to drop to one-tenth of its initial value. Studies performed with HRP and SBP have

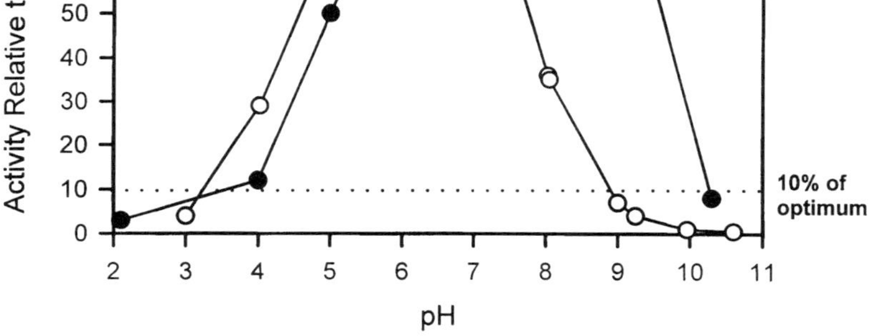

Figure 4 pH dependence of the catalytic activities of soybean peroxidase (SBP) and HRP. Activities were measured using an assay consisting of 2.0 mM phenol, 2.4 mM 4-aminoantipyrine, and 0.2 mM hydrogen peroxide buffered to the desired pH. (Adapted from Refs. 9 and 90.)

shown that the D value of SBP was several orders of magnitude larger than that of HRP. In addition, it has been shown that SBP can maintain its activity over reasonably long periods at temperatures above 70°C [90]. Therefore, the extreme stability of SBP under elevated temperature conditions points to its inherent advantage for the treatment of high-temperature wastes.

These studies have demonstrated that enzymes such as HRP and SBP can be used to transform certain contaminants of interest. They have also demonstrated restrictions on their application to certain types of substrates and conditions of pH and temperature. However, although enzymes are capable of carrying out the desired transformations, other criteria play an important role in judging the suitability of particular enzymes, as will be seen below.

C. Inactivation Mechanisms

One of the major challenges associated with enzyme-catalyzed phenol removal is the prohibitive cost of the enzyme resulting from inactivation processes that occur during treatment [24]. Enzymes are not only susceptible to inactivation by contaminants that are not directly involved in the catalytic reaction, but rather, they can also be inactivated, or at least inhibited, by some substrates and products of the catalytic reaction. Because such mechanisms can strongly influence the economics of enzymatic processes, they must be studied in reasonable detail. From such knowledge, engineering approaches can be found to minimize the occurrence of inactivation mechanisms.

Fig. 2 shows that peroxidase enzymes are subject to inactivation in the presence of excess hydrogen peroxide. That is, Compound II can be oxidized by hydrogen peroxide to form Compound III (HRP_{iii}), a catalytically inactive form of the enzyme, which only slowly reverts to the native state and reenters the catalytic cycle. For example, Fig. 5 shows the dependence of the activity of SBP and HRP on hydrogen peroxide concentration during the oxidation of phenol in the presence of 4-aminoantipyrine. The latter reagent reacts with the products of enzyme catalysis to produce a colored product that can be measured spectrophotometrically. Based on an understanding of the mechanisms and kinetics of peroxidase interactions with hydrogen peroxide and organic substrates, a model was derived to describe the relative dependence of activity on peroxide concentration. This model is as follows:

$$R = \left(\alpha + \frac{\beta}{[H_2O_2]} + \gamma[H_2O_2]\right)^{-1} \tag{8}$$

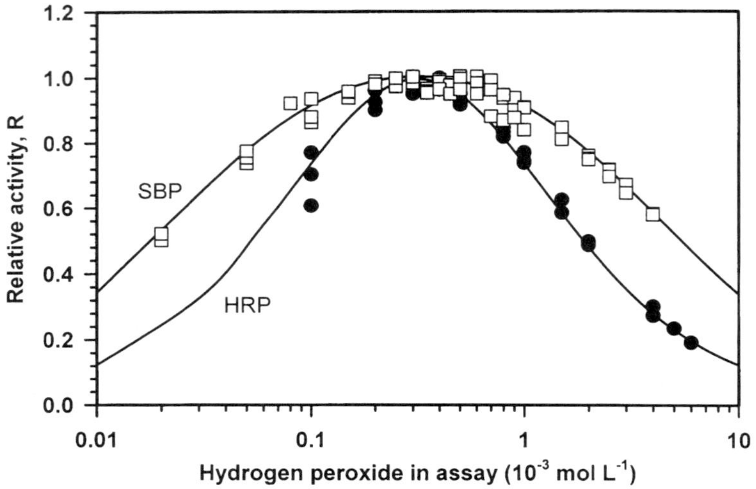

Figure 5 SBP and HRP catalytic activities as a function of peroxide concentration. The activity assay was buffered to pH 7.4 and consisted of 2.0 mM phenol, 2.4 mM 4-aminoantipyrine, and the appropriate quantity of hydrogen peroxide. (Adapted from Ref. 91.)

where R is the activity relative to the optimum ($0.0 < R < 1.0$), $[H_2O_2]$ is the peroxide concentration, and α, β, and γ are calibration constants. The fit of this model to the data of Fig. 5 demonstrates that the rate of reaction (represented by activity) is decreased in the presence of high concentrations of hydrogen peroxide (i.e., >0.3 mM) for both SBP and HRP. Thus, in cases where a high concentration of aromatic contaminants is being treated, it will be necessary to add the peroxide gradually to the treatment system to avoid inefficiencies associated with the production of Compound III. However, it should also be noted that if the peroxide concentration were lower than 0.3 mM, then the rate of transformation would be limited by the availability of peroxide. Thus, the reactor system must be optimized to maximize reaction rate, thereby minimizing capital costs, while also minimizing enzyme losses and reagent costs through inhibition by hydrogen peroxide. It should be noted that the optimum peroxide concentration is likely to be a function of the substrate treated and the reaction conditions used. Therefore, the specific value of 0.3 mM may or may not be germane to the turnover of phenol or other phenolic compounds, in the absence of 4-aminoantipyrine. In addition to reaction optimization, the model also serves a secondary purpose. A comparison of the constants (α, β, and γ) corresponding to various peroxidases allows these enzymes to be compared in terms of their susceptibility to inactivation by hydrogen peroxide [91]. This allows enzymes to be selected based on the conditions that are required during treatment. Such knowledge can only be gained based on an understanding of the mechanism and kinetics of enzyme interactions.

Permanent inactivation of peroxidases results from interactions between the enzyme and the free radical and polymer reaction products [92]. It has been postulated that the free radicals can form covalent bonds at or near the active site of the enzyme, resulting in interruption of its catalytic activity [20,93]. In addition, it has also been observed that the enzyme can become associated with polymeric products as they precipitate from solution [13]. To overcome these problems, which result in excessive losses of catalytic activity, researchers have focussed on the use of protective additives that interact with reaction products before they can inactivate the enzyme. For example, it has been shown that additives such as polyethylene glycol (PEG), gelatin, sodium borate, chitosan, and talc can significantly minimize enzyme inactivation through interaction with reaction products [13,15,94]. Sodium borate is not considered as a practical alternative due to its insecticidal properties. In addition, whereas talc does protect the enzyme, it may result in the production of significant quantities of solid residues following treatment. In contrast, PEG exerts a strong protective effect and is a nontoxic compound that has been declared fit for human consumption [95].

Of all additives, PEG has been studied the most due to the promise it has shown as a protective additive. For example, Fig. 6 shows differences in HRP concentrations that are required to treat a synthetic waste consisting of pure aqueous phenol and a foundry waste, both with and without PEG. In this case, sufficient PEG has been provided to provide maximal protection of the enzyme. That is, the addition of more PEG will not improve the quantity of phenols removed by a given dose of HRP. In this particular case, it is observed that the quantity of enzyme required to achieve over 98% conversion of total phenols has been reduced by approximately 22-fold. Experiments have demonstrated that the protective effect is best for high molecular weight PEGs [96] and that the quantity of PEG that is required for maximal protection of the enzyme is a function of the quantity of phenol being treated [14,97]. PEG is removed from solution together with the polymer products [96,98]. It has been shown that PEG is removed in the precipitate in fixed proportion to the phenol and, further, that if PEG is added in that proportion, there is essentially no residue of it once the phenol is removed [98]. However, any excess PEG that has been provided to the reaction mixture remains in solution and must be treated in a downstream process. Because of the low cost of PEG compared to HRP, the tremendous reduction in enzyme requirements will translate directly into major savings

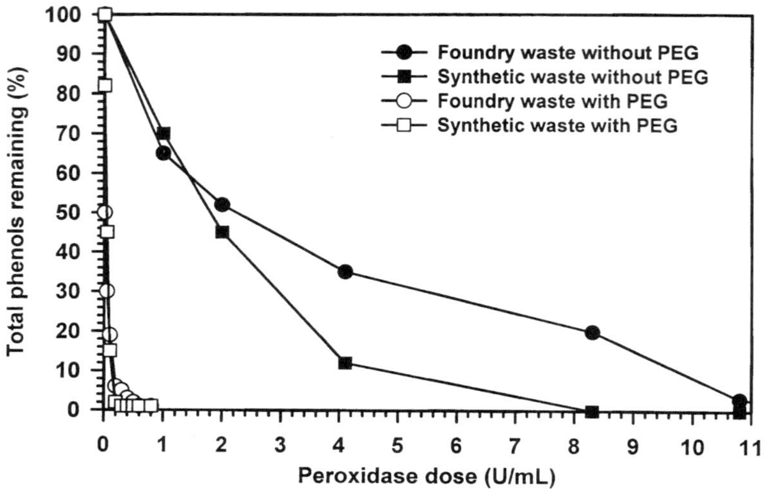

Figure 6 Phenol removal from a synthetic waste and a foundry waste as a function of HRP dose with and without the presence of 100 mg/L PEG. [Initial phenol] = 3.5 mM (330 mg/L as phenol), [initial peroxide] = 4.9 mM. (Adapted from Ref. 24.)

in reagent costs. Similar protective effects have been noted when using chitosan as an additive during the treatment of aqueous phenols with peroxidase [15].

D. Kinetics and Reactor Design

Economic considerations during the design of a treatment system include capital and operating costs. Operating costs may account for a large portion of the overall treatment costs and are therefore a major consideration when choosing among alternative designs. Capital costs often are closely related to hydraulic retention time (HRT), which directly influences reactor volume. Therefore, the system that provides the optimal combination of economical reactor size and efficient enzyme use will be preferred.

Considerable efforts have aimed at optimizing the reactor configuration in which the peroxidase-catalyzed removal of phenols can be carried out. Much of this work has focused on maximizing the efficiency of enzyme use since the majority of the operating cost associated with peroxidase treatment is likely to be that of the enzyme itself [24]. Improvements in the useful life of the enzyme, with a corresponding reduction in treatment cost, have been accomplished through the selection of an appropriate reactor configuration

[11,12]. This was accomplished based on an understanding of the mechanism of action of peroxidase and kinetic modeling of the reaction system.

Traditionally, steady-state kinetic models, such as the Michaelis–Menten model described earlier, have been developed to evaluate rate constants for enzyme systems. Such models have limited application to reactor development because they describe only the initial stages of a reaction. The conditions under which such a model holds true are not usually the conditions that are encountered in the milieu of a reactor. For example, the progress of a reaction may be modeled inaccurately if the time dependence of reactant concentrations and enzymes activities is neglected. The alternative is to explicitly establish the differential equations describing the system of reactions and to integrate them over the desired time interval. This allows for the simulation of reactions of much greater complexity than is possible by applying an overall steady-state approach. Such a model can then be used to simulate the treatment of wastes in different reactor configurations and can be used as an aid to reactor selection and design.

Using a similar approach to that used in the derivation of the Michaelis–Menten model, the kinetic equations of the peroxidase-catalyzed removal of an aromatic substrate were derived from the reaction pathways illustrated in Fig. 2 [92]. The rate of change of the aromatic compound concentration can be written as

$$\frac{\mathrm{d}[\mathrm{AH}_2]}{\mathrm{d}t} = -(E_o - E_{iii} - E_{inact})\left[\frac{1}{k_1[\mathrm{H}_2\mathrm{O}_2]} + \frac{1}{[\mathrm{AH}_2]}\left(\frac{k_2+k_3}{k_2k_3}\right)\right]^{-1} \tag{9}$$

where E_o, E_{iii}, E_{inact}, $[H_2O_2]$, and $[AH_2]$ equal the concentrations of the initial active enzyme, Compound III, inactive enzyme, hydrogen peroxide, and aromatic compound, respectively. The k values correspond to the rate constants of the individual reactions shown in Fig. 2. The following differential equation was also derived to describe the rate of enzyme inhibition by peroxide resulting from the formation of Compound III:

$$\frac{\mathrm{d}E_{iii}}{\mathrm{d}t} = \frac{k_{app}}{k_3}[\mathrm{H}_2\mathrm{O}_2](E_o - E_{iii} - E_{inact})\left[\frac{[\mathrm{AH}_2]}{k_1[\mathrm{H}_2\mathrm{O}_2]} + \frac{k_2+k_3}{k_2k_3}\right]^{-1} - k_{eff}E_{iii} \tag{10}$$

Finally, a differential equation was derived that described the permanent inactivation of the enzyme during the treatment process:

$$\frac{\mathrm{d}E_{inact}}{\mathrm{d}t} = k_r(E_o - E_{inact})\sqrt{-\frac{\mathrm{d}[\mathrm{AH}_2]}{\mathrm{d}t}} - k_e\frac{\mathrm{d}[\mathrm{AH}_2]}{\mathrm{d}t} \tag{11}$$

where k_r is a rate constant associated with inactivation by free radicals and k_e is a dimensionless proportionality constant associated with inactivation by polymer products. Note that in the latter equation, the rate of inactivation predicted by the equation increases with increasing reaction rate. This equation reveals that inactivation can be reduced by carrying the treatment reaction out at a reduced rate. Therefore, any increase in enzyme efficiency will come at the cost of a slower reaction rate and, thus, a larger system volume, as will be seen below.

The above equations have been calibrated and validated based on treatment experiments carried out with a range of concentrations of phenol, peroxide and HRP and for reactions conducted with and without the presence of PEG [10,92]. The simultaneous solution of these equations along with several mass balance equations and appropriate initial conditions yields the time-dependent concentrations of aromatic substrate, peroxide, and enzyme forms. Once these equations are calibrated for a given peroxidase, they can be used to simulate the behavior of various reactor systems. For example, the equations shown above have been used as a basis for the simulation of reactor systems including batch, plug-flow (PFR), continuous flow stirred tank reactors (CFSTR), and multistage CFSTRs, with each stage having equal volume [11].

Fig. 7 shows how these models can be used to aid in reactor selection. The figure illustrates model predictions of the quantity of HRP that is required to achieve 99% removal of phenol from an influent that initially contained 4 mM phenol (approximately 380 mg/L). This phenol concentration is typical of a high-strength foundry wastewater such as that shown in Fig. 6. A 99% phenol removal would yield an effluent that is within the range of typical discharge limits for total phenols [11]. Initial simulations revealed that equimolar amounts of influent hydrogen peroxide and phenol yield the lowest enzyme requirements for a given HRT and reactor configuration. This equimolar concentration satisfies the observed stoichiometry between peroxide and phenol and minimizes Compound III formation. Subsequent simulations shown in Fig. 7 indicate that the HRP required to achieve 99% phenol removal in two-stage or three-stage CFSTRs is less than the amount required by a PFR when the overall HRT is greater than approximately 4.8 and 2.5 hr, respectively. For an HRT less than 2.5 hr, a PFR or batch reactor is more efficient. The figure also shows that a single-stage CFSTR would require a significantly higher enzyme concentration to achieve the target level of phenol removal regardless of HRT. Additional simulations also revealed that the optimal choice of reactor configuration changes depending on the degree of treatment required. For example, for a 1-hr HRT, the optimal reactor configurations for 50%, 75%, and 99% removal of 4.0 mM phenol are two-stage CFSTRs, three-stage CFSTRs, and PFRs, respectively. Such observations are not necessarily intuitive. Rather, only through a good understanding of the reactions involved in a

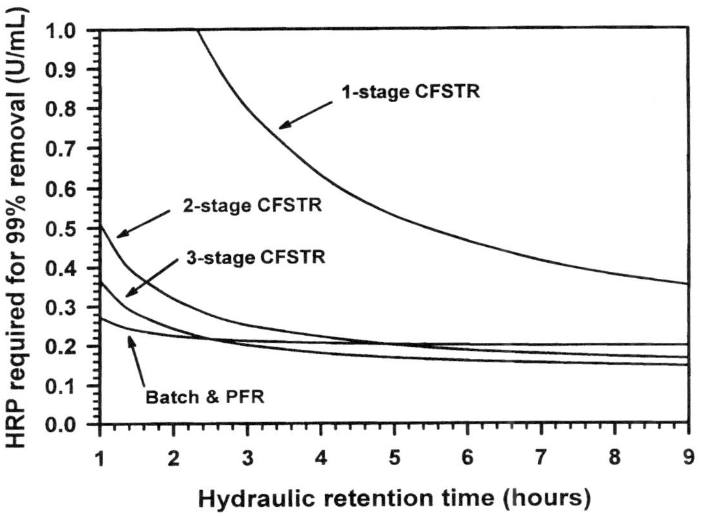

Figure 7 HRP concentration required for 99% removal of phenol in the presence of excess PEG as a function of hydraulic retention time and reactor configuration. pH 7.5, 25°C, [Initial phenol] = 4.0 mM (380 mg/L as phenol), [initial peroxide] = 4.0 mM. (Adapted from Ref. 11.)

given treatment processes may the most suitable reactor configuration and design be identified.

E. Assessment of Reaction Products

In waste-treatment applications, lowering the toxicity of the waste is at least as important as accomplishing the removal of specific contaminants. Therefore, perhaps the greatest challenge to the eventual application of enzymatic treatment is the reported tendency of the enzymatic reaction to yield soluble by-products that have been shown to exhibit toxicity [29,99,100,101]. Although the majority of the polymer products have very limited water solubility and tend to precipitate quite readily, some trace quantities can remain in the aqueous phase. In many cases, significant detoxification of solutions was achieved. For example, as shown in Fig. 8, solutions of 3-chlorophenol, 4-chlorophenol, 2,6-dichlorophenol, 3-methylphenol, and 4-methylphenol were substantially detoxified during treatment with HRP and SBP. In addition, solutions of pentachlorophenol were detoxified to a great extent by HRP. There were no significant differences in the toxicities of

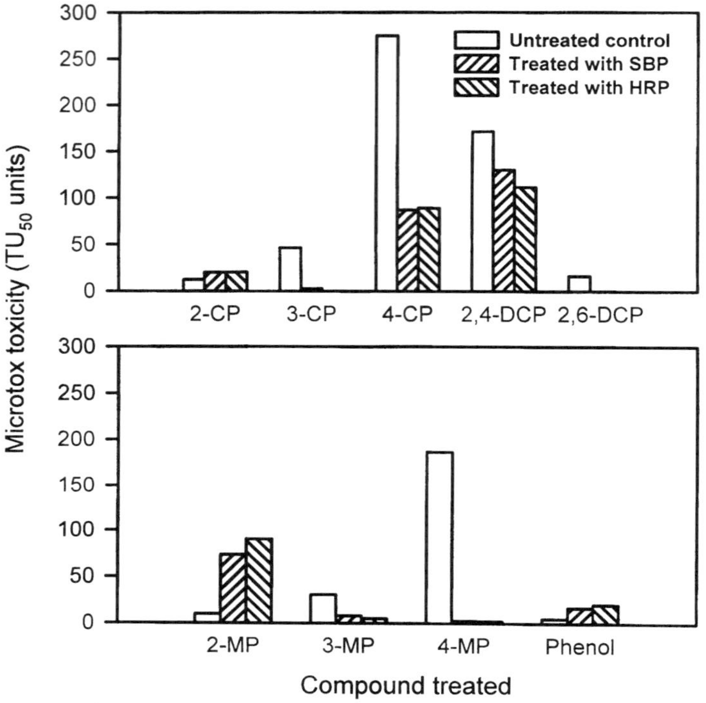

Figure 8 Toxicities of chlorophenols (CP), methylphenols (MP) and phenol before and after treatment with HRP and SBP. [Initial substrate] = 1 mM, [Initial peroxide] = 1.5 mM. All reactions were conducted at pH 7.0 and 25°C and sufficient enzyme was provided to achieve greater than 95% conversion of phenolic substrate. Residual peroxide was destroyed before toxicity was assessed. (Adapted from Ref. 101.)

reaction products arising from the use of SBP or HRP. Unfortunately, however, many of these solutions still exhibited higher than expected residual toxicities given the high degree of substrate conversion (i.e., >95%). In some cases, including phenol, 2-methylphenol, and 2-chlorophenol, the residual toxicity was higher than the toxicity before treatment (see Fig. 8).

Detailed experiments revealed that there is a gradual accumulation of dissolved toxic soluble products during the treatment of phenols [100]. Subsequent experiments showed that when phenolic substrates were treated in mixtures, the copolymerized products tended to be of reduced toxicity [101]. This is of practical importance given that many wastewaters are contaminated with a variety of phenols. The mutagenicity of reaction products resulting from the oxidation of phenols by enzymes has also been studied [29,102]. For example, 17 reaction products from the oxidation of phenols by chloroperoxidase, HRP, LiP and polyphenol oxidase were investigated [102]. It was reported that, in general, the formation of mutagenic products was not observed. However, the oxidation of nitrophenol by LiP did result in the formation of mutagenic products. Chloroperoxidase might also form some toxic products in the presence of chloride ions because of its capacity to oxidize the latter [29]. Hence, it would be advisable to use chloroperoxidase only in the absence of chloride ions.

In recent work, it has been shown that the toxicity of treated solutions decreases substantially over time [103]. It was suggested that toxic species could gradually form compounds that have a lower solubility and eventually precipitate from solution. Alternatively, they could react over time to produce nontoxic soluble products. The instability of the toxic products led to the idea that their destruction could be hastened. This was demonstrated when many solutions were detoxified through the application of an additional dose of hydrogen peroxide after the enzyme-catalyzed reaction was over. It was also shown that a moderate dose of ultraviolet light could accomplish the same effect and that the toxic products arising from the transformation of phenol could be removed using a small quantity of activated carbon in a polishing step [100]. Fortunately, these studies demonstrate that there are approaches that can be used to reduce residual toxicity.

It should be noted that all of these studies involved the treatment of phenols in pure aqueous solutions. However, in a waste treatment environment, there are potentially many other types of pollutants present that can interact with the free radicals that are generated by the enzyme. Thus, the products of the enzymatic reaction are likely to be quite different in a true waste. For example, during the treatment of a foul condensate wastewater (see the next section), there was a substantial toxicity drop during treatment with HRP [22]. A detailed analysis of the waste revealed that even though the total phenol content represented only a small fraction of the total organic content of the waste, they constituted a large fraction of the total toxicity. Thus, the enzymatic process not only targeted the phenols for treatment but, in effect, the most toxic components were targeted. Additional experiments showed that when a solution of pure phenol was treated in the presence of lignin, which is not a substrate of peroxidase, the toxicity was substantially lower than when phenol was treated alone. Thus, it was confirmed that the presence of nonsubstrates has a significant impact on the distribution of reaction products. Similar success was also noted during the treatment of a petroleum refinery wastewater [23]. Sufficient phenol transformation was accomplished to satisfy regulatory guidelines for total phenols. Along with phenol removal, the quality of the wastewater improved substantially. This was reflected in 60% and 80% reductions in COD and BOD_5, respectively, and a toxicity reduction of approximately 95%.

Based on these findings, it is clear that any future implementation of this process for industrial wastewater treatment will require that effluent toxicity be circumvented in a suitable manner.

F. Wastewater Treatment

Before any enzymatic treatment system can be implemented, it is necessary to confirm that the enzyme can carry out its catalytic action in the wastewater matrix. Losses in system efficiency may occur as a result of interaction of the enzyme with waste components (organics, oil and grease, suspended solids, metals, etc.) or due to the nonoptimal characteristics of the wastewater (pH, temperature, etc.). Peroxidase enzymes are one of the few enzymes whose capabilities have been demonstrated in actual waste matrices.

Brief studies demonstrated that industrial effluents including coal-gasification wastewater (2000 mg/L phenol, pH 9.0), flushing liquor from a coke plant (400 mg/L phenol, pH 8.6), and effluent from a triarylphosphate plant (100 mg/L total phenols, pH 7.8) could be treated successfully using peroxidase [20]. In particular, it was noted that the coal gasification wastewaters also contain significant quantities of ammonia, chloride, cyanide, thiocyanate, and other constituents that normally interfere with biological treatment. It was reported that some of these species appeared to interfere with treatment using peroxidase, but not when phenols were present in quantities that are often found in wastewaters. HRP and H_2O_2 were used to increase removal of low molecular mass color bodies (lignin fragments and derivatives) from biologically treated pulp mill effluents by 50% above that achieved using lime precipitation alone [21]. Treatment conditions were not optimized in the above studies and, consequently, the amount of enzyme required was excessive.

As shown in Fig. 6, foundry wastewater (330 mg/L total phenols, pH 6.5) was successfully treated resulting in more than 99% removal of total phenols [24]. Enzyme requirements were significantly reduced through optimization of treatment conditions and the use of PEG as an additive. It is notable, however, that the protective effect exerted by PEG was not as great as was observed in a synthetic waste. It was surmised that this could be due to interaction of PEG with other components of the waste, thereby reducing its availability for protection. Alternatively, the polymer products created in the foundry waste may be significantly different from those experienced during the treatment of aqueous solutions of pure phenols.

The phenol content of petroleum refinery wastewater (45 mg/L total phenols, pH 8.6, 300 mg/L COD, 90 mg/L BOD, 17 mg/L suspended solids) was also reduced below the discharge limit following treatment with HRP [23]. Approximately 58% of COD, 78% of BOD_5, and 95% of toxicity were removed along with the phenols. As a result of treatment, phenols were transformed into less biodegradable compounds that could be removed by coagulation. Optimization of the peroxide dose led to 20% enzyme savings. The use of polyethylene glycol and chitosan as protective additives resulted in 4-fold and 25-fold reductions in enzyme requirements, respectively. Phenol removal did not appear to be affected adversely by the presence of hydrocarbons that are frequently present in refinery wastewaters. In addition, a foul condensate from a Kraft mill (15 mg/L total phenols, pH 9.4, 1600 mg/L COD, 200 mg/L suspended solids) was treated with HRP resulting in the removal of phenols to a level that satisfied discharge limits [22]. Treatment also resulted in a drop in toxicity that was greater than could be accounted for by the transformation of phenols. However, only marginal COD removal was achieved, thereby indicating that the treatment process selectively targeted phenols in the waste. PEG was not effective as a protective additive in this waste matrix but results indicated that the condensate already contained species that protect the enzyme from inactivation by reaction products. These species were identified as lignin derivatives. Unexpectedly, peroxidase has consistently performed more efficiently in the foundry, petroleum, and Kraft mill wastes than when used to treat pure solutions of aqueous phenols [22–24]. However, this is not expected to be the case for all other enzymes that have been identified as candidates for environmental applications due to their fragility.

All of these studies demonstrate that peroxidase enzymes can catalyze the transformation of contaminants in a real wastewater matrix. This is an extremely important step before full-scale applications can be realized.

G. Commercial Availability of the Enzyme

A major hurdle that must be overcome before enzymatic treatment can become a reality is the price of the enzyme [24]. Recent research has focused on improving the economics of enzymatic systems through the direct use of plants or plant materials that contain enzymes, the use of crude enzyme extracts, the development of cloned cells that can be stimulated to produce enzymes efficiently in reactor systems, and transgenic manipulation of plants to stimulate enzyme production. Specific examples of some of these efforts are described below.

Plant material has been successfully used to decontaminate water polluted with phenolic compounds through enzymatic reactions [24,104].

Peroxidases from minced horseradish, potato, and white radish were used to remove 2,4-dichlorophenol up to concentrations of 850 mg/L and the removal rates achieved were comparable to those of purified HRP [104]. An exhaustive study was also conducted in which minced horseradish roots were used to treat 50 compounds that included eight USEPA priority pollutants and one suspected carcinogen. Most compounds were removed very effectively using the minced roots and their substrate specificity was comparable to that of purified HRP. It was concluded that the broad substrate specificity of horseradish made it suitable for the treatment of waters contaminated with a variety of phenols and anilines including at least six USEPA priority pollutants (phenol, 2-chlorophenol, 2,4-dichlorophenol, 2,4,6-trichlorphenol, 4-chloro-3-cresol, and pentachlorophenol). A crude juice extract from horseradish roots was also used to successfully treat a foundry waste [24]. As shown in Fig. 9, when no protective additive was used, the efficiency of the crude extract was better than that of the purified HRP. Presumably this is due to the significant quantity of organic matter that was present in the crude extract (with a COD of 180,000 mg O_2/L), which provided some degree of protection of the enzyme from inactivation. When PEG was used to protect the enzyme, significantly lower quantities of enzyme were required to achieve full treatment and, in this case, the efficiencies of the pure and crude enzyme preparations were very similar.

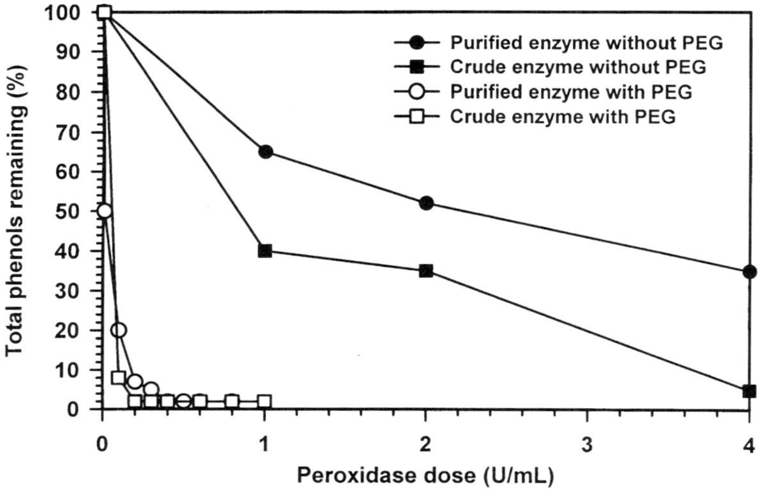

Figure 9 Treatment of a foundry waste with pure and crude preparations of HRP with and without the presence of polyethylene glycol (100 mg/L). [Initial phenol] = 3.5 mM (330 mg/L as phenol), [initial peroxide] = 4.9 mM. (Adapted from Ref. 24.)

It should be noted that full treatment of the waste could be accomplished with a retention time of approximately 5 min when only 0.8 U/mL of crude or pure enzyme were used in the presence of PEG. In contrast, an aerobic biological treatment system that was used to treat this particular foundry waste required a retention time of approximately 6 hr. This illustrates one of the most significant advantages of enzymatic treatment. On the other hand, the use of crude enzyme preparations and PEG contributes significant quantities of organic compounds to the waste, which ultimately must be removed in a subsequent treatment step.

Other plants such as potatoes, cauliflower, cherries, and soybeans and several fungi may also be used as sources of peroxidase enzymes. Soybeans, in particular, may represent a valuable source of peroxidase because the enzyme is found in the seed coat, which is a waste product from soybean-based industries [90]. In this case, it may be possible to use the solid waste from the soybean industry to treat the wastewaters of various chemical industries. In fact, the direct use of raw soybean hulls to accomplish the removal of phenol and 2-chlorophenol has been demonstrated [105]. However, it should be noted that this type of approach would result in an increase in the amount of solid residues that must be disposed following treatment. Peroxidases extracted from tomato and water hyacinth plants were also used to polymerize phenolic substrates [106]. Actual plant roots were also used for in vivo experiments of pollutant removal. The peroxidases studied accomplished good removal of the test substrate guaiacol and the plant roots precipitated the phenolic pollutants at the roots' surface. It was suggested that plant roots be used as natural immobilized enzyme systems to remove phenolic compounds from aquatic systems and soils. The direct use of plant material as an enzyme source represents a very interesting alternative to the use of purified enzymes due to its potentially lower cost. However, further studies are needed to confirm the feasibility of such a process.

Genetic engineering techniques are also expected to play a significant role in peroxidase production. For example, the successful development of a hairy root clone from horseradishes and its implementation in a bioreactor has been reported [107]. These cultures had peroxidase activity comparable to that of the original plant. This development of a peroxidase production system independent of field-grown horseradish is extremely advantageous in terms of the commercial production of peroxidase. Tobacco has also undergone transgenic manipulation to produce peroxidase. For example, cationic peanut (*Arachis hypogaea*) peroxidase was expressed in transgenic tobacco plants and represented about 0.3% of the total soluble proteins [108]. In addition, two peroxidases derived from sweet potatoes were expressed in high levels in tobacco [109]. Young, fully expanded leaves of transgenic plants showed higher peroxidase activity than those of non-transgenic plants.

The developments summarized above are quite significant. However, large-scale production of these enzymes is not yet a reality. Such full-scale applications must await the development and commercialization of new enzyme sources before enzymatic treatment can be considered feasible.

VI. CONCLUSION

A large number of enzymes from a variety of different plants and microorganisms have been reported to play an important role in an array of waste-treatment applications. Enzymes can act on specific recalcitrant pollutants to remove them by precipitation or transformation to other innocuous products. They can also change the characteristics of a given waste to render it more amenable to treatment or to aid in converting waste material to value-added products. Overall, there appears to be a great potential for enzymes in a large number of waste-treatment areas. Particularly notable is the possibility of using enzymes to convert wastes from the food processing and pulp and paper industries or simply municipal solid wastes. This has the double advantage of reducing the amount of waste materials that must be discarded, while at the same time creating valuable products such as fuels, feeds, and commodity chemicals. Enzymes seem to have a promising future. However, before their full potential can be realized, several issues remain to be addressed. These include, but are not limited to the following: (1) development of low-cost enzymes in quantities that are required at the industrial scale (2) demonstration of the feasibility of the enzymes under conditions experienced during wastewater treatment; (3) characterization of reaction products and assessment of their impact on downstream processes and the environment; and (4) identification of methods for the disposal of solid residues.

REFERENCES

1. Aitken MD. Waste treatment applications of enzymes: opportunities and obstacles. Chem Eng J 1993; 52:B49–B58.
2. Aitken MD, Irvine RL. Stability testing of ligninase and Mn-peroxidase from *Phanerochaete chrysosporium*. Biotechnol Bioeng 1989; 34:1251–1260.
3. Fan C-Y, Krishnamurthy S. Enzymes for enhancing bioremediation of petroleum-contaminated soils: A brief review. J Air Waste Manag Assoc 1995; 45:4453–4460.
4. Baily JE, Ollis DF. Biochemical Engineering Fundamentals. 2nd ed. McGraw-Hill, 1986.

5. Vieth WR, Venkatasubramanian K. Enzyme engineering: Part I. The utility of supported enzyme systems. Chemtech November 1973; 2:677–684.
6. Nicell JA, Al-Kassim L, Bewtra JK, Taylor KE. Wastewater treatment by enzyme catalysed polymerization and precipitation. Biodeterior Abst 1993; 7(1):1–8.
7. Klibanov AM, Alberti BN, Morris ED, Felshin LM. Enzymatic removal of toxic phenols and anilines from waste waters. J Appl Biochem 1980; 2:414–421.
8. Klibanov AM, Morris ED. Horseradish peroxidase for the removal of carcinogenic aromatic amines from water. Enzyme Microb Technol 1981; 3:119–122.
9. Nicell JA, Bewtra JK, Biswas N, St. Pierre CC, Taylor KE. Enzyme catalyzed polymerization and precipitation of aromatic compounds from aqueous solution. Can J Civ Eng 1993; 20:725–735.
10. Buchanan ID, Nicell JA. Kinetics of peroxidase interactions in the presence of a protective additive. J Chem Technol Biotechnol 1998; 72:23–32.
11. Buchanan ID, Nicell JA, Wagner M. Reactor models for horseradish-peroxidase-catalyzed aromatic removal. J Environ Eng 1998; 124(9):794–802.
12. Nicell JA, Bewtra JK, Biswas N, Taylor KE. Reactor development for peroxidase catalyzed polymerization and precipitation of phenols from wastewater. Water Res 1993; 27(11):1629–1639.
13. Nakamoto S, Machida N. Phenol removal from aqueous solutions by peroxidase-catalyzed reaction using additives. Water Res 1992; 26:49–54.
14. Wu J, Taylor KE, Bewtra JK, Biswas N. Optimization of the reaction conditions for enzymatic removal of phenol from wastewater in the presence of polyethylene glycol. Water Res 1993; 27(12):1701–1706.
15. Ganjidoust J, Tatsumi K, Wada S, Kawase M. Role of peroxidase and chitosan in removing chlorophenols from aqueous solution. Water Sci Technol 1996; 34(10):151–159.
16. Bodzek M, Bohdziewicz J, Kowalska M. Preparation of membrane-immobilised enzymes for phenol decomposition. J Chem Technol Biotechnol 1994; 61:231–239.
17. Coppella SJ, DelaCruz N, Payne GF, Pogell BM, Speedie MK, Karns JS, Sybert EM, Connor MA. Genetic engineering approach to toxic waste management: case study for organophosphate waste treatment. Biotech Prog 1990; 6:76–81.
18. Caldwell SR, Raushel FM. Detoxification of organophosphate pesticides using an immobilized phosphotriesterase from *Pseudomonas diminuta*. Biotechnol Bioeng 1991; 37:103–109.
19. Munnecke DM. Properties of an immobilized pesticide-hydrolyzing enzyme. Appl Environ Microbiol 1977; 33:503–507.
20. Klibanov AM, Tu T-M, Scott KP. Peroxidase-catalyzed removal of phenols from coal-conversion waste waters. Science 1983; 221:259–260.
21. Schmidt RL, Joyce TW. An enzymatic pretreatment to enhance the lime precipitability of pulp mill effluents. Tappi 1980; 63(12):63–67.

22. Wagner M, Nicell JA. Treatment of a foul condensate from Kraft pulping with horseradish peroxidase and hydrogen peroxide. Water Res 2001; 35(2): 485–495.
23. Wagner M, Nicell JA. Peroxidase-catalyzed removal of phenols from a petroleum refinery wastewater. Water Sci Technol 2001; 43(2):253–260.
24. Cooper VA, Nicell JA. Removal of phenols from a foundry wastewater using horseradish peroxidase. Water Res 1996; 30(4):954–964.
25. Glenn JK, Morgan MA, Mayfield MB. An extracellular H_2O_2-requiring enzyme preparation involved in lignin biodegradation by the white rot basidiomycete *Phanerochaete chrysosporium*. Biochem Biophys Res Commun 1983; 114:1077–1083.
26. Tien M, Kirk TK. Lignin-degrading enzyme from the hymenomycete *Phanerochaete chrysosporium* Burds. Science 1983; 221:661–663.
27. Venkatadri R, Irvine RL. Cultivation of *Phanerochaete chrysosporium* and production of lignin peroxidase in novel biofilm reactor systems: hollow fiber reactor and silicone membrane reactor. Water Res 1993; 27:591–596.
28. Joshi DK, Gold MH. Oxidation of dibenzo-*p*-dioxin by lignin peroxidase from the Basidiomycete *Phanerochaete chrysosporium*. Biochem 1994; 33:10969–10976.
29. Aitken MD, Massey IJ, Chen T, Heck PE. Characterization of reaction products from the enzyme catalyzed oxidation of phenolic pollutants. Water Res 1994; 28:1879–1889.
30. Hammel KE. Organopollutant degradation by ligninolyic fungi. Enzyme Microb Technol 1989; 11:776–777.
31. Cornwell RL, Tinland-Butez M-F, Tardone PJ, Cabasso I, Hammel KE. Lignin degradation and lignin peroxidase production in cultures of *Phanerochaete chrysosporium* immobilized on porous ceramic supports. Enzyme Microb Technol 1990; 12:916–920.
32. Kirk TK, Yang HH. Partial delignification of unbleached kraft pulp with ligninolytic fungi. Biotechnol Lett 1979; 1:347–352.
33. Royer G, Yerushalmi L, Rouleau D, Desrochers M. Continuous decolorization of bleached kraft effluents by *Coriolus versicolor* on the form of pellets. J Ind Microbiol 1991; 7:269–278.
34. Ferrer I, Dezotti M, Durán N. Decolorization of kraft effluent by free and immobilized lignin peroxidases and horseradish peroxidase. Biotechnol Lett 1991; 13:577–582.
35. Pellinen J, Yin C-F, Joyce TW, Chang H-M. Treatment of chlorine bleaching effluent using a white-rot fungus. J Biotechnol 1988; 8:67–76.
36. Peralta-Zamora P, Gomes de Moraes S, Esposito E, Antunes R, Ryes J, Duran N. Decolorization of pulp mill effluents with immobilized lignin and manganese peroxidase from *Phanerochaete chrysosporium*. Environ Technol 1998; 19:521–528.
37. van Gelder CWG, Flurkey WH, Wichers HJ. Sequence and structural features of plant and fungal tyrosinases. Phytochemistry 1997; 45(7):1309–1323.

38. Atlow SC, Bonadonna-Aparo L, Klibanov AM. Dephenolization of industrial wastewaters catalyzed by polyphenol oxidase. Biotechnol Bioeng 1984; 26:599–603.
39. Wada S, Ichikawa H, Tatsumi K. Removal of phenols from wastewater by soluble and immobilized tyrosinase. Biotechnol Bioeng 1993; 42:854–858.
40. Sun W-Q, Payne GF, Moas M, Chu JH, Wallace KK. Tyrosinase reaction/chitosan adsorption for removing phenols from wastewater. Biotech Prog 1992; 8:179–186.
41. Naidja A, Huang PM, Bollag J-M. Comparison of reaction products from the transformation of catechol catalyzed by birnessite or tyrosinase. Soil Sci Am J 1998; 62:188–195.
42. Ikehata K, Nicell JA. Characterization of tyrosinase for the treatment of aqueous phenols. Bioresour Technol 2000; 74:191–199.
43. Wada S, Ichikawa H, Tatsumi K. Removal of phenols and aromatic amines from wastewater by a combination treatment with tyrosinase and a coagulant. Biotechnol Bioeng 1995; 45:304–309.
44. Ikehata K, Nicell JA. Assessment of the products of tyrosinase-catalyzed oxidation of phenols. Biotech Prog 2000; 16(4):533–540.
45. Wada S, Ichikawa H, Tatsumi K. Removal of phenols with tyrosinase immobilized on magnetite. Water Sci Technol 1992; 26(9–11):2057–2059.
46. Bollag J-M, Shuttleworth KL, Anderson DH. Laccase-mediated detoxification of phenolic compounds. Appl Environ Microbiol 1988; 54:3086–3091.
47. Bollag J-M. Decontaminating soil with enzymes. Environ Sci Technol 1992; 26:1876–1881.
48. Majcherczyk A, Johannes C, Huttermann A. Oxidation of polycyclic aromatic hydrocarbons (PAH) by laccase of *Trametes versicolor*. Enzyme Microb Technol 1998; 22:335–341.
49. Wong Y, Yu J. Laccase-catalyzed decolorization of synthetic dyes. Water Res 1999; 33(16):3512–3520.
50. Milstein O, Haars A, Majcherczyk A, Trojanowski J, Tautz D, Zanker H, Hüttermann A. Removal of chlorophenols and chlorolignins from bleaching effluent by combined chemical and biological treatment. Water Sci Technol 1988; 20(1):161–170.
51. Lankinen VP, Inkeröinen MM, Pellinen J, Hatakka AI. The onset of lignin-modifying enzymes, decrease of AOX and color removal by white-rot fungi grown on bleach plant effluents. Water Sci Technol 1991; 24(3–4):189–198.
52. Forss K, Jokinen J, Savolainen M, Williamson H. Utilization of enzymes for effluent treatment in the pulp and paper industry. Tappi 1983; 179–183.
53. Munnecke DM. Detoxification of pesticides using soluble or immobilised enzymes. Proc Biochem 1978; 13:14–17.
54. Smith JM, Payne GF, Lumpkin JA. Enzyme-based strategy for toxic waste treatment and waste minimization. Biotechnol Bioeng 1982; 39:741–752.
55. Havens PL, Rase HF. Reusable immobilized enzyme/polyurethane sponge for removal and detoxification of localized organophosphate pesticide spills. Ind Eng Chem Res 1993; 32:2254–2258.

56. Basheer S, Kut ÖM, Prenosil JE, Bourne JR. Kinetics of enzymatic degradation of cyanide. Biotechnol Bioeng 1992; 41:465–473.
57. Basheer S, Kut ÖM, Prenosil JE, Bourne JR. Development of an enzyme membrane reactor for treatment of cyanide-containing wastewaters from the food industry. Biotechnol Bioeng 1993; 39:629–635.
58. Nazly N, Knowles CJ, Beardsmore AJ, Naylor WT, Corcoran EG. Detoxification of cyanide by immobilised fungi. J Chem Technol Biotechnol 1983; 33B:119–126.
59. Thomas L, Jungschaffer G, Sprössler B. Improved sludge dewatering by enzymatic treatment. Water Sci Technol 1993; 28(1):189–192.
60. Anazia I, Misra M. Enzymatic dewatering of Florida phosphate slimes. Miner Metall Process 1989; 6(2):93–95.
61. Hakulinen R. The use of enzymes for wastewater treatment in the pulp and paper industry—a new possibility. Water Sci Technol 1988; 20(1):251–262.
62. Khan LI, Sarker M. Enzyme enhanced stabilization of soil and fly ash. Fly ash for Soil Improvement Proceedings of the 1993 ASCE Annual Convention, Dallas, Texas, 1992, 43-50.
63. Shoemaker S. The use of enzymes for waste management in the food industry. Harlander SK, Labuza TP, ed., Biotechnology in Food Processing. Park Ridge, NY: Nayes Publications, 1986; 20(1):259–267.
64. Blasheck H. Approaches to making the food processing industry more environmentally friendly. Trends Food Sci Technol 1992; 3:107–110.
65. Leward DA, Lawrie RA. Recovery and utilisation of by-product proteins of the meat industry. J Chem Technol Biotechnol 1984; 34B:223–228.
66. Dalev PG. Utilisation of waste feathers from poultry slaughter for production of a protein concentrate. Bioresour Technol 1994; 48:265–267.
67. Venugopal V, Alur MD, Nerkar DP. Solubilization of fish proteins using immobilized microbial cells. Biotechnol Bioeng 1989; 33:1098–1103.
68. Coleman R. Biodegradable plastics from potato waste double savings to environment. Agric Eng 1990; 71(6):20–22.
69. Cosio IG, Fisher RA, Carroad PA. Bioconversion of shellfish chitin waste: waste pretreatment, enzyme production, process design and economic analysis. J Food Sci 1982; 47:901–905.
70. Bajza Z, Markovic I. Influence of enzyme concentration on leather waste hydrolysis kinetics. J Soc Leather Technol Chem 1999; 83(3):172–176.
71. Clanet M, Durand H, Tiraby G. Enzymatic saccharification of municipal wastes. Biotechnol Bioeng 1988; 32:930–934.
72. Coughlan MP. Enzymic hydrolysis of cellulose: An overview. Bioresour Technol 1992; 39:107–115.
73. Lagerkvist A, Chen H. Control of two-step anaerobic degradation of municipal solid waste (MSW) by enzyme addition. Water Sci Technol 1993; 27:47–56.
74. Rivers DB, Emert GH. Factors affecting the enzymatic hydrolysis of municipal-solid-waste components. Biotechnol Bioeng 1988; 3:278–281.

75. Duff SJ, Moritz JW, Andersen KL. Simultaneous hydrolysis and fermentation of pulp mill primary clarifier sludge. Can J Chem Eng 1994; 72:1013–1020.
76. Duff SJ, Moritz JW, Casavant TE. Effect of surfactant and particle size reduction on hydrolysis of deinking sludge and nonrecyclable newsprint. Biotechnol Bioeng 1995; 45:239–244.
77. Jobbins JM, Franks NE. Enzymatic deinking of mixed office waste: process condition optimization. Tappi 1997; 80(9):73–78.
78. Brennan MB. New age paper and textiles: Fungi, enzymes and closed-loop catalysis offer environmental, economic gains in manufacturing and recycling. CENEAR March, 1998; 23:39–47.
79. Grasshof A. Laboratory studies on cleaning milk-heaters with enzymatic cleaning solutions. Kiel Milchwirtsch Forschungsber 1999; 51(4):295–318.
80. Hemming D. Molecular farming: using transgenic plants to produce novel proteins and other chemicals. Agbiotech News Inf 1995; 7(1):19N–29N.
81. Worley J. UK's Tobacco and Health Institute branching out: Tobacco as a "factory" to grow new products. Odyssey: The Magazine of the University of Kentucky, Spring, 1998, 3–7.
82. O'Brien AM, Ó'Fágáin C. Dye bleaching and phenol precipitation by phthalic anhydride-modified horseradish peroxidase. J Chem Technol Biotechnol 2000; 75:363–368.
83. May S. Applications of oxidoreductases. Curr Opin Biotechnol 1999; 10:370–375.
84. Gillikin JW, Graham JS. Purification and developmental analysis of the major anionic peroxidase from the seed coat of *Glycine max*. Plant Physiol 1991; 96:214–220.
85. Al-Kassim L, Taylor KE, Nicell JA, Bewtra JK, Biswas N. Enzymatic removal of selected aromatic contaminants from wastewater by a fungal peroxidase from *Coprinus macrorhizus* in batch reactors. J Chem Technol Biotechnol 1994; 61:179–182.
86. Hammel KE, Kalyanaraman B, Kirk TK. Oxidation of polycyclic aromatic hydrocarbons and dibenzo-[*p*]-dioxins by *Phanerochaete chrysosporium* ligninase. J Biol Chem 1986; 261:16948–16952.
87. Hammel KE, Tardone PJ. The oxidative 4-dechlorination of polychlorinated phenols is catalyzed by extracellular fungal lignin peroxidases. Biochemistry 1988; 27:6563–6568.
88. Bumpus JA, Brock BJ. Biodegradation of crystal violet by the white rot fungus *Phanerochaete chrysosporium*. Appl Environ Microbiol 1988; 54: 1143–1150.
89. Schreiner RD, Stevens SE, Tien M. Oxidation of thianthrene by the ligninase of *Phanerochaete chrysosporium*. Appl Environ Microbiol 1990; 54: 1858–1860.
90. Wright H, Nicell JA. Characterization of soybean peroxidase for the treatment of aqueous phenols. Bioresour Technol 1999; 70:69–79.
91. Nicell JA, Wright H. A model of peroxidase activity with inhibition by hydrogen peroxide. Enzyme Microb Technol 1997; 21:302–310.

92. Buchanan ID, Nicell JA. Model development for horseradish peroxidase-catalyzed removal of aqueous phenol. Biotechnol Bioeng 1997; 54(3):251–261.
93. Ator MA, Ortiz de Montellano PR. Protein control of prosthetic heme reactivity: reaction of substrates with the heme edge of horseradish peroxidase. J Biol Chem 1987; 262(4):1542–1551.
94. Arseguel D, Baboulène M. Removal of phenol from coupling of talc and peroxidase. Application for depollution of wastewater containing phenolic compounds. J Chem Technol Biotechnol 1994; 61:331–335.
95. Harris JM. Introduction to biotechnical and biological applications of poly(ethylene glycol). In: Harris JM, ed. Poly(Ethylene Glycol) Chemistry: Biotechnical and Biomedical Applications. New York: Plenum Press, 1992: 1–14.
96. Kinsley C, Nicell JA. Treatment of aqueous phenol with soybean peroxidase in the presence of polyethylene glycol. Bioresour Technol 2000; 73(2): 139–146.
97. Nicell JA, Saadi KW, Buchanan ID. Phenol polymerization and precipitation by horseradish peroxidase enzyme and an additive. Bioresour Technol 1995; 54:5–16.
98. Wu Y, Taylor KE, Biswas N, Bewtra JK. A model for the protective effect of additives on the activity of horseradish peroxidase in the removal of phenol. Enzyme Microb Tech 1998; 22:315–322.
99. Heck PE, Massey IJ, Aitken MD. Toxicity of reaction products from enzymatic oxidation of phenolic pollutants. Water Sci Technol 1992; 26(9–11):2369–2371.
100. Ghioureliotis M, Nicell JA. Assessment of soluble products of peroxidase-catalyzed polymerization of aqueous phenol. Enzyme Microb Technol 1999; 25:185–193.
101. Ghioureliotis M, Nicell JA. Toxicity of soluble products from the peroxidase-catalyzed polymerization of substituted phenolic compounds. J Chem Technol Biotechnol 2000; 75:98–106.
102. Massey IJ, Aitken MD, Ball LM, Heck PE. Mutagenicity screening of reaction products from the enzyme-catalyzed oxidation of phenolic pollutants. Environ Toxi Chem 1994; 11:1743–1752.
103. Wagner M, Nicell JA. Detoxification of phenolic solutions with horseradish peroxidase and hydrogen peroxide. Wat Res 2002; 36(16):4041–4052.
104. Dec J, Bollag J-M. Use of plant material for the decontamination of water polluted with phenols. Biotechnol Bioeng 1994; 44:1132–1139.
105. Flock C, Bassi A, Gijzen M. Removal of aqueous phenol and 2-chlorophenol with purified soybean peroxidase and raw soybean hulls. J Chem Technol Biotechnol 1999; 74(4):303–309.
106. Adler PR, Arora R, El Ghaouth A, Glenn DM, Solar JM. Bioremediation of phenolic compounds from water with plant root surface peroxidases. J Environ Qual 1994; 23:1113–1117.
107. Taya M, Yoyama A, Nomura R, Kondo O, Matsui C, Kobayashi T.

Production of peroxidase hairy root cells in a two step culture system. J Ferment Bioeng 1989; 67:31–34.
108. Lige B, Ma S, Zhao D, Van Huystee RB. Cationic plant peroxidase: Expression and characterization in transgenic tobacco and purification of the histidine-tagged protein. Plant Sci 1998; 136(2):159–168.
109. Huh GH, Yun BW, Lee HS, Jo JK, Kwak SS. Overproduction of sweet potato peroxidases in transgenic tobacco plants. Phytochemistry 1998; 47(5): 695–700.
110. Kimura CC, Garcia EE, Martins AC, Nozaki J. Chemical and enzymatic hydrolysis of chrome shavings. Anal Asoc Quim Argen 1999; 87(3/4):97–103.
111. Iwanowski H, Linowska M. An attempt of application of enzymes to wastewater treatment. Environ Prot Eng 1981; 10:39–46.
112. Vanderberg LA, Foreman TM, Attrep M, Brainard JR, Sauer NN. Treatment of heterogeneous mixed wastes: enzyme degradation of cellulosic materials contaminated with hazardous organics and toxic and radioactive metals. Environ Sci Technol 1999; 33:1256–1262.
113. Swanson PE. Dehalogenases applied to industrial-scale biocatalysis. Curr Opin Biotechnol 1999; 10:365–369.
114. Mileski GJ, Bumpus JA, Jurek MA, Aust SD. Biodegradation of pentachlorophenol by the white rot fungus *Phanerochaete chrysosporium*. Appl Environ Microbiol 1988; 54:2885–2889.
115. Kilara R. Enzymes and their uses in the processed food industry: A review. Proc Biochem 1982; 17:36–41.
116. Jelen P. Reprocessing of whey and other dairy wastes for use as food ingredients. Food Technol 1983; 37:81–84.

Index